CONTROL ENGINEERING SERIES 68

Flexible Robot Manipulators

Other volumes in this series:

Flexible Robot Manipulators

Modelling, simulation and control

Edited by
M.O. Tokhi and A.K.M. Azad

The Institution of Engineering and Technology

Published by The Institution of Engineering and Technology, London, United Kingdom

© 2008 The Institution of Engineering and Technology

First published 2008

The Institution of Engineering and Technology
Michael Faraday House
Six Hills Way, Stevenage
Herts, SG1 2AY, United Kingdom

www.theiet.org

British Library Cataloguing in Publication Data
Flexible robot manipulators: modelling, simulation and control. - (IET control series)
   1. Manipulators (Mechanism) 2. Manipulators (Mechanism) - Automatic control
   I. Tokhi, M.O. II. Azad, Abul III. Institution of Engineering and Technology
   629.8'92

ISBN 978-0-86341-448-0

Typeset in India by Newgen Imaging Systems (P) Ltd, Chennai
Printed in the UK by Athenaeum Press Ltd, Gateshead, Tyne & Wear

Contents

Preface

The ever-increasing utilization of robotic manipulators in various applications in recent years has been motivated by the requirements and demands of industrial automation. Among the rigid and flexible manipulator types, attention is focused more towards flexible manipulators. This is owing to various advantages such manipulators offer as compared to their rigid counterparts. Exploitation of the potential benefits and capabilities of rigid and flexible manipulators introduces a further emerging line of research in which hybrid rigid–flexible manipulator structures are considered.

Flexural dynamics (vibration) in flexible manipulators has been the main research challenge in the modelling and control of such systems. Accordingly, research activities in flexible manipulators have looked into the development of methodologies to cope with the flexural motion dynamics of such systems.

A considerable amount of research on the development of dynamic models of flexible manipulators has been carried out. These have led to descriptions in the form of either partial differential equations, or finite-dimensional ordinary differential equations. From a control perspective, an input/output characterisation of the system is desired, which can be obtained through suitable online estimation and adaptation mechanisms. Given the dynamic nature of flexible manipulator systems, the practical realisation of such methodologies presents new challenges.

Numerical techniques using finite difference and finite element methods have been researched for dynamic characterisation of flexible manipulators. Accordingly, simulation algorithms characterising the dynamic behaviour of flexible manipulators have been developed that provide flexible means of analysis, test and verification of control techniques. With the widely available use of digital computing technology, such platforms are first-step favoured option in a wide range of applications.

Control structures adopted for flexible manipulators can broadly be separated into open loop and closed loop. Although the mathematical theory of open loop control is well established, only a limited number of successful applications in the control of distributed parameter flexible manipulator systems have been reported. A further research dimension, with this class of control structures, is online adaptation of the input shaping mechanism with the changing behaviour of the system and the environment. With closed-loop control techniques, a common trend that has been adopted by researchers is partitioning of the dynamics of the system into the slow (rigid-body) and fast (flexural motion) dynamics and accordingly devising separate control loops. An important consideration with this has been to adequately cope with the non-minimum phase behaviour exhibited by the system characterisation, which with optimal feedback control techniques leads to unstable control. Although this problem can be avoided with some traditional techniques, emerging intelligent control methodologies incorporating soft-computing paradigms offer a great deal of potential in solving such problems.

This book reports on recent and new developments in modelling, simulation and control of flexible robot manipulators, in light of the issues mentioned above. The contents of the book are divided into 19 chapters. Following a general overview of flexible manipulators from the perspective of modelling, simulation, control and applications in Chapter 1, the rest of the book may be grouped into four parts, although some overlap between the parts is allowed for reasons of completeness and coherency as far as required: (1) Chapters 2–4 provide a range of modelling approaches including classical techniques based on the Lagrange equation formulation and parametric approaches based on linear input–output models using system identification techniques and neuro-modelling approaches; (2) Chapters 5–7 present numerical modelling/simulation techniques for dynamic characterisation of flexible manipulators using the finite difference, finite element, symbolic manipulation and customised software techniques, with Chapter 7 dedicated to manipulators in space; (3) Chapters 8–17 present a range of open-loop and closed-loop control techniques based on classical and modern intelligent control methods including soft-computing and smart structures for flexible manipulators; (4) Chapters 18 and 19 are dedicated to software environments for analysis, design, simulation and control of flexible manipulators.

The book is intended for teaching in graduate courses on robotics, mechatronics, control, electrical and mechanical engineering. It can also serve as a source of reference for research in areas of modelling, simulation and control of dynamic flexible structures in general and, specifically, of flexible robotic manipulators.

The material presented in this book comprises contributions of worldwide researchers in the field, and the editors are grateful for their professional and scientific support. The editors would also like to thank Professor Derek Atherton for his encouragement and support during the initial planning of this project, and the IET publication team for their patience and support throughout this project.

M.O. Tokhi
University of Sheffield, UK

A.K.M. Azad
Northern Illinois University, USA

Contributors

M. Armada
Automatic Control Department
Industrial Automation Institute
Spanish National Research
   Council (CSIC)
Madrid, Spain

A.K.M. Azad
Department of Technology
Northern Illinois University
Illinois, USA

M.A. Botto
DEM–IDMEC
Instituto Superior Técnico
Technical University of Lisbon
Lisbon, Portugal

M.D. Brown
Lockheed Martin UK Integrated
   Systems
Havant, Hampshire, UK

G. Fernández
Electronic and Circuits Department
Universidad Simón Bolívar
Caracas, Venezuela

S.S. Ge
Department of Electrical and
   Computer Engineering
National University of Singapore
Singapore

S.P. Goh
GGS Systems (M) Sdn. Bhd.
Johor Bahru
Johor, Malaysia

J.C. Grieco
Electronic and Circuits Department
Universidad Simón Bolívar
Caracas, Venezuela

M. Gu
Space Technologies
Canadian Space Agency
Québec, Canada

G.W. Irwin
School of Electrical and Electronics
   Engineering
Queen's University of Belfast
Belfast, UK

J. Kövecses
Mechanical Engineering Department
McGill University
Montreal, Canada

C. Lange
Space Technologies
Canadian Space Agency
Québec, Canada

T.H. Lee
Department of Electrical and
    Computer Engineering
National University of Singapore
Singapore

S. Mahil
College of Engineering
Purdue University
Indiana, USA

J.M. Martins
DEM–IDMEC
Instituto Superior Técnico
Technical University of Lisbon
Lisbon, Portugal

Z. Mohamed
Faculty of Electrical Engineering
University Technology Malaysia
Malaysia

J.-C. Piedboeuf
Space Technologies
Canadian Space Agency
Québec, Canada

H. Poerwanto
PT PAL
Surabaya, Indonesia

H.R. Pota
Australian Defence Force Academy
Canberra, Australia

N.K. Poulsen
Department of Informatics and
    Mathematical Modelling
Technical University of Denmark
Lyngby, Denmark

O. Ravn
Automation, Ørsted. DTU
Technical University of Denmark
Lyngby, Denmark

J. Sá da Costa
DEM–IDMEC
Instituto Superior Técnico
Technical University of Lisbon
Lisbon, Portugal

W.P. Seering
Department of Mechanical Engineering
Massachusetts Institute of Technology
Cambridge, USA

K. Senda
Division of Mechanical Science and
    Engineering
Kanazawa University
Kanazawa, Japan

M.H. Shaheed
Department of Engineering
Queen Mary and Westfield College
University of London
London, UK

S.K. Sharma
School of Engineering
University of Plymouth
Plymouth, UK

B. Siciliano
Dipartimento di Informatica e Sistemistica
Università degli Studi di Napoli
    Federico II
Napoli, Italy

M.N.H. Siddique
School of Computing and Intelligent
   Systems
The University of Ulster
Londonderry, UK

W.E. Singhose
George W. Woodruff School of
   Mechanical Engineering
Georgia Institute of Technology
Atlanta, USA

M.O. Tokhi
Department of Automatic Control and
   Systems Engineering
The University of Sheffield
Sheffield, UK

L. Villani
Dipartimento di Informatica e
   Sistemistica
Università degli Studi di Napoli
   Federico II
Napoli, Italy

Z.P. Wang
Department of Electrical and
   Computer Engineering
National University of Singapore
Singapore

A.S. Yigit
Department of Mechanical
   Engineering
Kuwait University
Safat, Kuwait

Abbreviations

AB	Articulated body
ACC	Adaptive composite controller
ACU	Arm computer unit
A/D	Analogue/digital
ADAM	Aerospace dual-arm flexible manipulator
AMM	Assumed modes method
ANN	Artificial neural network
ARMAX	Autoregressive moving average with exogeneous inputs
ARX	Autoregressive with exogenous inputs
AVC	Active vibration control
BC	Boundary condition
CACE	Computer aided control engineering
CCD	Charge coupled device
CDT	Contact dynamics toolkit
CLIK	Closed-loop inverse kinematics
CMFC	Centralised model-free controller
CMM	Coordinate measuring machine
CI	Composite inertia
CRB	Composite rigid body
CSA	Canadian Space Agency
D/A	Digital/analogue
DFM	Duisburg flexible manipulator
DMFC	Decentralised model-free controller
DNA	Direct Nyquist array
DOF	Degree of freedom
DSP	Digital signal processing
EAP	Electroactive polymer
EBRC	Energy-based robust controller
ERLS	Equivalent rigid link system
EVA	Extravehicular activity
FD	Finite difference
FE	Finite element

FMA	Force moment accommodation
FMS	Flexible manipulator system
FRF	Frequency response function
GA	Genetic algorithm
GOCF	Generalized observability canonical form
GUI	Graphical user interface
HC	Hard computing
HLS	Hardware-in-the-loop simulation
IIR	Infinite impulse response
IMSC	Independent modal space control
INA	Inverse Nyquist array
I/O	Input/output
ISS	International Space Station
ISTE	Integral squared timed error
JBC	Joint-based collocated
LED	Light-emitting diode
LHP	Left half of s-plane
LMS	Least mean squares
LPDC	Local proportional, derivative control
LQ	Linear quadratic
LQG	Linear quadratic Gaussian
LQR	Linear quadratic regulator
LRMS	Long-reach manipulator system
MAM	Manual augmented mode
MBS	Mobile base system
MDR	MacDonald Dettwiler Space and Advanced Robotics Ltd.
MDSF	Manipulator development and simulation facility
MF	Membership function
MIMO	Multi-input multi-output
MIQ	Machine intelligence quotient
MLFM	Multi-link flexible manipulator
MLP	Multi-layered perceptron
MNN	Modular neural network
MOTS	MSS operation and training simulator
MPIPD	Multivariable PI–PD
MPO	Model predicted output
MPP	Multivariable pole-placement
MRO	MSS robotics operator
MSL	Mechatronic Simulink library
MSS	Mobile servicing system
MTF	Matrix transfer function
NARMAX	Non-linear autoregressive moving average with exogeneous inputs
NARX	Non-linear autoregressive with exogeneous inputs
NB	Negative big
NN	Neural network

NRT	Non-real-time
NS	Negative small
ODE	Ordinary differential equation
OLS	Orthogonal least squares
OPDE	Ordinary partial differential equation
ORU	Orbital replacement unit
OSA	One-step-ahead
OTCM	Orbital tool change-out mechanism
OTCME	ORU tool change-out mechanism emulator
PB	Positive big
PD	Proportional, derivative
PDE	Partial differential equation
PI	Proportional, integral
PID	Proportional, integral, derivative
PR	Probabilistic reasoning
PRBS	Pseudo-random binary sequence
PS	Positive small
PSD	Power spectral density
PZT	Lead Zirconate Titanate (piezoelectric ceramic material)
RAC	Resolved-acceleration control
RBF	Radial basis function
RC	Resistance–capacitance
RFM	Real flexible manipulator
RFR	Rigid–flexible–rigid
RH	Routh–Hurtwiz
RHP	Right half of s-plane
RLS	Recursive least squares
RMRC	Resolved-motion-rate control
RMS	Remote manipulator system
RTAI	Real-time application interface
RTW	Real-time workshop
RVDT	Rotary variable differential transformer
SC	Soft computing
SCEFMAS	Simulation and Control Environment for Flexible Manipulator Systems
SHM	Shared memory
SIM	Dynamic simulator
SISO	Single-input single-output
SLFM	Single-link flexible manipulator
SM	Symbolic manipulation
SMG	Symbolic model generator
SMP	System for monitoring and maintaining MSS robotics operators performance
SMT	STVF test-bed
SMT-SIM	SMT dynamic simulator

SPDM   Special purpose dextrous manipulator
SSRMS   Space station remote manipulator system
STVF   SPDM test verification facility
TLFM   Two-link flexible manipulator
USB   Universal Serial Bus
VR   Visual renderer
VRM   Virtual rigid manipulator
VSC   Variable structure control
ZN   Ziegler–Nichols
ZPETC   Zero phase error tracking controller
ZV   Zero vibration
ZVD   Zero vibration and derivative
1D   One-dimensional
1DOF   One-degree-of-freedom
2D   Two-dimensional
2DOF   Two-degrees-of-freedom
3D   Three-dimensional

Notations

a	Thickness of beam
b	Width of beam and of smart material
a_i, b_i, c_i	Constants, parameters, coefficients of polynomials
$\mathbf{a}_p$	Absolute linear acceleration vector of point p expressed in the body reference frame
$\mathbf{a}_n, \mathbf{a}_k$	Absolute linear acceleration vector of the body reference frame and of the cross section reference frame, expressed in the body reference frame
a_{ij}	Matrix element
A	Cross-sectional area
$\mathbf{A}_n$	Generalized acceleration vector of body reference frame n
$\mathbf{A}, \mathbf{B}, \mathbf{C}$	System state matrices
b_j^m	Bias on the jth neuron of the mth layer
$\mathcal{B}_{n_o}, \Sigma \mathcal{B}_{n_o}$	A body and its surface in the reference undeformed configuration
$\mathbf{B} \in R^3$	Magnetic flux density vector
$\mathbf{c}^M \in R^{6 \times 6}$	Symmetric matrix of elastic stiffness coefficients of the beam
$\mathbf{c}^S \in R^{6 \times 6}$	Symmetric matrix of elastic stiffness coefficients of the smart material
c_1	Thickness of upper surface smart material patch
c_{11}^M	Stiffness of the beam
c_{11}^S	Stiffness of the piezoelectric material
c_2	Thickness of lower surface smart material patch
c_{L1}	Stiffness per unit length of the beam
c_{L2}	Stiffness per unit length of the smart material
C, C_n	Kinetic energy, capacitance
C_a	Actuator voltage constant
C_s	Sensor voltage constant
d	Constant
d_i	Components of the displacement gradient strain vector
d_{31}	Piezoelectric charge constant
$\mathbf{D}$	Damping matrix
$\mathbf{D}$	Displacement gradient strain vector at a beam cross section

$\mathbf{D}(x,t) \in R^3$	Electrical displacement at location x and time t
e	Error
$\dot{e}$	Change in error
$\mathbf{e}_1, \mathbf{e}_2, \mathbf{e}_3$	Unit vectors along the axis of the cross section reference frame, expressed in the body reference frame
$\mathbf{E}_1, \mathbf{E}_2, \mathbf{E}_3$	Unit vectors along the axis of the body reference frame, expressed in the inertial reference frame
E	Young modulus
E_a	Actuating layer Young's modulus
E_K	System kinetic energy
E_P	System potential energy
$\mathbf{E} \in R^3$	Electrical field intensity vector
$E[.]$	Expectation
f_n	Natural frequency in Hz
$\mathbf{f}_s$	Force per unit area applied on the surface of a beam
$\mathbf{f}_n, \mathbf{f}_{\hat{n}}$	Force per unit area applied on the base and tip of a beam
$\mathbf{f}_{\Sigma \bar{B}_{no}}$	Force per unit area applied on the lateral faces of a beam excluding the edges of the first and last cross section
$F(t)$	Force function
$\mathbf{F}_n, \mathbf{M}_n$	Resulting external force and moment applied at the origin of the body reference frame
$\mathbf{F}_{\Sigma \bar{B}_{no}}, \mathbf{M}_{\Sigma \bar{B}_{no}}$	Resulting external force and moment per unit length applied at the origin of the cross section reference frame
$\mathbf{F}_{\hat{n}}, \mathbf{M}_{\hat{n}}$	Resulting external force and moment applied at the origin of the tip cross section reference frame
$\mathbf{F}^M \in R^6$	Simplified stress vector of the beam
$\mathbf{F}^S \in R^6$	Simplified stress vector of the smart material
g	Number of generation in genetic algorithms
$\mathbf{g}$	Gravitational acceleration vector expressed in the body reference frame
g_{max}	Maximum number of generation in genetic algorithms
g_{31}	Piezoelectric stress constant
$\mathbf{G}$	Green–Lagrange strain tensor
G	Half Young modulus, $G = E/2$
$\mathbf{G}_b$	Green–Lagrange strain vector at a beam cross section
$G(s)$	Transfer function (continuous)
$\mathbf{h} \in R^{6\times3}$	Coupling coefficients matrix
h_{12}	Coupling parameter per unit volume of the piezoelectric material
h_L	Coupling parameter per unit length of the smart material robot
$\mathbf{H}(x,t) \in R^3$	Magnetic field intensity at location x and time t
H	Depth/height of arm/link
$H(j\omega)$	Frequency response function
$\mathbf{H}_{n\omega}, \mathbf{H}_{nv}$	Rotation and translation Jacobian matrices of joint n
i	Constants, index, polynomial/model order

I	Area moment of inertia
$\mathbf{I}$	Identity matrix
I_h	Hub inertia
I_p	Inertia associated with payload
I_T	Total inertia $\left(I_h + l^3 \rho A / 3\right)$
j	Constants, index, polynomial/model order, unit imaginary number
J	Cost function
$\mathbf{J}$	$_k\mathbf{J}$ expressed in the body reference frame
J, I_2, I_3	Torsional and bending geometric moments of inertia.
$\mathbf{J}_{R_{ek}}$	Elastic rotation Jacobian of the kth cross section
$\mathbf{J}_n$	Second moment of inertia tensor of rigid body n expressed in the body reference frame
$\mathbf{J}_{T_k}$	Elastic translation Jacobian of the kth cross section
k	Constant, index
$_k\mathbf{J}$	Geometrical moments of inertia tensor of a cross section relative to and expressed in the cross section reference frame
K_c	Control output scaling factor
$\mathbf{K}$	Stiffness matrix
$\mathbf{K}_k$	Vector of bending curvatures expressed in the cross section reference frame
$\mathbf{K}_n$	Elemental stiffness matrix
k_{31}	Piezoelectric electromagnetic coupling constant
K_1, K_2, K_3	Components of $\mathbf{K}_k$. K_1 is the torsional strain, and K_2 and K_3 are the bending strains
K_p, K_i, K_d	Proportional, integral, derivative parameters in PID control
K_v	Derivative gain in PD control
l	Elemental length
L	Length of arm/link
L_g	Lagrangian
m	Total number of NNs in an MNN
m_3	Payload at end-point
$\mathbf{M}$	Mass matrix
$\mathbf{M}_n$	Elemental mass matrix
M_{e_n}	Mass matrix of flexible beam n
$M_a(x, t)$	Local moment induced in beam by piezoelectric actuating layer
M_p	Payload mass
m_{ij}	Elements of mass matrix
n	Constant, number of elements, number of modes
N	Constant, number of samples
$\mathbf{N}_n$	Coriollis and centrifugal force terms of body n
$\bar{\mathcal{N}}_{e_n}$	Generalized force vector contemplating non-linear inertial forces, linear elastic forces and the generalized external forces applied on the boundary of the beam
$\mathbf{N}(x), \mathbf{N}_a(x)$	Shape function vector
O	Origin of Cartesian coordinate system

$\{O_I, X_I Y_I Z_I\}$	Inertial reference frame
$\{O_n, X_n Y_n Z_n\}$	Body n reference frame
$\{O_k, X_k Y_k Z_k\}$	Beam cross section k reference frame
$OX_1 Y_1$	Local reference frame with axis OX_1 tangential to the beam at the base
$OX_0 Y_0$	Fixed base frame
p	Constant, pole
P_c	Crossover probability
P_m	Mutation probability
P_{cd}	Dynamic crossover probability
P_{md}	Dynamic mutation probability
P, P_n	Potential energy
$\mathbf{q}_{e_n}$	Global vector of the elastic generalized coordinates of a beam
$\mathbf{q}_{K_n}$	Vector of pure bending displacement and pure torsion angle generalized coordinates
$\mathbf{q}_{\varphi_n}$	Vector of pure shear displacement generalised coordinates
$\mathbf{q}_{n\omega}, \mathbf{q}_{nv}$	Vectors of angular and linear position parameters of joint n
$\mathbf{Q}(t), \mathbf{Q}_a(t)$	Nodal displacement vector
$Q(t)$	Total charge in sensing layer
$q(x, t)$	Charge distribution in sensing layer
$q_i(t)$	Time-dependent generalized coordinates
$\mathbf{r}$	Position vector of point P expressed in the fixed base frame
r	Reference input
$\mathbf{r}_n$	Position vector of body reference frame n relative to and expressed in the inertial reference frame
$\mathbf{r}_p$	Position vector of material point p relative to and expressed in the inertial reference frame
$\mathbf{r}_{n-1,n}$	Vector describing the position of body n relative to body $n-1$, expressed in body $n-1$ reference frame
$\mathbf{R}_n$	Rotation matrix from the inertial reference frame to body n reference frame
$\mathbf{R}_{e_k}$	Rotation matrix from body n reference frame to cross section k reference frame
$\mathbf{R}_{n/n-1}$	Orthogonal rotation matrix expressing the rotation of body n relative to body $n-1$
s	Laplace variable
$\mathbf{S} \in R^6$	Simplified strain vector
t	Time (continuous)
$\mathbf{t}, {}_k\mathbf{t}$	Tangent vector to the beam neutral fibre expressed in the body reference frame, and expressed in the cross section reference frame
t_a, t_b, t_c	Respective thicknesses of piezoelectric actuator, beam and piezoelectric sensor
T	Total elapsed time of the desired trajectory
T_n	Generalized control force vector at joint n
u	Plant input, control output

$\mathbf{v}(x,t)$	Voltage applied to the piezoelectric actuator
$V_a(\cdot)$	Actuator voltage
$V_s(\cdot)$	Sensor voltage
v	System state
$\mathbf{v}_p$	Absolute linear velocity vector of point p expressed in the body reference frame
$\mathbf{v}_n$, $\mathbf{v}_k$	Absolute linear velocity vector of the body reference frame and of the cross section reference frame, expressed in the body reference frame
$\mathbf{V}$	Generalized velocity vector of body reference frame n
w	Elastic deflection
w_{ij}^m	Connection weight between the ith neuron of the $(m-1)$th layer and the jth neuron of the mth layer
$w(x,t)$	Deflection at location x and time t
W	Width of arm/link, virtual work done by non-conservative forces
x	Distance from hub along the arm/link
$\mathbf{X}_{g_n}$	Centre of mass of rigid body n, relative to and expressed in the body reference frame
X, Y, Z	Moving coordinate system
X_o, Y_o, Z_o	Fixed coordinate system
$\mathbf{X}_p, \mathbf{x}_p, \mathbf{u}_p$	Reference position, displaced position and displacement vector of a material point p, expressed in the body reference frame
$\mathbf{X}_k, \mathbf{x}_k, \mathbf{u}_k$	Reference position, displaced position and displacement vector of a material point on the beam neutral axis, expressed in the body reference frame
X_1, X_2, X_3	Material coordinates of a body
y	Plant output, total displacement, actual output
$\mathbf{Y}_p$	Position vector of material point p in a given cross section, relative to and expressed in the cross section frame
$\hat{y}$	Estimated/predicted output
z	z-transform variable, zero
$\tilde{\mathbf{Z}}$	Skew symmetric matrix formed with the components of a given vector $\mathbf{Z}$
z^{-1}	Unit left-shift
α	End-point acceleration, pure torsion angle
α_k	Absolute angular acceleration vector of a cross section expressed in the body reference frame
β	Flexural rigidity
$\boldsymbol{\beta} \in R^{3\times 3}$	Symmetric matrix of impermittivity coefficients
β_{22}	Impermittivity per unit volume of the piezoelectric material
β_L	Impermittivity per unit length of the smart material robot
δ	Variational operator of the Principle of Virtual Powers
γ_i	Pure shear deflections
$\boldsymbol{\Gamma}_k$	Strain vector of the beam neutral axis expressed in the cross section reference frame

$\Gamma_1, \Gamma_2, \Gamma_3$	Components of $\mathbf{\Gamma}_k$. Γ_1 is the longitudinal strain, and Γ_2 and Γ_3 are the transverse Timoshenko shear strains
λ	Number of correctly classified pattern
ψ	Number of connections in the MNN
ε_c	Strain in sensing layer
ε_{ij}	Components of the Green–Lagrange strain tensor
$\phi_i(x)$	Admissible functions
ϕ_i	Element of shape function vector
$\varphi_{12}, \varphi_{13}$	Shear angles as defined in the classical theory of elasticity
$\mathcal{F}_n$	Generalized force vector applied at the origin of body reference frame n
$\mathbf{\Phi}^T_{n+1,n}$	Matrix transforming generalized forces applied at $n+1$ to generalized forces applied at n
$\theta, \theta(t)$	Hub/joint angle
$\dot{\theta}$	Hub/joint angular velocity
	Joint angle at the hub
$\theta_d(t)$	Desired joint angular trajectory
ρ	Mass density per unit volume
ρ_p	Specific mass of a beam at point p
ρ_A	Mass of a beam per unit length
ρ_1	Mass per unit volume of the beam
ρ_2	Mass per unit volume of the smart material
ρ_{L1}	Mass per unit length of the beam
ρ_{L2}	Mass per unit length of the smart material
σ	Piola–Kirchoff stresss vector at a beam cross section
δ	Variational operator of the Principle of Virtual Powers
$\mathbf{\mu} \in R^{3\times3}$	Permeability coefficients matrix
μ_{33}	Permeability of the piezoelectric material
μ_L	Permeability per unit length of the smart material robot
σ_a	Longitudinal stress in actuating layer
τ	Torque
τ_s	Sample period
$\tau(t)$	Torque applied to the base of the manipulator
υ	End-point residual
υ_i	Pure bending deflections
$\mathbf{v}_{ix}, \mathbf{\alpha}_x, \mathbf{\gamma}_{ix}$	Vectors of shape functions for the pure elastic deflections
$\mathbf{v}_{it}, \mathbf{\alpha}_t, \mathbf{\gamma}_{it}$	Vectors of the elastic generalised coordinates
ω	Frequency in radian
ω_c	Cut-off frequency in radian
ω_n	Natural frequency in radian
$\mathbf{\omega}_{n/n-1}$	Vector describing the angular velocity of body n relative to body $n-1$ expressed in body n reference frame
$\mathbf{\omega}_n, \mathbf{\omega}_k$	Absolute angular velocity vector of the body reference frame and of the cross section reference frame, expressed in the body reference frame

$\boldsymbol{\Omega}_k$	Angular velocity vector of the cross section reference frame relative to the body reference frame, expressed in the body reference frame
Λ_k	Absolute angular acceleration matrix of a cross section expressed in the body reference frame
ζ , ζ_i	Damping ratio
ξ_i	Slope parameter of the activation function of ith NN output
$\tanh(x)$	Activation function of hidden neurons
$\tanh(\xi_i x)$	Activation function of ith NN output
σ_e^2	Variance of variable e
$(-)_{\mathrm{rfr}_n}$	Vector or matrix, referring to rigid-flexible-rigid body n

Chapter 1

Flexible manipulators – an overview

M.O. Tokhi, A.K.M. Azad, H.R. Pota and K. Senda

This chapter presents a general overview of previously developed methodologies for modelling, simulation and control of flexible manipulators. A selection of currently available flexible manipulator experimental systems in various research laboratories and outside laboratory environments are introduced and their features and design merits described. A structured overview of common applications and future research prospects and applications of flexible and hybrid manipulators are provided.

1.1 Introduction

Flexible manipulator systems (FMSs) offer several advantages in contrast to their traditional rigid counterparts. These include faster system response, lower energy consumption, the requirement of relatively smaller actuators, reduced non-linearity owing to elimination of gearing, lower overall mass and, in general, lower overall cost. However, owing to the distributed nature of the governing equations describing dynamics of such systems, the control of flexible manipulators has traditionally involved complex processes (Aubrun, 1980; Book *et al.*, 1986; Plunkell and Lee, 1970). Moreover, to compensate for flexural effects and thus yield robust control the design focuses primarily on non-collocated controllers (Cannon and Schmitz, 1984; Harashima and Ueshiba, 1986).

Research on FMSs ranges from a single-link manipulator rotating about a fixed axis (Hastings and Book, 1987) to three-dimensional multi-link arms (Nagathan and Soni, 1986). However, experimental work, in general, is almost exclusively limited to single-link manipulators. This is because of the complexity of multi-link manipulator systems, resulting from more degrees of freedom and the increased interactions between gross and deformed motions. It is important for control purposes to recognise the flexible nature of the manipulator system and to build a suitable mathematical framework for modelling of the system. The use of dynamic models for FMSs is threefold: forward dynamics, inverse dynamics and controller design. Flexible manipulators are distributed parameter systems with rigid body as well as flexible movements. There are two physical limitations associated with the system:

1. The control torque can only be applied at the joint,
2. Only a finite number of sensors of bounded bandwidth can be used and at restricted locations along the length of the manipulator.

Such issues are considered in this chapter through a structured overview of techniques for modelling, dynamic simulation and control of flexible manipulators.

1.2   Modelling and simulation techniques

According to reported results, dynamic models of flexible manipulators are described either by partial differential equations (PDEs) or by finite-dimensional ordinary differential equations (ODEs) through some kind of approximation. Owing to the principles used, various types of model of flexible manipulator have been developed (Kanoh *et al.*, 1986). These can be classified as

- Lagrange's equation and modal expansion (Ritz–Kantrovitch)
- Lagrange's equation and finite element (FE) method
- Euler–Newton equation and modal expansion
- Euler–Newton equation and FE method
- Singular perturbation and frequency-domain techniques.

A commonly used approach for solving a PDE that represents the dynamics of a manipulator, sometimes referred to as the separation of variables method, is to utilize a representation of the PDE, obtained through a simplification process, by a finite set of ordinary differential equations. This model, however, does not always represent the fine details of the system (Hughes, 1987). A method in which the flexible manipulator is modelled as a massless spring with a lumped mass at one end and a lumped rotary inertia at the other end has previously been proposed (Feliú *et al.*, 1992; Oosting and Dickerson, 1988). In practice, dynamic models are mostly formulated on the basis of considering forward and inverse dynamics. In this manner, consideration is given to computational efficiency, simplicity and accuracy of the model. Here, a means of predicting changes in the dynamics of the manipulator resulting from changing configurations and loading is proposed, where predictions of changes in mode shapes and frequencies can be made without the need to solve the full determinantal equation of the system.

An alternative to modelling the manipulator in the time domain is to use a method based on frequency domain analysis (Book and Majette, 1983; Yuan *et al.*, 1989). This method develops a concise transfer matrix model using the Euler–Bernoulli beam equation for a uniform beam. The weakness of this method is that it makes no allowance for interaction between the gross motion and the flexible dynamics of the manipulator, nor can these effects be easily included in the model. As a result, the model can only be regarded as approximate. In another approach, a chain of flexible links is modelled by considering a flexible multi-body dynamic approach, based on an equivalent rigid link system (ERLS), where an ERLS (which is the closest possible to the deformed linkage) is defined, in order to match, at best, the requirements of a small displacement assumption. As the choice of ERLS is completely arbitrary, it could introduce artificial kinematic constraints, which in turn introduce modelling error (Giovagnoni, 1994).

Unfortunately, the solutions obtained through the above modelling processes are approximate and do not represent fine details of a system. To resolve this problem, numerical solution of the system's equation is performed allowing development of simulation environments. Dynamic simulation is important from a system design and evaluation viewpoint. It provides a characterisation of the system in the real sense as well as allowing online evaluation of controller designs. Commonly used simulation approaches involve finite element (FE), finite difference (FD) and symbolic manipulation (SM) methods. The FE method has been previously utilized to describe the flexible behaviour of manipulators (Dado and Soni, 1986; Usoro *et al.*, 1984). The steps involved in FE simulation are discretisation of the structure into small elements; selection of an approximating function to interpolate the result; derivation of an equation for these small elements; calculation of the system equation and solving the system equation considering the boundary conditions. The development of the algorithm can be divided into three main parts: the FE analysis, state-space representation and obtaining and analysing the system transfer function. The computational complexity and consequent software coding involved in the FE method is a major disadvantage of this technique. However, as the FE method allows irregularities in the structure and mixed boundary conditions to be handled, the technique is found to be suitable in applications involving irregular structures.

In applications involving uniform structures, such as manipulator systems, the FD method is found to be more appropriate. Simulation studies of flexible beam systems have demonstrated the relative simplicity of the FD method (Kourmoulis, 1990). The FD method is used to obtain an efficient numerical means of solving the PDE by developing a finite-dimensional simulation of the FMS through a discretisation, both, in time and space (distance) coordinates. The algorithm allows inclusion of distributed actuator and sensor terms in the PDE and modification of boundary conditions. The development of such an algorithm for a FMS has previously been reported (Tokhi and Azad, 1995*a*; Tokhi *et al.*, 1995). The algorithm thus developed has been implemented digitally and simulation results characterising the behaviour of the system under various loading conditions have been reported.

Investigations with symbolic manipulation have resulted in automated symbolic derivation of dynamic equations of motion of rigid and flexible manipulators utilizing Lagrangian formulation and assumed mode methods (Cetinkunt and Ittop, 1992; De Luca *et al.*, 1988; Lin and Lewis, 1994), Hamilton's principle and non-linear integro-differential equations (Low and Vidyasagar, 1988) and FD approximations (Tzes *et al.*, 1989). These methods have demonstrated that the approach has some advantages, such as allowing independent variation of flexure parameters. A study on utilizing the symbolic manipulation approach, for the modelling and analysis of a flexible manipulator using FE methods has also been reported (Mohammed and

Tokhi, 2002). It is argued that the effect of payload on the manipulator is important for modelling and control purposes, as successful implementation of a flexible
manipulator control is contingent upon achieving acceptable uniform performance in
the presence of payload variations. The developed model has been verified by using
an experimental rig to demonstrate the performance of the symbolic algorithm in
modelling and analysis of a flexible manipulator.

1.3   Control techniques

The dynamic behaviour of a flexible manipulator may be considered as a combination
of rigid-body and flexible dynamics. Accordingly, control strategies devised for such
systems are to take account of both rigid-body motion and flexible motion control.
The former corresponds to methods developed within the framework of conventional
rigid manipulator control. The latter, on the other hand, corresponds to approaches
developed within the framework of vibration control of flexible structures.

Vibration control techniques for flexible structures are generally classified into
two categories: passive and active control (Tokhi and Veres, 2002). Active control
utilizes the principle of wave interference. This is realised by artificially generating
anti-source(s) (actuator(s)) to destructively interfere with the unwanted disturbances
and thus result in reduction in the level of vibration. Active control of FMSs can in
general be divided into two categories: open-loop and closed-loop control. Open-loop
control involves altering the shape of actuator commands by considering the physical
and vibration properties of the FMS. The approach may account for changes in the
system after the control input is developed. Closed-loop control differs from open-
loop control in that it uses measurements of the system state and change the actuator
input accordingly to reduce the system response oscillation.

1.3.1   Passive control

Passive control utilizes the absorption property of matter and thus is realised by a fixed
change in the physical parameters of the structure, for example, adding viscoelastic
material to increase the damping properties of the flexible manipulator. It has been
reported that the control of vibration of a flexible manipulator by passive means is not
sufficient by itself to eliminate structural deflection (Book *et al.*, 1986). On the other
hand, if only active control is used then, owing to actuator and sensor dynamics,
destabilisation of modes near the bandwidth of the actuator or sensor may result
(Aubrun, 1980). To avoid such destabilisation a certain amount of passive damping
will be required to be employed, thus using hybrid control, that is, a combination of
active and passive control methods. Combined active/passive control strategies have
been proposed previously where low-frequency modes of vibration are controlled by
active means and the modes with frequencies just above the actively controlled modes
are controlled by passive means (Plunkell and Lee, 1970).

Several methods of passive vibration control of FMSs have been developed over
the years. These mainly include methods of implementation of a constrained viscoelastic damping layer to provide an energy dissipation medium (Kerwin, 1959) and

the utilization of composite materials in the construction of a flexible manipulator to provide higher strength and stiffness-to-weight ratio and larger structural damping than a metallic flexible manipulator Aubrun, 1980; Choi *et al.*, 1988; Thompson and Sung, 1986). Observations have shown that although passive damping provides a sharp increase in damping at higher frequency modes, the lower frequency modes still remain uncontrolled. Moreover, the addition of viscoelastic material and a constraining layer leads to an increase in the size and dynamic load of the system (Tzou, 1988).

1.3.2 Open-loop control

Open-loop control methods have been considered in vibration control where the control input is developed by considering the physical and vibrational properties of the FMS. Although, the mathematical theory of open-loop control is well established, only few successful applications in the control of distributed parameter systems, including flexible manipulator, have been reported (Dellman *et al.*, 1956; Singh *et al.*, 1989). The method involves the development of suitable forcing functions in order to reduce the vibration at resonance modes. The methods developed include shape command methods, the computed torque technique and bang-bang control. Shaped command methods attempt to develop forcing functions that minimise vibrations and the effect of parameters that affect the resonance modes (Aspinwall, 1980; Meckl and Seering, 1990; Swigert, 1980; Wang, 1986). Common problems of concern encountered in these methods include long move (response) time, instability owing to un-reduced modes and controller robustness in the case of a large change of the manipulator dynamics.

In the computed torque approach, depending on the detailed model of the system and desired output trajectory, the joint torque input is calculated using a model inversion process (Moulin and Bayo, 1991). The technique suffers from several problems, owing to, for instance, model inaccuracy, uncertainty over implementability of the desired trajectory, sensitivity to system parameter variations and response time penalties for a causal input.

Bang-bang control involves the utilization of single and multiple switched bang-bang control functions (Onsay and Akay, 1991). Bang-bang control functions require accurate selection of switching time, depending on the representative dynamic model of the system. A minor modelling error could cause switching error and thus result in a substantial increase in the residual vibrations. Although, utilization of minimum energy inputs has been shown to eliminate the problem of switching times that arise in the bang-bang input (Jayasuriya and Choura, 1991), the total response time, however, becomes longer (Meckl and Seering, 1990; Onsay and Akay, 1991).

1.3.3 Closed-loop control

Effective control of a system always depends on accurate real-time monitoring and the corresponding control effort. Initial discussions on feedback control of a flexible manipulator and the usefulness of optimal regulator as applied to this problem date back to the early 1970s. It is known in the conventional approach that compensation can alter the first vibrational mode by either adding some damping or extending the bandwidth of the system (Ogata, 2001). Compensation, however, will limit the performance of the manipulator because inputs with frequency contents above the first flexible mode could still cause vibration. Various modern control designs have been proposed during the last two decades for FMSs with different types of vibration measuring systems.

When free motion of a system consists mainly of a limited number of clearly separable modes, then it is possible to control these modes directly using the so-called independent modal space control (IMSC) method, where the controller is designed for each mode independent of other modes (Baz *et al.*, 1992; Sinha and Kao, 1991). Modal space control has been used for suppression of flexible motion in a three-link log loading manipulator with which considerable improvement has been achieved over conventional joint-based collocated controller. Although, initial investigations on the use of IMSC lack consideration of the location of the actuator (Meirovitch *et al.*, 1983), later investigations have shown that actuator placement is important for suppression of spillover and, thus, methods for optimal placement of sensors and actuators have been developed (Schulz and Heimbold, 1983).

Variable structure control (VSC) utilizes a viable high-speed switching feedback control law to drive the plant's state trajectory onto a specified and user-specified surface in the state-space, and to maintain the plant's state trajectory on this surface for all subsequent times. One of the first studies on the application of VSC to one-link flexible manipulators was reported by Qian and Ma (1992), where they controlled the end-point position in a non-collocated manner. In this study, a sliding surface (a line in the study) is constructed from the end-point position and its derivative is employed in the design. Qian and Ma (1992) claim that if the slope of this line is chosen positive and the system variables are made to stay on this line, these would converge to zero exponentially, thus yielding a stable system in sliding mode. The performance of the controller was evaluated through a series of simulations, followed by an analysis of the designed control system. Thomas and Bandyopadhyay (1997), however, have pointed out that the choice of a positive constant as the slope for this switching line would not guarantee the stability of the system in sliding mode. The switching line is in fact a switching hyper-surface in view of the functional relationship of the tip (end-point) position with the generalized coordinates of the system through mode shape functions. The stability of the system in sliding mode is guaranteed only if the motion on this hyper-surface is asymptotically stable (Young, 1977), whereas the positive value for the slope of the switching line employed by Qian and Ma (1992) will not guarantee this stability. Moreover, a stable VSC controller based on a state transformation has been designed in this study.

The application of VSC to multi-link flexible manipulators is very limited. There are difficulties in both modelling and controller design. Sira-Ramirez *et al.* (1992) have derived dynamical sliding mode regulators within the context of generalized observability canonical form (GOCF) (Fliess, 1989). The GOCF is obtained by means of a state elimination procedure, carried out on the system of differential equations describing the manipulator dynamics. Therefore the system can be considered as a linear system. Although simulation examples illustrate the performance of the proposed controller for a robotic manipulator with flexible joint, it is not easy to apply to general multi-link manipulators with flexible links. There are also applications of VSC to other plants similar to flexible manipulators, for example, a spacecraft with flexibility (Karray and Modi, 1995), a flexible structure on the ground (Iwamoto *et al.*, 2002), a disk drive actuator (Supino and Romano, 1997), and so on.

An appreciable amount of work carried out on the control of FMSs involves the utilization of strain gauges, mainly to measure mode shapes (Sakawa *et al.*, 1985). There are two essential components involved in measuring the modal response using strain gauges. The first is a method of measurement of the modes of vibration of the flexible manipulator. The second is the development of a computational technique for distinguishing different modes in the overall deflection of the flexible manipulator. Once modal information is available a control loop can be closed for each mode either to damp or to actively drive the manipulator in a manner that reduces the vibration. It appears that the strain gauge measurement is very simple and relatively inexpensive to use. However, the technique may place more stringent requirements on the dynamic modelling and control tasks. Strain gauges have the disadvantage of not giving a direct measurement of manipulator displacement, as they can only provide local information. Thus, displacement measurement by using strain gauges requires more complex and possibly time-consuming computations, which can lead to inaccuracies.

To solve the problem of displacement measurement, as encountered in using strain gauges only, attempts have been made to develop schemes that incorporate end-point measurements as well (Cannon and Schmitz, 1984; Kotnik *et al.*, 1988). Some researchers have proposed an approach that utilizes local or global measurement of flexible displacement of a manipulator to control system vibration (Harashima and Ueshiba, 1986; Wang *et al.*, 1989). In this method, the deflection of the manipulator is detected (measured), using, for example, a charge coupled device (CCD) camera or laser beam, relative to a rotating reference X–Y frame fixed to the hub of the manipulator. However, as an end-point position control system has a smaller stability margin than collocated control, it is necessary to include a collocated rate feedback (hub-velocity) to obtain acceptable performance of the closed-loop system. By using an end-point sensor, more accurate end-point positioning can be accomplished, but the resulting controller is less robust to plant uncertainties than the corresponding collocated design.

The difficulty in maintaining stability and performance robustness, owing to spillover effects from unmodelled modes that occur when a high-order system is controlled by a low-order controller, is of major concern in the control of flexible systems. To improve robustness it is typically required that the controller bandwidth be sufficiently reduced (Nesline and Zarchan, 1984). Studies have shown that most robust control techniques that ensure stability in the presence of parameter errors can only increase damping by a limited amount (Dorato, 1987). If the inherent damping is very low, this increase may be insufficient to adequately improve the response. Moreover, the controllers rely on accurate system models. This makes the controller very sensitive to modelling errors, leading to degradation in system performance and, in some cases, instability. It is evident that in using either global or local displacement measurement a device is required to be attached on the manipulator. This affects the behaviour of the manipulator (Mace, 1991).

Both feedforward and feedback control structures have been utilized in the control of vibration of FMSs (Shchuka and Goldenberg, 1989; Wells and Schueller, 1990). These include combined feedforward and feedback methods based on control law partitioning schemes, which use end-point position signal in an outer loop to control the flexible modes and the inner loop to control the rigid-body motion. Although, the pole-zero cancellation property of the feedforward control speeds up the system response, it increases overshoot and oscillation. However, it is found that, in contrast to many high-order compensators, systems with feedforward control incorporating proportional and derivative (PD) feedback are not highly sensitive to plant parameter variations.

In investigations carried out on control of FMSs the only non-collocated sensor/actuator pairs that have successfully been employed include motor torque with either the manipulator strain or global/local end-point position. However, practical realisation of both methods has associated short- and long-term drawbacks. If a state-space description of the closed-loop dynamics is available, it is possible to use acceleration feedback to stabilise a rigid manipulator (Stadenny and Belanger, 1986). Investigations on the control of a FMS using acceleration feedback to design the compensator and the end-point position feedback using a design based on a full-state feedback observer have shown that the controller using end-point position feedback exhibits a relatively slow and rough response in comparison with an acceleration feedback controller; the difference becoming more noticeable with increasing slewing angle (Kotnik *et al.*, 1988). Moreover, acceleration feedback produces relatively higher overshoot. The use of acceleration feedback appears to have intuitive appeal from an engineering design viewpoint, particularly because of the relative ease of implementation and low cost. Moreover, in sensing acceleration for control implementation, all sensing and actuation equipment is structure mounted. This implies that issues such as camera positioning or field of view are not of major concern, which is an important consideration, specifically, in large-scale applications such as telerobotics. Furthermore, applications to multi-link flexible manipulators (MLFMs) could benefit from such methods to a greater extent. Some researchers have also proposed adaptive control methods to compensate for parameter variations (Feliu *et al.*, 1990; Yang *et al.*, 1991). However, these approaches utilize optical methods of global/local end-point sensing for obtaining the feedback signal.

Many of the controllers have been designed on the basis of various input shaping mechanisms using both open- and closed-loop configurations. Zuo and Wang (1992) designed a closed-loop control mechanism based on shaped input filter, to reduce or eliminate vibrations and to reject external disturbances of a multi-link manipulator. An adaptive input shaping control scheme for vibration suppression in slewing flexible structures with particular application to flexible-link robotic manipulators has been reported by Tzes and Yurkovich (1993). The scheme combines a frequency-domain identification technique, with input shaping, in order to adjust critical parameters of the input shapers in the case of payload variation or other unmodelled dynamics. The scheme was realised through simulation and experimentation. Hillsley and Yurkovich (1993) have reported a composite control strategy for a two-link flexible robotic arm in conjunction with post-slew feedback scheme. In this work attention has been focused on end-point position control, for point-to-point movements assuming a fixed reference frame for the base with two rotary joints. Khorrami *et al.* (1994) addressed experimentation on rigid-body-based controllers with input preshaping for a two-link flexible manipulator (TLFM). The scheme is shown to be effective when the plant dynamics are linear and time invariant. It has also been shown that application of an inner-loop non-linear control to cancel some of the non-linearities and to reduce configuration dependence of structural frequencies enhances the performance of the input pre-shaping scheme. Borowiec and Tzes (1996) proposed a frequency-shaped explicit output feedback force control for a TLFM. In this work the frequency shaping dependence has been included to eliminate the undesirable effects associated with control and observation spillover. Magee *et al.* (1997) developed a control approach, combining command shaping and internal damping, to control a small robot attached to the end of a flexible manipulator. They also verified the proposed control system experimentally using two separate test-beds.

1.3.4 Artificial intelligence control

It is noted that the non-linear dynamics of rigid manipulators are compensated by an inverse-dynamic strategy, and use of such an approach for a flexible manipulator is restricted by non-minimum phase characteristics of the arm when end-point response is taken as output of the system (Talebi *et al.*, 1998b). Several conventional approaches have been proposed as solutions to this problem based on different methods such as non-causal torque, singular perturbation, integral manifold, transmission zero and redefined output (Bayo and Moulin, 1989; De Luca and Siciliano, 1989; Geniele *et al.*, 1992; Hashtrudi-Zaad and Khorasani, 1996; Kwon and Book, 1990; Madhavan and Singh, 1991; Moallem *et al.*, 1997; Schoenwald and Özgüner, 1990; Siciliano and Book, 1988; Wang and Vidaysagar, 1989a,b). However, performance of these control strategies may not be satisfactory in real-applications as it is difficult to accurately model a flexible manipulator.

In many cases, when it is difficult to obtain a model structure for a system with traditional system identification techniques, intelligent techniques are desired that can describe the system in the best possible way (Elanayar and Yung, 1994). Genetic algorithms (GAs) and artificial neural networks (ANNs) are commonly used for modelling dynamic systems. The main advantages of utilizing GAs for system identification are that they simultaneously evaluate many points in the parameter space and converge towards the global solution (Kargupta and Smith, 1991; Kristinsson and Dumont, 1992). The superiority of a GA over recursive least squares (RLS) in modelling a fixed-free flexible beam has been addressed by Hossain *et al.* (1995). In contrast, neural network (NN) approaches for system identification offer many advantages over traditional ones especially in terms of flexibility and hardware realisation (Ljung and Sjöberg, 1992). This technique is quite efficient in modelling non-linear systems or if the system possesses non-linearities to any degree.

Application of NNs for identification and control of dynamic systems has gained significant momentum in recent years. Narendra and Parthasarathy (1990) addressed system identification using the globally approximating characteristics of NNs. Neuro-modelling with different approaches, involving backpropagation, has been reported by various researchers (Nerrand *et al.*, 1994; Srinivasan *et al.*, 1994). The successful application of radial basis function (RBF) networks for modelling dynamic systems is also widely addressed in the literature (Casdagli, 1989; He and Lapedes, 1993; Sze, 1995). Chen *et al.* (1991) proposed orthogonal least square learning algorithm for RBF networks to model non-linear dynamic systems. Elanayar and Yung (1994) have addressed the use of RBF to approximate dynamic and state equations and to estimate state variables of stochastic systems.

A considerable amount of work has been carried out to develop and implement NN-based controllers for flexible manipulators. Cheng and Wen (1993) proposed a neuro-controller to drive a flexible arm to a desired trajectory along with using hub position and velocity measurement techniques for stabilising the system. Newton and Xu (1993) have addressed the joint tracking control problem for a space manipulator using feedback error learning technique. In this case, end-point position tracking cannot be guaranteed especially for high-speed desired trajectories. Control of a single-link flexible manipulator (SLFM) whose dynamics are partially known has been considered by Donne and Özgüner (1994). In this work, a model-based predictive control scheme is adopted for the known dynamics and unsupervised NN-based control scheme is utilized to control the unknown system dynamics. Identification and control are implemented as a two-stage process where identification of the unknown part of the system is done using a NN in supervised learning mode. Talebi *et al.* (1997) proposed a NN-based adaptive controller for a single flexible-link manipulator. Output redefined approach is used in designing the controller. They examined three different types of NN scheme. The controller has been realised both in simulation and experimental environments. The advantage of this controller over conventional PD type controllers has also been demonstrated. Gutiérrez *et al.* (1998) have reported implementation of a NN tracking controller for a single flexible link. In this work the practical implementation of a multi-loop non-linear NN tracking controller for a single flexible link has been tested and its performance compared to that of the standard PD and proportional, integral, derivative (PID) controllers. The controller includes an outer PD tracking loop and a singular perturbation inner loop for stabilisation of the fast dynamics, and a NN inner loop is used for feedback linearisation of the slow dynamics. Song and Koivo (1998) addressed NN-based control of a flexible manipulator, where a non-linear predictive control approach is presented using a discrete-time multi-layered perceptron network model for the plant. The predictive control framework allows variations in the model order, time delay and non-minimum phase effects in the plant. The method has been compared against a collocated passive PD controller. Development of a multi-loop non-linear NN tracking controller for a multi-link flexible arm using singular perturbation based fast control and outer loop slow control has been addressed by Yesildirek *et al.* (1994). Adaptive NN control of flexible manipulators based on singular perturbation has been reported by Ge *et al.* (1997). In this work the full model dynamics of the flexible manipulator were separated into the slow subsystem and the fast subsystem by applying singular perturbation techniques. Thus, an adaptive NN control based on direct adaptive techniques is designed to control the slow subsystem, and the fast control is designed as a simple linear quadratic regulator (LQR) control to stabilise the fast subsystem along the trajectory of the slow subsystem. Talebi *et al.* (1998*a*) addressed inverse-dynamic control of flexible-link manipulators using NNs, where a modified output redefined approach is utilized to overcome the problem caused by the non-minimum phase characteristics of the flexible-link system.

Neural network applications often incorporate a large number of neurons, thus requiring a great deal of computation for training and causing problems for error reduction (Bishop, 1995). A recent trend in NN design for large-scale problems is to split the original task into simpler subtasks, and use a subnetwork module for each one (Happel and Murre, 1994; Hodge *et al.*, 1999; Jacobs and Jordan, 1993; Kecman, 1996). This divide-and-conquer strategy then leads to super-linear speedup in training and one can improve the generalization ability over that of a single large network (Hanson and Salamon, 1990; Jacobs and Jordan, 1993). It is also easier to encode *a priori* knowledge in a modular framework. In general, a modular neural network (MNN) is constructed from two types of network, namely, expert networks and a gating network (Hodge *et al.*, 1999; Jacobs and Jordan, 1993). Expert networks compete to learn the training patterns and the gating network mediates this competition. During training, the weights of the expert and gating networks are adjusted simultaneously using the backpropagation algorithm (Rumelhart *et al.*, 1986). Sharma *et al.* (2003) reported work involving this strategy for modelling of a flexible manipulator. In this approach MNN learns to partition an input task into subtasks, allocating a different NN to learn each one. However, accuracy of the MNN depends greatly on accurate fusion of the individual networks as decided by a gating network (Hodge *et al.*, 1999). Researchers have presented a new method, using GAs (Goldberg, 1989; Holland, 1992), which removes the need for a gating network. Fusion of individual networks is decided by optimum slope selection of the activation function. The GA also optimises the structure and weights of the individual networks in the MNN.

1.4   Flexible manipulator systems

The first experimental SLFM systems were developed in the early 1980s (Cannon and Schmitz, 1984; Hastings and Book, 1987). Although research into the control of single-link setups continues, attention has now shifted to (MLFMs). There is a wide spectrum of MLFMs, starting from two-link experimental set-ups to 17.6 m, 7 degrees of freedom (DOF) remote manipulator system for space station assembly.

A good survey of initial efforts in flexible manipulator research is given in Hu (1993). Many of the manipulators discussed in Hu (1993) are now decommissioned. This section gives a brief description of the rich variety of flexible manipulators in current use, both for experimental and industrial use.

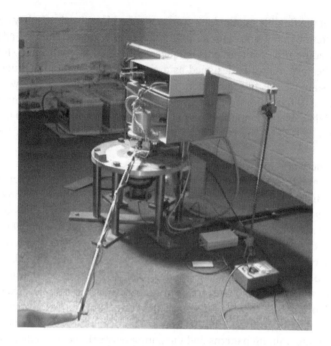

Figure 1.1   The Sheffield flexible manipulator system

1.4.1   Typical FMSs

Typical FMS configurations have not changed much since the early experimental systems (Book *et al.*, 1986; Cannon and Schmitz, 1984). The current commercial availability of sensor and actuator hardware has made it much easier to build an experimental FMS. The Sheffield manipulator at the University of Sheffield (UK) and the IST manipulator at the Technical University of Lisbon (Portugal) are good examples of the many experimental flexible manipulators used for research purposes (Martins *et al.*, 2003). A typical FMS has a flexible link, an actuator-gear mechanism to rotate the link, an optical encoder to measure joint rotation, accelerometers and strain gauges to sense flexible motion, an optical arrangement to measure the end-point position and an occasional force sensor attached to the end-point. There are variations in configuration among different FMS setups, for example, many setups have directly driven d.c. motors and others use harmonic drive gears with d.c. motors, some use only accelerometers or strain gauges to measure flexible deflections while others have cameras, some have semi-rigid flexible links while others have very flexible links. The sensor and actuator hardware is available in a wide variety and researchers make a selection to meet the needs of their experimental research. As a good example illustrating such variations in experimental FMS setups, first the Sheffield manipulator, shown in Figure 1.1 and second the IST manipulator, shown in Figure 1.2 are considered.

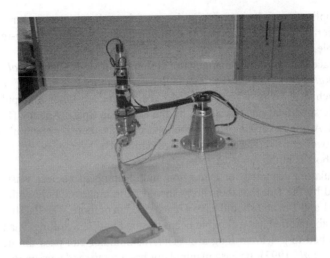

Figure 1.2   The IST flexible manipulator system

The Sheffield manipulator is a directly driven 0.9 m long aluminium alloy (mass density 2 710 kg/m^3, Young's modulus 71 × 10^9 Pa) flexible link (Tokhi and Azad, 1997). It has a width of 0.0032 m, a height of 0.019 m and a cross-section moment of inertia of 5.253 × 10^{-11} m^4. The actuator is a U9M4AT printed circuit armature motor driven by an Electro-Craft Corporation LA56 000 linear drive amplifier; the hub-inertia is 5.8598 × 10^{-11} kgm^2. The joint rotation is measured by a 2 048 pulse/rev encoder, a tachometer senses the angular velocity and a miniature accelerometer (sensitivity 1.02 mv/ms^2) measures the end-point acceleration. The manipulator is constructed so that the link is flexible in the horizontal plane and stiff in vertical bending and torsion.

The IST manipulator is a very flexible manipulator designed to experiment with force control methods. It is a 0.5 m long link made of spring-steel (mass density 7 850 kg/m^3, Young's modulus 209 × 10^9 Pa) driven by a harmonic drive (gear ratio 50) coupled servo system. A 12A8 servo amplifier from Advanced Motion Control drives the RH-14-6002 harmonic drive servo system. The link has 0.001 m width, 0.02 m height and hub-radius of 0.075 m. A 2 000 pulse/rev shaft encoder measures the angular displacement and a SR-series Kodak motion corder analyser (camera) with an acquisition rate of 1 000 frames/s senses the end-point position.

1.4.2   Flexible manipulators for industrial applications

In many applications a long-reach manipulator is essential. The structure of these long-reach robots is designed to minimise vibrations. In spite of careful design, their response time is governed by structural vibration. Active damping control is essential to improve response times of these long-reach manipulators. Since the response time is a critical performance parameter there is a great deal of interest in experimental work on these robots. Space robotics and inspection of nuclear waste storage tanks are two applications where prototype long-reach application robots are in service, although, it may be a few years before commercial mass-produced long-reach robots are available.

The space station remote manipulator system (SSRMS) (Stieber *et al.*, 1999) has a 116 000 kg/1 500 kg, payload-to-manipulator mass ratio. It is a 7-DOF manipulator with a reach of 17.6 m when fully stretched. There are other manipulators with comparable specifications developed by the Canadian Space Agency (Canadarm2) and NASA. Among a variety of long-reach space robots, there is a manipulator that can move end-over-end to cover the entire space station. The average power consumption of these robots is in the range of 350–450 W.

Manipulators designed for inspection and cleaning of nuclear waste tanks are constrained by the fact that their body should fit in a narrow hole with a diameter between 0.1 and 1 m and should have a reach of 25 m or more. The Pacific Northwest National Laboratories has a flexible beam test-bed with a Schilling Titan II manipulator attached to a 4.17 m long arm. This manipulator can pass through a 0.3 m hole (Kress *et al.*, 1997). Its base manipulator has a rotary and a prismatic joint. This gives it 2 DOF. The rotary motion orients the arm and the prismatic joint gives it a telescopic motion to set its reach. Another manipulator in this class of long-reach arm with a dexterous manipulator at the end is the MIT spatial long-reach manipulator system (LRMS). The so-called Shaky II experiment uses a 1.5 m arm with a PUMA 250 at its end.

1.4.3 Multi-link flexible manipulators

Figure 1.3 shows an experimental aerospace dual-arm flexible manipulator (ADAM) at the Tohoku University, Sendai, Japan (Miyabe *et al.*, 2001). The manipulator has two arms with seven joints and two flexible links in each arm. The length of each link is 0.5 m, elbow mass is 6 kg, and the combined wrist and end-effector mass is 2.9 kg. Force/torque sensors are attached to the end-effectors to measure contact force/torque. Link deflections are measured by strain gauges. Laser displacement sensors measure the distance between the end-effector and the surface of the object to be inspected (Miyabe *et al.*, 2001; Space Machines Laboratory, 2004a).

Sandia National Laboratories (2004) have several MLFM experimental test facilities. These manipulators are mostly driven by hydraulic actuators and are designed for handling heavy payloads.

A four-link manipulator facility is available at the Control Engineering Laboratory, Ruhr-University Bochum, Germany (Wang *et al.*, 2002). Each link, 0.6 m long, is a composite of a rigid and a 0.24 m long flexible part. All the four joints are driven by d.c. motors with harmonic gear drives. Elastic deflections of the links are measured by strain gauges. The mass at joints one and two is 9.16 kg and at three and four is 6.14 kg.

1.4.4 Two-link flexible manipulators

Many two-link flexible manipulator (TLFM) experimental test-beds have been set up over the last couple of decades but with a few exceptions the overall structure is not

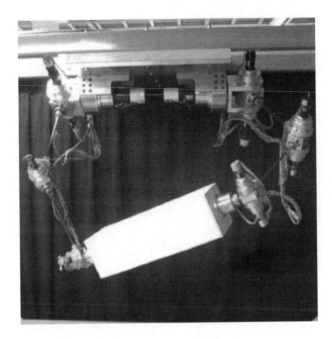

Figure 1.3   Aerospace dual-arm flexible manipulator

much different from the early setups. Most TLFMs are constrained to the horizontal plane. One of the earliest experimental TLFM was built at Stanford University for the purposes of demonstrating non-collocated control (Oakley and Cannon, 1990). Both links in this manipulator are driven by directly connected d.c. motors in the horizontal plane, the joint rotation is measured by rotary variable differential transformers (RVDTs), a CCD television camera is used to measure the end-point position. The manipulator is made from very flexible links and the highest modelled mode is at 12 Hz.

Figure 1.4 shows a 3-DOF, TLFM system named as FLEBOT II (Space Machines Laboratory, 2004*b*). It has two flexible links and three joints and is designed for research in space robotics applications. This setup is different from other manipulators; its workspace includes both vertical and horizontal planes.

The TLFM at the University of Washington (Bossert *et al.*, 1996) is for position and force control research. Its motion is limited to the horizontal plane. It is driven by two direct drive d.c. motors (continuous torques 3.39 Nm and 0.424 Nm), joints angles are measured by 1 024 cycles/rev optical encoders, the end-point position is sensed by a three-dimensional dynaSight bi-optic infrared tracking system, and it has a 6-DOF force/torque sensor attached to its end-point.

A TLFM setup as an experimental space robot simulator has been reported in Romano *et al.* (2002), where both flexible links (0.5 m and 0.51 m) are actuated by d.c. motors with 50 times gear ratio harmonic drives. The manipulator is supported by four airpads on a granite table. Joint angular position is obtained with 1 800 pulse/rev

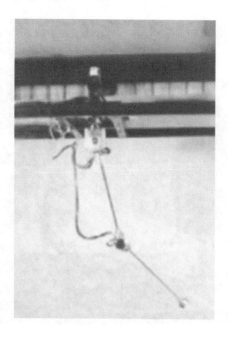

Figure 1.4   A 3-DOF two-link flexible manipulator

optical encoders mounted on motor shafts. End-point position is measured by a monochrome 640×480 pixel PLUNIX TM-6701 (60 Hz refresh frequency) camera. The resolution and accuracy of this camera is poor and to back it up an accelerometer is used to measure end-point oscillations.

The Duisburg flexible manipulator (DFM01) is another TLFM system reported in Bai *et al.* (1998). Its two links (0.571 m and 0.465 m), driven by geared d.c. motors, are made of single spring steel strip and have a limited range ($\pm 35°$ link 1 and $\pm 45°$ link 2). The manipulator has only horizontal motion and the end-effector is supported by an airpad on a plastic table. Optical encoders are used to measure angular rotation, deformation of links is measured by two strain gauges mounted on each link and an optical three-dimensional sensor is used to measure the end-point position.

The TLFM at Keio University, Japan (Kino *et al.*, 1998) drives the first link (0.45 m) with a directly driven d.c. motor and the second link (0.43 m) with a harmonic geared motor, the end-point position is measured by a position sensitive detector (something like a camera) mounted above the TLFM.

Most flexible manipulator facilities use external optical sensing device for end-point position measurement. But there are many experimental facilities, which use strain gauges for flexibility measurements. For example, the TLFMs at IRCCyN (Benosman and Le Vey, 2002), PSTECH (Cheong *et al.*, 2001), University of Missouri (Xu *et al.*, 2001), and The National Taiwan University (Lin and Fu, 1998) use strain gauges and optical encoders to measure the end-point elastic displacement.

Figure 1.5   A two-link joint flexible macro manipulator carrying a two-armed micro manipulator with a 'satellite' 3DOF two-link flexible manipulator

Achieving high-performance control of very flexible structures is a challenging task, but one that is critical to the success of many important applications. Manipulation with very flexible structures, such as the Space Shuttle RMS, is a very challenging task. Figure 1.5 demonstrates a two-link joint flexible macro/micro-manipulator performing this kind of task. The system was developed by the Aerospace Robotics Laboratory of Stanford University. (Stevensand and How, 1996). The macro-part of this manipulator features two 1.5 m long links, each with built-in joint flexibility. The two micro-arms are similar to the ones on the free-flying space robots. The arm operates in a two-dimensional space-like environment: the end-point floats on an air bearing over a large, flat granite table. Also shown is a small free floating object that the robot can catch and manipulate. An overhead vision system senses the end-point position of the arms as well as the position of the object.

The goal of this research is to remove the human from the low-level control of moving each arm. Instead, the human should give tasks for the robot to complete autonomously. It is desired to have a low-level control loop that can accurately control the end-point of each micro-arm. While much research has been done on controlling redundant robotic systems, many of these techniques break down when there is flexibility between the motor and the robot. Current research is on using techniques such as back-stepping to account for the joint flexibility in combination with rigid robot redundancy management schemes.

1.4.5   Single-link flexible manipulators

Most SLFMs have similar configuration to the Sheffield and IST manipulators described earlier. SLFMs reported in Nagarkatti *et al.* (2001) and Li and Chen (2001) each use a camera to obtain the end-point position. The system reported in Su and Khorasani (2001) uses a photodiode, with a camera lens to focus light on

Figure 1.6   A flexible 2-DOF single-link manipulator

the photodiode and an infrared emitting diode, for sensing end-point deflection. Two SLFMs, which have different configurations from most other setups, are discussed next.

The very flexible arm at IPNG-CNRS-UJF (Landau *et al.*, 1996) is made of two aluminium strips (1 m long, 0.1 m wide and 0.0008 m thick) coupled by ten regularly spaced rigid frames. A d.c. motor is used an as actuator and 1 250 pulses/rev encoder with a one-turn potentiometer (4.7 kΩ) measures the angular rotation. The SLFM can rotate 360°. The sensor to measure the end-point deflection is rather unique. It is a combination of an LED at the end-point and a mirror-detector at the hub. The mirror is centred on the hub axis and is rotated by a d.c. motor. The light from the LED can fall on the detector only at one position of the mirror (when the incidence and reflected angles are equal) and that gives the end-point deflection.

Another interesting SLFM is the experimental single-link 2-DOF flexible manipulator shown in Figure 1.6 (Goh *et al.*, 2000*b*). It has been constructed according to the specifications shown in Table 1.1. The 2 DOF allow the manipulator to move in both horizontal and vertical planes, allowing any coupling between the motions to be studied. The flexible link is a homogeneous, cylindrical, aluminium rod with constant properties along its length. Unlike many other experimental setups, the motion of the link is not artificially constrained in any plane within the safe working envelope. The link is symmetrical about its longitudinal axis, and this axis intersects both horizontal and vertical drive axes at the hub. Horizontal and vertical end-point displacements of the link are calculated based on two measurements. First, the position of the link end-point in relation to the hub axis (θ_{flex}) is measured using a precision potentiometer. This is combined with the angle of the hub relative to horizontal or vertical axis

Table 1.1   Specifications for a flexible 2-DOF single-link manipulator

Flexible link	Length (L)	1.00 m
	Mass (M_l)	0.35 kg
	Offset from motor axes	0.06 m
	Flexural rigidity (β)	72.2 Nm^2
Payload	Mass (M_p)	0.79 kg
	Centre of gravity offset	0.25 m
Hub	Horizontal inertia (I_{hy})	0.468 kgm^2
	Vertical inertia (I_{hz})	0.249 kgm^2
Torque motors	Maximum torque	33 Nm
	Rotor inertia	0.013 kgm^2
	Torque constant	1.2 Nm/A
	Maximum control signal	$\pm$ 8.5 V
Encoders	Line count	3 600
(incremental)	Resolution ($\times$4 counting)	4.4×10^{-4} rad
Potentiometers	Linearity	$\pm$ 0.2%
	Linearity with respect to end-point position	$\pm 3 \times 10^{-3}$ rad
Interface	A/D (0–10 V differential input)	14 bit
	D/A ($\pm$ 10 V outputs)	16 bit
	Encoder interface card	24 bit

(θ) measured by an incremental optical encoder. These signals are interfaced to a PC that is used to implement the control algorithms.

1.5   Applications

The properties and capabilities provided by flexible manipulators stand for a clear challenge in opening new applications for robots. Situations where the workspace is constrained, or when it is required to perform operations such as assembly in space, prevents the use of classical, rigid-link, industrial robot configurations. For these applications structural mass and stiffness must be reduced, to allow entering very confined workspaces and/or to permit cost-effective launching, and to enlarge manipulator reach out and dexterity. This could be of interest not only in space applications, but also in the industrial sector. Flexible manipulators, equipped with an active vibration control system, can reach quite the same accuracy of traditional industrial robots with low mass of moving parts and reduced cost and power consumption. Some known examples are the application of fast, flexible manipulators in the food industry (robotic packing and palletising) and in assembly tasks.

Flexibility is also becoming an important issue for other fields such as machine tools and civil engineering machinery, for example, tunnel boring machines, excavators, and so on, where requirements for extending tools life, increasing accuracy and speeding up overall performance entail making control systems fully aware of true system dynamics. Existing robotic systems tend to overkill, they are too complex, which makes them expensive to purchase and maintain.

As the potential of flexible manipulator technology is being demonstrated in laboratories, and some results are moving to industry, new ideas on their applicability are arising. One very important aspect of flexible robot technology is their intrinsic capability to accommodate forces with the environment, which could be a main issue when developing robots for direct cooperation with humans. Thus, a number of foreseen robot applications regarding safety and dependability could take clear advantage of structural flexibility. The emerging humanoid robots, where a clear need in mass reduction is mandatory for their operation, constitute undoubtedly an area where flexible robot technology is to play a major role in the future.

Although numerous potential industrial applications of flexible robotic manipulator systems have been identified, there are a number of technological issues, which need to be addressed before the industry can accept flexible robotic manipulator systems. These are development and study of flexible manipulator construction material, efficient actuation and sensing technologies, and simple and effective controller designs. Developments in these areas require assistance from government agencies and investment from industries.

Despite all of these, flexible robotic manipulators are in use to some extent in space applications. This is because of the weight restriction for a spacecraft. One such manipulator is made of a composite rod that is lifted by a longitudinal rope actuator and has an end-effecter gripper with bending electroactive polymer (EAP) driven fingers allowing to grab and hold an object (Bar-Cohen *et al.*, 1998). The EAP surface wiper operates like a human finger and can be used to remove dust from windows and solar cells. Other potential areas of application are manipulation in nuclear and other hazardous environments, car/vehicle painting, manufacturing of electronic hardware and food industry.

1.6 Summary

This chapter has presented a general overview of flexible robot manipulators. Research on FMSs ranges from single-link manipulators rotating about fixed axes to three-dimensional multi-link arms. However, experimental work, in general, is almost exclusively limited to single-link manipulators. This is owing to the complexity of multi-link manipulator systems, resulting from more DOF and the increased interactions between gross and deformed motions. It is noted that the first experimental SLFM systems were developed in the early 1980s. Although research in the control of single-link setups still continues, a shift of attention to MLFMs is currently noted.

The control of flexible manipulators is a challenge and researchers have faced up to such challenges through numerous investigations at devising various modelling, simulation and control strategies. Such investigations, which are still in progress at various research laboratories worldwide, have led to the realisation of the potential of flexible robotic manipulators in a number of existing and emerging applications including space exploration, nuclear industries and manufacturing industries. Moreover, opportunities for future applications are opening up.

It is noted that flexible manipulators have several advantages as compared to their rigid counterparts and accordingly are potential candidates in a variety of industrial and service sector applications. Despite all of this, flexible robotic manipulators are only in use to some extent in space applications because of the tight weight restriction for a spacecraft.

Although numerous further potential industrial and service sector applications of flexible robotic manipulators have been identified, the realisation of these will depend on further research and development work at structural design of manipulators and further efficient modelling and control approaches. Developments in these areas also require institutional collaboration, assistance from government agencies and investment from the industry.

Chapter 2

Modelling of a single-link flexible manipulator system: Theoretical and practical investigations

A.K.M. Azad and M.O. Tokhi

In this chapter, an analytical model of a single-link flexible manipulator, characterised by a set of infinite number of natural modes, is first developed. This is used to develop state-space and equivalent frequency-domain models of the system. These models can further be used for controller design exercises. An experimental flexible manipulator system is used for identifying model parameters. The model parameter identification procedure involves spectral analysis of collected input–output data from the experimental system. The identified parameters are then used with the developed model and the model response is verified with the experimental system.

2.1   Introduction

In practice, dynamic models are mostly formulated on the basis of considering forward and inverse dynamics. In this manner, consideration is given to computational efficiency, simplicity and accuracy of the model. Here, a means of predicting changes in the dynamics of the manipulator resulting from changing configurations and loading is proposed, where predictions of changes in mode shapes and frequencies can be made without the need to solve the full determinantal equation of the system. It is important for control purposes to recognise the flexible nature of the manipulator system and to build a suitable mathematical framework for modelling of the system. The flexible manipulator under consideration is a distributed parameter system with rigid body as well as flexible movements. There are two physical limitations associated with the system: (a) the control torque can only be applied at the joint (hub), and (b) only a finite number of sensors of bounded bandwidth can be used and at restricted locations along the length of the manipulator.

Owing to the principles used various types of models of flexible manipulator have been developed (Kanoh *et al.*, 1986). As indicated in Chapter 1, these include the Lagrange's equation and modal expansion (Ritz–Kantrovitch) or assumed mode method, the Lagrange's equation and finite element method, the Euler–Newton equation and modal expansion, the Euler–Newton equation and finite element, and the singular perturbation and frequency-domain techniques.

In the Lagrange's equation and modal expansion (Ritz–Kantrovitch method), the deflection of the manipulator is represented as a summation of modes. Each mode is assumed as a product of two functions; one dependent on the distance along the length of the manipulator, and the other, a generalized coordinate, dependent on time. In principle, the summation amounts to an infinite number of modes. However, for practical purposes, a small number of modes are used. The Lagrange's equation and finite element method is conceptually similar to the above (assumed modes) method. Here the generalized coordinates are the displacement and/or slope at specific points (nodes) along the manipulator.

The Euler–Newton's method is a more direct means of calculating system dynamics. The rate of change of linear and angular momentum is derived explicitly, rather than via Lagrange's equation. Newton's second law is used to balance these terms with the applied forces (Raksha and Goldenberg, 1986). In simulation, or forward dynamics, the linear and angular momentums of the manipulator are unknown while the actuator forces are known. Expressing the former in terms of a set of assumed modes of finite elements leads to a dynamic model relating the time dependency of the modes/elements to the external forces. The basic approach in the Euler–Newton and assumed modes method is to divide the manipulator into a number of elements and carry out a dynamic balance on each element. For a large number of elements this is a very tedious process. On the other hand, it is far easier to include non-linear effects without complicating the basic model.

In the singular perturbation technique, the system characteristic modes are separated into two distinct groups: a set of low frequency or slow modes and a set of high frequency or fast modes. In the case of flexible manipulators, the rigid-body modes are the slow modes and the flexible modes are the fast modes. The dynamics of the system can then be divided into two subsystems. The slow subsystem is of the same order as that of the equivalent rigid manipulator. The slow variables are considered as constant parameters for the fast subsystem (Khorrami and Özgüner, 1988)

An alternative to modelling the manipulator in the time domain is to use a method based on frequency-domain analysis (Book and Majette, 1983; Yuan *et al.*, 1989). This method develops a concise transfer matrix model using the Euler–Bernoulli beam equation for a uniform beam. The weakness of this method is that it makes no allowance for interaction between the gross motion and the flexible dynamics of the manipulator, nor can these effects be easily included in the model. As a result, the model can only be regarded as approximate. In another approach a chain of flexible links is modelled by considering flexible multi-body dynamic approach, based on an equivalent rigid link system (ERLS), where an ERLS, which is the closest possible to the deformed linkage, is defined, in order to match at best the requirements for small displacement assumption. As the choice of ERLS is completely arbitrary, it could introduce artificial kinematic constraints, which in turn introduces modelling error (Giovagnoni, 1994).

In this chapter an analytical model of a SLFM, characterised by a set of infinite numbers of natural modes, is first developed. This is used to develop state-space and equivalent frequency-domain models of the system. Experimental model identification involves estimating a transfer function or some equivalent mathematical description of the system from measurements of the system input and output. To ensure the acquisition of high-quality data, an experimental set-up involving a single-link flexible manipulator system is designed so that the essential aspects of the measurement process requiring particular attention are considered. These include proper excitation of the structure, choice and location of suitable transducers, selection of conditioning amplifiers and filters and method of signal processing suitable for the system under consideration.

The manipulator is excited and the corresponding input and output signals measured and used to determine the dynamic characteristics of the system. As the manipulator is very lightly damped, two independent methods are used to extract the model parameters. The characteristics thus obtained are then used to validate the system model by comparing these with the corresponding analytical results. Investigations are also carried out to study the effect of payload on the characteristics of the manipulator. Experimental results, thus, obtained are presented and discussed.

2.2 Dynamic equations of the system

2.2.1 The flexible manipulator system

The flexible manipulator system under consideration is modelled as a pinned-free flexible beam, with a mass at the hub, which can bend freely in the horizontal plane but is stiff in vertical bending and torsion. The model development utilizes the Lagrange equation and modal expansion method (Hastings and Book, 1987; Korolov and Chen, 1989). To avoid the difficulties arising owing to time varying length, the length of the manipulator is assumed to be constant. Moreover, shear deformation, rotary inertia and the effect of axial force are neglected.

A schematic representation of the SLFM is shown in Figure 2.1 where a manipulator with a moment of inertia I_b, hub inertia I_h, a linear mass density ρ and length l is considered. The payload mass is M_p and I_p is the inertia associated with the payload. A control torque $\tau(t)$ is applied at the hub of the manipulator by an actuator motor. The angular displacement of the manipulator, moving in the *POQ*-plane, is denoted by $\theta(t)$. The height of the link is assumed to be much greater than its width, thus, allowing the manipulator to vibrate (be flexible) dominantly in the horizontal direction. The shear deformation and rotary inertia effects are also ignored.

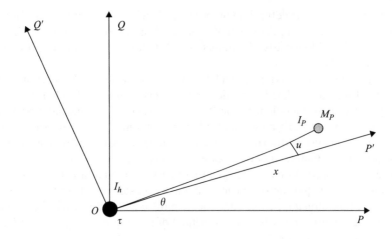

Figure 2.1   Schematic representation of the flexible manipulator system

For an angular displacement θ and an elastic deflection u, the total (net) displacement $y(x, t)$ of a point along the manipulator at a distance x from the hub can be described as a function of both the rigid-body motion $\theta(t)$ and elastic deflection $u(x, t)$ measured from the line OX.

$$y(x, t) = x\theta(t) + u(x, t) \tag{2.1}$$

To obtain equations of motion of the manipulator, the associated energies have to be obtained. These include the kinetic, potential and dissipated energies.

2.2.2   Energies associated with the system

The energies associated with the system include the kinetic, potential and dissipated energy. These are considered for the FMS in this section.

As the contribution of the rotational moment of inertia is neglected, the kinetic energy of the system can be written as

$$E_K = \frac{1}{2}I_h\dot{\theta}^2 + \frac{1}{2}\int_0^l \left(\frac{\partial u}{\partial t} + x\dot{\theta}\right)^2 \rho dx + \frac{1}{2}M_p \left(\frac{\partial u}{\partial t} + x\dot{\theta}\right)^2_{x=1} \tag{2.2}$$

Note in the above equation that the first term on the right-hand side is because of the hub inertia, the second term is because of the rotation of the manipulator with respect to the origin and the third term is because of the payload mass. Moreover, note that only small elastic deflection and small angular velocity are considered.

The potential energy is related to the bending of the manipulator. Since the height of the manipulator under consideration is assumed to be significantly larger that its thickness, the effects of shear displacements can be neglected. In this manner, the potential energy of the manipulator can be written as

$$E_P = \frac{1}{2}EI \int_0^l \left[\frac{\partial^2 u}{\partial x^2}\right]^2 dx \tag{2.3}$$

where E and I are the Young's modulus of the manipulator material and the moment of inertia of the cross-sectional area (second moment of inertia) of the manipulator, respectively. Note that, in general, there will be motion of the manipulator in the vertical plane as well, in the form of permanent elastic deflections caused by gravitational forces. This effect, however, is ignored as the flexibility of the manipulator has been minimised in the vertical direction.

To consider the mechanism by which energy is absorbed from the structure during its dynamic operation, let the resistance to transverse velocity be represented by $D(x)$, the resistance to angular velocity at the hub by D_0 and the resistance to strain velocity by D_s. In this manner, the energy dissipated by the damping moment and force can be written as

$$E_F = \frac{1}{2}\int_0^l D(x)\left(\frac{\partial u}{\partial t}\right)^2 dx + \frac{1}{2}\int_0^l D_s I \left(\frac{\partial^3 u}{\partial x^2 \partial t}\right)^2 dx + \frac{1}{2}D_0\left(\frac{\partial^2 u}{\partial x \partial t}\right)^2_{x=0} \tag{2.4}$$

It has been shown that the damping matrix satisfies the orthogonality conditions and thus can be uncoupled in a similar manner as the inertia and stiffness matrices (Clough and Penzien, 1975). To satisfy the mode superposition analysis, it is assumed that $D(x) = b_0\rho A$, $D_s = b_1 E$ and $D_0 = b_0 I_h$, with b_0 and b_1 representing proportionality constants and A the cross-sectional area of the manipulator.

2.2.3   The dynamic equations of motion

The non-conservative work for the input torque τ can be written as

$$W = \tau\theta \tag{2.5}$$

To obtain the equations of motion of the manipulator, the Hamilton's extended principle described by (Meirovitch, 1970)

$$\int_{t_1}^{t_2} (\delta L + \delta W)\, dt = 0 \tag{2.6}$$

can be used, subject to $\delta\theta = \delta u = 0$ at t_1 to t_2, where, t_1 and t_2 are two arbitrary times $(t_1 < t_2)$ and $L = E_K - E_P$ is the system Lagrangian. δW represents the virtual work, $\delta\theta$ represents a virtual rotation and δu represents a virtual elastic displacement. Using equations (2.2), (2.3) and (2.5) the integral in equation (2.6) can be written as

$$\delta \int_{t_2}^{t_1} (E_K - E_P + W)dt = 0$$

The rotary inertia and shear deformation are more pronounced at high frequencies and more influential on the higher modes (Tse *et al.*, 1978). Since investigations have shown that the first two modes are sufficient in modelling the manipulator the rotary inertia and shear deformation effects can be ignored. Manipulation of the above equation yields the equation of motion of the manipulator as

$$EI\frac{\partial^4 u\,(x,t)}{\partial x^4} + \rho\frac{\partial^2 u\,(x,t)}{\partial t^2} = -\rho x \ddot{\theta} \qquad (2.7)$$

with the corresponding boundary and initial conditions as

$$u\,(0,t) = 0$$

$$I_h\frac{\partial^3 u\,(0,t)}{\partial t^2 \partial x} - EI\frac{\partial^2 u\,(0,t)}{\partial x^2} = \tau\,(t)$$

$$M_p\frac{\partial^2 u\,(l,t)}{\partial x^2} - EI\frac{\partial^3 u\,(l,t)}{\partial x^3} = 0 \qquad (2.8)$$

$$I_p\frac{\partial^3 u\,(l,t)}{\partial t^2 \partial x} + EI\frac{\partial^2 u\,(l,t)}{\partial x^2} = 0$$

$$u(x,0) = 0, \qquad \frac{\partial u(x,0)}{\partial x} = 0$$

Substituting for $u\,(x,t)$ from equation (2.1) into equations (2.7) and (2.8) and simplifying yields the governing equation of motion of the manipulator in terms of $y\,(x,t)$ as

$$EI\frac{\partial^4 y\,(x,t)}{\partial x^4} + \rho\frac{\partial^2 y\,(x,t)}{\partial t^2} = 0 \qquad (2.9)$$

with the corresponding boundary and initial conditions as

$$y\,(0,t) = 0$$

$$I_h\frac{\partial^3 y\,(0,t)}{\partial t^2 \partial x} - EI\frac{\partial^2 y\,(0,t)}{\partial x^2} = \tau\,(t)$$

$$M_p\frac{\partial^2 y\,(l,t)}{\partial x^2} - EI\frac{\partial^3 y\,(l,t)}{\partial x^3} = 0 \qquad (2.10)$$

$$I_p\frac{\partial^3 y\,(l,t)}{\partial t^2 \partial x} + EI\frac{\partial^2 y\,(l,t)}{\partial x^2} = 0$$

$$y(x,0) = 0, \qquad \frac{\partial y(x,0)}{\partial x} = 0$$

Equation (2.9) gives the fourth-order partial differential equation (PDE), which represents the dynamic equation describing the motion of the flexible manipulator. This equation could also be directly obtained by using Newton's law (Breakwell, 1980). The Hamilton's principle, however, is more convenient to use because it automatically generates appropriate boundary conditions.

2.3   Mode shapes

Using the assumed modes method (Meirovitch, 1970; Meyer, 1971), a solution of the dynamic equation of motion of the manipulator can be obtained as a linear combination of the product of admissible functions $\phi_i(x)$ and time-dependent generalized coordinates $q_i(t)$:

$$y(x,t) = \sum_{i=0}^{n} \phi_i(x)\, q_i(t) \quad \text{for} \quad i = 0,1,...,n \tag{2.11}$$

where, the admissible function, ϕ_i, also called the mode shape, is purely a function of the displacement along the length of the manipulator and q_i is purely a function of time and includes an arbitrary, multiplicative constant. The zeroth mode is the rigid-body mode of the manipulator, characterising the so-called rigid manipulator as considered without elastic deflection.

Substituting for $y(x, t)$ from equation (2.11) into equation (2.9) and manipulating yields two ordinary differential equations as

$$\frac{d^4\phi_i(x)}{dx^4} - \beta_i^4 \phi_i(x) = 0, \qquad \frac{d^2 q_i(t)}{dt^2} + \omega_i^2 q_i(t) = 0 \tag{2.12}$$

where

$$\omega_i^2 = \frac{EI}{\rho}\beta_i^4 \tag{2.13}$$

and β_i is a constant. Let two constants λ and ε be defined as

$$\lambda_i = \beta_i l, \quad \varepsilon = \frac{I_h}{Ml^2} = \frac{3I_h}{I_b} \tag{2.14}$$

where, M is the mass of the manipulator. The first relation in equation (2.12) is a fourth-order ordinary differential equation with a solution of the form,

$$\phi_i(x) = A_i \sin \beta_i x + B_i \sinh \beta_i x + C_i \cos \beta_i x + D_i \cosh \beta_i x \tag{2.15}$$

To find the natural frequencies and mode shapes of the system, the values of λ_i satisfying the boundary conditions of the undriven manipulator, $\tau = 0$, are determined together with the corresponding values of the coefficients A_i, B_i, C_i and D_i in equation (2.15). This requires utilization of the orthogonality properties and the mode shapes (Meirovitch, 1970). Using equation (2.15) and the boundary conditions in equation (2.10) yields

$$\int_0^l M\phi_i(x)\, \phi_j(x)\, dx + I_h \phi_i'(0)\, \phi_j'(0) + M_p \phi_i(l)\, \phi_j(l) + I_p \phi_i'(l)\, \phi_j'(l) = I_T \delta_{ij} \tag{2.16}$$

where δ_{ij} is the Kronecker delta and the normalization constant $I_T = I_h + I_b + I_p$ is the total inertia about the motor armature. Equation (2.15) uniquely defines A_i and, thus, the magnitude of the mode $\phi_i(x)$. From the properties of self-adjoint systems, the mode shapes must also satisfy the orthogonality condition

$$\int_0^l EI\phi_i''(x)\,\phi_j''(x)\,dx = I_T\omega_i^2\delta_{ij} \tag{2.17}$$

where $\phi_i'' = d^2\phi_i/dx^2$. The analytical values of natural frequencies ω_i can then be obtained using equation (2.14). ε determines the vibration frequencies of the manipulator; a small ε corresponds to the manipulator with lower vibration frequencies. For a very large ε the vibration frequencies correspond to those of a cantilever beam. The effect of a payload mass, on the other hand, is significant on the vibration frequencies. By considering the boundary conditions, the mode shape function $\phi_i(x)$ of the manipulator in equation (2.15) can thus be obtained.

2.4   State-space model

In the absence of an external torque, equation (2.7) describes the behaviour of the manipulator in free transverse vibration with a solution $u(x,t)$ as

$$u(x,t) = \sum_{i=1}^{n} q_i(t)\,\phi_i(x) \quad \text{for } i = 1, 2, \ldots, n \tag{2.18}$$

Utilizing equations (2.2), (2.3) and (2.18), and using the orthogonality properties in equations (2.16) and (2.17), the kinetic energy E_K and potential energy E_P of the system, in terms of the natural modes, can be obtained as

$$E_K = \frac{1}{2}\delta_{ij}I_b \sum_{i=1}^{n} \dot{q}_i^2$$

$$E_P = \frac{1}{2}\sum_{i=1}^{n}\sum_{j=1}^{n} q_i q_j \int_0^l EI\phi_i''\phi_j''\,dx = \frac{1}{2}\delta_{ij}I_T \sum_{i=1}^{n} \omega_i^2 q_i^2 \tag{2.19}$$

Similarly, using equations (2.4) and (2.5) the dissipated energy E_F and the work W can be obtained as

$$E_F = \frac{1}{2}(I_h + I_b)\,2\xi_i\omega_i^2 q_i^2$$

$$W = \tau\theta = \tau \sum_{i=0}^{n} \phi'(0)\,q_i \tag{2.20}$$

where $\xi_i = \frac{b_0}{2\omega_i} + \frac{b_1\omega_i}{2}$ is the damping ratio.

The dynamic equation of the system can now be formed using the kinetic energy, potential energy and dissipated energy in the Lagrangian of the energy expression given as (Tse *et al.*, 1978)

$$\frac{d}{dt}\left(\frac{\partial L}{\partial \dot{q}_i}\right) - \frac{\partial L}{\partial \dot{q}_i} + \frac{\partial E_F}{\partial \dot{q}_i} = W_i \tag{2.21}$$

where $L = E_K - E_P$, q_i represents the time-dependent generalized coordinates, W_i represents the work done by the input torque at the joint in each coordinate.

Substituting for E_K, E_P, E_F and W from equations (2.19) and (2.20) into equation (2.21) and using the orthogonality relations in equations (2.16) and (2.17), an infinite set of decoupled ordinary differential equations are obtained as

$$\ddot{q}_0 = \frac{\tau}{I_T}$$

$$\ddot{q}_1 + 2\xi_1 \dot{q}_1 + \omega_1^2 q_1 = \frac{d\phi_1(0)}{dx}\frac{\tau}{I_T}$$

$$\ddot{q}_2 + 2\xi_2 \dot{q}_2 + \omega_2^2 q_2 = \frac{d\phi_2(0)}{dx}\frac{\tau}{I_T} \tag{2.22}$$

$$\vdots$$

where, I_T is the total inertia of the system, $(I_T = I_b + I_h + I_m)$. In this manner, owing to the distributed nature of the system, there will be an infinite number of modes of vibration of the flexible manipulator that can be represented. However, in practice, it is observed that the contribution of higher modes to the overall movement is negligible. Therefore, a reduced-order model incorporating the lower (dominant) modes can be assumed. This assumption is justified by the fact that the dynamics of the system are dominantly governed by a finite number of lower modes (Hughes, 1987). Retaining the first $n + 1$ modes of interest, equation (2.22) can be written in a state-space form as

$$\dot{X} = \mathbf{A}X + \mathbf{B}\tau \tag{2.23}$$

where, $\dot{X} = dX/dt$,

$$X^T = \{\begin{array}{cccccc} q_0 & \dot{q}_0 & q_1 & \dot{q}_1 & \cdots & q_n & \dot{q}_n \end{array}\}$$

$$\mathbf{A} = \begin{bmatrix}
0 & 1 & 0 & 0 & 0 & 0 & \cdots & 0 & 0 \\
0 & 0 & 0 & 0 & 0 & 0 & \cdots & 0 & 0 \\
0 & 0 & 0 & 1 & 0 & 0 & \cdots & 0 & 0 \\
0 & 0 & -\omega_1^2 & -2\xi_1\omega_1 & 0 & 0 & \cdots & 0 & 0 \\
0 & 0 & 0 & 0 & 0 & 1 & \cdots & 0 & 0 \\
0 & 0 & 0 & 0 & -\omega_2^2 & -2\xi_2\omega_2 & \cdots & 0 & 0 \\
\vdots & \vdots & \vdots & \vdots & \vdots & \vdots & \cdots & \vdots & \vdots \\
0 & 0 & 0 & 0 & 0 & 0 & \cdots & 0 & 1 \\
0 & 0 & 0 & 0 & 0 & 0 & \cdots & \omega_n^2 & -2\xi_n\omega_n
\end{bmatrix}$$

$$\mathbf{B}^T = \frac{1}{I_T}\left\{\begin{array}{ccccccc} 0 & 1 & 0 & \dfrac{d\phi_1(0)}{dx} & \cdots & 0 & \dfrac{d\phi_n(0)}{dx} \end{array}\right\}$$

Let the manipulator be facilitated with three sensors: end-point acceleration sensor, hub-velocity sensor and hub-angle sensor. The output vector Y of these sensors is related to the state vector by

$$Y = \mathbf{C}X \tag{2.24}$$

where,

$$\mathbf{C} = \begin{bmatrix} 1 & 0 & \dfrac{d\phi_1^2\,(l)}{dx^2} & 0 & \cdots & \dfrac{d\phi_n^2\,(l)}{dx^2} & 0 \\[2mm] 1 & 0 & \dfrac{d\phi_1\,(0)}{dx} & 0 & \cdots & \dfrac{d\phi_n\,(0)}{dx} & 0 \\[2mm] 0 & 1 & 0 & \dfrac{d\phi_1\,(0)}{dx} & \cdots & 0 & \dfrac{d\phi_n\,(0)}{dx} \end{bmatrix}$$

with $\phi_i''(l)$ representing the end-point acceleration sensor modal gain and $\phi_i'(0)$ the actuator (motor) modal gain.

2.5 Transfer function model

For frequency-domain control design input/output relationships are usually expressed in a transfer function form. This allows the use of classical design methods such as Bode plots, Nyquist diagrams and root loci. The open-loop transfer function of the system, $G\,(s)$, is given as the ratio $Y\tau^{-1}$. Using the state-space equations, (2.23) and (2.24), this can be obtained as

$$G\,(s) = \mathbf{C}\,(s\mathbf{I} - \mathbf{A})^{-1}\,\mathbf{B} \tag{2.25}$$

where, $\mathbf{I}$ is the identity matrix of the same dimension as $\mathbf{A}$, and s is the Laplace transform variable. Alternatively, using a method developed by Breakwell (1980), the transfer function can be obtained directly using equation (2.13). Taking the Laplace transform of this equation yields the ordinary differential equation:

$$EI\frac{d^4\bar{y}\,(x,s)}{dx^4} + \rho s^2\bar{y}\,(x,s) = 0 \tag{2.26}$$

with the transformed boundary conditions as

$$\bar{y}\,(0,s) = 0$$
$$I_h s^2 \bar{y}'\,(0,s) - EI\bar{y}''\,(0,s) = \tau\,(s)$$
$$M_p\bar{y}''\,(l,s) - EI\bar{y}'''\,(l,s) = 0$$
$$I_p s^2 \bar{y}'\,(l,s) + EI\bar{y}''\,(l,s) = 0$$

where $\bar{y}$ denotes the Laplace transform of y and $\bar{y}''' = d^3\bar{y}/dx^3$. Equation (2.26) has the general solution

$$\bar{y}\,(x,s) = A_1 \sin \beta x + B_1 \cos \beta x + C_1 \sinh \beta x + D_1 \cosh \beta x$$

where $\beta^4 = -\rho s^2/EI$. Once the constants A_1, B_1, C_1 and D_1, and hence $\bar{y}(x, s)$ are known, it is possible to derive the transfer function from input torque to a particular output when the latter is expressed as a function of y. All the transfer functions that can be derived in this manner share a common denominator. In practice, the resulting expressions are complex transcendental functions of β. For the single-link manipulator with end-point mass and hub inertia the numerator function $N(\lambda)$ and the denominator function $D(\lambda)$ can be represented by their Maclaurin expressions as

$$D(\lambda) = \sum_{n=0}^{\infty} \frac{\lambda^n d^n D(0)}{n! d\lambda^n}, \qquad N(\lambda) = \sum_{n=0}^{\infty} \frac{\lambda^n d^n N(0)}{n! d\lambda^n}$$

As a consequence of the distribution of roots the Maclaurin expansions can be expressed as a product of quartic factors:

$$D(\lambda) = P_d(\lambda) \sum_{i=0}^{\infty} \left(1 - \frac{\lambda_i^4}{\lambda_{di}^4}\right), \qquad N(\lambda) = P_n(\lambda) \sum_{i=0}^{\infty} \left(1 - \frac{\lambda_i^4}{\lambda_{ni}^4}\right)$$

where P is a polynomial in λ of degree 3 or less. Using the relationship $\lambda^4 = (-\rho s^2 l^4/EI)$, the transfer function can be expressed in terms of the Laplace transform variable s. The final transfer function from the input torque to hub-angle θ, input torque to hub-velocity $\dot{\theta}$ and input torque to end-point acceleration α can be written respectively as

$$\frac{\theta(s)}{\tau(s)} = \frac{1}{I_T s^2} \sum_{i=1}^{\infty} \frac{\left(1 + s^2/\omega_{\theta i}^2\right)}{\left(1 + s^2/\omega_i^2\right)}$$

$$\frac{\dot{\theta}(s)}{\tau(s)} = \frac{1}{I_T s} \sum_{i=1}^{\infty} \frac{\left(1 + s^2/\omega_{vi}^2\right)}{\left(1 + s^2/\omega_i^2\right)} \qquad (2.27)$$

$$\frac{\alpha(s)}{\tau(s)} = \frac{1}{I_T s} \sum_{i=1}^{\infty} \frac{\left(1 + s^2/\omega_{\alpha i}^2\right)}{\left(1 + s^2/\omega_i^2\right)}$$

where $\omega_{ci}, \omega_{\alpha i}$ $\quad \omega_{ci}, \omega_{\alpha i}$ and α_{ti} are real constants corresponding to the system zeros.

2.6 Experimentation

The set-ups used for experimental identification process is shown in Figure 2.2. This consists of an aluminium type flexible manipulator of dimensions and characteristics given in Table 2.1, driven by a high torque printed-circuit armature type motor. The measurement sensors consist of an accelerometer at the end-point of the manipulator, a shaft encoder and a tachometer, both at the hub of the manipulator and four strain gauges located along the manipulator length at uniform spacing.

The outputs of these sensors as well as a voltage proportional to the current applied to the motor are fed to a personal computer through a signal conditioning circuit and an anti-aliasing filter for further analysis. The cut-off frequency of the anti-aliasing

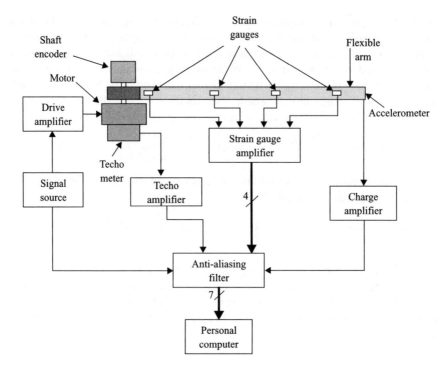

Figure 2.2   Experimental set-up for model parameter identification

Table 2.1   Parameters of the flexible manipulator system

Parameter	Symbol	Formula	Value
Length	l	—	960 mm
Width	b	—	19.008 mm
Thickness	d	—	3.2004 mm
Material	—	—	Aluminium
Density	ρ	Constant	2710 kg/m^3
Cross-sectional area	a	$b \times d$	6.08332×10^{-5} m^2
Mass	m	$a \times l$	0.158264 kg
Second moment of inertia	I	$(1/12)\, b \times d^3$	5.19238×10^{-11} m^4
Young modulus	E	Constant	71×10^9 N/m^2
Rigidity	EI	$E \times I$	3.6866 NM^2
Moments of inertia	I_b	$(1/3)\, ml^2$	0.04862 kgm^2
Hub inertia	I_h	Measured	5.86×10^{-4} kgm^2

filter is set appropriately to satisfy the sampling requirements. In this process the measurement chain is adjusted to achieve a good signal-to-noise ratio; the amplitude of the noise is kept within 5 mV with the signal level within 10 V. An RTI-815 multi-function analogue/digital (A/D) and digital/analogue (D/A) board is used as an input/output (I/O) hardware unit.

2.6.1 Natural frequencies

In the case of a flexible manipulator system only the first few modes dominantly characterise the system behaviour. Considering this, only the parameters associated with the first two modes are extracted. A sampling period of 5 ms is used which accommodates comfortably the first and second modes within the measurements (Ziemer *et al.*, 1998). In the experimental investigations to follow, two approaches for obtaining the required information are used:

- The manipulator system is excited by a random signal from a noise generator as a signal source over a frequency range of 0–50 Hz, which covers the first two flexible modes of the system. The responses at various points are measured with a resolution of 0.448 Hz and fed to the computer through a set of amplifiers and filters. The collected data is then analysed to obtain the overall frequency response function (FRF).
- The manipulator system is excited by a stepped sine wave from a Solartron 1 170 spectrum analyser and the response around the pole and zero frequencies is measured. Here the measurement is made only around the pole and zero frequencies with a resolution of 0.001 Hz. The total FRF plot is obtained by combining the results of the two methods of measurement.

To obtain the natural frequencies of vibration of the flexible manipulator system, two methods are introduced and their utilization explored. The first method is based on the measurement of the autopower spectral density of the response of the system. This is referred to as the spectral density method. The second method is based on the measurements of the FRF and coherency function of the system. This is referred to as the FRF method. The former method is used extensively and demonstrated throughout this book (see Chapters 8, 12 and 15) in determining the level of vibration at resonance frequencies of flexible structures.

The autopower spectral density $S_{xx}(\omega)$ of a signal x is defined as

$$S_{xx}(\omega) = S_x(j\omega) S_x^*(j\omega) \tag{2.28}$$

where $S_{xx}(\omega)$ is a real valued function containing the magnitude information only, $S_x(j\omega)$ is the linear spectrum of x given by the Fourier transform of the time signal $x(t)$. $S_x^*(j\omega)$ is the complex conjugate of $S_x(j\omega)$.

The response of the system can alternatively be described by the frequency response function. The equations relating to the response of a system in random vibrations to the excitation are given as (Newland, 1996)

$$S_{yy}(\omega) = |H(j\omega)|^2 S_{xx}(\omega)$$
$$S_{xy}(j\omega) = H(j\omega) S_{xx}(\omega) \qquad (2.29)$$
$$S_{yy}(\omega) = H(j\omega) S_{yx}(j\omega)$$

where $S_{xx}(\omega)$ and $S_{yy}(\omega)$ are the autopower spectral densities of the excitation signal x and the response signal y, respectively. $S_{xy}(j\omega)$ and $S_{yx}(j\omega)$ are the cross-spectral densities between these two signals and $H(j\omega)$ is the FRF of the system. Let $H_1(j\omega)$ and $H_2(j\omega)$ denote two estimates of the FRF obtained according to the relations in equation (2.29) as

$$H_1(j\omega) = \frac{S_{xy}(j\omega)}{S_{xx}(\omega)} \quad \text{and} \quad H_2(j\omega) = \frac{S_{yy}(\omega)}{S_{yx}(j\omega)} \qquad (2.30)$$

The error between the two functions is given by the coherency function defined as

$$\gamma^2(\omega) = \frac{H_1(j\omega)}{H_2(j\omega)} \qquad (2.31)$$

In this manner, the coherency function gives a measure of the estimation error and indicates the level of coherence between the input and output. If γ^2 is unity at some frequency ω, this means that the output is entirely owing to the input at that frequency. However, a value of γ^2 less than unity means that either the output is owing to the input as well as other inputs or the output is corrupted with noise. When analysing the extracted signal, it is not sufficient to compute the Fourier transform of the signal. Instead an estimate for the spectral densities and correlation functions should be obtained, which are used to characterise the extracted signal. Although these are computed from the Fourier transform, there are additional considerations concerning their accuracy and statistical reliability. It is, additionally, necessary to perform an averaging process, involving several samples of the measurement, before a result is obtained. The two major factors, which determine the number of averages required, are the statistical reliability of the results and the removal of random noise from the signal (Bendat and Piersol, 1986; Newland, 1996). To overcome the leakage problem, a hanning type window function is employed before processing each block of data. MATLAB is used here to analyse the data and obtain the FRF, autopower spectral density and the coherency function (Little and Shure, 1988). This incorporates the use of an averaging method to remove the noise associated with the signal.

The system was excited using a random signal as the signal source, and the time response of the system was measured at the hub, at the four strain gauge locations and at the end-point. These were used to obtain autopower spectral densities of the signals using equation (2.28) and the FRFs and coherency functions between the input torque and the system response at these points using equations (2.29), (2.30) and (2.31). These are shown for hub-angle, strain at location-1 and end-point acceleration, in Figures 2.3–2.6. The pole and zero frequencies of the flexible manipulator system for the first two modes, as obtained by identifying the peaks (maximum amplitudes) and valleys (minimum amplitudes) in the autopower spectral density

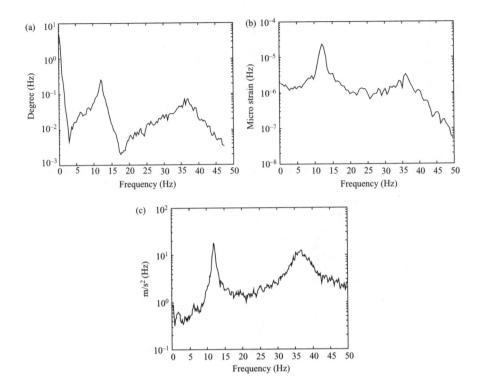

Figure 2.3   Autopower spectrum of signals at various measurement locations: (a) hub-angle, (b) strain at location-3, (c) end-point acceleration

functions and the FRFs are shown in Table 2.2 along with the corresponding analytical values.

As seen, a reasonable level of accuracy in the extraction of pole and zero frequencies from autopower spectrum and the FRF is achieved. It is noted in Figures 2.4–2.6 that the coherence shows some error around the pole frequencies with the hub-angle measurement and at the lower frequency range with the strain gauge location-3 and end-point acceleration measurement. There are various reasons for the coherence not to be unity. These include the presence of noise on one or other of the two signals, improper coupling between the structure and the excitation, non-linearity and low frequency resolution.

In the measurements above low coherence is mostly owing to the lower frequency resolution of the analysis programme and partly owing to the non-linearity of the structure. As the frequency response is measured near the pole and zero regions with better accuracy using the spectrum analyser, it is observed that there is a small variation in pole and zero frequencies with the location of measurement. It is also observed that these frequencies increase with the distance between the excitation of the structure and the response measurement location. Note that the error in pole frequencies decreases with increasing the distance between the excitation and the

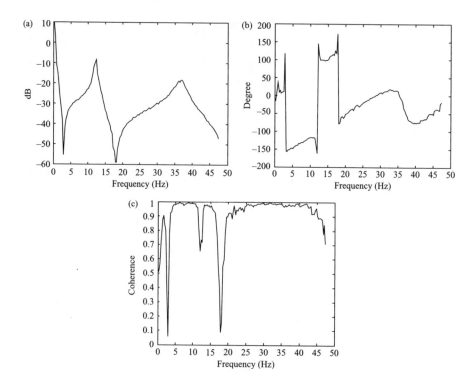

Figure 2.4   Transfer function from torque input to shaft encoder output

response measurement location. This is because the two corresponding vibration modes of the structure become more pronounced in amplitude along the length of the manipulator with distance from the hub. The error trend in zero frequencies, on the other hand, is variable with varying the distance between the excitation and the measurement point.

For better accuracy to be achieved near the pole and zero frequencies, the system was excited with a stepped sinusoid signal from a spectrum analyser. The input and output measurements of the system were used to obtain the FRF around the pole/zero regions. The complete FRF was then obtained by combining the response with random excitation and the stepped sinusoid excitation. The pole and zero frequencies thus obtained are shown in Table 2.3. The difference between the values obtained for poles and zeros using random excitation and stepped sinusoid excitation demonstrate the need for higher resolution around the pole and zero frequencies for the measurement of FRF. The pole and zero frequencies thus obtained using the spectrum analyser will be used in subsequent calculations.

It follows from the transfer functions in equation (2.27) that the poles and zeros of the manipulator are functions of the loading conditions. These indicate that the system poles or natural frequencies are functions of both the hub inertia and payload whereas the system zeros are functions of payload mass. To investigate this further,

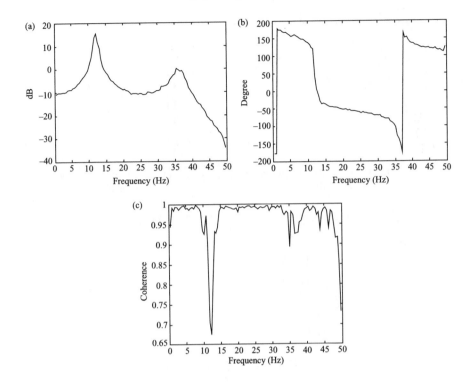

Figure 2.5   Transfer function from torque input to strain at location-3: (a) magnitude, (b) phase, (c) coherence

the effect of payload on the pole and zeros of the system at the hub was studied. This was done by exciting the system with a random torque input and measuring the response at the shaft encoder with various payloads at the end-point. The response was analysed to obtain the FRF and, thus the pole and zero frequencies. Figure 2.7 shows the first two poles and zeros of the manipulator as a function of the payload. As noted, the variation in each case is predominant for small payloads.

Figure 2.7 demonstrates that as the mass increases there will be a noticeable change in the response of the manipulator; increased inertia owing to additional load leads to a reduction in the overall displacement of the manipulator in a fixed time period. It is noted that as the payload increases the system zeros migrate from cantilever beam frequencies and coverage to the corresponding natural frequencies of a clamped beam where the latter represents the theoretical limit as the payload mass tends to infinity. These observations are important in the development of suitable control strategies for flexible manipulator systems.

2.6.2   Damping ratios

There are several possible forms of damping within the system. Depending on the source, these can be classified into three groups: (a) viscous damping and Coulomb

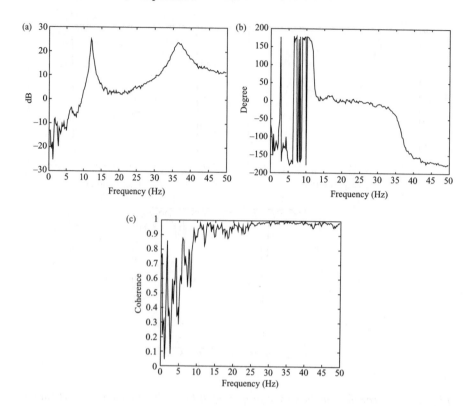

Figure 2.6   Transfer function from torque to end-point acceleration signal: (a) magnitude, (b) phase (c) coherence

damping (stiction/friction) associated with driving motor, (b) external effects such as primarily air resistance as the manipulator rotates and (c) structural damping owing to dissipation of energy within the manipulator material. The Coulomb damping associated with the motor is a constant retarding torque, which always acts in the opposite direction to the velocity of the manipulator. This is overcome by introducing a constant voltage bias in the input torque to result in a torque equal and opposite to the frictional torque. The sign of the bias voltage will vary according to the direction of motion of the hub. The damping because of air resistance was not considered in the experiment. However, a comparison of the model with the system reveals that this damping does not affect the total damping of the system significantly.

To measure the damping coefficients for the flexible modes a stepped sine input was used from the spectrum analyser to excite the system and the frequency response of the system was measured. The frequency response data was then used to draw a Nyquist plot from which the damping ratio can be obtained. The method is based on the principle that, in the vicinity of a resonance, the behaviour of a system is dominantly determined by the corresponding resonance mode being observed (Silva

Table 2.2 Comparison of the analytical and experimental pole and zero frequencies

Method	Measurement location	Pole 1 (12.499 Hz)		Pole 2 (12.499 Hz)		Zero 1 (12.499 Hz)		Zero 2 (12.499 Hz)	
		Measured (Hz)	Error (%)	Measured (Hz)	Error (%)	Measured (Hz)	Error (%)	Measured (Hz)	Error (%)
Spectral density	Shaft encoder	11.719	6.24	37.109	2.06	2.93	2.06	17.08	5.08
	Tachometer	12.207	2.34	36.133	0.76	2.93	2.06	17.568	2.37
	Location-1	11.719	6.24	35.156	3.31	3.418	19.05	13.184	26.73
	Location-2	12.207	2.34	35.645	1.97	3.906	36.05	20.508	13.97
	Location-3	12.207	2.34	36.133	0.76	—	—	20.461	13.71
	Location-4	12.207	2.34	36.133	0.76	—	—	20.461	13.71
	Accelerometer	12.207	2.34	35.645	1.97	3.418	19.05	20.019	11.25
	Average	12.068	3.35	35.993	1.01	3.027	5.43	18.754	4.22
FRF	Shaft encoder	12.207	2.34	36.133	0.76	2.93	2.06	17.09	5.02
	Tachometer	12.207	2.34	36.133	0.76	2.93	2.06	18.066	0.40
	Location-1	11.719	6.24	36.133	0.76	3.418	19.05	13.184	26.73
	Location-2	12.207	2.34	36.133	0.76	3.906	36.05	20.508	13.97
	Location-3	12.207	2.34	35.645	1.97	2.93	2.06	21.973	22.11
	Location-4	12.207	2.34	35.612	0.69	2.93	2.06	21.973	22.11
	Accelerometer	12.207	2.34	36.133	0.76	3.418	19.05	20.02	11.26
	Average	12.137	2.9	36.132	0.63	3.209	11.77	18.973	5.44

Table 2.3   Pole and zero frequencies obtained using the spectrum analyser

	Pole 1 (12.499 Hz)		Pole 2 (36.35 Hz)		Zero 1 (2.871 Hz)		Zero 2 (17.994 Hz)	
Measurement location	Measured (Hz)	Error (%)	Measured (Hz)	Error (%)	Measured (Hz)	Error (%)	Measured (Hz)	Error (%)
Shaft encoder	12.00	3.99	35.18	3.25	2.90	1.01	17.92	0.41
Tachometer	12.00	3.99	35.19	3.22	2.91	1.36	17.919	0.42
Location-1	12.00	3.99	35.19	3.22	2.91	1.36	17.923	0.39
Location-2	12.00	3.99	35.20	3.19	2.924	1.85	17.925	0.38
Location-3	12.025	3.79	35.42	2.59	2.931	2.09	17.928	0.37
Location-4	12.037	3.7	35.60	2.09	2.935	2.23	17.930	0.36
Accelerometer	12.05	3.59	36.00	0.99	2.96	3.1	17.942	0.39
Average	12.016	3.86	35.397	2.65	2.924	1.85	17.927	0.37

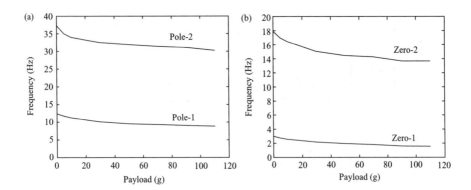

Figure 2.7   Variation of system poles and zeros with a change in payload: (a) poles, (b) zeros

and Maia, 1988). The damping ratios thus obtained using frequency response measurements of the system from the torque input to hub-angle, hub-velocity, end-point acceleration and strain locations 1–4 are shown in the first two modes in Table 2.4. Similar to the natural frequencies, the damping ratios also show small variations with the measurement location. It follows from Table 2.4 that the damping ratio decreases as the measurement location moves further away from the hub of the manipulator, where excitation is applied. This is also an indication of the non-linear behaviour of the system. The average values given in Table 2.4 provide estimates of the overall damping ratios for the two resonance modes of the system and are useful for global analysis and control of the system.

Table 2.4   Damping ratios obtained from responses at different locations

Response location	Damping ratios (ξ_i)	
	Mode-1	Mode-2
Shaft encoder	0.029	0.170
Tachometer	0.027	0.168
Location-1	0.027	0.162
Location-2	0.026	0.162
Location-3	0.020	0.159
Location-4	0.018	0.1317
Accelerometer	0.018	0.084
Average	0.024	0.148

2.6.3   Modal gain

To obtain the hub modal slope coefficient, a method based on the construction of the system transfer function is utilized (Gevarter, 1970; Martin, 1978). The transfer function of a linear elastic structure can be built up as a set of (alternative) poles and zeros, the values of which can be obtained experimentally. Thus, the open-loop transfer function from torque input $\tau(s)$ to the hub-angle $\theta(s)$ of the flexible manipulator can be expressed as

$$\frac{\theta(s)}{\tau(s)} = \frac{1}{I_T s^2} \prod_{i=1}^{n} \frac{\left[(s^2/\Omega_i^2) + 2\xi_i(s/\Omega_i) + 1\right]}{\left[(s^2/\omega_i^2) + 2\xi_i(s/\omega_i) + 1\right]} \tag{2.32}$$

where ξ_i is the damping ratio for mode i, Ω_i is the frequency of the zero and ω_i is the frequency of the pole corresponding to this mode with the zero frequency falling between two consecutive pole frequencies. Using equations (2.23) and (2.24) the open-loop transfer function from the input torque to the hub-angle can be written as

$$\frac{\theta(s)}{\tau(s)} = \frac{1}{I_T s^2} + \frac{1}{I_T} \sum_{i=1}^{n} \frac{[\phi_i'(0)]^2}{s^2 + 2\xi_i\omega_i s + \omega_i^2} \tag{2.33}$$

As the values of ξ_i, ω_i and Ω_i can be obtained using the procedure outlined earlier, the value of $\phi_i'(0)$ can be computed using equations (2.32) and (2.33).

Evaluating the hub-angle and end-point acceleration at a response frequency ω_i using equation (2.33) and the system open-loop transfer function from input torque to end-point acceleration and manipulating yields the end-point acceleration modal gain $\phi_i''(l)$ as

$$\phi_i''(l) = \frac{\alpha(j\omega_i)}{\theta(j\omega_i)} \frac{\phi_i'(0)}{\omega_i^2} \tag{2.34}$$

Table 2.5   Model parameters of the flexible manipulator system

Parameter	Symbol	Mode-1	Mode-2
Zero	Ω_i	2.924Hz	17.927Hz
Pole	ω_i	12.016Hz	35.397Hz
Hub modal gain	$\phi_i'(0)$	4.03	2.36
Acceleration modal gain	$\phi_i''(l)$	−2.19	3.47
Damping ratio	ξ_i	0.024	0.148

Thus, the steady-state peak-to-peak ratio of end-point acceleration to hub-angle can be measured experimentally at each resonance frequency and used in equation (2.17) to obtain the end-point acceleration sensor modal gain. The sign of the modal gain $\phi_i''(0)$ is determined by observing whether the end-point acceleration signal is in phase or out of phase with the shaft encoder signal. Using the estimated system parameters and the measured frequency response data in equations (2.32), (2.33) and (2.34) the corresponding end-point acceleration modal gains can be obtained. These are shown in Table 2.5.

2.7   Model validation

To evaluate the model reliability the output of the model is compared in this section with the response of the experimental system using a bang-bang input torque. The system was first excited by the input torque and the system response, consisting of the hub-angle, hub-velocity and end-point acceleration, was measured.

Figure 2.8(a) shows a comparison between the experimental system and model inputs. The torque inputs applied to the model and the system are principally the same. However, owing to the motor drive amplifier behaviour and motor dynamics, the shape of the torque input to the experimental system is slightly changed. The corresponding hub-angle is shown in Figure 2.8(b). It is noted that the overall behaviour is similar for both the model and the system. However, some abruptness in the system behaviour at half way and at the end of the movement is observed. As seen in Figures 2.8(c) and 2.8(d), the system and model response for the hub-velocity and end-point acceleration in general agree with one another very closely, during the transient as well as steady-state periods. A number of factors contribute to the occasional slight disagreements in magnitudes of the responses noticed at some points. The model, for instance, is a reduced-order one in which only the first two modes are included. The torsion and vertical motion as well as higher modes are not accounted for in the model. The effects of these, however, are present in the response of the experimental system. Moreover, friction losses and dynamics of the motor, which can considerably affect the system response, are not accounted for in the model. Note further that the input torque appearing at the hub of the experimental system is not exactly the same as the model input torque; a ringing effect and time delay during the

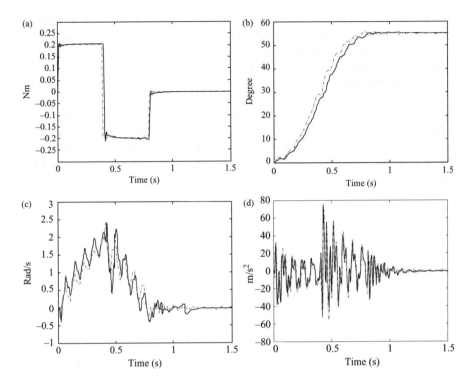

Figure 2.8   Model validation plots (broken line = model, solid line = experimental system): (a) torque profile, (b) hub angle, (c) hub velocity, (d) end-point acceleration

change of state of the system torque input is observed. However, the model response appears to agree with the system response reasonably well.

2.8   Summary

A procedure for development of a suitable model of a single-link flexible manipulator system has been presented. The Lagrange's equation and modal expansion method has been utilized in obtaining an analytical model of the system characterised by an infinite number of modes. It has been shown that for practical implementation of the model, it is helpful to convert the infinite dimensional model to one with a finite number of natural modes. This leads to a matrix differential form of the model, which can be readily converted to a state-space form. Careful choice of the truncation level based on the physical attributes of the system has been shown to yield a very good approximation of the behaviour of the full-order model over a limited frequency range.

Transfer function models have also been obtained. These are useful in frequency-domain controller designs for the system. It has been shown and experimentally verified that these transfer functions vary with the payload condition of the manipulator. A small change in the payload can result in a significant change in the natural frequencies of the manipulator. This, in a control context, can lead to excessive deterioration in the performance of a highly tuned control system.

An experimental procedure for identification of a model of flexible manipulator systems has been presented and verified. This involves obtaining the FRF at various response locations and using this to extract the model parameters. A reasonable degree of coherence has been obtained for the measured FRFs. A reduced-order linear model of the flexible manipulator, including the rigid-body mode and two flexible modes has been identified. A close agreement between the model and the system output has been achieved.

Chapter 3

Classical mechanics approach of modelling multi-link flexible manipulators

J. Sá da Costa, J.M. Martins and M.A. Botto

This chapter presents a dynamics modelling approach for flexible robotic manipulators. The driving motivation for this paper is to understand the dynamics of such systems from a control point of view. A general framework is presented that provides an adequate basis for analysis and control of flexible manipulators aimed at terrestrial or orbital applications. To this end, the generalized coordinates chosen to describe the elastic deformation are based on the Ritz approximation, and the approach to arrive at the dynamics model of a single flexible link is the principle of virtual powers, or Jourdain's principle, leading to a Lagrangian description for the elastic deformation and a Eulerian description for the rigid-body motion. A rigid–flexible–rigid body is modelled. The evolution to the multi-link case is then natural considering the topology of a serial link manipulator. Two procedures are presented, the first is the $O(N^3)$ composite inertia method and the second is the $O(N)$ articulated body method.

3.1 Introduction

The study of robotic manipulator arms is currently, and has been for the last 30 years, an intensively active field of research. From a technological point of view, manipulator arms have been found to be highly practical for both terrestrial and spatial applications, and from a theoretical point of view, they represent highly complex dynamic systems probably the predilect mechanical dynamic system for analysis. The introduction of elastic deformation of the manipulator structure, either willingly or unwillingly, has intensified the field of research even further.

The advantages of lightweight manipulators, as referred to in Chapter 1, over the traditional heavy and rigid manipulators have been recognised for a long time.

However, there is still a setback in the use of these machines. This resides in the fact that lightweight manipulators taken to a certain level will give way to a natural loss of stiffness allowing the manipulator to vibrate due to elastic deformation, highly deteriorating machine precision. A promising approach to compensate for this setback is to incorporate into these machines more sophisticated control algorithms, with an involved actuating and sensing network. A basis for this approach is a manipulator model capable of reproducing the fundamental system dynamics in a given application and amenable for real-time computation.

The first problem one faces in the modelling of flexible manipulators is at the link level: how large are the elastic displacements? This problem has been a topic of research for many years, and one answer is that the displacements should not be assumed only as first order. If one requires a dynamics model capable of capturing centrifugal stiffening and a kinematic model capable of reproducing the link fore-shortening, at least second-order strain–displacement relations are needed. In terms of beam kinematics or rigid cross-sections, this implies assuming at least a second-order rotation matrix of a cross-section due to elasticity. Parameterisation of higher-order rotation matrices, on the other hand, require additional care in the choice of parameters to use. One possible choice is the beam bending curvature.

The second problem one faces is the global dynamics formulation and solution. Two approaches are available for dealing with the dynamics of rigid multi-body systems: the composite rigid body (CRB) method and the articulated body (AB) method (Featherstone, 1987). For systems of smaller dimension, the CRB method provides solution faster than the AB method; as the size of the system increases, the AB method becomes more efficient. In robotics applications, both approaches should be available, since the size of the system changes frequently (e.g. cooperating manipulators). The formulation and solution problem is designated as the formulation stiffness problem in Ascher *et al.* (1997). These methods may be extended to flexible manipulator systems (FMSs) and the CRB method is currently designated as the composite inertia (CI) method.

Following the works of D'Eleuterio (1992), Jain and Rodriguez (1992), Bremer and Pfeiffer (1992), Piedboeuf (1998) and Pai *et al.* (2000) a systematic modelling approach leading to a general modelling environment is presented here, which may be used either in analysis or in the development of real-time control schemes for flexible manipulator arms. The cross-section rotations may be of higher order as needed, and the formulation method may be either the global dynamics CI method or the recursive AB method. To this end, Section 3.2 describes the reference frames and the kinematic equations of a flexible beam. In Section 3.3 the non-linear strain–displacement relations are presented. In Section 3.4 the dynamic model of a single flexible link is deduced, and in Section 3.5 this model is used in the formulation of the dynamics of a flexible link manipulator. Finally, Section 3.6 concludes the chapter.

3.2   Kinematics: the reference frames

The problem of where to place reference frames in order to describe the motion of flexible robot links modelled as beams has two major solutions. One is designated as the inertial (also designated as fixed) reference frame approach and the other is designated as the body (moving or shadow) reference frame approach. The distinction between these two approaches is well established in the work of Simo and Vu-Quoc (1986). The body reference frame approach is especially motivated by the assumption of small displacements. On the other hand, the inertial reference frame approach requires the use of beam theories capable of accounting for large rotations of the beam, that is, the strain measures must be invariant under superposed rigid body motion.

The main advantages of the body reference frame approach is that it yields a formulation of conceptual closeness to the flexible robotics control problem and it allows the use of shape functions of the structure for spatial discretisation of the partial differential equations (PDEs) of motion. The inertial reference frame approach, on the other hand, requires the use of the finite element method yielding a finite dimensional model of much larger dimension. Simo and Vu-Quoc (1986) argued that, from the computational standpoint, in the general case, the use of shape functions does not pose a significant advantage due to the non-linear and highly coupled inertia terms emanating from the body reference frame approach. In the inertial reference frame approach, the non-linearity of the problem is shifted to the stiffness part of the equations of motion leading to a simpler structure of the same equations. This aspect has not been investigated here on account of the introductory statement of the paragraph; in robotic manipulators, joint actuator torque is applied along a body reference frame.

A pitfall in the body reference frame approach appears when small displacements are assumed. Under this assumption, the resulting equations of motion predict an incorrect behaviour in relation to the effects of centrifugal force on the deformation of the link. As was eloquently shown in Simo and Vu-Quoc (1987), appropriate account of the influence of centrifugal force on the bending stiffness of a flexible beam requires the use of a geometrically non-linear (at least second-order) beam theory. This does not necessarily imply assuming large strains. Even with infinitesimal strains, the geometry of the beam may allow for large displacements that necessarily must be treated with a non-linear beam theory. Small displacements imply small strains but the converse does not necessarily apply. This leads to capturing the kinematic effect known as beam foreshortening (Kane *et al.*, 1987; Smith and Baruh, 1991).

Under these settings the kinematic description of a flexible link is achieved here through the use of a body reference frame. The inertial reference frame is designated as $\{O_I, X_I Y_I Z_I\}$, and the body reference frame is designated as $\{O_n, X_n Y_n Z_n\}$ (see Figure 3.1). The latter is rigidly attached to the first point of neutral axis of the beam, and describes the rigid body motion of the beam. It gives the orientation of the beam in space, in its reference, un-deformed configuration. In an inertial reference frame formulation this reference frame does not exist.

3.2.1 Deformation assumptions

The adoption of the body reference frame allows the separation of rigid-body motion from elastic deformation. The former is described by the body reference frame and the latter is described through the introduction of a third reference frame. The choice

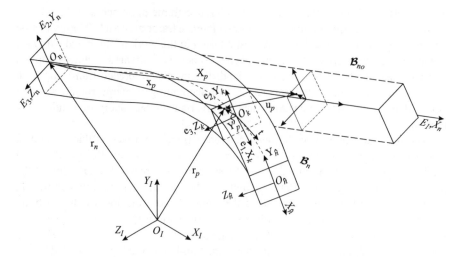

Figure 3.1   Kinematic description of a flexible link

of this last reference frame is based on the following deformation assumptions for the flexible beam:

(i)   Plane beam cross-sections before deformation, remain plane after deformation (warping is not considered).
(ii)   Bending and shear deformation are considered (Timoshenko beam theory).
(iii)   The beam neutral fibre does not suffer extension.
(iv)   The beam neutral axis in the un-deformed configuration is a straight line.
(v)   The beam cross-section is of constant specific mass, and is symmetrical relative to its principal axis.
(vi)   The shear strains and bending strains are considered to be small, that is, they are accounted for up to the first order, $O(1)$.

Assumptions (i) and (ii) fall in the theory of Timoshenko beams (Bayo, 1989; Timoshenko *et al.*, 1974). This constitutes a general case of applications, where the slenderness of robot links are relaxed beyond the Euler–Bernoulli beam theory assumptions to a slenderness ratio of length/width >10.

Assumptions (iii), (iv) and (v) are motivated by the physical reality of flexible manipulator arms. The stiffness of the manipulator links in the direction of its axis is significantly larger than in the transverse directions, therefore, the beam axis is considered inextensible. Assumptions (iv) and (v) are common design criteria. Although these can be exploited to improve the design, this is not attempted here.

The motivation for assumption (vi) is twofold. First, if local deformation is large, most probably the material is not adequate to build a robotic manipulator! Second, and more determinant, common materials used in applications retain their linearity for extremely small strain. It is worth stressing that what is enforced in this assumption is small strain, and not small displacement.

The last reference frame is now placed, taking advantage of the assumption that the beam cross-section is rigid during deformation. This reference frame is designated as cross section reference frame $\{O_k, X_k Y_k Z_k\}$. Its origin is placed at the point where the beam neutral axis intersects the beam cross-section (Figure 3.1), and its axes are directed along the axes of symmetry of the cross section. The assumption of rigid cross sections has also implications on the reference frame $\{O_n, X_n Y_n Z_n\}$. This reference frame is now not just rigidly attached to the first point of the neutral axis, but rigidly attached to the first cross section of the beam. For this cross section, reference frames $\{O_n, X_n Y_n Z_n\}$ and $\{O_k, X_k Y_k Z_k\}$ coincide with one another. Similarly, when referring to the end cross section of the beam, the cross-section reference frame becomes $\{O_{\hat{n}}, X_{\hat{n}} Y_{\hat{n}} Z_{\hat{n}}\}$. Notice that these reference frames are of the co-rotational type and not of the convected basis type (Simo, 1985). In the latter, the axis normal to the cross section is removed, and placed tangent to the neutral axis, rendering the reference frame non-orthogonal.

In summary, a flexible beam is seen as a group of rigid cross sections distributed along an inextensible line, kept together by elastic forces and moments due to small local shear deformation and bending curvature, respectively. Attached to each cross section there is a reference frame $\{O_k, X_k Y_k Z_k\}$ whose position and orientation relative to reference frame $\{O_n, X_n Y_n Z_n\}$ describes the beam deformation. The position and orientation of the reference frame $\{O_n, X_n Y_n Z_n\}$ relative to the inertial reference frame, $\{O_I, X_I Y_I Z_I\}$, describes the rigid-body motion.

3.2.2   Kinematics of a flexible link

The kinematic description presented in this section is similar to that of (Géradin and Cordona, 2001) with the exception of introducing the moving reference frame. The orthogonal matrix expressing the orientation of $\{O_k, X_k Y_k Z_k\}$ relative to $\{O_n, X_n Y_n Z_n\}$ is designated as $\mathbf{R}_{e_k}$, and is given by

$$\mathbf{R}_{e_k} = \begin{bmatrix} \mathbf{e}_1 & \mathbf{e}_2 & \mathbf{e}_3 \end{bmatrix} \tag{3.1}$$

where $\mathbf{e}_1$, $\mathbf{e}_2$ and $\mathbf{e}_3$ are the unit vectors along the axes of the cross-section reference frame, written in the body reference frame. The orthogonal matrix expressing the orientation of $\{O_n, X_n Y_n Z_n\}$ relative to $\{O_I, X_I Y_I Z_I\}$ is designated as $\mathbf{R}_n$, and is given by

$$\mathbf{R}_n = \begin{bmatrix} \mathbf{E}_1 & \mathbf{E}_2 & \mathbf{E}_3 \end{bmatrix} \tag{3.2}$$

where $\mathbf{E}_1$, $\mathbf{E}_2$ and $\mathbf{E}_3$ are unit vectors along the axes of the body reference frame, expressed in the inertial reference frame.

The position vector of a material point p of the beam belonging to a certain cross section, relative to the origin of the cross-section reference frame and with components along the same reference frame is defined as $\mathbf{Y}_p$. Owing to the assumption of rigid cross section, the components of this vector remain constant during deformation. However, relative to the moving reference frame, the position coordinates of point p change because of deformation. The reference position of point p is designated as $\mathbf{X}_p$ and the displaced position of point p is defined as $\mathbf{x}_p$. The components of $\mathbf{X}_p$ are the material coordinates of the beam. The displacement vector carrying point p from position $\mathbf{X}_p$ to position $\mathbf{x}_p$ is defined as $\mathbf{u}_p$. The components of the displacement vector, just like the components of $\mathbf{X}_p$ and $\mathbf{x}_p$ are along the reference frame $\{O_n, X_n Y_n Z_n\}$. In component form

$$\mathbf{Y}_p = \begin{bmatrix} 0 & X_2 & X_3 \end{bmatrix}^T \tag{3.3}$$

$$\mathbf{X}_p = \begin{bmatrix} X_1 & X_2 & X_3 \end{bmatrix}^T \tag{3.4}$$

A material point in the beam neutral fibre is described by setting $\mathbf{Y}_p = \begin{bmatrix} 0 & 0 & 0 \end{bmatrix}^T$. Accordingly, its reference position, displaced position and displacement vector become $\mathbf{X}_k = \begin{bmatrix} X_1 & 0 & 0 \end{bmatrix}^T$, $\mathbf{x}_k$ and $\mathbf{u}_k$, respectively. The tangent vector to the neutral fibre is given by

$$\mathbf{t} = \frac{d\mathbf{x}_k}{dX_1} \tag{3.5}$$

This vector, also of unit magnitude due to assumption (iii), does not coincide with vector $\mathbf{e}_1$ owing to the shear deformation of Timoshenko beam theory.

In order to specify the position of point p in the inertial reference frame, $\{O_I, X_I Y_I Z_I\}$, defined as $\mathbf{r}_p$, the position of the body reference frame relative to the inertial reference frame, $\mathbf{r}_n$, must be known. These two vectors are considered here with components along the inertial reference frame. Finally, the basic kinematic equation for a material point p of the flexible link is

$$\begin{aligned} \mathbf{r}_p &= \mathbf{r}_n + \mathbf{R}_n \mathbf{x}_p \\ &= \mathbf{r}_n + \mathbf{R}_n \left(\mathbf{x}_k + \mathbf{R}_{e_k} \mathbf{Y}_p \right) \\ &= \mathbf{r}_n + \mathbf{R}_n \left(\mathbf{X}_k + \mathbf{u}_k + \mathbf{R}_{e_k} \mathbf{Y}_p \right) \end{aligned} \tag{3.6}$$

where because of deformation alone one has

$$\mathbf{x}_p = \mathbf{X}_k + \mathbf{u}_k + \mathbf{R}_{e_k} \mathbf{Y}_p \tag{3.7}$$

In this equation, the only terms that are owing to the deformation of the link are the displacement vector of the material points of the neutral axis, $\mathbf{u}_k$, and the rotation matrix of a cross section $\mathbf{R}_{e_k}$. These are deduced in the next section.

3.3   The strain–displacement relations

The strain measures that are most commonly applied to the treatment of flexible beams are the Green–Lagrange strain measures (Sharf, 1996) and the displacement gradient measure of deformation (Géradin and Cordona, 2001). The former provides a general means of measuring the deformation inside a three-dimensional continuum whereas the latter is a simplification specific to beam kinematics. Both of these are invariant under rigid-body motion, therefore, they can be calculated from vectors expressed in the body reference frame or in the inertial reference frame. Here, the body reference frame is adopted. Considering Figure 3.1 they are respectively given as

$$G\left(\mathbf{X}_p\right) = \frac{1}{2}\left(\frac{d\mathbf{u}_p}{d\mathbf{X}_p} + \frac{d\mathbf{u}_p^T}{d\mathbf{X}_p} + \frac{d\mathbf{u}_p^T}{d\mathbf{X}_p}\frac{d\mathbf{u}_p}{d\mathbf{X}_p}\right) \tag{3.8}$$

and

$$D\left(\mathbf{X}_p\right) = \mathbf{R}_{e_k}^T \frac{d\mathbf{x}_p}{dX_1} - \frac{d\mathbf{X}_p}{dX_1} \tag{3.9}$$

The Green–Lagrange strain tensor, G, is calculated by considering the change during deformation of the squared length of an infinitesimal fibre and expressing the difference in the body reference frame (the reference un-deformed configuration). On the other hand, the displacement gradient strain vector, D, is obtained by calculating the position gradient with respect to the parameter describing the neutral axis, X_1, in the cross-section reference frame, and subtracting from it the position gradient in the reference un-deformed configuration.

G is a 3×3 symmetric matrix, and it expresses the strain state of a small material volume. D is a 3×1 vector, and it expresses the strain state at a specific cross section. Defining the components of the Green–Lagrange strain tensor as

$$G = \begin{bmatrix} \varepsilon_{11} & \varepsilon_{21} & \varepsilon_{31} \\ \varepsilon_{12} & \varepsilon_{22} & \varepsilon_{32} \\ \varepsilon_{13} & \varepsilon_{23} & \varepsilon_{33} \end{bmatrix} \tag{3.10}$$

and considering the rigidity of cross sections one has that $\varepsilon_{22} = \varepsilon_{33} = \varepsilon_{32} = \varepsilon_{23} = 0$, and the remaining strain components are reduced to three, namely ε_{11}, $\varepsilon_{12} = \varepsilon_{21}$ and $\varepsilon_{13} = \varepsilon_{31}$ since the tensor must be symmetric. The Green–Lagrange strains at a beam cross section may now be obtained by multiplying the strain tensor by the unit vector normal to the cross section expressed in the same reference frame as the tensor. Multiplying the shear components by 2 to give engineering strains, the result is

$$G_b = \begin{bmatrix} \varepsilon_{11} & 2\varepsilon_{12} & 2\varepsilon_{13} \end{bmatrix}^T \tag{3.11}$$

The component form of the displacement gradient strain vector, on the other hand, is defined as

$$D = \begin{bmatrix} d_1 & d_2 & d_3 \end{bmatrix}^T \tag{3.12}$$

At this point, assuming small displacement vectors in equations (3.8) and (3.9) in order to lead to small strain components may be premature and lead to incorrect models as mentioned in Section 3.2. Therefore, it is best to rewrite these strain vectors in terms of the variables that express deformation in the basic kinematic equation (3.7), namely, the displacement vector of the material points on the neutral axis, $\mathbf{u}_k$ (or their displaced positions $\mathbf{x}_k$), and the rotation matrix of a cross section $\mathbf{R}_{e_k}$. Substituting for $\mathbf{x}_p$ from equation (3.7) into equation (3.9) the displacement gradient strains become

(Géradin and Cardona, 2001)

$$\mathbf{D}\left(\mathbf{X}_p\right) = \mathbf{R}_{e_k}^T \left(\mathbf{t} - \mathbf{e}_1\right) + \mathbf{R}_{e_k}^T \frac{d\mathbf{R}_{e_k}}{dX_1} \mathbf{Y}_p$$

$$= \boldsymbol{\Gamma}_k + \tilde{\mathbf{K}}_k \mathbf{Y}_p \tag{3.13}$$

where, equation (3.5) has been used. The vector $\boldsymbol{\Gamma}_k = \begin{bmatrix} \Gamma_1 & \Gamma_2 & \Gamma_3 \end{bmatrix}^T$ represents the strains of neutral axis of the beam. Γ_1 contemplates two phenomena, the extensional strain of the neutral axis owing to a longitudinal force, and the longitudinal strain induced by transverse shear (Simo and Vu-Quoc, 1986, 1987);

$$\Gamma_1 = \mathbf{e}_1^T \mathbf{t} - 1 \tag{3.14}$$

The former has been neglected by setting $\|\mathbf{t}\| = 1$ in equation (3.5) owing to assumption (iii), and the latter will be neglected in Section 3.3.1 on account of assumption (vi). Γ_2 and Γ_3 are the transverse Timoshenko shears,

$$\Gamma_i = \mathbf{e}_i^T \mathbf{t}, \qquad i = 2, 3 \tag{3.15}$$

and will also be considered according to assumption (vi).

The skew symmetric matrix $\tilde{\mathbf{K}}_k$ represents the curvature of the beam cross sections from which the curvature vector with components along the cross-section reference frame may be extracted, $\mathbf{K}_k = \mathrm{vect}(\tilde{\mathbf{K}}_k) = \begin{bmatrix} K_1 & K_2 & K_3 \end{bmatrix}^T$ with

$$\tilde{\mathbf{K}}_k = \begin{bmatrix} 0 & -K_3 & K_2 \\ K_3 & 0 & -K_1 \\ -K_2 & K_1 & 0 \end{bmatrix} = \mathbf{R}_{e_k}^T \frac{d\mathbf{R}_{e_k}}{dX_1} \tag{3.16}$$

K_1 is the torsional strain of the neutral axis and K_2 and K_3 are the bending strains, all expressed in the cross-section reference frame. Similar to the shear strains, these bending strains will also be considered according to assumption (vi). Note that the beam curvature is not given by this vector owing to shear deformation.

The above quantities provide an alternative representation of the strain vector that is more adequate for expansion than the representation in equation (3.9). Assumptions at the local deformation level are more realistic than at the displacement level. The latter depend on the length of the beam whereas the former do not.

Similar to the displacement gradient strains, it is shown in Géradin and Cardona (2001) that the Green–Lagrange strains are given by

$$\mathbf{G}_b\left(\mathbf{X}_p\right) = \mathbf{D}\left(\mathbf{X}_p\right) + \begin{bmatrix} \frac{1}{2}\mathbf{D}\left(\mathbf{X}_p\right)^T \mathbf{D}\left(\mathbf{X}_p\right) \\ 0 \\ 0 \end{bmatrix} \tag{3.17}$$

It is verified that the Green–Lagrange strains yield the same shear strains but different longitudinal strains than the displacement gradient. Expanding the second term in equation (3.17) yields longitudinal strain dependence on quadratic bending curvature terms $\mathbf{K}_k$, quadratic shear terms $\boldsymbol{\Gamma}_k$ and coupling shear and bending terms. All these terms are neglected on the basis of assumption (vi). Therefore, the two strain measures are approximate, and are explicitly written as

$$\mathbf{G_b}\left(\mathbf{X}_p\right) \approx \mathbf{D}\left(\mathbf{X}_p\right) = \begin{bmatrix} \Gamma_1 \\ \Gamma_2 \\ \Gamma_3 \end{bmatrix} + \begin{bmatrix} -K_3X_2 + K_2X_3 \\ -K_1X_3 \\ K_1X_2 \end{bmatrix} \tag{3.18}$$

3.3.1 Parameterisation of the rotation matrix

Equation (3.13) implies that for infinitesimal strain measures one has infinitesimal shear strains, Γ_k, and infinitesimal bending strains, $\mathbf{K}_k$. However, equation (3.16) implies that infinitesimal bending strains along the beam, $\mathbf{K}_k$, do not necessarily yield infinitesimal cross-section rotations. Similar to Bremer and Pfeiffer (1992) and Schwertassek and Wallrapp (1999), expanding $\mathbf{R}_{e_k}$ and $\mathbf{K}_k$ in Taylor series in equation (3.16), and retaining only the first-order term of the bending strains results

$$\underbrace{\frac{d\mathbf{R}_{ek1}}{dX_1}}_{O(1)} + \underbrace{\frac{d\mathbf{R}_{ek2}}{dX_1}}_{O(2)} + \cdots = \left(\underbrace{\mathbf{I}}_{o(0)} + \underbrace{\mathbf{R}_{ek1}}_{o(1)} + \underbrace{\mathbf{R}_{ek2}}_{o(2)} \cdots \right) \underbrace{\tilde{\mathbf{K}}_{k_1}}_{o(1)} \tag{3.19}$$

where $\mathbf{R}_{e_{ki}}$ is the term of order i of the rotation matrix $\mathbf{R}_{e_k}$ and $\mathbf{K}_{k_1}$ is the first-order term of the bending strains. The rotation matrix can thus be calculated up to order m through integration of the same-order terms in equation (3.19):

$$\mathbf{R}_{e_k} = \mathbf{I} + \mathbf{R}_{ek1} + \mathbf{R}_{ek2} + \cdots + \mathbf{R}_{ekm} \tag{3.20}$$

where

$$\mathbf{R}_{ek1} = \int_0^{X_1} \mathbf{I}\tilde{\mathbf{K}}_{k_1} d\xi$$

$$\mathbf{R}_{eki} = \int_0^{X_1} \mathbf{R}_{eki-1}\tilde{\mathbf{K}}_{k_1} d\xi \tag{3.21}$$

3.3.2 Parameterisation of the neutral axis tangent vector

On introducing the shear angles φ_{12} and φ_{13}, as defined by the classical theory of elasticity (Novozhilov, 1961; Schwertassek and Wallrapp, 1999), equation (3.15) becomes

$$\Gamma_i = \mathbf{e}_i^T \mathbf{t} = \cos\left(\frac{\pi}{2} - \varphi_{1i}\right) = \sin(\varphi_{1i}), \quad i = 2, 3 \tag{3.22}$$

and equation (3.14) becomes

$$\Gamma_1 = \mathbf{e}_1^T \mathbf{t} - 1 = \sqrt{1 - \sin^2(\varphi_{12}) - \sin^2(\varphi_{13})} - 1 \tag{3.23}$$

where assumption (iii), $\|\mathbf{t}\| = 1$, has been considered. Furthermore, vector $\mathbf{t}$ may be written with components along the cross-section reference frame as

$$_k\mathbf{t} = \sqrt{1 - \sin^2{(\varphi_{12})} - \sin^2{(\varphi_{13})}}\, \mathbf{e}_1 + \sin{(\varphi_{12})}\, \mathbf{e}_2 + \sin{(\varphi_{13})}\, \mathbf{e}_3 \qquad (3.24)$$

Considering now assumption (vi), yields

$$\Gamma_i \approx \varphi_{1i}, \quad i = 2, 3 \qquad (3.25)$$

and

$$\Gamma_1 \approx 0 \qquad (3.26)$$

and finally

$$_k\mathbf{t} = \mathbf{e}_1 + \varphi_{12}\mathbf{e}_2 + \varphi_{13}\mathbf{e}_3 \qquad (3.27)$$

The resulting vector of infinitesimal shear strains is finally

$$\Gamma_k \approx \Gamma_{k_1} = \begin{bmatrix} 0 & \varphi_{12} & \varphi_{13} \end{bmatrix}^T \qquad (3.28)$$

The strain state of the beam is described by the variables K_1, K_2, K_3, φ_{12} and φ_{13}. With the Rayleigh–Ritz approach to spatially discretise the system equations these are the variables for which trial functions have to be chosen that obey the imposed boundary conditions at the ends of the beam.

3.3.3   Displacement of the neutral axis

To complete the kinematic description of equation (3.7), the displacement vector of the points on the neutral axis must be specified. To this end, note that

$$\frac{d\mathbf{x}_k}{dX_1} = \mathbf{R}_{e_k k}\mathbf{t} \qquad (3.29)$$

and also

$$\frac{d\mathbf{x}_k}{dX_1} = \frac{d}{dX_1}(\mathbf{X}_k + \mathbf{u}_k) = \begin{bmatrix} 1 \\ 0 \\ 0 \end{bmatrix} + \frac{d\mathbf{u}_k}{dX_1} \qquad (3.30)$$

$\mathbf{u}_k$ can now be calculated through integration resulting

$$\mathbf{u}_k = \int_0^{X_1} \mathbf{R}_{e_k k}\mathbf{t}\,d\xi - \begin{bmatrix} X_1 \\ 0 \\ 0 \end{bmatrix} \qquad (3.31)$$

Even though the bending strains are infinitesimal, equation (3.19), and the shear strains are infinitesimal, equation (3.28), the strain–displacement relations, equation (3.9), may be non-linear if the rotation matrix $\mathbf{R}_{e_k}$ is expanded up to non-linear terms, equation (3.21), in the kinematics equation (3.7).

3.4   The dynamic model of a single flexible link

In Martins *et al.* (2002), an analysis was performed on continuous dynamic models obtained through the Extended Hamilton Principle in order to identify the terms crucial for capturing the dynamic stiffening of a flexible beam. These continuous equations were then discretised through a modal expansion in order to allow simulation and eventually controller analysis. Here, the strategy is slightly changed since the subtleness that causes incorrect modelling has been uncovered. The intent is to deduce a discrete dynamic model of the Newton–Euler type capturing the fundamental dynamics required for flexible manipulator analysis. A Eulerian formulation is used for the rigid-body motion and a total Lagrangian formulation is used for the deformation (Boyer and Coiffet, 1996). To this end, Jourdain's Principle or the Principle of Virtual Powers is adopted (Schwertassek and Wallrapp, 1999), assuming a Rayleigh–Ritz expansion of the elastic variables.

Considering the flexible link in Figure 3.1, Jourdain's Principle is then stated as follows:

$$\int_{\mathcal{B}_{no}} \delta\mathbf{v}_p^T \mathbf{a}_p \rho_p d\mathcal{B}_{no} + \int_{\mathcal{B}_{no}} \delta\dot{\mathbf{G}}_p^T \boldsymbol{\sigma} d\mathcal{B}_{no} = \int_{\mathcal{B}_{no}} \delta\mathbf{v}_p^T \mathbf{g} \rho_p d\mathcal{B}_{no} + \int_{\Sigma\mathcal{B}_{no}} \delta\mathbf{v}_p^T \mathbf{f}_s d\Sigma\mathcal{B}_{no}$$

(3.32)

The first term on the left-hand side represents the virtual power of the inertial force, where $\mathbf{v}_p$ and $\mathbf{a}_p$ are the absolute (relative to the inertial frame) velocity and acceleration vector, respectively, expressed in the body reference frame. The second term of the left-hand side represents the virtual power of the internal elastic stress where $\boldsymbol{\sigma}$ is the vector of Piola–Kirchoff stresses acting on a cross section. The first term on the right-hand side represents the virtual power of the gravitational force, where ρ_p is the specific mass of the beam at point p in the reference configuration and $\mathbf{g}$ is the gravitational acceleration vector expressed in the body reference frame. The last term represents the virtual power of the external forces that are applied on the surface of the body. $\mathbf{f}_s$ is the external force vector (force per unit of un-deformed surface area) applied on the beam surface. $\mathcal{B}_{no}$ and $\Sigma\mathcal{B}_{no}$ represent the body and its surface in the un-deformed configuration.

Owing to the kinematic assumptions for the beam deformation, equation (3.32) may be written in a more practical form where linear and angular motions of a cross section are separated (Boyer and Coiffet, 1996; Bremer and Pfeiffer, 1992).

3.4.1   The inertial force term

The absolute velocity and acceleration vectors of material point p expressed in the body reference frame are obtained from equation (3.6) as

$$\mathbf{v}_p = \mathbf{R}_n^T \dot{\mathbf{r}}_p = \underbrace{\mathbf{v}_n + \widetilde{\boldsymbol{\omega}}_n (\mathbf{X}_k + \mathbf{u}_k) + \dot{\mathbf{u}}_k}_{\mathbf{v}_k} + \underbrace{(\widetilde{\boldsymbol{\omega}}_n + \widetilde{\boldsymbol{\Omega}}_k)}_{\bar{\boldsymbol{\omega}}_k} \mathbf{R}_{e_k} \mathbf{Y}_p$$

(3.33)

and

$$\mathbf{a}_p = \mathbf{R}_n^T \ddot{\mathbf{r}}_p = \underbrace{\mathbf{a}_n + \left(\dot{\tilde{\boldsymbol{\omega}}}_n + \tilde{\boldsymbol{\omega}}_n \tilde{\boldsymbol{\omega}}_n\right)(\mathbf{X}_k + \mathbf{u}_k) + 2\tilde{\boldsymbol{\omega}}_n \dot{\mathbf{u}}_k + \ddot{\mathbf{u}}_k}_{\mathbf{a}_k}$$

$$+ \underbrace{\left(\dot{\tilde{\boldsymbol{\omega}}}_n + \tilde{\boldsymbol{\omega}}_n \tilde{\boldsymbol{\omega}}_n + \dot{\tilde{\boldsymbol{\Omega}}}_k + \tilde{\boldsymbol{\Omega}}_k \tilde{\boldsymbol{\Omega}}_k + 2\tilde{\boldsymbol{\omega}}_n \tilde{\boldsymbol{\Omega}}_k\right)}_{\boldsymbol{\Lambda}_k} \mathbf{R}_{e_k} \mathbf{Y}_p \tag{3.34}$$

respectively, where on account of the orthogonality of $\mathbf{R}_n$ and $\mathbf{R}_{e_k}$ the following equalities have been used (Géradin and Cordona, 2001):

$$\mathbf{R}_n^T \mathbf{R}_n = \mathbf{I} \tag{3.35}$$

$$\mathbf{R}_n^T \dot{\mathbf{R}}_n = \tilde{\boldsymbol{\omega}}_n \tag{3.36}$$

$$\mathbf{R}_n^T \ddot{\mathbf{R}}_n = \frac{d}{dt}\left(\mathbf{R}_n^T \dot{\mathbf{R}}_n\right) - \dot{\mathbf{R}}_n^T \dot{\mathbf{R}}_n = \dot{\tilde{\boldsymbol{\omega}}}_n + \tilde{\boldsymbol{\omega}}_n \tilde{\boldsymbol{\omega}}_n \tag{3.37}$$

and

$$\mathbf{R}_{e_k} \mathbf{R}_{e_k}^T = \mathbf{I} \tag{3.38}$$

$$\dot{\mathbf{R}}_{e_k} \mathbf{R}_{e_k}^T = \tilde{\boldsymbol{\Omega}}_k \tag{3.39}$$

$$\ddot{\mathbf{R}}_{e_k} \mathbf{R}_{e_k}^T = \frac{d}{dt}\left(\dot{\mathbf{R}}_{e_k} \mathbf{R}_{e_k}^T\right) - \dot{\mathbf{R}}_{e_k} \dot{\mathbf{R}}_{e_k}^T = \dot{\tilde{\boldsymbol{\Omega}}}_k + \tilde{\boldsymbol{\Omega}}_k \tilde{\boldsymbol{\Omega}}_k \tag{3.40}$$

$\mathbf{v}_n = \mathbf{R}_n^T \dot{\mathbf{r}}_n$ and $\mathbf{a}_n = \mathbf{R}_n^T \ddot{\mathbf{r}}_n$ are the absolute linear velocity and absolute linear acceleration vectors of the body reference frame, respectively, $\boldsymbol{\omega}_n$ is the absolute angular velocity vector of the body reference frame and $\boldsymbol{\Omega}_k$ is the angular velocity vector of the cross-section reference frame relative to the body reference frame. All these vectors are expressed in the body reference frame. $\mathbf{v}_n$ and $\boldsymbol{\omega}_n$ form the set of *quasi-coordinates* that describe the rigid body motion of the beam in a *Eulerian description*. $\mathbf{u}_k$ and $\mathbf{R}_{e_k}$ contain the coordinates that describe the elastic deformation in a *total Lagrangean description*. Furthermore, in equation (3.33), $\mathbf{v}_k$ and $\boldsymbol{\omega}_k$ are the absolute linear and angular velocity vectors of the kth cross section expressed in the body reference frame, and in equation (3.34), $\mathbf{a}_k$ and $\boldsymbol{\Lambda}_k$ are the absolute linear acceleration vector and absolute angular acceleration matrix of a cross section expressed in the body reference frame. From equations (3.33) and (3.34) one can verify the well-known relation,

$$\mathbf{a}_j = \dot{\mathbf{v}}_j + \tilde{\boldsymbol{\omega}}_n \mathbf{v}_j, \qquad j = n, k, p \tag{3.41}$$

The absolute angular acceleration matrix in equation (3.34) may be rearranged as

$$\boldsymbol{\Lambda}_k = \left(\dot{\boldsymbol{\omega}}_n + \dot{\boldsymbol{\Omega}}_k + \tilde{\boldsymbol{\omega}}_n \boldsymbol{\Omega}_k\right)^{\sim} + \left(\tilde{\boldsymbol{\omega}}_n + \tilde{\boldsymbol{\Omega}}_k\right)^2 \tag{3.42}$$

where the first part is skew symmetric and contains the absolute angular acceleration vector, and the second part is symmetric and contains the centrifugal acceleration terms. The angular acceleration vector, α_k, also relates to the angular velocity vector, ω_k, as in equation (3.41) as

$$\alpha_k = \dot{\omega}_k + \widetilde{\omega}_n \omega_k = \dot{\omega}_n + \dot{\Omega}_k + \widetilde{\omega}_n \Omega_k \tag{3.43}$$

The inertial term in equation (3.32) may be simplified by first applying the variational operator δ (Jourdain's variation) (Piedboeuf, 1993) to the velocity expression:

$$\begin{aligned}
\delta v_p &= \delta v_n + \delta \widetilde{\omega}_n (X_k + u_k) + \delta \dot{u}_k + \left(\delta \widetilde{\omega}_n + \delta \widetilde{\Omega}_k \right) R_{e_k} Y_p \\
&= \delta v_k + \delta \widetilde{\omega}_k R_{e_k} Y_p
\end{aligned} \tag{3.44}$$

and then by substituting for a_p and δv_p from equations (3.34) and (3.44) and solving. This yields

$$\int_{\mathcal{B}_{no}} \delta v_p^T a_p \rho_p d\mathcal{B}_{no} = \int_0^L \rho_A \delta v_k^T a_k + \rho_A \delta \omega_k^T \left(J\alpha_k + \widetilde{\omega}_k J\omega_k \right) dX_1 \tag{3.45}$$

where

$$\rho_A = \int_A \rho_p d_A = \rho_p A \tag{3.46}$$

$$\int_A Y_p d_A = \begin{bmatrix} 0 & 0 & 0 \end{bmatrix}^T \tag{3.47}$$

$$J = R_{e_k k} JR_{e_k}^T \tag{3.48}$$

with

$$_k J = \int_A \widetilde{Y}_p \widetilde{Y}_p^T d_A = \begin{bmatrix} J & 0 & 0 \\ 0 & I_2 & 0 \\ 0 & 0 & I_3 \end{bmatrix} \tag{3.49}$$

In equation (3.46) the beam specific mass, ρ_p, is considered constant throughout the cross section, and ρ_A is the beam mass per unit length. In equation (3.37) the beam cross sections are assumed symmetric. In equations (3.48) and (3.49) $_k J$ represents the geometrical moments of inertia tensor of a cross section relative to, and expressed in the cross-section reference frame and J represents the same tensor expressed in the moving reference frame. The components of $_k J$ are constant where

$$I_2 = \int_A X_3^2 d_A \tag{3.50}$$

$$I_3 = \int_A X_2^2 d_A \tag{3.51}$$

and

$$J = I_2 + I_3 \tag{3.52}$$

are the bending and torsional geometric moments of inertia respectively. The components of $\mathbf{J}$ on the other hand depend on the rotation matrix due to elastic deformation, $\mathbf{R}_{e_k}$.

3.4.2   The elastic force term

The elastic term is expanded into an integral along the beam length, similar to the inertial term, by first taking the variation of the time derivative to the strains given in the form of equation (3.13):

$$\delta \dot{\mathbf{G}}_b = \delta \dot{\boldsymbol{\Gamma}}_k + \delta \tilde{\dot{\mathbf{K}}}_k \mathbf{Y}_p \tag{3.53}$$

and then by applying Hook's law while neglecting Poisson's ratio

$$\boldsymbol{\sigma} = \mathbf{E} \mathbf{G}_b \tag{3.54}$$

where

$$\mathbf{E} = \begin{bmatrix} E & 0 & 0 \\ 0 & G & 0 \\ 0 & 0 & G \end{bmatrix} \tag{3.55}$$

with E representing Young's modulus and $G = E/2$.

Substituting for $\delta \dot{\mathbf{G}}_b$ and $\boldsymbol{\sigma}$ from equations (3.53) and (3.54) into the elastic term in equation (3.32), and taking into account the symmetry of cross sections, equation (3.47), yields

$$\int_{\mathcal{B}_{no}} \delta \dot{\mathbf{G}}_p^T \boldsymbol{\sigma} d\mathcal{B}_{no} = \int_0^L \delta \dot{\boldsymbol{\Gamma}}_k^T A \mathbf{E} \boldsymbol{\Gamma}_k + \delta \dot{\mathbf{K}}_k^T \mathbf{C} \mathbf{K}_k dX_1 \tag{3.56}$$

with

$$\mathbf{C} = \int_A \tilde{\mathbf{Y}}_p \mathbf{E} \tilde{\mathbf{Y}}_p^T d_A = \begin{bmatrix} GJ & 0 & 0 \\ 0 & EI_2 & 0 \\ 0 & 0 & EI_3 \end{bmatrix} \tag{3.57}$$

Carrying out the multiplications in equation (3.56) yields

$$\int_{\mathcal{B}_{no}} \delta \dot{\mathbf{G}}_p^T \boldsymbol{\sigma} d\mathcal{B}_{no} = \int_0^L \delta \dot{\varphi}_{12} GA\varphi_{12} + \delta \dot{\varphi}_{13} GA\varphi_{13} + \delta \dot{K}_1 GJK_1$$
$$+ \delta \dot{K}_2 EI_2 K_2 + \delta \dot{K}_3 EI_3 K_3 dX_1 \tag{3.58}$$

The elastic forces due to deformation that are applied on a cross section are given by $GA\varphi_{12}$ and $GA\varphi_{13}$, and the elastic moments due to deformation that are applied on a cross section are given by GJK_1, $EI_2 K_2$ and $EI_3 K_3$.

3.4.3 The gravitational force term

Using equation (3.44) and considering the symmetry of cross sections, equation (3.47), the gravitational force term reduces to

$$\int_{\mathcal{B}_{no}} \delta \mathbf{v}_p^T \mathbf{g} \rho_p d\mathcal{B}_{no} = \int_0^L \rho_A \delta \mathbf{v}_k^T \mathbf{g} dX_1 \qquad (3.59)$$

Due to the placement of the cross-section reference frame at the centre of mass of the cross section, the gravity term only affects its linear motion. Comparing equations (3.45) and (3.59), one notices that the contribution of the gravitational acceleration to the dynamic equations may be accounted for by simply subtracting the gravitational acceleration vector $\mathbf{g}$ from the linear acceleration of a cross section.

3.4.4 The external force term

The boundary of the beam may be separated into three parts: the first cross section (attached to $\{O_n, X_n Y_n Z_n\}$), the lateral faces of the beam parallel to the beam axis and the last cross section (attached to $\{O_{\hat{n}}, X_{\hat{n}} Y_{\hat{n}} Z_{\hat{n}}\}$). The external load virtual power term is correspondingly separated into three terms:

$$\int_{\Sigma \mathcal{B}_{no}} \delta \mathbf{v}_p^T \mathbf{f}_s d\Sigma \mathcal{B}_{no} = \int_{A_n} \delta \mathbf{v}_p^T \mathbf{f}_n dA_n + \int_{\Sigma \overline{\mathcal{B}}_{no}} \delta \mathbf{v}_p^T \mathbf{f}_{\Sigma \overline{\mathcal{B}}_{no}} d\Sigma \overline{\mathcal{B}}_{no} + \int_{A_{\hat{n}}} \delta \mathbf{v}_p^T \mathbf{f}_{\hat{n}} dA_{\hat{n}}$$
$$(3.60)$$

where A_n represents the first cross section, $A_{\hat{n}}$ represents the last cross section and $\Sigma \overline{\mathcal{B}}_{no}$ represents the boundary of the beam excluding the first and last cross sections, that is, the lateral faces of the beam excluding the edges of the first and last cross section. This separation is consensual with the kinematic assumptions introduced in Section 3.2. Basically, the borders of the beam are the first cross section, the last cross section and the edges of the in-between cross sections. In a typical application, the first and last cross sections would be attached to the neighbouring bodies, and the lateral faces of the beam may have forces applied on them either from external disturbances or for control purposes. The latter may be achieved for example through the use of control-moment-gyros or proof-mass actuators (Preumont, 1997).

Using equation (3.44), the lateral faces term of the beam becomes

$$\int_{\Sigma \overline{\mathcal{B}}_{no}} \delta \mathbf{v}_p^T \mathbf{f}_{\Sigma \overline{\mathcal{B}}_{no}} d\Sigma \overline{\mathcal{B}}_{no} = \int_0^{\overline{L}} \delta \mathbf{v}_k^T \mathbf{F}_{\Sigma \overline{\mathcal{B}}_{no}} + \delta \boldsymbol{\omega}_k^T \mathbf{M}_{\Sigma \overline{\mathcal{B}}_{no}} d\Sigma \overline{\mathcal{B}}_{no} \qquad (3.61)$$

where

$$\mathbf{F}_{\Sigma \overline{\mathcal{B}}_{no}} = \int_{\Sigma A_k} \mathbf{f}_{\Sigma \overline{\mathcal{B}}_{no}} d\Sigma A_k \qquad (3.62)$$

and

$$\mathbf{M}_{\Sigma \overline{\mathcal{B}}_{no}} = \int_{\Sigma A_k} \left(\mathbf{R}_{e_k} \mathbf{Y}_p \right)^{\sim} \mathbf{f}_{\Sigma \overline{\mathcal{B}}_{no}} d\Sigma A_k \qquad (3.63)$$

where $\overline{0}$ and $\overline{L}$ represent the integration excluding the edges of the first and last cross sections and ΣA_k represents the edge of cross section A_k. The first term on the right-hand side of equation (3.61) represents the contribution of the external forces to the linear motion of a cross section and the second term represents the contribution of the external forces to the angular motion of a cross section. Equations (3.62) and (3.63) represent the resulting external force and moment, respectively, applied at the centre of the cross section.

The contribution of the loads applied on the first and last cross sections are determined in a similar manner to the above term. For the first cross-section one has that $\Omega_k = \begin{bmatrix} 0 & 0 & 0 \end{bmatrix}^T$, $X_1 = 0$, $\mathbf{u}_k = \begin{bmatrix} 0 & 0 & 0 \end{bmatrix}^T$ and $\mathbf{R}_{e_k} = \mathbf{R}_{e_n} = \mathbf{I}$. Therefore,

$$\int_{A_n} \delta\mathbf{v}_p^T \mathbf{f}_n dA_n = \delta\mathbf{v}_n^T \mathbf{F}_n + \delta\boldsymbol{\omega}_n^T \mathbf{M}_n \tag{3.64}$$

where $\mathbf{F}_n$ and $\mathbf{M}_n$ are the resulting external force and moment applied at the centre of the cross section expressed in the body reference frame.

For the last cross-section one has $X_1 = L$, and therefore the rotation matrix due to deformation, the reference position vector, the displacement vector of the centre of the cross section, and the velocity vector due to elasticity become $\mathbf{R}_{e_{\hat{n}}}$, $\mathbf{X}_{\hat{n}}$, $\mathbf{u}_{\hat{n}}$ and $\Omega_{\hat{n}}$, respectively. The virtual power expression becomes

$$\int_{A_{\hat{n}}} \delta\mathbf{v}_p^T \mathbf{f}_{\hat{n}} dA_{\hat{n}} = \underbrace{(\delta\mathbf{v}_n + \delta\widetilde{\boldsymbol{\omega}}_n (\mathbf{X}_{\hat{n}} + \mathbf{u}_{\hat{n}}) + \delta\dot{\mathbf{u}}_{\hat{n}})^T}_{\delta\mathbf{v}_{\hat{n}}} \mathbf{F}_{\hat{n}} + \underbrace{(\delta\boldsymbol{\omega}_n + \delta\Omega_{\hat{n}})^T}_{\delta\boldsymbol{\omega}_{\hat{n}}} \mathbf{M}_{\hat{n}} \tag{3.65}$$

where $\mathbf{F}_{\hat{n}}$ and $\mathbf{M}_{\hat{n}}$ represent the resulting external force and moment, respectively, applied at the centre of the cross section and written in the body reference frame. In a robotic manipulator, the last cross section of a flexible beam is connected to another body, either rigid or another flexible beam. Therefore, reference frames $\{O_{\hat{n}}, X_{\hat{n}}Y_{\hat{n}}Z_{\hat{n}}\}$ and $\{O_{n+1}, X_{n+1}Y_{n+1}Z_{n+1}\}$ are placed coincident and equation (3.65) is rewritten as

$$\int_{A_{\hat{n}}} \delta\mathbf{v}_p^T \mathbf{f}_{\hat{n}} dA_{\hat{n}} = -\delta\mathbf{v}_{\hat{n}}^T \mathbf{R}_{e_{\hat{n}}} \mathbf{F}_{n+1} - \delta\boldsymbol{\omega}_{\hat{n}}^T \mathbf{R}_{e_{\hat{n}}} \mathbf{M}_{n+1} \tag{3.66}$$

3.4.5   Rayleigh–Ritz discretisation

The system of ordinary differential equations that describe the dynamics of the flexible beam is obtained by taking equations (3.45), (3.58), (3.59), (3.61), (3.64) and (3.65), and performing a spatial discretisation of the elastic variables. In these equations, the terms that are dependent on the elastic variables, $\mathbf{K}_k$ and Γ_k, are the displacement vector $\mathbf{u}_k$ and its first- and second-time derivatives, the angular velocity vector Ω_k and its first time derivative, and the cross-section inertia tensor $\mathbf{J}$. The elastic variables, $\mathbf{K}_k$ and Γ_k, appear in $\mathbf{u}_k$ in the form of the rotation matrix $\mathbf{R}_{e_k}$ and of vector $_k\mathbf{t}$ respectively, and in Ω_k and $\mathbf{J}$ in the form of the rotation matrix $\mathbf{R}_{e_k}$.

The shear angles have been assumed to be of the first order and the rotation matrix has been left expanded up to an undefined order m. The calculation of the rotation matrix, according to equations (3.20) and (3.21), leads to a notation involving a large amount of spatial integrals of the curvatures. Therefore, to simplify the notation, the following integral expressions are defined (Bremer and Pfeiffer, 1992):

$$\upsilon_2 = \int_0^{X_1} \int_0^{\xi} K_3 d\eta d\xi = \mathbf{\upsilon}_{2x} (X_1)^T \mathbf{\upsilon}_{2t} (t) \tag{3.67}$$

$$\upsilon_3 = \int_0^{X_1} \int_0^{\xi} -K_2 d\eta d\xi = \mathbf{\upsilon}_{3x} (X_1)^T \mathbf{\upsilon}_{3t} (t) \tag{3.68}$$

$$\alpha = \int_0^{X_1} K_1 d\xi = \mathbf{\alpha}_x (X_1)^T \mathbf{\alpha}_t (t) \tag{3.69}$$

$$\gamma_2 = \int_0^{X_1} \varphi_{12} d\xi = \mathbf{\gamma}_{2x} (X_1)^T \mathbf{\gamma}_{2t} (t) \tag{3.70}$$

$$\gamma_3 = \int_0^{X_1} \varphi_{13} d\xi = \mathbf{\gamma}_{3x} (X_1)^T \mathbf{\gamma}_{3t} (t) \tag{3.71}$$

where υ_2 and υ_3 represent pure bending deflections, α represents a pure torsion angle and γ_2 and γ_3 represent pure shear deflections. $\mathbf{\upsilon}_{2x}$, $\mathbf{\upsilon}_{3x}$, $\mathbf{\alpha}_x$, $\mathbf{\gamma}_{2x}$ and $\mathbf{\gamma}_{3x}$ are the vectors of shape functions for the elastic deflections, and $\mathbf{\upsilon}_{2t}$, $\mathbf{\upsilon}_{3t}$, $\mathbf{\alpha}_t$, $\mathbf{\gamma}_{2t}$ and $\mathbf{\gamma}_{3t}$ are the corresponding vectors of generalized coordinates. The global vector of elastic coordinates is defined as

$$\mathbf{q}_{en} = \begin{bmatrix} \mathbf{q}_{Kn}^T & \vdots & \mathbf{q}_{\varphi n}^T \end{bmatrix}^T = \begin{bmatrix} \mathbf{\upsilon}_{2t}^T & \mathbf{\upsilon}_{3t}^T & \mathbf{\alpha}_t^T & \vdots & \mathbf{\gamma}_{2t}^T & \mathbf{\gamma}_{3t}^T \end{bmatrix}^T \tag{3.72}$$

Using the above discretisation, the terms that depend on the elastic variables may be written in compact form as

$$\mathbf{R}_{e_k} = \mathbf{R}_{e_k} \left(\mathbf{q}_{K_n} \right) \tag{3.73}$$

$$\mathbf{\Omega}_k = \mathbf{J}_{\mathbf{R}_{ek}} \left(\mathbf{q}_{K_n} \right) \dot{\mathbf{q}}_{K_n} \tag{3.74}$$

$$\dot{\mathbf{\Omega}}_k = \dot{\mathbf{J}}_{\mathbf{R}_{ek}} \dot{\mathbf{q}}_{K_n} + \mathbf{J}_{\mathbf{R}_{ek}} \ddot{\mathbf{q}}_{K_n} \tag{3.75}$$

$$\mathbf{x}_k = \mathbf{X}_k + \mathbf{u}_k \left(\mathbf{q}_{en} \right) \tag{3.76}$$

$$\dot{\mathbf{u}}_k = \mathbf{J}_{T_k} \left(\mathbf{q}_{en} \right) \dot{\mathbf{q}}_{en} = \begin{bmatrix} \mathbf{J}_{TK_k} & \mathbf{J}_{T\varphi_k} \end{bmatrix} \begin{bmatrix} \dot{\mathbf{q}}_{K_n} \\ \dot{\mathbf{q}}_{\varphi n} \end{bmatrix} \tag{3.77}$$

$$\ddot{\mathbf{u}}_k = \dot{\mathbf{J}}_{T_k} \dot{\mathbf{q}}_{en} + \mathbf{J}_{T_k} \ddot{\mathbf{q}}_{en} = \begin{bmatrix} \dot{\mathbf{J}}_{TK_k} & \dot{\mathbf{J}}_{T\varphi_k} \end{bmatrix} \begin{bmatrix} \dot{\mathbf{q}}_{K_n} \\ \dot{\mathbf{q}}_{\varphi n} \end{bmatrix} + \begin{bmatrix} \mathbf{J}_{TK_k} & \mathbf{J}_{T\varphi_k} \end{bmatrix} \begin{bmatrix} \ddot{\mathbf{q}}_{K_n} \\ \ddot{\mathbf{q}}_{\varphi n} \end{bmatrix} \tag{3.78}$$

In these equations $\mathbf{J}_{\mathbf{R}_{ek}}$ is the elastic rotation Jacobian and $\mathbf{J}_{T_k}$ is the elastic translation Jacobian of the kth cross section of beam n. The former depends only on the curvatures and the latter depends on the curvatures and the shear angles. The discretised dynamic equations of a flexible link are finally obtained by: first applying the variational operator of the Principle of Virtual Powers to $\dot{\mathbf{u}}_k$ and $\mathbf{\Omega}_k$, yielding

$$\delta \dot{\mathbf{u}}_k = \mathbf{J}_{T_k}(\mathbf{q}_{en}) \, \delta \dot{\mathbf{q}}_{en} = \begin{bmatrix} \mathbf{J}_{TK_k} & \mathbf{J}_{T\varphi_k} \end{bmatrix} \begin{bmatrix} \delta \dot{\mathbf{q}}_{K_n} \\ \delta \dot{\mathbf{q}}_{\varphi_n} \end{bmatrix} \tag{3.79}$$

and

$$\delta \mathbf{\Omega}_k = \mathbf{J}_{R_{ek}}(\mathbf{q}_{K_n}) \, \delta \dot{\mathbf{q}}_{K_n} \tag{3.80}$$

and then, according to the same principle, by gathering the coefficients of $\delta \mathbf{v}_n$, $\delta \boldsymbol{\omega}_n$, $\delta \dot{\mathbf{q}}_{K_n}$, and $\delta \dot{\mathbf{q}}_{\varphi_n}$ in equations (3.45), (3.58), (3.59), (3.61), (3.64) and (3.65),

$$\int_0^L \rho_A \begin{bmatrix} \mathbf{I} & \tilde{\mathbf{x}}_k^T & \mathbf{J}_{TK_k} & \mathbf{J}_{T\varphi_k} \\ \tilde{\mathbf{x}}_k & \tilde{\mathbf{x}}_k \tilde{\mathbf{x}}_k^T + \mathbf{J} & \tilde{\mathbf{x}}_k \mathbf{J}_{TK_k} + \mathbf{J}\mathbf{J}_{R_{ek}} & \tilde{\mathbf{x}}_k \mathbf{J}_{T\varphi_k} \\ \mathbf{J}_{TK_k}^T & (\tilde{\mathbf{x}}_k \mathbf{J}_{TK_k} + \mathbf{J}\mathbf{J}_{R_{ek}})^T & \mathbf{J}_{TK_k}^T \mathbf{J}_{TK_k} + \mathbf{J}_{R_{ek}}^T \mathbf{J}\mathbf{J}_{R_{ek}} & \mathbf{J}_{TK_k}^T \mathbf{J}_{T\varphi_k} \\ \mathbf{J}_{T\varphi_k}^T & (\tilde{\mathbf{x}}_k \mathbf{J}_{T\varphi_k})^T & (\mathbf{J}_{TK_k}^T \mathbf{J}_{T\varphi_k})^T & \mathbf{J}_{T\varphi_k}^T \mathbf{J}_{T\varphi_k} \end{bmatrix} dX_1 \begin{bmatrix} \mathbf{a}_n - \mathbf{g} \\ \dot{\boldsymbol{\omega}}_n \\ \ddot{\mathbf{q}}_{K_n} \\ \ddot{\mathbf{q}}_{\varphi_n} \end{bmatrix}$$

$$+ \int_0^L \rho_A \begin{bmatrix} 0 & \tilde{\boldsymbol{\omega}}_n \tilde{\mathbf{x}}_k^T & 2\tilde{\boldsymbol{\omega}}_n \mathbf{J}_{TK_k} & 2\tilde{\boldsymbol{\omega}}_n \mathbf{J}_{T\varphi_k} \\ 0 & \tilde{\boldsymbol{\omega}}_n \tilde{\mathbf{x}}_k \tilde{\mathbf{x}}_k^T & 2\tilde{\mathbf{x}}_k \tilde{\boldsymbol{\omega}}_n \mathbf{J}_{TK_k} + \mathbf{J}\tilde{\boldsymbol{\omega}}_n \mathbf{J}_{R_{ek}} & 2\tilde{\mathbf{x}}_k \tilde{\boldsymbol{\omega}}_n \mathbf{J}_{T\varphi_k} \\ & +(\boldsymbol{\omega}_n + \mathbf{J}_{R_{ek}} \dot{\mathbf{q}}_{K_n})^{\sim} \mathbf{J} & +(\boldsymbol{\omega}_n + \mathbf{J}_{R_{ek}} \dot{\mathbf{q}}_{K_n})^{\sim} \mathbf{J}\mathbf{J}_{R_{ek}} & \\ 0 & \mathbf{J}_{TK_k}^T \tilde{\boldsymbol{\omega}}_n \tilde{\mathbf{x}}_k^T & 2\mathbf{J}_{TK_k}^T \tilde{\boldsymbol{\omega}}_n \mathbf{J}_{TK_k} + \mathbf{J}_{R_{ek}}^T \mathbf{J}\tilde{\boldsymbol{\omega}}_n \mathbf{J}_{R_{ek}} & 2\mathbf{J}_{TK_k}^T \tilde{\boldsymbol{\omega}}_n \mathbf{J}_{T\varphi_k} \\ & +\mathbf{J}_{R_{ek}}^T (\boldsymbol{\omega}_n + \mathbf{J}_{R_{ek}} \dot{\mathbf{q}}_{K_n})^{\sim} \mathbf{J} & +\mathbf{J}_{R_{ek}}^T (\boldsymbol{\omega}_n + \mathbf{J}_{R_{ek}} \dot{\mathbf{q}}_{K_n})^{\sim} \mathbf{J}\mathbf{J}_{R_{ek}} & \\ 0 & \mathbf{J}_{T\varphi_k}^T \tilde{\boldsymbol{\omega}}_n \tilde{\mathbf{x}}_k^T & 2\mathbf{J}_{T\varphi_k}^T \tilde{\boldsymbol{\omega}}_n \mathbf{J}_{TK_k} & 2\mathbf{J}_{T\varphi_k}^T \tilde{\boldsymbol{\omega}}_n \mathbf{J}_{T\varphi_k} \end{bmatrix} dX_1 \begin{bmatrix} \mathbf{v}_n \\ \boldsymbol{\omega}_n \\ \dot{\mathbf{q}}_{K_n} \\ \dot{\mathbf{q}}_{\varphi_n} \end{bmatrix}$$

$$+ \int_0^L \rho_A \begin{bmatrix} 0 & 0 & \dot{\mathbf{J}}_{TK_k} & \dot{\mathbf{J}}_{T\varphi_k} \\ 0 & 0 & \tilde{\mathbf{x}}_k \tilde{\boldsymbol{\omega}}_n \mathbf{J}_{TK_k} + \mathbf{J}\dot{\mathbf{J}}_{R_{ek}} & \tilde{\mathbf{x}}_k \dot{\mathbf{J}}_{T\varphi_k} \\ 0 & 0 & \mathbf{J}_{TK_k}^T \dot{\mathbf{J}}_{TK_k} + \mathbf{J}_{R_{ek}}^T \mathbf{J}\dot{\mathbf{J}}_{R_{ek}} & \mathbf{J}_{TK_k}^T \dot{\mathbf{J}}_{T\varphi_k} \\ 0 & 0 & \mathbf{J}_{T\varphi_k}^T \dot{\mathbf{J}}_{TK_k} & \mathbf{J}_{T\varphi_k}^T \dot{\mathbf{J}}_{T\varphi_k} \end{bmatrix} dX_1 \begin{bmatrix} \mathbf{v}_n \\ \boldsymbol{\omega}_n \\ \dot{\mathbf{q}}_{K_n} \\ \dot{\mathbf{q}}_{\varphi_n} \end{bmatrix}$$

$$+ \int_0^L \begin{bmatrix} 0 & 0 & 0 & 0 \\ 0 & 0 & 0 & 0 \\ 0 & 0 & \mathrm{diag}(EI_3 \boldsymbol{\upsilon}''_{2x} \boldsymbol{\upsilon}''^T_{2x}, EI_2 \boldsymbol{\upsilon}''_{3x} \boldsymbol{\upsilon}''^T_{3x}, GJ\boldsymbol{\alpha}'_x \boldsymbol{\alpha}'^T_x) & 0 \\ 0 & 0 & 0 & \mathrm{diag}(GA\boldsymbol{\gamma}'_{2x} \boldsymbol{\gamma}'^T_{2x}, GA\boldsymbol{\gamma}'_{3x} \boldsymbol{\gamma}'^T_{3x}) \end{bmatrix} dX_1 \begin{bmatrix} 0 \\ 0 \\ \mathbf{q}_{K_n} \\ \mathbf{q}_{\varphi_n} \end{bmatrix}$$

$$= \begin{bmatrix} \mathbf{I} & 0 \\ 0 & \mathbf{I} \\ 0 & 0 \\ 0 & 0 \end{bmatrix} \begin{bmatrix} \mathbf{F}_n \\ \mathbf{M}_n \end{bmatrix} + \int_0^{\bar{L}} \begin{bmatrix} \mathbf{I} & 0 \\ \tilde{\mathbf{x}}_k & \mathbf{I} \\ \mathbf{J}_{TK_k}^T & \mathbf{J}_{R_{ek}}^T \\ \mathbf{J}_{T\varphi_k}^T & 0 \end{bmatrix} \begin{bmatrix} \mathbf{F}_{\Sigma \bar{B}_{no}} \\ \mathbf{M}_{\Sigma \bar{B}_{no}} \end{bmatrix} dX_1 - \begin{bmatrix} \mathbf{R}_{e\hat{n}} & 0 \\ \tilde{\mathbf{x}}_{\hat{n}} \mathbf{R}_{e\hat{n}} & \mathbf{R}_{e\hat{n}} \\ \mathbf{J}_{TK_{\hat{n}}}^T \mathbf{R}_{e\hat{n}} & \mathbf{J}_{R_{e\hat{n}}}^T \mathbf{R}_{e\hat{n}} \\ \mathbf{J}_{T\varphi_{\hat{n}}}^T \mathbf{R}_{e\hat{n}} & 0 \end{bmatrix} \begin{bmatrix} \mathbf{F}_{n+1} \\ \mathbf{M}_{n+1} \end{bmatrix}$$

$$\tag{3.81}$$

According to the blocks delineated, this equation may be written in a compact form as

$$
\begin{bmatrix} \mathcal{M}_{en,rr} & \mathcal{M}_{en,re} \\ \mathcal{M}_{en,re}^T & \mathcal{M}_{en,ee} \end{bmatrix} \begin{bmatrix} \mathbf{A}_n \\ \ddot{\mathbf{q}}_{en} \end{bmatrix} + \mathcal{N}_{en} \left(\mathbf{V}_n, \dot{\mathbf{q}}_{en} \right) + \mathcal{K}_{en} \left(\mathbf{q}_{en} \right)
$$

$$
= \begin{bmatrix} \mathbf{I} \\ 0 \end{bmatrix} \mathcal{F}_n + \mathcal{F}_{\Sigma n} - \begin{bmatrix} \mathbf{\Phi}_{r\,n+1,n}^T \\ \mathbf{\Phi}_{e\,n+1,n}^T \end{bmatrix} \mathcal{F}_{n+1}
$$

(3.82)

or

$$
\begin{bmatrix} \mathcal{M}_{en,rr} & \mathcal{M}_{en,re} \\ \mathcal{M}_{en,re}^T & \mathcal{M}_{en,ee} \end{bmatrix} \begin{bmatrix} \mathbf{A}_n \\ \ddot{\mathbf{q}}_{en} \end{bmatrix} + \begin{bmatrix} \overline{\mathcal{N}}_{en,r} \\ \overline{\mathcal{N}}_{en,e} \end{bmatrix} \left(\mathbf{V}_n, \dot{\mathbf{q}}_{en}, \mathbf{q}_{en}, \mathcal{F}_{\Sigma n} \right)
$$

$$
= \begin{bmatrix} \mathbf{I} \\ 0 \end{bmatrix} \mathcal{F}_n - \begin{bmatrix} \mathbf{\Phi}_{r\,n+1,n}^T \\ \mathbf{\Phi}_{e\,n+1,n}^T \end{bmatrix} \mathcal{F}_{n+1}
$$

(3.83)

where the non-linear inertial generalized forces $\mathcal{N}_{en}$, the generalized linear elastic forces $\mathcal{K}_{en}$ and the generalized forces and moments applied on the boundary of the beam, $\mathcal{F}_{\Sigma n}$, have been grouped under the same vector $\overline{\mathcal{N}}_{en}$.

3.5 The dynamic model of a multi-link manipulator

A flexible manipulator consists of rigid components and flexible beams connected by joints in a serial chain topology as shown in Figure 3.2. The body at one of the extremities of the chain is designated as the base body $\mathcal{B}_1$ and the body at the other end of the chain is designated as the tip body $\mathcal{B}_N$. The in-between bodies of the chain are connected to only two joints. Body zero, $\mathcal{B}_0$, is the designation reserved for the inertial body.

The joints are numbered in a similar fashion as the bodies. For the nth body, $\mathcal{B}_n$, the inboard joint is designated as $\mathcal{J}_n$, and the outboard joint is designated as $\mathcal{J}_{n+1}$. The designation of inboard (outboard) refers to the topological position of the body or joint in the manipulator, which is closer to/further from the inertial body. The first joint, $\mathcal{J}_1$, is the joint connecting the chain to the inertial body, $\mathcal{B}_0$.

The joints may have from 0 to 6 degrees of freedom (DOF). If the manipulator is floating in space for example, then J_1 is set to allow the 6-DOF. Similarly, if the system consists of a mobile robot to which a flexible manipulator is attached, then $\mathcal{J}_1$ must possess translation degrees of freedom. For a typical case as in industrial manipulators, $\mathcal{J}_1$ only includes rotational degrees of freedom.

A joint with 0-DOF serves the purpose of rigidly connecting flexible beams with flexible beams, as in $\mathcal{J}_N$, for example, and flexible beams with rigid bodies, as in $\mathcal{J}_2$,

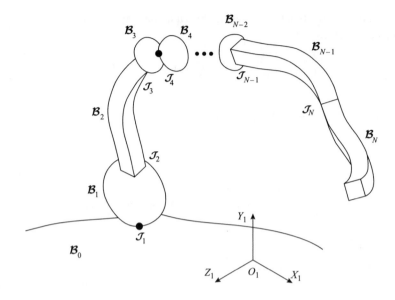

Figure 3.2   Flexible manipulator

$\mathcal{J}_3$ and $\mathcal{J}_{N-1}$, for example. By rigidly connecting flexible beams, flexible bodies of greater complexity may be modelled; higher displacements may be achieved and the complexity of the shape functions reduced by assuming a flexible beam composed of several flexible beams connected base to tip. The dimension of the model, however, is increased.

3.5.1   Joint kinematics

Joint $\mathcal{J}_n$ connecting rigid bodies $\mathcal{B}_{n-1}$ and $\mathcal{B}_n$ is represented in Figure 3.3. Here a joint representation similar to that of Rodriguez *et al.* (1992), but with the increasing body numbering from base to tip as in Hardt (1999) is used. The reference position of $\mathcal{J}_n$ on $\mathcal{B}_{n-1}$ is designated as the inboard point of $\mathcal{J}_n$, O_{n-1}. Similarly, the reference position of $\mathcal{J}_n$ on $\mathcal{B}_n$ is designated as the outboard point of $\mathcal{J}_n$, O_n, the origin of the body reference frame $\{\mathcal{O}_n, X_n Y_n Z_n\}$. The inter-body position vector, which describes the position of O_n relative to O_{n-1} expressed in $\{O_{n-1}, X_{n-1} Y_{n-1} Z_{n-1}\}$, is defined as $\mathbf{r}_{n-1,n}$.

The rotation of $\mathcal{B}_n$ relative to $\mathcal{B}_{n-1}$ is defined through the orthogonal rotation matrix $\mathbf{R}_{n/n-1}$. The columns of $\mathbf{R}_{n/n-1}$ are the projections of the basis vectors of reference frame $\{O_n, X_n Y_n Z_n\}$ on $\{O_{n-1}, X_{n-1} Y_{n-1} Z_{n-1}\}$. Similarly to equation (3.36), the angular velocity of $\mathcal{B}_n$ relative to $\mathcal{B}_{n-1}$ expressed in $\{O_n, X_n Y_n Z_n\}$ is obtained from

$$\widetilde{\omega}_{n/n-1} = \mathbf{R}_{n/n-1}^T \dot{\mathbf{R}}_{n/n-1} \qquad (3.84)$$

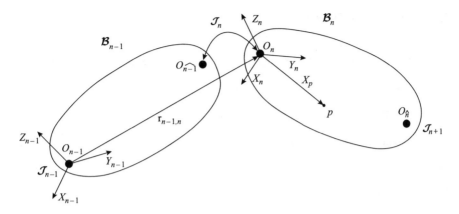

Figure 3.3   Joint kinematics

The absolute angular velocity of body $\mathcal{B}_n$ expressed in $\{O_n, X_n Y_n Z_n\}$ may then be written as

$$\omega_n = \mathbf{R}_{n/n-1}^T \omega_{n-1} + \omega_{n/n-1} \tag{3.85}$$

The relative angular velocity between the two rigid bodies may be expressed in terms of the time derivative of Euler angles (Sincarsin *et al.*, 1993) (the case of typical robotic manipulators), therefore, $\omega_{n/n-1}$ may be written as a product of a rotational joint matrix, $\mathbf{H}_{n\omega}(\mathbf{q}_{n\omega})$ (rotation Jacobian matrix of joint n), multiplied by the vector of angular velocity parameters, $\dot{\mathbf{q}}_{n\omega}$;

$$\omega_{n/n-1} = \mathbf{H}_{n\omega}(\mathbf{q}_{n\omega})\dot{\mathbf{q}}_{n\omega} \tag{3.86}$$

The absolute linear velocity of $\mathcal{B}_n$, expressed in $\{O_n, X_n Y_n Z_n\}$, is given by

$$
\begin{aligned}
\mathbf{v}_n &= \mathbf{R}_{n/n-1}^T \left(\mathbf{v}_{n-1} + \widetilde{\omega}_{n-1} \mathbf{r}_{n-1,n} + \dot{\mathbf{r}}_{n-1,n} \right) \\
&= \mathbf{R}_{n/n-1}^T \left(\mathbf{v}_{n-1} + \widetilde{\omega}_{n-1} \mathbf{r}_{n-1,n} \right) + \mathbf{H}_{nv}(\mathbf{q}_{n\omega}) \dot{\mathbf{q}}_{nv}
\end{aligned} \tag{3.87}
$$

where $\mathbf{H}_{nv}(\mathbf{q}_{n\omega})$ is the translation joint matrix (translation Jacobian matrix of joint n), and $\dot{\mathbf{q}}_{nv}$ is the vector of linear velocity parameters. $\mathbf{H}_{nv}(\mathbf{q}_{n\omega})$ is dependent on $\mathbf{q}_{n\omega}$ if joint translation occurs in $\mathcal{B}_{n-1}$. If translation occurs in $\mathcal{B}_n$ it is a constant matrix. Writing equation (3.85) together with equation (3.87) yields (Jain and Rodriguez, 1992)

$$
\begin{bmatrix} \mathbf{v}_n \\ \omega_n \end{bmatrix} = \begin{bmatrix} \mathbf{R}_{n/n-1}^T & \mathbf{R}_{n/n-1}^T \tilde{\mathbf{r}}_{n-1,n}^T \\ 0 & \mathbf{R}_{n/n-1}^T \end{bmatrix} \begin{bmatrix} \mathbf{v}_{n-1} \\ \omega_{n-1} \end{bmatrix} + \begin{bmatrix} \mathbf{H}_{nv} & 0 \\ 0 & \mathbf{H}_{n\omega} \end{bmatrix} \begin{bmatrix} \dot{\mathbf{q}}_{nv} \\ \dot{\mathbf{q}}_{n\omega} \end{bmatrix}
$$

$$\Leftrightarrow \mathbf{V}_n = \mathbf{\Phi}_{rn,n-1}\mathbf{V}_{n-1} + \mathbf{H}_n \dot{\mathbf{q}}_n \tag{3.88}$$

The absolute linear and angular acceleration vectors of $\mathcal{B}_n$, expressed in $\{O_n, X_n Y_n Z_n\}$, may be obtained as in equations (3.41) and (3.43), or by analogy to equations (3.34) and (3.43), as

$$
\begin{aligned}
\mathbf{a}_n &= \mathbf{R}_{n/n-1}^T \left(\mathbf{a}_{n-1} + \dot{\tilde{\boldsymbol{\omega}}}_{n-1}\mathbf{r}_{n-1,n} + \tilde{\boldsymbol{\omega}}_{n-1}\tilde{\boldsymbol{\omega}}_{n-1}\mathbf{r}_{n-1,n} + 2\tilde{\boldsymbol{\omega}}_{n-1}\dot{\mathbf{r}}_{n-1,n} + \ddot{\mathbf{r}}_{n-1,n} \right) \\
&= \mathbf{R}_{n/n-1}^T \left(\mathbf{a}_{n-1} + \dot{\tilde{\boldsymbol{\omega}}}_{n-1}\mathbf{r}_{n-1,n} \right) + \mathbf{H}_{nv}\ddot{\mathbf{q}}_{nv} \\
&\quad + \mathbf{R}_{n/n-1}^T \tilde{\boldsymbol{\omega}}_{n-1}\tilde{\boldsymbol{\omega}}_{n-1}\mathbf{r}_{n-1,n} + \left(2\left(\mathbf{R}_{n/n-1}^T \boldsymbol{\omega}_{n-1} \right)^{\sim} \right. \\
&\quad \left. + \tilde{\boldsymbol{\omega}}_{n/n-1} \right) \mathbf{H}_{nv}\dot{\mathbf{q}}_{nv} + \dot{\mathbf{H}}_{nv}\dot{\mathbf{q}}_{nv}
\end{aligned}
\tag{3.89}
$$

and

$$
\begin{aligned}
\boldsymbol{\alpha}_n &= \dot{\boldsymbol{\omega}}_n = \dot{\mathbf{R}}_{n/n-1}^T \boldsymbol{\omega}_{n-1} + \mathbf{R}_{n/n-1}^T \dot{\boldsymbol{\omega}}_{n-1} + \dot{\mathbf{H}}_{n\omega}\dot{\mathbf{q}}_{n\omega} + \mathbf{H}_{n\omega}\ddot{\mathbf{q}}_{n\omega} \\
&= \mathbf{R}_{n/n-1}^T \dot{\boldsymbol{\omega}}_{n-1} + \mathbf{H}_{n\omega}\ddot{\mathbf{q}}_{n\omega} + \left(\mathbf{R}_{n/n-1}^T \boldsymbol{\omega}_{n-1} \right)^{\sim} \mathbf{H}_{n\omega}\dot{\mathbf{q}}_{n\omega} + \dot{\mathbf{H}}_{n\omega}\dot{\mathbf{q}}_{n\omega}
\end{aligned}
\tag{3.90}
$$

Writing the above two equations together, similar to equation (3.88), yields

$$
\begin{bmatrix} \mathbf{a}_n \\ \dot{\boldsymbol{\omega}}_n \end{bmatrix} = \begin{bmatrix} \mathbf{R}_{n/n-1}^T & \mathbf{R}_{n/n-1}^T \tilde{\mathbf{r}}_{n-1,n}^T \\ 0 & \mathbf{R}_{n/n-1}^T \end{bmatrix} \begin{bmatrix} \mathbf{a}_{n-1} \\ \dot{\boldsymbol{\omega}}_{n-1} \end{bmatrix} + \begin{bmatrix} \mathbf{H}_{nv} & 0 \\ 0 & \mathbf{H}_{n\omega} \end{bmatrix} \begin{bmatrix} \ddot{\mathbf{q}}_{nv} \\ \ddot{\mathbf{q}}_{n\omega} \end{bmatrix}
$$

$$
+ \begin{bmatrix} \mathbf{R}_{n/n-1}^T \tilde{\boldsymbol{\omega}}_{n-1}\tilde{\boldsymbol{\omega}}_{n-1}\mathbf{r}_{n-1,n} + \left(2\left(\mathbf{R}_{n/n-1}^T \boldsymbol{\omega}_{n-1} \right)^{\sim} + \tilde{\boldsymbol{\omega}}_{n/n-1} \right) \mathbf{H}_{nv}\dot{\mathbf{q}}_{nv} + \dot{\mathbf{H}}_{nv}\dot{\mathbf{q}}_{nv} \\ \left(\mathbf{R}_{n/n-1}^T \boldsymbol{\omega}_{n-1} \right)^{\sim} \mathbf{H}_{n\omega}\dot{\mathbf{q}}_{n\omega} + \dot{\mathbf{H}}_{n\omega}\dot{\mathbf{q}}_{n\omega} \end{bmatrix}
$$

$$
\Leftrightarrow \mathbf{A}_n = \boldsymbol{\Phi}_{rn,n-1}\mathbf{A}_{n-1} + \mathbf{H}_n\ddot{\mathbf{q}}_n + \mathbf{N}_n
\tag{3.91}
$$

In the case of a rigid connection where $\mathcal{B}_{n-1}$ is a flexible beam and $\mathcal{B}_n$ is either a flexible beam or a rigid-body, the reference frames $\{O_{\wedge_{n-1}}, X_{\wedge_{n-1}} Y_{\wedge_{n-1}} Z_{\wedge_{n-1}}\}$ and $\{O_n, X_n Y_n Z_n\}$ coincide, and equations (3.88) and (3.91) are obtained by evaluating $\mathbf{v}_k, \boldsymbol{\omega}_k, \mathbf{a}_k$ and $\boldsymbol{\alpha}_k$ at $O_{\wedge_{n-1}}$:

$$
\begin{bmatrix} \mathbf{v}_n \\ \boldsymbol{\omega}_n \end{bmatrix} = \begin{bmatrix} \mathbf{R}_{n/n-1}^T & \mathbf{R}_{n/n-1}^T \tilde{\mathbf{r}}_{n-1,n}^T & \mathbf{R}_{n/n-1}^T \mathbf{J}_{TK_{\wedge_{n-1}}} & \mathbf{R}_{n/n-1}^T \mathbf{J}_{T\varphi_{\wedge_{n-1}}} \\ 0 & \mathbf{R}_{n/n-1}^T & \mathbf{R}_{n/n-1}^T \mathbf{J}_{R_{e\wedge_{n-1}}} & 0 \end{bmatrix} \begin{bmatrix} \mathbf{v}_{n-1} \\ \boldsymbol{\omega}_{n-1} \\ \dot{\mathbf{q}}_{K_{n-1}} \\ \dot{\mathbf{q}}_{\varphi_{n-1}} \end{bmatrix}
$$

$$
\Leftrightarrow \mathbf{V}_n = \begin{bmatrix} \boldsymbol{\Phi}_{rn,n-1} & \boldsymbol{\Phi}_{en,n-1} \end{bmatrix} \begin{bmatrix} \mathbf{V}_{n-1} \\ \dot{\mathbf{q}}_{en-1} \end{bmatrix}
\tag{3.92}
$$

and

$$\begin{bmatrix} \mathbf{a}_n \\ \dot{\boldsymbol{\omega}}_n \end{bmatrix} = \begin{bmatrix} \mathbf{R}^T_{n/n-1} & \mathbf{R}^T_{n/n-1}\tilde{\mathbf{r}}^T_{n-1,n} & \mathbf{R}^T_{n/n-1}\mathbf{J}_{TK_{\hat{\wedge}_{n-1}}} & \mathbf{R}^T_{n/n-1}\mathbf{J}_{T\varphi_{\hat{\wedge}_{n-1}}} \\ 0 & \mathbf{R}^T_{n/n-1} & \mathbf{R}^T_{n/n-1}\mathbf{J}_{R_{e\hat{\wedge}_{n-1}}} & 0 \end{bmatrix} \begin{bmatrix} \mathbf{a}_{n-1} \\ \dot{\boldsymbol{\omega}}_{n-1} \\ \ddot{\mathbf{q}}_{K_{n-1}} \\ \ddot{\mathbf{q}}_{\varphi_{n-1}} \end{bmatrix}$$

$$+ \begin{bmatrix} \mathbf{R}^T_{n/n-1}\left(\tilde{\boldsymbol{\omega}}_{n-1}\tilde{\boldsymbol{\omega}}_{n-1}\mathbf{r}_{n-1,n} + 2\tilde{\boldsymbol{\omega}}_{n-1}\mathbf{J}_{T_{\hat{\wedge}_{n-1}}}\dot{\mathbf{q}}_{en-1} + \dot{\mathbf{J}}_{T_{\hat{\wedge}_{n-1}}}\dot{\mathbf{q}}_{en-1} \right) \\ \mathbf{R}^T_{n/n-1}\left(\tilde{\boldsymbol{\omega}}_{n-1}\mathbf{J}_{R_{e\hat{\wedge}_{n-1}}}\dot{\mathbf{q}}_{K_{n-1}} + \dot{\mathbf{J}}_{R_{e\hat{\wedge}_{n-1}}}\dot{\mathbf{q}}_{K_{n-1}} \right) \end{bmatrix}$$

$$\Leftrightarrow \mathbf{A}_n = \begin{bmatrix} \boldsymbol{\Phi}_{rn,n-1} & \boldsymbol{\Phi}_{en,n-1} \end{bmatrix} \begin{bmatrix} \mathbf{A}_{n-1} \\ \ddot{\mathbf{q}}_{en-1} \end{bmatrix} + \mathbf{N}_{en} \tag{3.93}$$

where $\mathbf{R}_{n/n-1} = \mathbf{R}_{e\hat{\wedge}_{n-1}}$.

3.5.2 Dynamics of a rigid body

The dynamic equations of a rigid body expressed in the body reference frame may be obtained through the Principle of Virtual Powers, similar to the case of a flexible beam. Considering body $\mathcal{B}_n$ of Figure 3.3 one has

$$\int_{\mathcal{B}_n} \delta\mathbf{v}^T_p \mathbf{a}_p \rho_p d\mathcal{B}_n = \int_{\mathcal{B}_n} \delta\mathbf{v}^T_p \mathbf{g}\rho_p d\mathcal{B}_n + \int_{\Sigma\mathcal{B}_n} \delta\mathbf{v}^T_p \mathbf{f}_s d\Sigma\mathcal{B}_n \tag{3.94}$$

where

$$\mathbf{v}_p = \mathbf{v}_n + \tilde{\boldsymbol{\omega}}_n \mathbf{X}_p \tag{3.95}$$

and

$$\mathbf{a}_p = \mathbf{a}_n + \dot{\tilde{\boldsymbol{\omega}}}_n \mathbf{X}_p + \tilde{\boldsymbol{\omega}}_n\tilde{\boldsymbol{\omega}}_n \mathbf{X}_p \tag{3.96}$$

Applying the variational operator, δ, to equation (3.95), and solving equation (3.94) yields

$$\begin{bmatrix} m_b\mathbf{I} & m_b\tilde{\mathbf{X}}^T_{gn} \\ m_b\tilde{\mathbf{X}}_{gn} & \mathbf{J}_n \end{bmatrix} \begin{bmatrix} \mathbf{a}_n - \mathbf{g} \\ \dot{\boldsymbol{\omega}}_n \end{bmatrix} + \begin{bmatrix} m_b\tilde{\boldsymbol{\omega}}_n\tilde{\mathbf{X}}^T_{gn}\boldsymbol{\omega}_n \\ \tilde{\boldsymbol{\omega}}_n\mathbf{J}_n\boldsymbol{\omega}_n \end{bmatrix}$$

$$= \begin{bmatrix} \mathbf{F}_n \\ \mathbf{M}_n \end{bmatrix} - \begin{bmatrix} \mathbf{R}_{n+1/n} & 0 \\ \tilde{\mathbf{r}}_{n,n+1}\mathbf{R}_{n+1/n} & \mathbf{R}_{n+1/n} \end{bmatrix} \begin{bmatrix} \mathbf{F}_{n+1} \\ \mathbf{M}_{n+1} \end{bmatrix} \tag{3.97}$$

which may be written in compact form as

$$\mathcal{M}_{rn}\mathbf{A}_n + \mathcal{N}_{rn}(\boldsymbol{\omega}_n) = \mathcal{F}_n - \boldsymbol{\Phi}^T_{r\,n+1,n}\mathcal{F}_{n+1} \tag{3.98}$$

The control force and the control torque at the joints may be obtained by performing a virtual power balance across joint n. Since a joint has no mass, the linear relative motion is expressed as

$$\delta \, (\mathbf{H}_{nv}\dot{\mathbf{q}}_{nv})^T \, \mathbf{F}_n = \delta \dot{\mathbf{q}}_{nv}^T \mathbf{H}_{nv}^T \mathbf{F}_n \tag{3.99}$$

and the angular relative motion as

$$\delta \, (\mathbf{H}_{n\omega}\dot{\mathbf{q}}_{n\omega})^T \, \mathbf{M}_n = \delta \dot{\mathbf{q}}_{n\omega}^T \mathbf{H}_{n\omega}^T \mathbf{M}_n \tag{3.100}$$

The externally applied generalized force vector (torques and forces) is then written as

$$\begin{bmatrix} \mathbf{T}_{nv} \\ \mathbf{T}_{n\omega} \end{bmatrix} = \begin{bmatrix} \mathbf{H}_{nv}^T & 0 \\ 0 & \mathbf{H}_{n\omega}^T \end{bmatrix} \begin{bmatrix} \mathbf{F}_n \\ \mathbf{M}_n \end{bmatrix}$$

$$\Leftrightarrow \mathcal{T}_n = \mathbf{H}_n^T \mathcal{F}_n \tag{3.101}$$

3.5.3 *Dynamics of a rigid–flexible–rigid body*

A flexible beam attached to two rigid bodies, one at each end, may be seen as a building block for a flexible manipulator. The rigid bodies are the supporting structures for actuators and sensors at the joints, and are typically of significant mass. However, if that is not the case, their mass may be easily neglected, and what remains are their geometric properties. To this end, consider the following:

Rigid body: $\mathcal{B}_n$, from equation (3.97), and setting $\dot{\mathbf{q}}_n = 0$ and $\ddot{\mathbf{q}}_n = 0$ in equations (3.88) and (3.91),

$$\mathcal{M}_{rn}\mathbf{A}_n + \mathcal{N}_{rn} = \mathcal{F}_n - \mathbf{\Phi}_{r\,n+1,n}^T \mathcal{F}_{n+1} \tag{3.102}$$

$$\mathbf{V}_{n+1} = \mathbf{\Phi}_{r\,n+1,n}\mathbf{V}_n \tag{3.103}$$

$$\mathbf{A}_{n+1} = \mathbf{\Phi}_{r\,n+1,n}\mathbf{A}_n + \mathbf{N}_{n+1} \tag{3.104}$$

$$\mathbf{N}_{n+1} = \begin{bmatrix} \mathbf{R}_{n+1/n}^T \widetilde{\boldsymbol{\omega}}_n \widetilde{\boldsymbol{\omega}}_n \mathbf{r}_{n,n+1} \\ 0 \end{bmatrix} \tag{3.105}$$

Flexible beam: $\mathcal{B}_{n+1}$, from equations (3.83), (3.92) and (3.93),

$$\begin{bmatrix} \mathcal{M}_{e\,n+1,rr} & \mathcal{M}_{e\,n+1,re} \\ \mathcal{M}_{e\,n+1,re}^T & \mathcal{M}_{e\,n+1,ee} \end{bmatrix} \begin{bmatrix} \mathbf{A}_{n+1} \\ \ddot{\mathbf{q}}_{e\,n+1} \end{bmatrix} + \begin{bmatrix} \mathcal{N}_{e\,n+1,r} \\ \mathcal{N}_{e\,n+1,e} \end{bmatrix}$$

$$= \begin{bmatrix} \mathbf{I} \\ 0 \end{bmatrix} \mathcal{F}_{n+1} - \begin{bmatrix} \mathbf{\Phi}_{r\,n+2,n+1}^T \\ \mathbf{\Phi}_{e\,n+2,n+1}^T \end{bmatrix} \mathcal{F}_{n+2} \tag{3.106}$$

$$\mathbf{V}_{n+2} = \begin{bmatrix} \mathbf{\Phi}_{r\,n+2,n+1} & \mathbf{\Phi}_{e\,n+2,n+1} \end{bmatrix} \begin{bmatrix} \mathbf{V}_{n+1} \\ \dot{\mathbf{q}}_{e\,n+1} \end{bmatrix} \tag{3.107}$$

$$\mathbf{A}_{n+2} = \begin{bmatrix} \mathbf{\Phi}_{r\,n+2,n+1} & \mathbf{\Phi}_{e\,n+2,n+1} \end{bmatrix} \begin{bmatrix} \mathbf{A}_{n+1} \\ \ddot{\mathbf{q}}_{e\,n+1} \end{bmatrix} + \mathbf{N}_{e\,n+2} \tag{3.108}$$

Rigid body: $\mathcal{B}_{n+2}$, from equation (3.97),

$$\mathcal{M}_{r\,n+2}\mathbf{A}_{n+2} + \mathcal{N}_{r\,n+2} = \mathcal{F}_{n+2} - \mathbf{\Phi}^T_{rn+3,\,n+2}\mathcal{F}_{n+3} \tag{3.109}$$

The generalized coordinates of the rigid–flexible–rigid (RFR) body are the absolute linear and angular velocity vectors $\mathcal{B}_n$, $\mathbf{v}_n$ and $\boldsymbol{\omega}_n$ respectively, and the elastic bending displacements, torsion angle and shear angles of body $\mathcal{B}_{n+1}$, $\mathbf{q}_{Kn+1}$ and $\mathbf{q}_{\varphi n+1}$. The dynamic model of the RFR body may then be obtained by reapplying the Principle of Virtual Powers. The generalized virtual velocities are $\delta\mathbf{V}_n$ and $\delta\dot{\mathbf{q}}_{e\,n+1}$, therefore,

$$\delta\mathbf{V}_{n+1} = \mathbf{\Phi}_{r\,n+1,\,n}\delta\mathbf{V}_n \tag{3.110}$$

and

$$\delta\mathbf{V}_{n+2} = \left[\, \mathbf{\Phi}_{r\,n+2,\,n+1}\mathbf{\Phi}_{r\,n+1,\,n} \,\middle|\, \mathbf{\Phi}_{e\,n+2,\,n+1} \,\right] \begin{bmatrix} \delta\mathbf{V}_n \\ \delta\dot{\mathbf{q}}_{e\,n+1} \end{bmatrix} \tag{3.111}$$

The acceleration of the flexible beam is given in relation to the acceleration of $\mathcal{B}_n$ by equation (3.104), and the acceleration of $\mathcal{B}_{n+2}$ is obtained from equations (3.104) and (3.108) as

$$\mathbf{A}_{n+2} = \left[\, \mathbf{\Phi}_{r\,n+2,\,n+1}\mathbf{\Phi}_{r\,n+1,\,n} \,\middle|\, \mathbf{\Phi}_{e\,n+2,\,n+1} \,\right] \begin{bmatrix} \mathbf{A}_n \\ \ddot{\mathbf{q}}_{e\,n+1} \end{bmatrix} + \mathbf{\Phi}_{r\,n+2,\,n+1}\mathbf{N}_{n+1} + \mathbf{N}_{e\,n+2} \tag{3.112}$$

Adding the contributions of the three bodies yields

$$\delta\mathbf{V}_n^T \left(\mathcal{M}_{rn}\mathbf{A}_n + \mathcal{N}_{rn} - \mathcal{F}_n + \mathbf{\Phi}^T_{r\,n+1,\,n}\mathcal{F}_{n+1} \right)$$

$$+ \left[\, \delta\mathbf{V}_n^T \;\; \delta\dot{\mathbf{q}}_{e\,n+1}^T \,\right] \begin{bmatrix} \mathbf{\Phi}^T_{r\,n+1,\,n} & 0 \\ 0 & \mathbf{I} \end{bmatrix} \left(\begin{bmatrix} \mathcal{M}_{e\,n+1,\,rr} & \mathcal{M}_{e\,n+1,\,re} \\ \mathcal{M}^T_{e\,n+1,\,re} & \mathcal{M}_{e\,n+1,\,ee} \end{bmatrix} \begin{bmatrix} \mathbf{A}_{n+1} \\ \ddot{\mathbf{q}}_{e\,n+1} \end{bmatrix} \right.$$

$$+ \begin{bmatrix} \mathcal{N}_{e\,n+1,\,r} \\ \mathcal{N}_{e\,n+1,\,e} \end{bmatrix} - \begin{bmatrix} \mathbf{I} \\ 0 \end{bmatrix}\mathcal{F}_{n+1} + \left. \begin{bmatrix} \mathbf{\Phi}^T_{r\,n+2,\,n+1} \\ \mathbf{\Phi}^T_{e\,n+2,\,n+1} \end{bmatrix}\mathcal{F}_{n+2} \right)$$

$$+ \left[\, \delta\mathbf{V}_n^T \;\; \delta\dot{\mathbf{q}}_{e\,n+1}^T \,\right] \begin{bmatrix} \left(\mathbf{\Phi}_{r\,n+2,\,n+1}\mathbf{\Phi}_{r\,n+1,\,n}\right)^T \\ \mathbf{\Phi}^T_{e\,n+2,\,n+1} \end{bmatrix}$$

$$\left(\mathcal{M}_{r\,n+2}\mathbf{A}_{n+2} + \mathcal{N}_{r\,n+2} - \mathcal{F}_{n+2} + \mathbf{\Phi}^T_{rn+3,\,n+2}\mathcal{F}_{n+3} \right) = 0 \tag{3.113}$$

and adding the coefficients of $\delta\mathbf{V}_n^T$ and $\delta\dot{\mathbf{q}}_{e\,n+1}^T$ and writing in matrix form results in the following system of dynamic equations:

$$\mathcal{M}_{\text{rfr}_n} \begin{bmatrix} \mathbf{A}_n \\ \ddot{\mathbf{q}}_{e\,n+1} \end{bmatrix} + \mathcal{N}_{\text{rfr}_n} = \begin{bmatrix} \mathbf{I} \\ 0 \end{bmatrix}\mathcal{F}_n - \begin{bmatrix} \left(\mathbf{\Phi}_{rn+3,\,n+2}\mathbf{\Phi}_{r\,n+2,\,n+1}\mathbf{\Phi}_{r\,n+1,\,n}\right)^T \\ \left(\mathbf{\Phi}_{rn+3,\,n+2}\mathbf{\Phi}_{e\,n+2,\,n+1}\right)^T \end{bmatrix}\mathcal{F}_{n+3} \tag{3.114}$$

where

$$M_{\text{rfr}_n} = \begin{bmatrix} M_{rn} & 0 \\ 0 & 0 \end{bmatrix} + \begin{bmatrix} \boldsymbol{\Phi}^T_{rn+1,n} M_{en+1,rr} \boldsymbol{\Phi}_{rn+1,n} & \boldsymbol{\Phi}^T_{rn+1,n} M_{en+1,re} \\ M^T_{en+1,re} \boldsymbol{\Phi}_{rn+1,n} & M_{en+1,ee} \end{bmatrix}$$

$$+ \begin{bmatrix} \left(\boldsymbol{\Phi}_{rn+2,n+1} \boldsymbol{\Phi}_{rn+1,n}\right)^T M_{rn+2} \boldsymbol{\Phi}_{rn+2,n+1} \boldsymbol{\Phi}_{rn+1,n} & \left(\boldsymbol{\Phi}_{rn+2,n+1} \boldsymbol{\Phi}_{rn+1,n}\right)^T M_{rn+2} \boldsymbol{\Phi}_{en+2,n+1} \\ \boldsymbol{\Phi}^T_{en+2,n+1} M_{rn+2} \boldsymbol{\Phi}_{rn+2,n+1} \boldsymbol{\Phi}_{rn+1,n} & \boldsymbol{\Phi}^T_{en+2,n+1} M_{rn+2} \boldsymbol{\Phi}_{en+2,n+1} \end{bmatrix}$$

$$(3.115)$$

and

$$N_{\text{rfr}_n} = \begin{bmatrix} N_{rn} \\ 0 \end{bmatrix} + \begin{bmatrix} \boldsymbol{\Phi}^T_{rn+1,n} \left(M_{en+1,rr} \mathbf{N}_{n+1} + \overline{N}_{en+1,r}\right) \\ M^T_{en+1,re} \mathbf{N}_{n+1} + \overline{N}_{en+1,e} \end{bmatrix}$$

$$+ \begin{bmatrix} \left(\boldsymbol{\Phi}_{rn+2,n+1} \boldsymbol{\Phi}_{rn+1,n}\right)^T \left(M_{rn+2} \left(\boldsymbol{\Phi}_{rn+2,n+1} \mathbf{N}_{n+1} + \mathbf{N}_{en+2}\right) + N_{rn+2}\right) \\ \boldsymbol{\Phi}^T_{en+2,n+1} \left(M_{rn+2} \left(\boldsymbol{\Phi}_{rn+2,n+1} \mathbf{N}_{n+1} + \mathbf{N}_{en+2}\right) + N_{rn+2}\right) \end{bmatrix}$$

$$(3.116)$$

3.5.4   Dynamics of a serial multi-RFR body system

Consider an articulated chain of RFR bodies representing a flexible manipulator. It follows from equations (3.88), (3.107) and (3.103) that

$$\mathbf{V}_{n+3} = \begin{bmatrix} \boldsymbol{\Phi}_{rn+3,n+2} \boldsymbol{\Phi}_{rn+2,n+1} \boldsymbol{\Phi}_{rn+1,n} & | & \boldsymbol{\Phi}_{rn+3,n+2} \boldsymbol{\Phi}_{en+2,n+1} \end{bmatrix} \begin{bmatrix} \mathbf{V}_n \\ \dot{\mathbf{q}}_{en+1} \end{bmatrix}$$

$$+ \mathbf{H}_{n+3} \dot{\mathbf{q}}_{n+3} \qquad (3.117)$$

and from equations (3.91) and (3.112) that

$$\mathbf{A}_{n+3} = \begin{bmatrix} \boldsymbol{\Phi}_{rn+3,n+2} \boldsymbol{\Phi}_{rn+2,n+1} \boldsymbol{\Phi}_{rn+1,n} & | & \boldsymbol{\Phi}_{rn+3,n+2} \boldsymbol{\Phi}_{en+2,n+1} \end{bmatrix} \begin{bmatrix} \mathbf{A}_n \\ \ddot{\mathbf{q}}_{en+1} \end{bmatrix}$$

$$+ \mathbf{H}_{n+3} \ddot{\mathbf{q}}_{n+3}$$

$$+ \mathbf{N}_{n+3} + \boldsymbol{\Phi}_{rn+3,n+2} \left(\boldsymbol{\Phi}_{rn+2,n+1} \mathbf{N}_{n+1} + \mathbf{N}_{en+2}\right) \qquad (3.118)$$

Renumbering equations (3.114), (3.117) and (3.118) in terms of the nth RFR body, and rewriting the generalized velocity, acceleration and force vectors in order to include the elastic contributions results in the dynamic equation,

$$M_{\text{rfr}_n} \begin{bmatrix} \mathbf{A}_n \\ \ddot{\mathbf{q}}_{en+1} \end{bmatrix} + N_{\text{rfr}_n} = \begin{bmatrix} \mathcal{F}_n \\ 0 \end{bmatrix} - \begin{bmatrix} \left(\boldsymbol{\Phi}_{rn+3,n+2} \boldsymbol{\Phi}_{rn+2,n+1} \boldsymbol{\Phi}_{rn+1,n}\right)^T & 0 \\ \left(\boldsymbol{\Phi}_{rn+3,n+2} \boldsymbol{\Phi}_{en+2,n+1}\right)^T & 0 \end{bmatrix} \begin{bmatrix} \mathcal{F}_{n+3} \\ 0 \end{bmatrix}$$

$$\Leftrightarrow M_{\text{rfr}_n} \mathbf{A}_{\text{rfr}_n} + N_{\text{rfr}_n} = \mathcal{F}_{\text{rfr}_n} - \boldsymbol{\Phi}^T_{\text{rfr}_{n+1,n}} \mathcal{F}_{\text{rfr}_{n+1}} \qquad (3.119)$$

the velocity equation,

$$\begin{bmatrix} \mathbf{V}_n \\ \dot{\mathbf{q}}_{e\,n+1} \end{bmatrix} = \begin{bmatrix} \boldsymbol{\Phi}_{rn,\,n-1}\boldsymbol{\Phi}_{rn-1,\,n-2}\boldsymbol{\Phi}_{rn-2,\,n-3} & \boldsymbol{\Phi}_{rn,\,n-1}\boldsymbol{\Phi}_{en-1,\,n-2} \\ 0 & 0 \end{bmatrix} \begin{bmatrix} \mathbf{V}_{n-3} \\ \dot{\mathbf{q}}_{en-2} \end{bmatrix}$$

$$+ \begin{bmatrix} \mathbf{H}_n & 0 \\ 0 & \mathbf{I} \end{bmatrix} \begin{bmatrix} \dot{\mathbf{q}}_n \\ \dot{\mathbf{q}}_{e\,n+1} \end{bmatrix}$$

$$\mathbf{V}_{\mathrm{rfr}_n} = \boldsymbol{\Phi}_{\mathrm{rfr}_n,\,n-1} \mathbf{V}_{\mathrm{rfr}_{n-1}} + \mathbf{H}_{\mathrm{rfr}_n}\dot{\mathbf{q}}_{\mathrm{rfr}_n} \qquad (3.120)$$

and the acceleration equation,

$$\begin{bmatrix} \mathbf{A}_n \\ \ddot{\mathbf{q}}_{e\,n+1} \end{bmatrix} = \begin{bmatrix} \boldsymbol{\Phi}_{rn,\,n-1}\boldsymbol{\Phi}_{rn-1,\,n-2}\boldsymbol{\Phi}_{rn-2,\,n-3} & \boldsymbol{\Phi}_{rn,\,n-1}\boldsymbol{\Phi}_{en-1,\,n-2} \\ 0 & 0 \end{bmatrix} \begin{bmatrix} \mathbf{A}_{n-3} \\ \ddot{\mathbf{q}}_{en-2} \end{bmatrix}$$

$$+ \begin{bmatrix} \mathbf{H}_n & 0 \\ 0 & \mathbf{I} \end{bmatrix} \begin{bmatrix} \ddot{\mathbf{q}}_n \\ \ddot{\mathbf{q}}_{e\,n+1} \end{bmatrix}$$

$$+ \begin{bmatrix} \mathbf{N}_n + \boldsymbol{\Phi}_{rn,\,n-1}\left(\boldsymbol{\Phi}_{rn-1,\,n-2}\mathbf{N}_{n-2} + \mathbf{N}_{en-1}\right) \\ 0 \end{bmatrix}$$

$$\Leftrightarrow \mathbf{A}_{\mathrm{rfr}_n} = \boldsymbol{\Phi}_{\mathrm{rfr}_n,\,n-1}\mathbf{A}_{\mathrm{rfr}_{n-1}} + \mathbf{H}_{\mathrm{rfr}_n}\ddot{\mathbf{q}}_{\mathrm{rfr}_n} + \mathbf{N}_{\mathrm{rfr}_n} \qquad (3.121)$$

Furthermore, the externally applied generalized force vector in equation (3.101) may be rewritten as

$$\begin{bmatrix} \mathcal{T}_n \\ 0 \end{bmatrix} = \begin{bmatrix} \mathbf{H}_n^T & 0 \\ 0 & \mathbf{I} \end{bmatrix} \begin{bmatrix} \mathcal{F}_n \\ 0 \end{bmatrix}$$

$$\Leftrightarrow \mathcal{T}_{\mathrm{rfr}_n} = \mathbf{H}_{\mathrm{rfr}_n}^T \mathcal{F}_{\mathrm{rfr}_n} \qquad (3.122)$$

Equations (3.119), (3.121) and (3.122) are in a form compatible with the problem formulation and solution methods for rigid manipulators presented in (Ascher *et al.* 1997) and Pai *et al.* (2000). The solution method consists of writing a large algebraic system as in Lubich *et al.* (1992) or von Schwerin (1999), which is solved through elimination methods leading to either the global dynamic CI method (the CRB method for rigid multi-body systems), or the AB method. The former has complexity $O(N^3)$ due to the need for inverting the global system mass matrix, whereas the latter has complexity $O(N)$. To illustrate, following (Ascher *et al.* 1997), the algebraic system for an unconstrained ($F_{\text{rfr4}} = 0$) three-RFR body manipulator is written as

$$
\begin{bmatrix}
\mathcal{M}_{\text{rfr}_3} & 0 & \mathbf{I} & & & & & & \\
0 & 0 & \mathbf{H}_{\text{rfr}_3}^T & & & & & & \\
\mathbf{I} & \mathbf{H}_{\text{rfr}_3} & 0 & -\boldsymbol{\Phi}_{\text{rfr}_{3,2}} & & & & & \\
& & -\boldsymbol{\Phi}_{\text{rfr}_{3,2}}^T & \mathcal{M}_{\text{rfr}_2} & 0 & \mathbf{I} & & & \\
& & & 0 & 0 & \mathbf{H}_{\text{rfr}_2}^T & & & \\
& & & \mathbf{I} & \mathbf{H}_{\text{rfr}_2} & 0 & -\boldsymbol{\Phi}_{\text{rfr}_{2,1}} & & \\
& & & & & -\boldsymbol{\Phi}_{\text{rfr}_{2,1}}^T & \mathcal{M}_{\text{rfr}_1} & 0 & \mathbf{I} \\
& & & & & & 0 & 0 & \mathbf{H}_{\text{rfr}_1}^T \\
& & & & & & \mathbf{I} & \mathbf{H}_{\text{rfr}_1} & 0
\end{bmatrix}
\begin{bmatrix}
-\mathbf{A}_{\text{rfr}_3} \\
\ddot{\mathbf{q}}_{\text{rfr}_3} \\
\mathcal{F}_{\text{rfr}_3} \\
-\mathbf{A}_{\text{rfr}_2} \\
\ddot{\mathbf{q}}_{\text{rfr}_2} \\
\mathcal{F}_{\text{rfr}_2} \\
-\mathbf{A}_{\text{rfr}_1} \\
\ddot{\mathbf{q}}_{\text{rfr}_1} \\
\mathcal{F}_{\text{rfr}_1}
\end{bmatrix}
=
\begin{bmatrix}
\mathcal{N}_{\text{rfr}_3} \\
\mathcal{T}_{\text{rfr}_3} \\
-\mathbf{N}_{\text{rfr}_3} \\
\mathcal{N}_{\text{rfr}_2} \\
\mathcal{T}_{\text{rfr}_2} \\
-\mathbf{N}_{\text{rfr}_2} \\
\mathcal{N}_{\text{rfr}_1} \\
\mathcal{T}_{\text{rfr}_1} \\
-\mathbf{N}_{\text{rfr}_1}
\end{bmatrix}
$$

(3.123)

The global dynamics CI method consists of rearranging the block rows and columns of the above system into the form

$$
\begin{bmatrix}
\mathcal{M}_{\text{rfr}_3} & & & \mathbf{I} & & & & & \\
& \mathcal{M}_{\text{rfr}_2} & & -\boldsymbol{\Phi}_{\text{rfr}_{3,2}}^T & \mathbf{I} & & & & \\
& & \mathcal{M}_{\text{rfr}_1} & & -\boldsymbol{\Phi}_{\text{rfr}_{2,1}}^T & \mathbf{I} & & & \\
& & & \mathbf{H}_{\text{rfr}_3}^T & & & & & \\
& & & & \mathbf{H}_{\text{rfr}_2}^T & & & & \\
& & & & & \mathbf{H}_{\text{rfr}_1}^T & & & \\
\mathbf{I} & -\boldsymbol{\Phi}_{\text{rfr}_{3,2}} & & \mathbf{H}_{\text{rfr}_3} & & & & & \\
& \mathbf{I} & -\boldsymbol{\Phi}_{\text{rfr}_{2,1}} & & \mathbf{H}_{\text{rfr}_2} & & & & \\
& & \mathbf{I} & & & \mathbf{H}_{\text{rfr}_1} & & &
\end{bmatrix}
\begin{bmatrix}
-\mathbf{A}_{\text{rfr}_3} \\
-\mathbf{A}_{\text{rfr}_2} \\
-\mathbf{A}_{\text{rfr}_1} \\
\ddot{\mathbf{q}}_{\text{rfr}_3} \\
\ddot{\mathbf{q}}_{\text{rfr}_2} \\
\ddot{\mathbf{q}}_{\text{rfr}_1} \\
\mathcal{F}_{\text{rfr}_3} \\
\mathcal{F}_{\text{rfr}_2} \\
\mathcal{F}_{\text{rfr}_1}
\end{bmatrix}
=
\begin{bmatrix}
\mathcal{N}_{\text{rfr}_3} \\
\mathcal{N}_{\text{rfr}_2} \\
\mathcal{N}_{\text{rfr}_1} \\
\mathcal{T}_{\text{rfr}_3} \\
\mathcal{T}_{\text{rfr}_2} \\
\mathcal{T}_{\text{rfr}_1} \\
-\mathbf{N}_{\text{rfr}_3} \\
-\mathbf{N}_{\text{rfr}_2} \\
-\mathbf{N}_{\text{rfr}_1}
\end{bmatrix}
$$

$$
\Leftrightarrow
\begin{bmatrix}
\mathcal{M} & 0 & \boldsymbol{\Phi}^{-T} \\
0 & 0 & \mathbf{H}^T \\
\boldsymbol{\Phi}^{-1} & \mathbf{H} & 0
\end{bmatrix}
\begin{bmatrix}
-\mathbf{A} \\
\ddot{\mathbf{q}} \\
\mathcal{F}
\end{bmatrix}
=
\begin{bmatrix}
\mathcal{N} \\
\mathcal{T} \\
-\mathbf{N}
\end{bmatrix}
$$

(3.124)

and solving through block-row elimination in order to arrive at

$$
\left(\mathbf{H}^T \boldsymbol{\Phi}^T \mathcal{M} \boldsymbol{\Phi} \mathbf{H} \right) \ddot{\mathbf{q}} + \mathbf{H}^T \boldsymbol{\Phi}^T \left(\mathcal{M} \boldsymbol{\Phi} \mathbf{N} + \mathcal{N} \right) = \mathcal{T}
$$

(3.125)

which is the joint space dynamic equation of the flexible manipulator. This is the solution that can be obtained through a reapplication of the Principle of Virtual Powers to the chain of RFR bodies. From equation (3.120) the following may be written:

$$
\begin{bmatrix}
\mathbf{V}_{\text{rfr}_3} \\
\mathbf{V}_{\text{rfr}_2} \\
\mathbf{V}_{\text{rfr}_1}
\end{bmatrix}
=
\begin{bmatrix}
\mathbf{I} & \boldsymbol{\Phi}_{\text{rfr}_{3,2}} & \boldsymbol{\Phi}_{\text{rfr}_{3,2}} \boldsymbol{\Phi}_{\text{rfr}_{2,1}} \\
0 & \mathbf{I} & \boldsymbol{\Phi}_{\text{rfr}_{2,1}} \\
0 & 0 & \mathbf{I}
\end{bmatrix}
\begin{bmatrix}
\mathbf{H}_{\text{rfr}_3} & 0 & 0 \\
0 & \mathbf{H}_{\text{rfr}_2} & 0 \\
0 & 0 & \mathbf{H}_{\text{rfr}_1}
\end{bmatrix}
\begin{bmatrix}
\dot{\mathbf{q}}_{\text{rfr}_3} \\
\dot{\mathbf{q}}_{\text{rfr}_2} \\
\dot{\mathbf{q}}_{\text{rfr}_1}
\end{bmatrix}
$$

$$
\Leftrightarrow \mathbf{V} = \boldsymbol{\Phi} \mathbf{H} \dot{\mathbf{q}}
$$

(3.126)

thus identifying the product $\boldsymbol{\Phi}\mathbf{H}$ as a Jacobian matrix.

The AB method on the other hand, solves the algebraic system in equation (3.123), taking advantage of its block diagonal structure. Each block corresponds to a RFR body, which is coupled to the next body in the chain through matrix $\boldsymbol{\Phi}^T_{\text{rfr}_{n+1,n}}$. The solution procedure consists of eliminating these matrices in order to decouple the diagonal blocks. Therefore, for $n = N, N-1, \ldots, 1$, eliminate the middle row of each block using the first row and then the last row of the block. Then, using the resulting nth block rows, eliminate matrix $\boldsymbol{\Phi}^T_{\text{rfr}_{n,n-1}}$, coupling block $n-1$ with block n. The resulting system is

$$
\left[
\begin{array}{ccc|ccc|ccc}
\mathcal{M}_{\text{rfr}_3} & 0 & \mathbf{I} & & & & & & \\
0 & \mathcal{D}_{\text{rfr}_3} & 0 & -\mathbf{H}^T_{\text{rfr}_3}\mathcal{M}_{\text{rfr}_3}\boldsymbol{\Phi}_{\text{rfr}_{3,2}} & & & & & \\
\mathbf{I} & \mathbf{H}_{\text{rfr}_3} & 0 & -\boldsymbol{\Phi}_{\text{rfr}_{3,2}} & & & & & \\
\hline
& & & \widehat{\mathcal{M}}_{\text{rfr}_2} & 0 & \mathbf{I} & & & \\
& & & 0 & \mathcal{D}_{\text{rfr}_2} & 0 & -\mathbf{H}^T_{\text{rfr}_2}\widehat{\mathcal{M}}_{\text{rfr}_2}\boldsymbol{\Phi}_{\text{rfr}_{2,1}} & & \\
& & & \mathbf{I} & \mathbf{H}_{\text{rfr}_2} & 0 & -\boldsymbol{\Phi}_{\text{rfr}_{2,1}} & & \\
\hline
& & & & & & \widehat{\mathcal{M}}_{\text{rfr}_1} & 0 & \mathbf{I} \\
& & & & & & 0 & \mathcal{D}_{\text{rfr}_1} & 0 \\
& & & & & & \mathbf{I} & \mathbf{H}_{\text{rfr}_1} & 0
\end{array}
\right]
\left[
\begin{array}{c}
-\mathbf{A}_{\text{rfr}_3} \\
\ddot{\mathbf{q}}_{\text{rfr}_3} \\
\mathcal{F}_{\text{rfr}_3} \\
\hline
-\mathbf{A}_{\text{rfr}_2} \\
\ddot{\mathbf{q}}_{\text{rfr}_2} \\
\mathcal{F}_{\text{rfr}_2} \\
\hline
-\mathbf{A}_{\text{rfr}_1} \\
\ddot{\mathbf{q}}_{\text{rfr}_1} \\
\mathcal{F}_{\text{rfr}_1}
\end{array}
\right]
=
\left[
\begin{array}{c}
\mathcal{N}_{\text{rfr}_3} \\
\widehat{\mathcal{T}}_{\text{rfr}_3} \\
-\mathbf{N}_{\text{rfr}_3} \\
\hline
\widehat{\mathcal{N}}_{\text{rfr}_2} \\
\widehat{\mathcal{T}}_{\text{rfr}_2} \\
-\mathbf{N}_{\text{rfr}_2} \\
\hline
\widehat{\mathcal{N}}_{\text{rfr}_1} \\
\widehat{\mathcal{T}}_{\text{rfr}_1} \\
-\mathbf{N}_{\text{rfr}_1}
\end{array}
\right]
$$

$$(3.127)$$

where

$$
\widehat{\mathcal{T}}_{\text{rfr}_n} = \mathcal{T}_{\text{rfr}_n} - \mathbf{H}^T_{\text{rfr}_n}\left(\widehat{\mathcal{N}}_{\text{rfr}_n} + \widehat{\mathcal{M}}_{\text{rfr}_n}\mathbf{N}_{\text{rfr}_n}\right) \tag{3.128}
$$

$$
\mathcal{D}_{\text{rfr}_n} = \mathbf{H}^T_{\text{rfr}_n}\widehat{\mathcal{M}}_{\text{rfr}_n}\mathbf{H}_{\text{rfr}_n} \tag{3.129}
$$

$$
\widehat{\mathcal{N}}_{\text{rfr}_n} = \mathbf{N}_{\text{rfr}_n} + \boldsymbol{\Phi}^T_{\text{rfr}_{n+1,n}}\left(\widehat{\mathcal{N}}_{\text{rfr}_{n+1}} + \widehat{\mathcal{M}}_{\text{rfr}_{n+1}}\mathbf{N}_{\text{rfr}_{n+1}}\right)
$$

$$
+ \boldsymbol{\Phi}^T_{\text{rfr}_{n+1,n}}\widehat{\mathcal{M}}_{\text{rfr}_{n+1}}\mathbf{H}_{\text{rfr}_{n+1}}D^{-1}_{\text{rfr}_{n+1}}\widehat{\mathcal{T}}_{\text{rfr}_{n+1}} \tag{3.130}
$$

and

$$
\widehat{\mathcal{M}}_{\text{rfr}_n} = \mathcal{M}_{\text{rfr}_n} + \boldsymbol{\Phi}^T_{\text{rfr}_{n+1,n}}\widehat{\mathcal{M}}_{\text{rfr}_{n+1}}\boldsymbol{\Phi}_{\text{rfr}_{n+1,n}}
$$

$$
- \boldsymbol{\Phi}^T_{\text{rfr}_{n+1,n}}\widehat{\mathcal{M}}_{\text{rfr}_{n+1}}\mathbf{H}_{\text{rfr}_{n+1}}D^{-1}_{\text{rfr}_{n+1}}\mathbf{H}^T_{\text{rfr}_{n+1}}\widehat{\mathcal{M}}_{\text{rfr}_{n+1}}\boldsymbol{\Phi}_{\text{rfr}_{n+1,n}} \tag{3.131}
$$

with $\widehat{\mathcal{M}}_{\text{rfr}_N} = \mathcal{M}_{\text{rfr}_N}$. Equation (3.127) represents a linear system for $\ddot{\mathbf{q}}_{\text{rfr}_n}$ and $\mathbf{A}_{\text{rfr}_n}$. For $n = 1, \ldots, N$, $\ddot{\mathbf{q}}_{\text{rfr}_n}$ may be calculated. Equation (3.131) is the CRB inertia introduced in Featherstone (1987) in the context of rigid multi-body dynamics.

3.6 Summary

A systematic modelling approach leading to a general modelling environment, which may be used either in analysis or in the development of real-time control schemes for flexible manipulator arms, has been presented. A flexible beam attached to two rigid bodies, one at each end, has been assumed as a building block for a flexible manipulator. The rigid bodies are the supporting structures for actuators and sensors at the joints, and are typically of significant mass. However, if that is not the case, their mass may be easily neglected, and what remains are their geometric properties. The cross-section rotations due to elasticity of the flexible beam may be of higher order as needed and shear deformation according to the Timoshenko beam theory is included. The method of formulation for the dynamics of a flexible manipulator may be either the global dynamics CI method or the recursive AB method.

Chapter 4

Parametric and non-parametric modelling of flexible manipulators

M.H. Shaheed and M.O. Tokhi

This chapter presents the development of parametric and non-parametric approaches for dynamic modelling of a single-link flexible manipulator system. The least mean squares, recursive least squares and genetic algorithms are used to obtain linear parametric models of the system. Non-parametric models of the system are developed using a non-linear autoregressive process with exogeneous input model structure with multi-layered perceptron and radial basis function neural networks. The system is in each case modelled from the input torque to hub-angle, hub-velocity and end-point acceleration outputs. The models are validated using several validation tests. Finally, a comparative assessment of the approaches used is presented and discussed in terms of accuracy, efficiency and estimation of vibration modes of the system.

4.1   Introduction

System identification is extensively used as a fundamental requirement in many engineering and scientific applications. The objective of system identification is to find exact or approximate models of dynamic systems based on observed inputs and outputs. Once a model of a physical system is obtained, it can be used for solving various problems; for example, to control the physical system or to predict its behaviour under different operating conditions (Tabrizi *et al.*, 1990; Worden *et al.*, 1994). Researchers have devised a number of techniques to determine models that best describe the input–output behaviour of a system. Parametric and non-parametric identification are two major classes of such techniques.

Parametric identification of a system comprises two main steps. The first step is qualitative operation, which defines the structure of the system, for example, type and order of the (differential/difference) equation relating the input to the output. This is known as *characterisation,* which means selection of a suitable model structure, for example, autoregressive with exogeneous inputs (ARX), autoregressive moving average with exogeneous inputs (ARMAX), Box–Jenkins, and so on. The second step namely *identification,* consists of determination of the numerical values of the structural parameters that minimise the distance between the system to be identified and its model. Common estimation methods include least mean squares (LMS), recursive least squares (RLS), instrumental-variables, maximum-likelihood and prediction-error. In simple terms, this is a curve-fitting exercise. Lately, genetic algorithm (GA) based parametric identification techniques have been utilized in many applications. Kristinsson and Dumont (1992) used a GA for system identification and control. Fonseca *et al.* (1993) have addressed a non-linear model term selection, where GA has been employed as an alternative to orthogonal least square (OLS) regression to find a smaller set of non-linear model terms from a broader set of possible terms. Caponetto *et al.* (1995) have utilized GAs to determine the parameters of the Chua's oscillator.

In the case of (non-linear) non-parametric models, neural networks (NNs) are commonly utilized. NNs possess various attractive features such as massive parallelism, distributed representation and computation, generalization ability, adaptability and inherent contextual information processing (Jain and Muhiuddin, 1996). Owing to the efficient nature of their working principles and other attractive characteristics, attempts are made to use NNs extensively in various identification and control applications, including robotics.

Among the various types of NNs, the multi-layered perceptron (MLP) and radial basis function (RBF) are commonly utilized in the identification and control of dynamic systems. Narendra and Parthasarathy (1990) have addressed system identification using the globally approximating characteristics of NNs, and have suggested a number of identification structures using NNs for adaptive control of unknown non-linear dynamic systems. Qin *et al.* (1992) have reported a comparative study of four NN learning methods for dynamic system identification. Srinivasan *et al.* (1994) have addressed backpropagation through adjoints for the identification of non-linear dynamic systems using recurrent neural models. They re-investigated the backpropagation for an efficient evaluation of the gradient with arbitrary interconnections of recurrent systems. They also proposed the accelerated backpropagation to eliminate the delay in obtaining the gradient. Nerrand *et al.* (1994) have addressed training methodologies of recurrent NNs for dynamic process modelling. Olurotimi (1994) has addressed recurrent network training methodologies with feedforward complexity to learn Bessel's difference equation, thereby generating Bessel functions within as well as outside the training set. Hagan and Menhaj (1994) have proposed modification of the backpropagation algorithm with the Marquardt algorithm to train feedforward networks.

The successful application of RBF networks for modelling dynamic systems has also been widely addressed in the literature (Casdagli, 1989; He and Lapedes, 1993; Sze, 1995). Chen *et al.* (1991) proposed the OLS learning algorithm for RBF networks to model non-linear dynamic systems. Elanayar and Yung (1994) have addressed the use of RBF networks to approximate the dynamic and state equations of stochastic systems and to estimate state variables. Identification of robotic manipulators with NNs has also been reported in the literature (Karakasoglu *et al.*, 1993).

In this chapter dynamic modelling of a single-link flexible manipulator described in Chapter 2 is considered using parametric and non-parametric identification techniques. The resulting models are subjected to several validation methods, and a comparative assessment of the results is presented and discussed.

4.2   Parametric identification techniques

The characteristics of flexible structure systems are generally of distributed nature amounting to resonance modes of vibration. The primary interest in this work lies in locating frequencies of these *resonance modes*, which ultimately dictate the behaviour of the system. With a view to find an accurate model of the system a comparative assessment of the LMS, RLS and GA methods is carried out in the parametric dynamic modelling of the flexible manipulator.

4.2.1   LMS algorithm

The LMS algorithm is based on the steepest descent method (Widrow *et al.*, 1975). The computational procedure of the LMS algorithm can be summarised as follows:

Initially, set each weight $w_i(k)$; $i = 0, 1, \ldots, N - 1$, to an arbitrary fixed value, such as 0. For each subsequent sampling instant $k = 1, 2, \ldots$ execute the relations below in the order given

$$\hat{y}(k) = \sum_{i=0}^{N-1} w_i(k)x(k-1) \tag{4.1}$$

$$e(k) = y(k) - \hat{y}(k) \tag{4.2}$$

$$w_i(k+1) = w_i(k) + 2\mu e(k)x(k-i) \tag{4.3}$$

where $x(k)$ is the observation vector at time step k, $e(k)$ is the error between the actual output $y(k)$ and the predicted output $\hat{y}(k)$, and μ represents the learning rate and is a constant.

4.2.2   RLS algorithm

The standard RLS estimation process at a time step k is described as (Tokhi and Leith, 1991)

$$\varepsilon(k) = \Theta(k)\Psi(k) - y(k) \tag{4.4}$$

$$\Theta(k) = \Theta(k-1) - P(k-1)\Psi^{\mathrm{T}}(k)\left[1 + \Psi(k)P(k-1)\Psi^{\mathrm{T}}(k)\right]^{-1}\varepsilon(k) \tag{4.5}$$

$$P(k) = \lambda^{-1}P(k-1) - \frac{\lambda^{-1}P(k-1)\Psi^{T}(k)\Psi(k)P(k-1)}{\lambda + \Psi(k)P(k-1)\Psi^{T}(k)} \tag{4.6}$$

where $\Psi(k)$ represents the observation matrix, $y(k)$ is the system output, $\Theta(k)$ is the model parameter vector, $P(k)$ is the covariance matrix and λ represents the forgetting factor. Thus, the RLS estimation process is to implement and execute the relations in equations (4.4)–(4.6) in the order given. The convergence of the algorithm is determined by the magnitude of the modelling error $\varepsilon(k)$ reaching a minimum or by the estimated set of parameters reaching a steady level. The purpose of using forgetting factor is to help the algorithm converge to the global minimum. However, the use of a forgetting factor could cause the predicted values of parameters to fluctuate rather than to converge to a certain value. The level of fluctuation depends on the value of λ; the smaller the value of λ the larger the fluctuation in the parameter values.

4.2.3   Genetic algorithms

Genetic algorithms form one of the prominent members of the broader class of evolutionary algorithms, inspired by the mechanism of natural biological evolution, that is, the principles of survival of the fittest (Holland, 1975). The operating mechanism of a GA can be described through the stages shown in Figure 4.1. These comprise the following:

1. *Creation of initial set of potential solutions (population) as strings*: An initial population of potential solutions is created. Each element of the population is mapped onto a set of strings (the chromosome) to be manipulated by the genetic operators.
2. *Evaluation of each solution and selection of the best ones*: The performance of each member of the population is assessed through an objective function

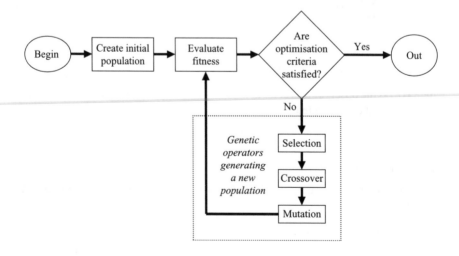

Figure 4.1   Genetic algorithm-simple working principles

imposed by the problem. This establishes the basis for selection of pairs of individuals that will mate during reproduction. For reproduction, each individual is assigned a fitness value derived from its raw performance measure, given by the objective function. This value is used in the selection to bias towards fitter individuals. Highly fit individuals, relative to the whole population, have a high probability of being selected for mating whereas less fit individuals have a correspondingly low probability of being selected (Chipperfield *et al.*, 1994).

3. *Genetic manipulation to create new population*: In this phase, genetic operators such as crossover and mutation are used to produce a new population of individuals (offspring) by manipulating the 'genetic information' usually called genes, possessed by the members (parents) of the current population. The crossover operator is used to exchange genetic information between pairs, or larger groups, of individuals. Mutation is generally considered to be a background operator, which ensures that the search process is not trapped at a local minimum, by introducing new genetic structures.

After manipulation by the crossover and mutation operators, the individual strings are, if necessary, decoded, the objective function evaluated, a fitness value assigned to each individual and individuals selected for mating according to their fitness, and so the process continues through subsequent generations. In this way, the average performance of individuals in a population is expected to increase, as good individuals are preserved and breed with one another and the less fit individuals die out. The GA is terminated when some criteria are satisfied, for example, a certain number of generations completed or when a particular point in the search space is reached.

In this chapter, randomly selected parameters are optimised for different, arbitrarily chosen order to fit to the system by applying the working mechanism of GAs as described above. The fitness function utilized is the sum-squared error between the actual output, $y(i)$, of the system and the predicted output, $\hat{y}(i)$;

$$f(e) = \sum_{i=1}^{N} \left(\left| y(i) - \hat{y}(i) \right| \right)^2 \tag{4.7}$$

where n is the number of input/output samples. With the fitness function given above, the global search technique of the GA is utilized to obtain the best set of parameters among all the attempted orders for the system.

4.3 Non-parametric identification techniques

Various modelling techniques can be used with NNs for identification of non-linear dynamic systems. These include state-output model, recurrent state model and non-linear autoregressive moving average with exogeneous input (NARMAX) model. It is evident from the literature that if the input and output data of the plant are available, the NARMAX model is a suitable choice for modelling non-linear systems with suitable neuro-learning algorithms. Mathematically the model is given as (Luo and Unbehauen, 1997)

$$\hat{y}(t) = f[(y(t-1), y(t-2), \ldots, y(t-n_y),$$
$$u(t-1), u(t-2), \ldots, u(t-n_u),$$
$$e(t-1), e(t-2), \ldots, e(t-n_e)] + e(t) \tag{4.8}$$

where $\hat{y}(t)$ is the output vector determined by past values of the system input vector, output vector and noise with maximum lags n_y, n_u and n_e, respectively, $f(\cdot)$ is the system mapping constructed through MLP or RBF NNs with an appropriate learning algorithm. The model is also known as the NARMAX equation error model. However, if the model is good enough to identify the system without incorporating the noise term or considering the noise as additive at the output, the model can be represented in a non-linear autoregressive with exogeneous input (NARX) model form as (Luo and Unbehauen, 1997)

$$\hat{y}(t) = f[(y(t-1), y(t-2), \ldots, y(t-n_y),$$
$$u(t-1), u(t-2), \ldots, u(t-n_u)] + e(t) \tag{4.9}$$

This is described in Figure 4.2.

4.3.1   *Multi-layered perceptron neural networks*

The MLP NNs are extensively used in numerous applications including pattern recognition, function approximation, system identification, prediction and control, speech and natural language processing. A MLP NN is capable of forming arbitrary decision boundaries and representing Boolean functions (Minsky and Papert, 1969). The network can be made up of any number of layers with reasonable number of neurons in each layer, based on the nature of the application. The layer to which the input data is supplied is called the input layer and the layer from which the output is taken is known

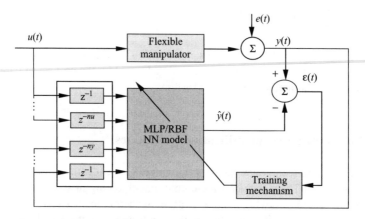

Figure 4.2   NARX model identification with neural networks

as the output layer. All other intermediate layers are called hidden layers. The layers are fully interconnected, which means that each processing unit (neuron) is connected to every neuron in the previous and succeeding layers. However, the neurons in the same layer are not connected to each other. A neuron performs two functions, namely, combining and activation. Different types of function such as threshold, piecewise linear, sigmoid, tansigmoid and Gaussian are used for activation.

The backpropagation learning algorithm is commonly used with MLP NN. According to the backpropagation algorithm, the connection weights between the layers of the multi-layered NN are adapted in accordance with the following rules:

$$\Delta w_{kj} = \eta \delta_k O_j \qquad (4.10)$$

$$\Delta w_{ji} = \eta \delta_j O_i \qquad (4.11)$$

where, for tansigmoid function,

$$\delta_k = O_k(1 - O_k^2)(\tau_k - O_k) \qquad (4.12)$$

$$\delta_j = O_j(1 - O_j^2) \sum_k \delta_k w_{kj} \qquad (4.13)$$

O_k, O_j and O_i are output values at the output, hidden and input layers, respectively. w_{kj} is a connection weight from neuron j in the hidden layer to neuron k in the output layer. Similarly, w_{ji} is a connection weight from neuron i at the input layer to neuron j in the hidden layer and τ_k is the target value. Derivation of the algorithm can be found in a number of books (Luo and Unbehauen, 1997; Omatu *et al.*, 1996). The NN training may get stuck in a shallow local minimum with standard backpropagation. This can be solved by using the Marquardt–Levenberg (Marquardt, 1963) modified version of the backpropagation. While backpropagation is a steepest descent algorithm, the Marquardt–Levenberg algorithm is an approximation to Newton's method (Battiti, 1992).

4.3.2 *Radial basis function neural networks*

The RBF NN is a special class of multilayer feedforward networks, widely studied and applied with supervised learning to solve engineering problems. The hidden neurons of the network provide a set of 'functions' that constitute an arbitrary 'basis' for the input vectors when they are expanded into the hidden-neuron space. These functions are called 'radial basis' functions. The distance between the input vector and a prototype vector (centre vector) determines the activation of the hidden neuron. In contrast to the MLP network, the RBF network has one hidden layer. The input layer has neurons with a linear function that simply feeds input signals to the hidden layer. Moreover, the connections between the input layer and the hidden layer are not weighted, that is, each hidden neuron receives the corresponding input value unaltered. The hidden neurons are processing units that perform the RBF. The transfer function of the hidden neurons in the RBF network can be local or global. Local RBFs that are widely studied include Gaussian and inverse multiquardic. Among the global RBFs, thin plate spline, linear and multiquardic functions are widely used.

Consider a RBF NN with the input at time t denoted by $\mathbf{u}(t) = [u_1(t), u_2(t), \ldots, u_n(t)]^T$ and let the connection vector (centre) of each hidden neuron be denoted by $\mathbf{c}_j$, for $j = 1, 2, \ldots, n_h$, where n_h represents the number of neurons in the hidden layer. Then, the output of each neuron in the hidden layer is given as

$$h_j(t) = f_j(\|\mathbf{u}(t) - \mathbf{c}_j\|) \quad (j = 1, 2, \ldots, n_h) \tag{4.14}$$

The connections between the output layer and the hidden layer are weighted. Each neuron of the output layer has a linear input–output relationship so that they perform simple summations; that is, the output of the kth neuron in the output layer at time t is

$$\hat{y}_k(t) = \sum_{j=1}^{n_h} w_{jk} h_j(t) = \sum_{j=1}^{n_h} w_{jk} f_j(\|\mathbf{u}(t) - \mathbf{c}_j\|), \qquad \text{for } k = 1, 2, \ldots, M \tag{4.15}$$

where M represents the number of neurons in the output layer and w_{jk} is the connection weight between the jth neuron in the hidden layer and kth neuron in the output layer. Including the bias parameter w_{k0} in the linear sum, equation (4.15) becomes

$$\hat{y}_k(t) = w_{k0} + \sum_{j=1}^{n_h} w_{jk} h_j(t) = w_{k0} + \sum_{j=1}^{n_h} w_{jk} f_j(\|\mathbf{u}(t) - \mathbf{c}_j\|) \tag{4.16}$$

The bias parameter compensates for the difference between the average value over the data set of the basis function activation and the corresponding average value of the target.

It follows from equations (4.14) and (4.16) that, in general, a RBF NN is specified by two sets of parameters: the connection weights and the vector of centres. These parameters can in principle be determined from the available sample vectors (training data) by solving the optimisation problem (Luo and Unbehauen, 1997):

$$E = \sum_{t=1}^{S} \|\mathbf{y}(t) - \hat{\mathbf{y}}(t)\|^2 \tag{4.17}$$

where S represents the number of available sample vectors, $\hat{\mathbf{y}}(t) = [\hat{y}_1(t), \hat{y}_2(t), \ldots, \hat{y}_M(t)]^T$ is the output vector computed from the sample input vector using equations (4.14) and (4.16), and $\mathbf{y}(t)$ is the corresponding desired output vector.

Since an RBF-NN has only one layer of weighted connections and the output neurons are simple summation units, equation (4.17) will become a linear least-squares problem once the vector of centres and the scaling parameters have been determined. That is,

$$\min_{\mathbf{w}} \sum_{t=1}^{S} \|\mathbf{y}(t) - \hat{\mathbf{y}}(t)\|^2 = \min_{\mathbf{w}} \|\mathbf{w}\mathbf{F} - \hat{\mathbf{y}}\|^2 \tag{4.18}$$

where $\mathbf{w} = \{w_{ij}\}$ is an $M \times n_h$ matrix of connection weights, $\mathbf{F}$ is an $n_h \times S$ matrix consisting of the outputs of the hidden neurons and whose elements are computed with

$$F_{jt} = f_j(\|\mathbf{u}(t) - \mathbf{c}_j\|) \text{ for } j = 1, 2, \ldots, n_h; t = 1, 2, \ldots, S \tag{4.19}$$

and $\hat{\mathbf{y}} = \left[\hat{\mathbf{y}}(1), \hat{\mathbf{y}}(2), \ldots, \hat{\mathbf{y}}(S)\right]$ is an $M \times S$ matrix of desired outputs. The connection weight matrix $\mathbf{w}$ in equation (4.14) can be found in an explicit form as

$$\mathbf{w} = \hat{\mathbf{y}}\mathbf{F}^+ \tag{4.20}$$

where, $\mathbf{F}^+$ is the pseudo-inverse of $\mathbf{F}$ (Luo and Unbehauen, 1997).

The task of a learning algorithm (optimisation scheme) for the RBF network is to select the centres and find a set of weights that make the network perform the desired mapping. A number of learning algorithms are commonly used for this purpose, including

- Non-linear optimisation of all the parameters (centres and output weights, or other free parameters)
- Random centre selection and a least square algorithm
- Clustering and a least square algorithm
- The OLS algorithm.

Among these the OLS algorithm (Chen *et al.*, 1991) is commonly used, and adopted in this chapter.

4.4 Model validation

Once a model of the system is obtained, it is required to validate whether the model is good enough to represent the system. A number of such validation tests are available in the literature, some of which are described below (Shaheed, 2000).

A common measure of predictive accuracy used in control and system identification is to compute the one-step-ahead (OSA) prediction of the system output. This is expressed as

$$\hat{y}(t) = f\left(u(t), u(t-1), \ldots, u(t-n_u), y(t-1), \ldots, y(t-n_y)\right) \tag{4.21}$$

where $f(\cdot)$ is a non-linear function, u and y are the inputs and outputs, respectively. The residual or prediction is given by

$$\varepsilon(t) = y(t) - \hat{y}(t) \tag{4.22}$$

Often $\hat{y}(t)$ will be a relatively good prediction of $y(t)$ over the estimation set, even if the model is biased, because the model was estimated by minimising the prediction error.

Another method to evaluate the predictive capability of the fitted model is to compute the model predicted output (MPO). This is defined by

$$\hat{y}_d(t) = f\left(u(t), u(t-1), \ldots, u(t-n_u), \hat{y}_d(t-1), \ldots, \hat{y}_d(t-n_y)\right) \tag{4.23}$$

and the deterministic error or deterministic residual is

$$\varepsilon_d(t) = y(t) - \hat{y}_d(t) \tag{4.24}$$

If only lagged inputs are used to assign network input nodes, then

$$\hat{y}(t) = \hat{y}_d(t) \tag{4.25}$$

If the fitted model behaves well for the OSA and MPO, this does not necessarily imply that the model is unbiased. The prediction over a different set of data often reveals that the model could be significantly biased. One way to overcome this problem is by splitting the data set into two sets, estimation set and test set (prediction set). Normally, the data is divided into two halves. The first half is used to train the NN and the output computed. The NN usually tracks the system output well and converges to a suitable error minimum. New inputs are presented to the trained NN and the predicted output is observed. If the fitted model is correct, that is, correct assignment of lagged *us* and *ys* then the network will predict well for the prediction set.

A more convincing method of model validation is to use correlation tests. If the model is adequate, then the residuals or prediction errors $\varepsilon(t)$ should be unpredictable from all linear and non-linear combinations of past inputs and outputs. The derivation of simple tests, which can detect these conditions, is complex, but it can be shown that the following conditions should hold (Billings and Voon, 1986).

$$\phi_{\varepsilon\varepsilon}(\tau) = E[\varepsilon(t-\tau)\varepsilon(t)] = \delta(\tau)$$

$$\phi_{u\varepsilon}(\tau) = E[u(t-\tau)\varepsilon(t)] = 0 \quad \forall \tau$$

$$\phi_{u^2\varepsilon}(\tau) = E[(u^2(t-\tau) - \bar{u}^2(t))\varepsilon(t)] = 0 \quad \forall \tau \tag{4.26}$$

$$\phi_{u^2\varepsilon^2}(\tau) = E[(u^2(t-\tau) - \bar{u}^2(t))\varepsilon^2(t)] = 0 \quad \forall \tau$$

$$\phi_{\varepsilon(\varepsilon u)}(\tau) = E[\varepsilon(t)\varepsilon(t-1-\tau)u(t-1-\tau)] = 0 \quad \tau \geq 0$$

where $\phi_{u\varepsilon}(\tau)$ indicates the cross-correlation function between $u(t)$ and $\varepsilon(t)$, $\varepsilon u(t) = \varepsilon(t+1)u(t+1)$, and $\delta(\tau)$ is an impulse function.

Ideally the model validity tests should detect all the deficiencies in the performance of the NN including bias due to internal noise. In practice normalized correlations are computed. The sampled correlation function between two sequences $\psi_1(t)$ and $\psi_2(t)$ is given by

$$\hat{\phi}_{\psi_1\psi_2}(\tau) = \frac{\sum_{t=1}^{N-\tau} \psi_1(t)\psi_2(t+\tau)}{\sqrt{\sum_{t=1}^{N} \psi_1^2(t) \sum_{t=1}^{N} \psi_2^2(t)}} \tag{4.27}$$

Normalization ensures that all the correlation functions lie in the range $-1 \leq \hat{\phi}_{\psi_1\psi_2}(\tau) \leq 1$ irrespective of the signal strengths. The correlations will never be exactly zero for all lags and the 95 per cent confidence bands defined as $1.96/\sqrt{N}$ are used to indicate if the estimated correlations are significant or not, where N is the data length. Therefore, if the correlation functions are within the confidence intervals, the model is regarded as adequate.

The OSA prediction and MPO tests are normally used to determine the model validity in the case of non-linear modelling. Estimation set and test set could be used in the cases of both linear and non-linear modelling. Among the five correlation tests, the first two in equation (4.26) are generally used to determine model validity in the case of linear modelling whereas all five are used in the case of non-linear modelling.

4.5　Data pre-processing

Data pre-processing is required to reduce the difference of magnitude of input variables used to fit to the model. This leads to faster convergence. One of the widely used forms of pre-processing is linear re-scaling of the input variables. Naturally, input variables presented to the model might have values, which differ by several orders of magnitude. Linear transformation is applied to arrange for all the inputs to have similar values. This is usually done by treating each of the input variables independently and calculating the mean $\bar{x}_i$ and variance σ_i^2 for each variable with respect to the presented data set (Bishop, 1995):

$$\bar{x}_i = \frac{1}{N} \sum_{n=1}^{N} x_i^n \tag{4.28}$$

$$\sigma_i^2 = \frac{1}{N-1} \sum_{n=1}^{N} (x_i^n - \bar{x}_i)^2 \tag{4.29}$$

where $n = 1, \ldots, N$ labels the patterns. Thus, a set of re-scaled variables is defined as

$$\tilde{x}_i^n = \frac{x_i^n - \bar{x}_i}{\sigma_i} \tag{4.30}$$

The transformed variables given by $\tilde{x}_i^n$ have zero mean and unity standard deviation over the transformed data set.

4.6　Experimentation and results

4.6.1　Parametric modelling

For identification with the LMS and RLS algorithms the ARMAX structure was considered. This is given as

$$y(t) + a_1 y(t-1) + \cdots + a_{n_a} y(t-n_a)$$
$$= b_1 u(t-1) + \cdots + b_{n_b} u(t-n_b)$$
$$+ e(t) + c_1 e(t-1) + \cdots + c_{n_c} e(t-n_c) \tag{4.31}$$

where a_i, b_i, c_i are parameters to be identified. This accounts for both the true system and noise models.

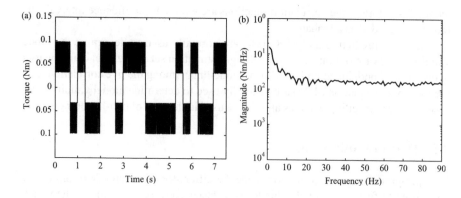

Figure 4.3   The composite PRBS torque input: (a) time domain, (b) spectral density

Table 4.1   Modes found from system responses to the composite PRBS input

Response	First mode (Hz)	Second mode (Hz)	Third mode (Hz)
Hub-angle	11.112	34.26	63.89
End-point acceleration	11.112	36.11	63.89

For the input to be persistently exciting, a combination of bang-bang and pseudo-random binary sequence (PRBS) signal was utilized to excite the system in the range 0–100 Hz covering the rigid-body motion and the first three resonance modes. The level of the bang-bang torque was chosen as ±0.065 Nm, which is enough to deliver the required amount of energy to excite the rigid-body mode of the system without driving it beyond its maximum angular displacement. A PRBS signal with a level of ±0.0325 Nm was used to excite the plant at higher modes. Figure 4.3 shows the combined bang-bang and PRBS signal, referred to as the composite PRBS signal.

The system response was observed over a duration of 7.5 s with no payload. It was noted with the hub-angle, hub-velocity and end-point acceleration responses that the composite PRBS sufficiently excited the first three vibration modes of the system. The first three resonance modes found from these results are shown in Table 4.1.

The manipulator was modelled with the LMS algorithm from torque input to the hub-angle and end-point acceleration output with model orders 8 and 6, respectively. The simulated outputs of hub-angle and end-point acceleration models with the actual system output are shown in Figure 4.4.

It was noted with the corresponding correlation tests of the hub-angle model that none of the correlation functions were within the 95 per cent confidence interval indicating that the model was inadequate to represent the system, and as noted in all the three cases the first mode was not identified.

In modelling the system with the RLS algorithm model orders of 8 and 6 were chosen for the hub-angle and end-point acceleration respectively. Figure 4.5 shows the

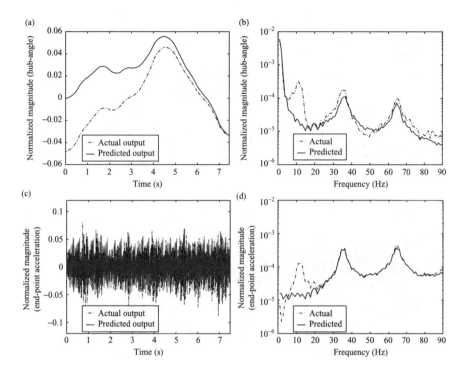

Figure 4.4 *System model with the LMS algorithm: (a) hub-angle, (b) spectral density of hub-angle, (c) end-point acceleration, (d) spectral density of end-point acceleration*

responses of the hub-angle and end-point acceleration models. It was noted with the corresponding correlation tests of the hub-angle model that all the correlation functions were within the 95 per cent confidence interval indicating adequate model fit.

In modelling the manipulator with the GA different initial values and operator rates were investigated, and satisfactory results were achieved with the following set of parameters:

Generation gap:	0.9
Mutation rate (hub-angle model):	0.00313
Mutation rate (end-point acceleration model):	0.00417

The GA was designed with 100 individuals in each generation. The maximum number of generations was set to 500. Different model orders were investigated, and the best results were achieved with orders 8 and 6 with the sum-squared error levels of 0.000163 and 0.18944 in the 500th generation for the hub-angle and end-point acceleration models respectively. It was noted that a reasonable level of accuracy was achieved with as smaller number as 50 generations. Figure 4.6 shows the responses of

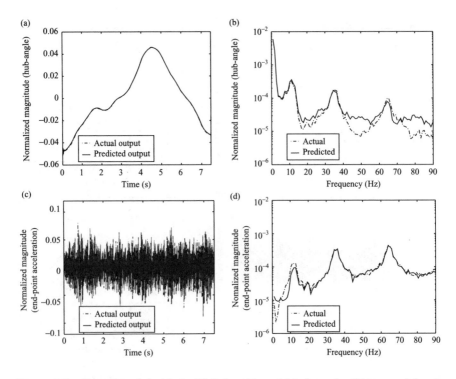

Figure 4.5   System model with the RLS algorithm: (a) hub-angle, (b) spectral density of hub-angle, (c) end-point acceleration, (d) spectral density of end-point acceleration

the hub-angle and end-point acceleration models with the best parameter set resulting in the 500th generation. It was noted with the corresponding correlation tests of the hub-angle model that all the correlation functions were within the 95 per cent confidence interval indicating an adequate model fit.

4.6.2   Non-parametric modelling

The composite PRBS signal described earlier has four amplitude levels, which might not be sufficient to capture the non-linear dynamics present in the system. Accordingly in the experiments to follow a uniformly distributed white noise signal in the range of 0−100 Hz, which covers the first three vibration modes of the system, was used to excite the system. The level of the signal was chosen within ±0.3 Nm. Figure 4.7 shows the noise input in the time and frequency domains thus used. The corresponding hub-angle and end-point acceleration responses of the system are shown in Figure 4.8. The first three vibration modes of the system found from these results are listed in Table 4.2. It is noted that the value of the first resonance frequency is consistent with the three responses. This is also reflected in a single resonance peak in each as noted in Figure 4.8. The discrepancy in the frequency of each of the second and

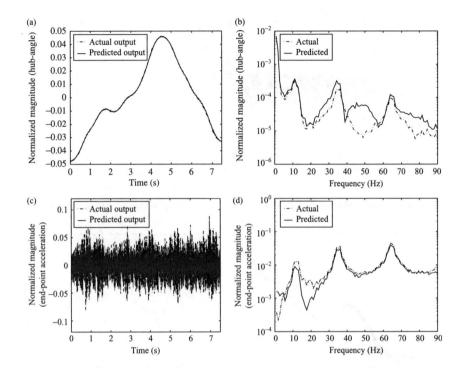

Figure 4.6   *System model with the GA: (a) hub-angle, (b) spectral density of hub-angle, (c) end-point acceleration, (d) spectral density of end-point acceleration*

third modes is due to multiple resonance peaks appearing about these modes as noted in Figure 4.8. In obtaining the frequencies in Table 4.2 for the second and third modes, an average of the frequencies corresponding to these peaks was taken in each case.

In this section results of modelling the manipulator from input torque to hub-angle and end-point acceleration outputs with MLP- and RBF-NNs are presented. In these investigations both the OSA prediction and the MPO models were obtained. The data set, in each case, comprised 1 500 data points. This was divided into two sets of 750 data points each. The first set was used to train the network and the whole 1 500 points, including the 750 points that were not used in the training process, were used to test the model.

4.6.2.1 Modelling with MLP NN

Figure 4.9 shows the OSA prediction of hub-angle response using a three layered MLP NN with five neurons in the hidden layer and $n_u = n_y = 8$. The model reached a sum-squared error level of 0.0025 with 100 training passes. It is noted with the corresponding correlation tests performed on the predicted data set that all the tests were largely within the 95 per cent confidence interval indicating

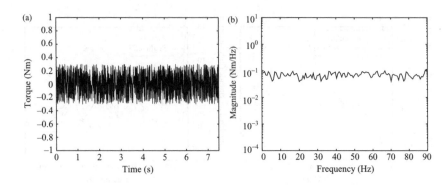

Figure 4.7   The white noise input to the system: (a) time domain, (b) spectral density

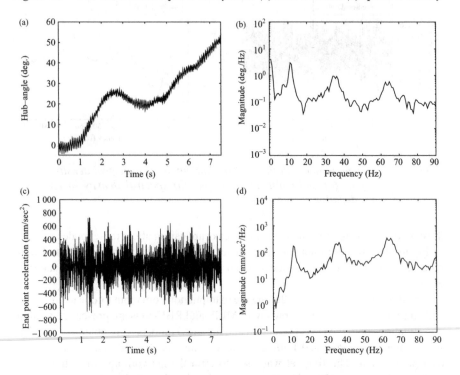

Figure 4.8   The system response to the white noise input: (a) hub-angle, (b) spectral density of hub-angle, (c) end-point acceleration, (d) spectral density of end-point acceleration

acceptable performance of the model in characterising the hub-angle behaviour of the manipulator.

A neuro-model constructed with 5 neurons in the input layer and 15 in the hidden layer was realised fit for the OSA prediction end-point acceleration model of the manipulator. The model reached a sum-squared error level of 0.0135 with 500 training

Table 4.2 *Resonance frequencies found from the responses to the white noise input*

Response	First mode (Hz)	Second mode (Hz)	Third mode (Hz)
Hub-angle	11.112	34.03	64.58
End-point acceleration	11.112	36.11	62.5

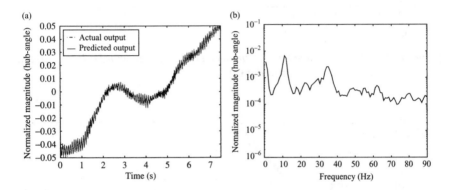

Figure 4.9 *One-step-ahead hub-angle prediction with MLP NN: (a) time domain, (b) spectral density*

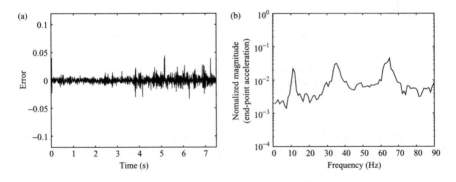

Figure 4.10 *One-step-ahead end-point acceleration prediction with MLP NN: (a) time domain, (b) spectral density*

passes. Figure 4.10 shows the performance of the model. It was noted in the corresponding correlation tests that all the test results were within the 95 per cent confidence interval indicating adequate model fit.

The model predicted end-point acceleration output with an MLP network, consisting of 5 neurons in the input layer and 15 neurons in the hidden layer, is shown in Figure 4.11. The model reached a sum-squared error level of 0.0097 with

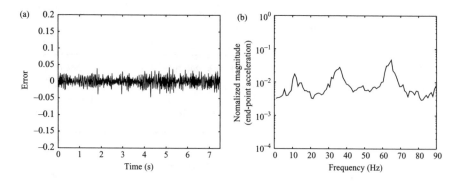

Figure 4.11   Model predicted end-point acceleration output with MLP NN: (a) time domain, (b) spectral density

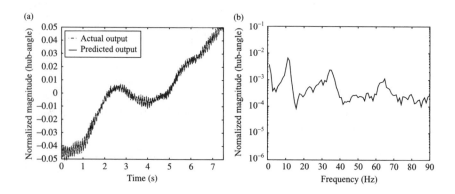

Figure 4.12   One-step-ahead hub-angle prediction with RBF NN: (a) time domain, (b) spectral density

5 000 training passes. The results of all the corresponding correlation tests were largely within the 95 per cent confidence interval indicating acceptable model.

4.6.2.2   Modelling with RBF NN

Figure 4.12 shows the results of OSA hub-angle prediction of the manipulator using 5 RBF terms with a spread constant of unity and lag of 8. As noted the network has predicted the system output very well. The model reached a sum-squared error level of 0.0034 within only four training passes. It was noted that the corresponding correlation test functions were within the 95 per cent confidence interval indicating an adequate hub-angle model fit.

A NN designed with 250 RBF terms, spread constant of unity and $n_u = n_y = 6$ was found fit for the OSA prediction of end-point acceleration. The model reached a sum-squared error level of 0.0115 with 249 training passes, as shown in Figure 4.13. It

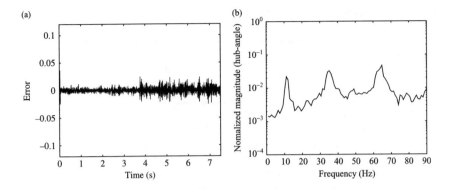

Figure 4.13   *One-step-ahead end-point acceleration prediction with RBF NN: (a) time domain, (b) spectral density*

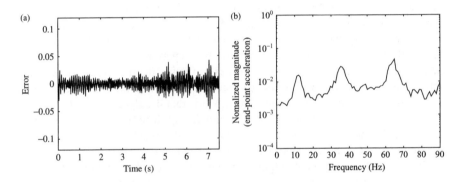

Figure 4.14   *Model predicted end-point acceleration output with RBF NN: (a) time domain, (b) spectral density*

was noted that the corresponding correlation test results were all within the 95 per cent confidence interval indicating adequate model fit.

Figure 4.14 shows the training and test set validation of the model predicted end-point acceleration output of the network constructed with 250 RBF terms, $n_u = n_y = 6$ and spread constant of unity. The model reached a sum-squared error level of 0.0115 with 249 training passes. It was noted that the corresponding correlation test results were all within the 95 per cent confidence interval indicating an adequate model fit.

4.7   Comparative assessment

Table 4.3 shows a comparative assessment of the three parametric modelling approaches in terms of estimation of the resonance modes of the system. It is noted

Table 4.3   Performance of parametric models in characterising the system

Excitation	Model	First mode error (%)	Second mode error (%)	Third mode error (%)
C-PRBS input	LMS-based hub-angle model	*	5.40	0.93
	LMS-based end-point acceleration model	*	0	0
	RLS-based hub-angle model	0.93	0	0
	RLS-based end-point acceleration model	7.99	0	0
	GA-based hub-angle model	0	0	0
	GA-based end-point acceleration model	0	0	0

* Mode did not appear

that the performances with the GA- and RLS-based modelling approaches were better than that with the LMS-based modelling. The LMS failed to detect the first vibration mode of the system in both the hub-angle and end-point acceleration responses. Among the GA and RLS modelling approaches, the performance was better with the GA, as the GA uses a global search process in finding the parameters. It was also noted in the correlation test results that the GA and RLS performed better than the LMS. However, a major advantage of the LMS is that the algorithm is very simple. The LMS is faster as compared to the RLS and the GA. The GA is extremely slower than the LMS and the RLS and thus, with the current computing technology it is not easy to utilize a GA-based identification process in real time.

It is noted from the input/output mappings presented above that the NN-based models have performed well. All the models were validated through the steps of training and test. It was noted from the results of the correlation tests that the NN-based modelling techniques considered in this study performed adequately well.

Comparative performance of the NN-based modelling of the manipulator is summarised in Table 4.4. It is noted that with all the techniques the vibration modes of the system were successfully determined with minor or no error except in very few cases. It was also noted that the RBF NNs were faster in convergence as compared to MLP networks. Comparing the results between OSA prediction and MPO, it is noted that the OSA prediction performed well in approximating the system responses. This agrees with the theory as the OSA prediction uses actual data for prediction whereas the MPO uses the predicted value for the same purpose. However, the MPO is more reliable as compared to the OSA prediction and in this investigation the results of the MPO were good enough to be relied upon.

4.8   Summary

Parametric and non-parametric modelling of a flexible manipulator have been carried out. Among the parametric modelling approaches, the LMS is extensively used in

Table 4.4   *Performance of non-parametric models in characterising the system*

Model type	Number of epochs	Sum-squared error	Error in first mode (%)	Error in second mode (%)	Error in third mode (%)
MLP NN one-step-ahead hub-angle prediction	100	0.0025	0	0.68	4
MLP NN one-step-ahead end-point acceleration prediction	500	0.0135	0	2.56	3.70
MLP NN model predicted end-point acceleration output	5 000	0.0097	0	2.56	3.70
RBF NN one-step-ahead hub-angle prediction	4	0.0034	0	0.68	0.36
RBF NN one-step-ahead end-point acceleration prediction	249	0.0115	0	2.56	3.70
RBF NN model predicted end-point acceleration output	249	0.0115	0	2.56	3.70

many real-time applications. However, its performance in this case was worse as compared to the RLS and the GA. These algorithms are suitable to different applications, but a closer match depends on the nature of the application. In this particular application the RLS and GA performed very well in detecting the vibration modes of the system and their performances were comparable.

Investigations in the case of non-parametric modelling of the manipulator reveal that both MLP and RBF NNs are suitable for modelling such systems. Results of non-linear modelling techniques adopted have been validated through various tests including input/output mapping, training and test validation and correlation tests. It has been noted that all the neuro-modelling techniques have performed well in approximating the system response, where in the time-domain the developed models have predicted the system output closely and in the frequency-domain the resonance frequencies of the model have matched those of the actual system very well.

Chapter 5

Finite difference and finite element simulation of flexible manipulators

A.K.M. Azad, M.O. Tokhi, Z. Mohamed, S. Mahil and H. Poerwanto

This chapter presents numerical approaches based on finite difference (FD) and finite element (FE) techniques for dynamic simulation of single-link flexible manipulator systems. A finite-dimensional simulation of the flexible manipulator system is developed using an FD discretisation of the dynamic equation of motion of the manipulator. A methodology is then presented for obtaining the dynamic model of a lightweight flexible manipulator using FE/Lagrangian technique. Structural damping, hub inertia and payload are incorporated in the dynamic model, which is then represented in a state-space form. Simulation results characterising the dynamic behaviour of the manipulator are presented and discussed for both FD and FE methods. A comparative study of the FD and the FE methods of dynamic modelling of flexible manipulators on the basis of computational accuracy, efficiency and demand are then considered. The performance of the algorithms is assessed with experimental results in time and frequency domains.

5.1 Introduction

Flexible manipulator systems (FMSs) offer several advantages in contrast to the traditional rigid ones. These include faster system response, lower energy consumption, requiring relatively smaller actuators, reduced non-linearity due to elimination of gearing, less overall mass and, in general, less overall cost. However, owing to the distributed nature of the governing equations describing system dynamics, the control of flexible manipulators has traditionally involved complex processes (Aubrun, 1980; Balas, 1978; Omatu and Seinfeld, 1986). Moreover, to compensate for flexure effects and thus yield robust control the design focuses primarily on non-collocated controllers (Cannon and Schmitz, 1984; Harashima and Ueshiba, 1986).

It is important initially to recognise the flexible nature of the manipulator and construct a mathematical model for the system that accounts for interactions with actuators and payload. Such a model can be constructed using partial differential equations (PDEs). A commonly used approach for solving a PDE representing the dynamics of a manipulator, sometimes referred to as the separation of variables method, is to utilize a representation of the PDE, obtained through a simplification process, by a finite set of ordinary differential equations. Such a model, however, does not always represent the fine details of the system (Hughes, 1987). A method in which the flexible manipulator is modelled as a massless spring with a lumped mass at one end and lumped rotary inertia at the other end has previously been proposed (Feliu *et al.*, 1992; Oosting and Dickerson, 1988). Unfortunately, the solution obtained through this method is also not accurate and suffers from similar problems as in the case of the separation of variables method. The finite element (FE) method has also been previously utilized to describe the flexible behaviour of manipulators (Dado and Soni, 1986; Usoro *et al.*, 1984). The computational complexity and consequent software coding involved in the FE method is a major disadvantage of this technique. However, as the FE method allows irregularities in the structure and mixed boundary conditions to be handled, the technique is found suitable in applications involving irregular structures. In applications involving uniform structures, such as the manipulator system considered here, the finite difference (FD) method is found to be more appropriate. Previous simulation studies of flexible beam systems have demonstrated the relative simplicity of the FD method (Kourmoulis, 1990). Dynamic simulation is important from a system design and evaluation viewpoint. It provides a characterisation of the system in the real sense as well as allowing on-line evaluation of controller designs.

This chapter presents an investigation into the dynamic simulation of a flexible manipulator system using FD and FE methods. The FD method is used to obtain an efficient numerical method of solving the PDE by developing a finite-dimensional simulation of the flexible manipulator system through a discretisation, both, in time and space (distance) coordinates. The algorithm proposed here allows the inclusion of distributed actuator and sensor terms in the PDE and modification of boundary conditions incorporating mode frequency dependent damping (Tokhi and Azad, 1995a). The FE/Lagrangian approach is used to obtain the dynamic equations of an actual experimental test-rig. Structural damping, hub inertia and payload are incorporated in the dynamic model, which is then represented in a state-space form. The algorithms thus developed are implemented digitally and simulation results verifying their performances in characterising the behaviour of the system under various loading conditions are presented and discussed. The results are also compared with the outputs from an actual experimental test-rig.

5.2   The flexible manipulator system

A description of a single-link flexible manipulator system is shown in Figure 5.1, where X_0OY_0 and XOY represent the stationary and moving coordinates respectively,

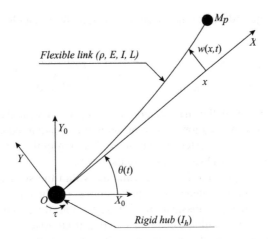

Figure 5.1   Mechanical model of the flexible manipulator system

$\tau(t)$ represents the applied torque at the hub, M_P is the payload mass and I_h is the hub inertia. E, I and ρ represent the Young's modulus, second moment of inertia and mass density per unit length of the manipulator, respectively. Gravity effects are neglected and the motion of the manipulator is confined to the XOY plane.

The flexible manipulator system can be modelled as a pinned-free flexible beam, incorporating inertia at the hub and payload mass at the end-point. The model is developed through the utilisation of Lagrange equation and modal expansion method (Hastings and Books, 1987; Korolov and Chen, 1989). For an angular displacement θ and an elastic deflection w, the total (net) displacement $y(x,t)$ of a point along the manipulator at a distance from the hub can be described as a function of both the rigid-body motion $\theta(t)$ and elastic deflection $w(x,t)$ measured from the line OX;

$$y(x,t) = x\theta(t) + w(x,t) \tag{5.1}$$

The dynamic equations of motion of the manipulator can be obtained using the Hamilton's extended principle (Azad, 1994; Meirovitch, 1967) with the associated kinetic, potential and dissipated energies of the system. Ignoring the effects of rotary inertia and shear deformation, a fourth-order PDE representing the manipulator motion can, thus, be obtained as (Azad, 1994)

$$EI\frac{\partial^4 w(x,t)}{\partial x^4} + \rho\frac{\partial^2 w(x,t)}{\partial t^2} = -\rho x\ddot{\theta} \tag{5.2}$$

To obtain the corresponding boundary conditions, the following must hold:

- The displacement at the hub ($w(0,t)$) must be zero.
- The total forces at the hub must be the same with the applied torque.
- The shear force at the end-point must be equal to $M_P\left(\partial^2 w(x,t)/\partial t^2\right)$ (Tse *et al.*, 1978).

- The stress at the end-point must be zero, that is, no force should be present at the free end.

$$w\,(0,t) = 0, \qquad\qquad I_h \frac{\partial^3 w(0,t)}{\partial t^2 \partial x} - EI \frac{\partial^2 w(0,t)}{\partial x^2} = \tau\,(t)$$

$$M_p \frac{\partial^2 w(l,t)}{\partial t^2} - EI \frac{\partial^3 w(l,t)}{\partial x^3} = 0, \qquad\qquad EI \frac{\partial^2 w(l,t)}{\partial x^2} = 0 \tag{5.3}$$

Where l is the length of the manipulator. Equation (5.2) with the corresponding boundary conditions in equation (5.3) represents the dynamic equation of motion of the flexible manipulator system assuming no damping in the system. In practice, however, such an effect is always present in the system.

There are several possible forms of damping within the system. These can be classified into three groups, depending on the source: (a) the manipulator itself has structural damping owing to dissipation of energy within the manipulator material; (b) viscous damping and Coulomb damping (stiction/friction) associated with driving motor and (c) external effects such as primarily air resistance as the manipulator rotates.

To incorporate damping in the governing dynamic equation of the system, a mode frequency dependent damping term proportional to $\partial^3 w\,(x,t)/\partial x^2 \partial t$ can be introduced (Davis and Hirschorn, 1988). Equation (5.2) can thus be modified to yield

$$EI \frac{\partial^4 w(x,t)}{\partial x^4} + \rho \frac{\partial^2 w(x,t)}{\partial t^2} - D_S \frac{\partial^3 w(x,t)}{\partial x^2 \partial t} = -\rho x \ddot{\theta} \tag{5.4}$$

where D_S is the resistance to strain velocity, that is, rate of change of strain and $D_S \left[\partial^3 w\,(x,t)/\partial x^2 \partial t \right]$ represents the resulting damping moment dissipated in the manipulator structure during its dynamic operation. The corresponding boundary conditions can, thus, be written as

$$w\,(0,t) = 0, \qquad\qquad I_h \frac{\partial^3 w(0,t)}{\partial t^2 \partial x} - EI \frac{\partial^2 w(0,t)}{\partial x^2} = \tau\,(t)$$

$$M_p \frac{\partial^2 w(l,t)}{\partial t^2} - EI \frac{\partial^3 w(l,t)}{\partial x^3} = 0, \qquad\qquad EI \frac{\partial^2 w(l,t)}{\partial x^2} = 0 \tag{5.5}$$

Note in Figure 5.1 that as line OX is tangential to the manipulator at the hub, point O, the following holds:

$$\frac{\partial w(0,t)}{\partial x} = 0 \tag{5.6}$$

Thus, using equations (5.1) and (5.6) yield

$$\frac{\partial y(0,t)}{\partial x} = \theta(t)$$

Substituting for $w\,(x,t)$ from equation (5.1) into equations (5.4) and (5.5), manipulating and simplifying yields the governing equation of motion of the manipulator in terms of $y\,(x,t)$ as

$$EI \frac{\partial^4 y\,(x,t)}{\partial x^4} + \rho \frac{\partial^2 y\,(x,t)}{\partial t^2} - D_S \frac{\partial^3 y\,(x,t)}{\partial x^2 \partial t} = \tau\,(t) \tag{5.7}$$

with the corresponding boundary conditions as

$$y(0, t) = 0$$

$$I_h \frac{\partial^3 y(0, t)}{\partial t^2 \partial x} - EI \frac{\partial^2 y(0, t)}{\partial x^2} = \tau(t)$$

$$M_p \frac{\partial^2 y(l, t)}{\partial t^2} - EI \frac{\partial^3 y(l, t)}{\partial x^3} = 0 \qquad (5.8)$$

$$EI \frac{\partial^2 y(l, t)}{\partial x^2} = 0$$

and initial conditions as

$$y(x, 0) = 0, \qquad \frac{\partial y(x, 0)}{\partial x} = 0 \qquad (5.9)$$

Equation (5.7) gives the fourth-order PDE, which represents the dynamic equation describing the motion of the flexible manipulator.

5.3   The FD method

To solve the PDE in equation (5.7) and thus develop a suitable simulation environment characterising the behaviour of the system, the FD method can be used. Thus, a set of equivalent difference equations defined by the central FD quotients of the FD method are obtained by discretising the PDE in equation (5.7) with its associated boundary and initial conditions in equations (5.8) and (5.9). The process involves dividing the manipulator into n sections each of length Δx and considering the deflection of each section at sample times Δt, see Figure 5.2. In this manner, a solution of the PDE is obtained by generating the central difference formulae for the partial derivative terms of the response $y(x, t)$ of the manipulator at points $x = i\Delta x$, $t = j\Delta t$ (Azad, 1994; Burden and Faires, 1989; Lapidus, 1982):

$$\frac{\partial^2 y(x, t)}{\partial t^2} = \frac{y_{i,j+1} - 2y_{i,j} + y_{i,j-1}}{\Delta t^2}$$

$$\frac{\partial^2 y(x, t)}{\partial x^2} = \frac{y_{i+1,j} - 2y_{i,j} + y_{i-1,j}}{\Delta x^2}$$

$$\frac{\partial^3 y(x, t)}{\partial x^3} = \frac{y_{i+2,j} + 2y_{i+1,j} + 2y_{i-1,j} - y_{i-2,j}}{2\Delta x^3}$$

$$\frac{\partial^4 y(x, t)}{\partial x^4} = \frac{y_{i+2,j} - 4y_{i+1,j} + 6y_{i,j} - 4y_{i-1,j} + y_{i-2,j}}{\Delta x^4}$$

$$\frac{\partial^3 y(x, t)}{\partial t^2 \partial x} = \frac{y_{i,j+1} - 2y_{i,j} + y_{i,j-1} - y_{i-1,j+1} + 2y_{i-1,j} - y_{i-1,j-1}}{\Delta x \Delta t^2}$$

$$\frac{\partial^3 y(x, t)}{\partial x^2 \partial t} = \frac{y_{i+1,j} - 2y_{i,j} + y_{i-1,j} - y_{i+1,j-1} + 2y_{i,j-1} - y_{i-1,j-1}}{\Delta t \Delta x^2} \qquad (5.10)$$

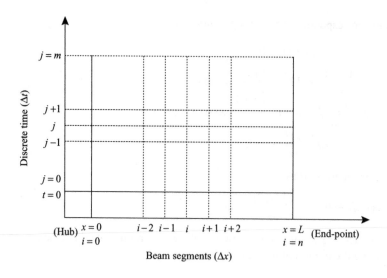

Figure 5.2   Finite difference discretisation in time and space variables

where $y_{i,j}$ represents the response $y(x, t)$ at $x = i\Delta x$ and $t = j\Delta t$ or $y(x_i, t_j)$. Note that, a time-space discretisation is adopted in the evaluation of the response of the manipulator.

5.3.1   Development of the simulation algorithm

A solution of the PDE in equation (5.7) can be obtained by substituting for $\partial^2 y/\partial t^2$, $\partial^4 y/\partial x^4$ and $\partial^3 y/\partial x^2 \partial t$ from equation (5.10) and simplifying to yield

$$\frac{EI}{\Delta x^4}\left[y_{i+2,j} - 4y_{i+1,j} + 6y_{i,j} - 4y_{i-1,j} + y_{i-2,j}\right] + \frac{\rho}{\Delta t^2}\left[y_{i,j+1} - 2y_{i,j} + y_{i,j-1}\right]$$
$$-\frac{D_S}{\Delta x^2 \Delta t}\left[y_{i+1,j} - 2y_{i,j} + y_{i-1,j} - y_{i+1,j-1} + 2y_{i,j-1} - y_{i-1,j-1}\right] = 0$$

or

$$y_{i,j+1} = -c\left[y_{i+2,j} + y_{i-2,j}\right] + b\left[y_{i+1,j} + y_{i-1,j}\right] + ay_{i,j} - y_{i,j-1}$$
$$+ d\left[y_{i+1,j} - 2y_{i,j} + y_{i-1,j} - y_{i+1,j-1} + 2y_{i,j-1} - y_{i-1,j-1}\right] \qquad (5.11)$$

where $a = 2 - \frac{6EI\Delta t^2}{\rho\Delta x^4}$, $b = \frac{4EI\Delta t^2}{\rho\Delta x^4}$, $c = \frac{EI\Delta t^2}{\rho\Delta x^4}$, $d = \frac{D_S\Delta t}{\rho\Delta x^2}$.

Equation (5.11) gives the displacement of section i of the manipulator at time step $j + 1$. It follows from this equation that, to obtain the displacements $y_{n-1,j+1}$ and $y_{n,j+1}$, the displacements of the fictitious points $y_{n+2,j}$, $y_{n+1,j}$, and $y_{n+1,j-1}$ are required. These can be obtained using the boundary conditions related to the dynamic equation of the flexible manipulator. The discrete form of the corresponding boundary conditions, obtained in similar manner as above, are

$$y_{0,j} = 0 \tag{5.12}$$

$$y_{-1,j} = y_{1,j} + \frac{\Delta x I_h}{EI \Delta t^2} \left[y_{1,j+1} - 2y_{1,j} + y_{1,j-1}\right] + \frac{\Delta x^2}{EI} \tau(j) \tag{5.13}$$

$$y_{n+2,j} = 2y_{n+1,j} - 2y_{n-1,j} + y_{n-2,j} + \frac{2\Delta x^3 M_p}{\Delta t^2 EI} \left[y_{n,j+1} - 2y_{n,j} + y_{n,j-1}\right] \tag{5.14}$$

$$y_{n+1,j} = 2y_{n,j} - y_{n-1,j} \tag{5.15}$$

5.3.2   The hub displacement

Note that the torque is applied at the hub of the flexible manipulator. Thus, $\tau(i,j) = 0$ for $i \geq 1$. Using equations (5.11) and (5.12), the displacement $y_{1,j+1}$ can be obtained as

$$y_{1,j+1} = -c\left[y_{3,j} + y_{-1,j}\right] + by_{2,j} + ay_{1,j} - y_{1,j-1} + d\left[y_{2,j} - 2y_{1,j} - y_{2,j-1} + 2y_{1,j-1}\right] \tag{5.16}$$

Substituting for $y_{-1,j}$ from equation (5.13) into equation (5.16) and simplifying yields

$$y_{1,j+1} = K_1 y_{1,j} + K_2 y_{2,j} + K_3 y_{3,j} + K_4 y_{1,j-1} + K_5 y_{2,j-1} + K_6 \tau(j) \tag{5.17}$$

where

$$K_1 = \frac{c\Delta t^2 EI + 2c\Delta x I_h + (a - 2d)\,\Delta t^2 EI}{\Delta t^2 EI + c\Delta x I_h}, \quad K_4 = -\frac{c\Delta x I_h + (1 - 2d)\,\Delta t^2 EI}{\Delta t^2 EI + c\Delta x I_h}$$

$$K_2 = \frac{(b + d)\,\Delta t^2 EI}{\Delta t^2 EI + c\Delta x I_h}, \quad K_5 = -\frac{d\Delta t^2 EI}{\Delta t^2 EI + c\Delta x I_h}$$

$$K_3 = -\frac{c\Delta t^2 EI}{\Delta t^2 EI + c\Delta x I_h}, \quad K_6 = \frac{c\Delta x^2 \Delta t^2}{\Delta t^2 EI + c\Delta x I_h}$$

5.3.3   The end-point displacement

Using equation (5.11) for $i = n - 1$, yields the displacement $y_{n-1,j-1}$ as

$$y_{n-1,j+1} = -c\left[y_{n-3,j} + y_{n+1,j}\right] + b\left[y_{n,j} + y_{n-2,j}\right] + ay_{n-1,j} - y_{n-1,j-1}$$
$$+ d\left[y_{n,j} - 2y_{n-1,j} + y_{n-2,j} - y_{n,j-1} + 2y_{n-1,j-1} - y_{n-2,j-1}\right] \tag{5.18}$$

Similarly, using equation (5.11) for $i = n$, yields the displacement $y_{n,j+1}$ as

$$y_{n,j+1} = -c\left[y_{n-2,j} + y_{n+2,j}\right] + b\left[y_{n+1,j} + y_{n-1,j}\right] + ay_{n,j} - y_{n,j-1}$$
$$+ d\left[y_{n+1,j} - 2y_{n,j} + y_{n-1,j} - y_{n+1,j-1} + 2y_{n,j-1} - y_{n-1,j-1}\right] \tag{5.19}$$

The fictitious displacements $y_{n+1,j}$ and $y_{n+2,j}$, appearing in equations (5.18) and (5.19), can be obtained using the boundary conditions in equations (5.14) and (5.15). $y_{n+1,j-1}$ can easily be obtained by shifting $y_{n+1,j}$ from time step j to time step $j-1$. Substituting for $y_{n+1,j}$ from equation (5.15) into equation (5.18) yields the displacement $y_{n-1,j+1}$ as

$$y_{n-1,j+1} = K_7 y_{n-3,j} + K_8 y_{n-2,j} + K_9 y_{n-1,j} + K_{10} y_{n,j} + K_{11} y_{n-2,j-1}$$
$$+ K_{12} y_{n-1,j-1} + K_{13} y_{n,j-1} \qquad (5.20)$$

where $K_7 = -c$, $K_8 = b + d$, $K_9 = a + c - 2d$, $K_{10} = -(2c - b - d)$, $K_{11} = -d$, $K_{12} = -(1 - 2d)$ and $K_{13} = -d$.

Similarly, substituting for $y_{n+2,j}$ and $y_{n+1,j}$ from equations (5.14) and (5.15) into equation (5.19), and simplifying yields the displacement $y_{n,j+1}$ as

$$y_{n,j+1} = K_{14} y_{n-2,j} + K_{15} y_{n-1,j} + K_{16} y_{n,j} + K_{17} y_{n,j-1} \qquad (5.21)$$

where

$$K_{14} = \frac{-2c\Delta t^2 EI}{\Delta t^2 EI + 2c\Delta x^3 M_P}$$

$$K_{15} = \frac{4c\Delta t^2 EI}{\Delta t^2 EI + 2c\Delta x^3 M_P}$$

$$K_{16} = \frac{\Delta t^2 EI}{\Delta t^2 EI + 2c\Delta x^3 M_P} \left\{ a + 2b - 4c + \frac{4c\Delta x^3 M_P}{\Delta t^2 EI} \right\}$$

$$K_{17} = \frac{-\Delta t^2 EI}{\Delta t^2 EI + 2c\Delta x^3 M_P} \left\{ \frac{2c\Delta x^3 M_P}{\Delta t^2 EI} + 1 \right\}$$

Equations (5.11), (5.17), (5.20) and (5.21) represent the dynamic equation of the manipulator for all the grid points (stations) at specified instants of time t in the presence of hub inertia and payload.

5.3.4   *Matrix formulation*

Using matrix notation, equations (5.11), (5.17), (5.20) and (5.21) can be written in a compact form as

$$\mathbf{Y}_{i,j+1} = \mathbf{A}\mathbf{Y}_{i,j} + \mathbf{B}\mathbf{Y}_{i,j-1} + \mathbf{CF} \qquad (5.22)$$

where $\mathbf{Y}_{i,j+1}$ is the displacement of grid points $i = 1, 2, \ldots, n$ of the manipulator at time step $j + 1$, $\mathbf{Y}_{i,j}$ and $\mathbf{Y}_{i,j-1}$ are the corresponding displacements at time steps j and $j - 1$, respectively. $\mathbf{A}$ and $\mathbf{B}$ are constant $n \times n$ matrices whose entries depend on the flexible manipulator specification and the number of sections the manipulator is divided into, $\mathbf{C}$ is a constant matrix related to the given input torque and $\mathbf{F}$ is an $n \times 1$ matrix related to the time step Δt and mass per unit length of the flexible manipulator:

$$
\mathbf{Y}_{i,j+1} = \begin{bmatrix} y_{1,j+1} \\ y_{2,j+1} \\ \vdots \\ y_{n,j+1} \end{bmatrix}, \qquad
\mathbf{Y}_{i,j} = \begin{bmatrix} y_{1,j} \\ y_{2,j} \\ \vdots \\ y_{n,j} \end{bmatrix}, \qquad
\mathbf{Y}_{i,j-1} = \begin{bmatrix} y_{1,j-1} \\ y_{2,j-1} \\ \vdots \\ y_{n,j-1} \end{bmatrix}
$$

$$
\mathbf{A} = \begin{bmatrix}
K_1 & K_2 & K_3 & 0 & 0 & \cdots & 0 & 0 \\
(b+d) & (a-2d) & (b+d) & -c & 0 & \cdots & 0 & 0 \\
-c & (b+d) & (a-2d) & (b+d) & -c & \cdots & 0 & 0 \\
\vdots & \ddots & \ddots & \ddots & \ddots & \ddots & \ddots & \vdots \\
0 & 0 & \cdots & -c & b+d & a-2d & b+d & -c \\
0 & 0 & \cdots & 0 & K_7 & K_8 & K_9 & K_{10} \\
0 & 0 & \cdots & 0 & 0 & K_{14} & K_{15} & K_{16}
\end{bmatrix}
$$

$$
\mathbf{B} = \begin{bmatrix}
K_4 & K_5 & 0 & 0 & 0 & \cdots & 0 & 0 \\
-d & 2d-1 & -d & 0 & 0 & \cdots & 0 & 0 \\
0 & -d & 2d-1 & -d & 0 & \cdots & 0 & 0 \\
\vdots & \ddots & \ddots & \ddots & \ddots & \ddots & \vdots \\
0 & 0 & \cdots & 0 & -d & 2d-1 & -d & 0 \\
0 & 0 & \cdots & 0 & 0 & K_{11} & K_{12} & K_{13} \\
0 & 0 & \cdots & 0 & 0 & 0 & 0 & K_{17}
\end{bmatrix}
$$

$$
\mathbf{C} = \tau(j), \qquad \mathbf{F} = [K_6 \quad 0 \cdots \quad 0]^{\mathrm{T}}
$$

5.3.5 State-space formulation

A state-space formulation of the dynamic equation of the manipulator can be constructed by referring to the matrix formulation. Using the notation for simulation of discrete-time linear systems, the dynamic equations of the flexible manipulator can be written as

$$
x(n+1) = \mathbf{P}x(n) + \mathbf{Q}u
$$
$$
y(n) = \mathbf{R}x(n) + \mathbf{S}u
$$

where

$$
\mathbf{P} = \begin{bmatrix} \mathbf{A} & \mathbf{B} \\ I_{N \times N} & 0_{N \times N} \end{bmatrix}, \quad
\mathbf{Q} = \begin{bmatrix} \mathbf{C} \\ 0_{N \times 1} \end{bmatrix}, \quad
\mathbf{R} = \begin{bmatrix} I_N & 0_N \end{bmatrix}, \quad
\mathbf{S} = [0_{2N}]
$$

$$
u = \begin{bmatrix} \tau & 0 & \cdots & 0 \end{bmatrix}^{\mathrm{T}}, \quad
y(n) = \begin{bmatrix} x(1,n) \cdots x(N,n), \; x(1, n-1) \cdots x(N, n-1) \end{bmatrix}
$$

Note that N represents the number of sections.

5.4   The FE/Lagrangian method

The dynamic model of a physical structure (including rigid robotic manipulators) is obtainable by using Lagrange's method (Crandall *et al.*, 1968). For robotic manipulators, the Lagrangian technique is simplified by first formulating certain (generalized) inertia matrices (Mahil, 1982). This approach can be combined with the FE method (Cook, 1981; Rao, 1989) to model the flexible links of robotic manipulators (Usoro *et al.*, 1983, 1984).

The flexible link is considered as an assemblage of a finite number of small elements. The elements are assumed interconnected at certain points, known as nodes. For each FE, the scalar kinetic and potential energy functions are formulated as functions of the generalized coordinates. A Lagrangian is formulated, and the dynamic model is obtained by applying Lagrange's equations. By reducing the element size, that is, by increasing the number of elements, the overall solution of the system equations can be made to converge to the exact solution as precisely as desired. The FE/Lagrangian approach is general but, for reasons of simplicity, it is demonstrated for two cases: (a) a manipulator with single flexible link, and (b) a manipulator with two flexible links. A two-link manipulator is shown in Figure 5.3.

5.4.1   *Elemental matrices*

The first step is to find a 'generalized inertia matrix' and a stiffness matrix for a single FE. Bernoulli–Euler beam theory is used in this process. For an arbitrary link, an FE j is considered to possess 2 degrees of freedom (DOF), a transverse flexural deflection and a flexural slope, at each end of the element. The flexural displacement y can be described in terms of these DOF and certain shape functions (Hermitian polynomials)

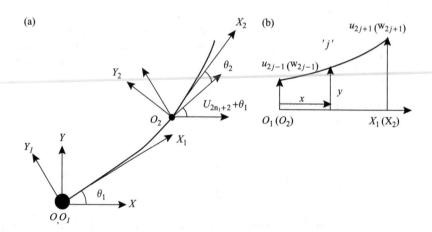

Figure 5.3   A two-link manipulator: (a) the manipulator, (b) element j of link 1(2)

$\phi_k(x)$ (Ross, 1996) as

$$y(x,t) = \sum_{k=1}^{4} \phi_k(x) u_{2j-2+k}(t) \tag{5.23}$$

where, for an element of length l, the shape functions are defined by

$$\phi_1(x) = 1 - 3\frac{x^2}{l^2} + 2\frac{x^3}{l^3}, \qquad \phi_2(x) = x - 2\frac{x^2}{l} + \frac{x^3}{l^2}$$

$$\phi_3(x) = 3\frac{x^2}{l^2} - 2\frac{x^3}{l^3}, \qquad \phi_4(x) = -\frac{x^2}{l} + \frac{x^3}{l^2} \tag{5.24}$$

The displacement variables x and $y(x)$, see Figure 5.3(b), satisfy the boundary conditions, that is, $y(0) = u_{2j-1}$, $\partial y(0)/\partial x = u_{2j}$, $y(l) = u_{2j+1}$, $\partial y(l)/\partial x = u_{2j+2}$.

5.4.1.1   Scalar energy functions

Let $\mathbf{r}$ be the position vector, in inertial coordinates, of a small differential element on the FE 'j' under consideration. The kinetic energy, T_j, of the (finite) element is obtained by integrating (over the element length l) the corresponding energy function of the differential element. Thus

$$T_j = \frac{1}{2} \int_0^l m \left[\frac{\partial \mathbf{r}^\mathrm{T}}{\partial t} \cdot \frac{\partial \mathbf{r}}{\partial t} \right] dx \tag{5.25}$$

which, after some algebraic manipulations, has the form

$$T_j = \frac{1}{2} \dot{\mathbf{z}}_j^\mathrm{T} \mathbf{M}_j \dot{\mathbf{z}}_j \tag{5.26}$$

The matrix $\mathbf{M}_j$ is a symmetric matrix, which behaves similar to inertia, and is called the 'generalized inertia matrix' of element 'j'. The (i,k) th element of this inertia matrix is

$$M_j(i,k) = \int_0^l m \left[\frac{\partial T_j^\mathrm{T}}{\partial z_{ji}} \cdot \frac{\partial T_j}{\partial z_{ji}} \right] dx; i, k = 1, 2, \ldots \tag{5.27}$$

The potential energy, V_j, due to elasticity of the FE is obtained, in a similar manner, by integrating the differential (potential) energy function over the element length l. Thus,

$$V_j = \frac{1}{2} \int_0^l EI \left[\frac{\partial^2 y}{\partial x^2} \right] dx \tag{5.28}$$

(For motion in planes other than horizontal, potential energy due to gravity need be included.) After some manipulation, the potential energy of the element can be determined as

$$V_j = \frac{1}{2} \boldsymbol{\psi}_j^\mathrm{T} \mathbf{K}_{j\omega\omega} \boldsymbol{\psi}_j \tag{5.29}$$

where

$$\mathbf{K}_{j\omega\omega}(l) = \left(\frac{EI}{l^3}\right) \begin{bmatrix} 12 & 6l & -12 & 6l \\ 6l & 4l^2 & -6l & 2l^2 \\ -12 & -6l & 12 & -6l \\ 6l & 2l^2 & -6l & 4l^2 \end{bmatrix} \tag{5.30}$$

is a 4×4 stiffness matrix of the jth FE corresponding to the elastic motion, and is a function of the element length 'l'. Summing over all FEs, the total *kinetic* and total *potential energies* T and V of the arbitrary link can be obtained as

$$T = \sum_{j=1}^{n} T_j = \sum_{j=1}^{n} \frac{1}{2} \dot{\mathbf{z}}_j^T \mathbf{M}_j \dot{\mathbf{z}}_j = \frac{1}{2} \tilde{\dot{\mathbf{q}}}^T \tilde{\mathbf{M}} \tilde{\dot{\mathbf{q}}} \tag{5.31}$$

$$V = \sum_{j=1}^{n} V_j = \sum_{j=1}^{n} \frac{1}{2} \boldsymbol{\psi}_j^T \mathbf{K}_{j\omega\omega} \dot{\mathbf{z}}_j = \frac{1}{2} \tilde{\boldsymbol{\psi}}^T \tilde{\mathbf{K}}_{\omega\omega} \tilde{\boldsymbol{\psi}} \tag{5.32}$$

The elements of the 'intermediate' 'generalized inertia matrix' $\tilde{\mathbf{M}}$ and the 'intermediate' 'stiffness matrix' $\tilde{\mathbf{K}}_{\omega\omega}$ can be easily obtained by simple manipulations. A matrix $\mathbf{P}(l)$, which is also a function of the element length 'l', is defined below. This is handy in the procedure development.

$$\mathbf{P}(l) = \begin{bmatrix} 156 & 22l & 54 & -13l \\ 22l & 4l^2 & 13l & -3l^2 \\ 54 & 13l & 156 & -22l \\ -13l & -3l^2 & -22l & 4l^2 \end{bmatrix} \tag{5.33}$$

5.4.2   A single-link flexible manipulator

In this case, the position of a small differential element located on the jth FE of link '1' is given by the vector

$$\mathbf{r}_1 = \begin{bmatrix} (j-1)l_1 + x \\ y \end{bmatrix}$$

in the body-fixed system of coordinates. This vector can be transformed to the inertial system of coordinates by using a transformation matrix $\mathbf{T}_0^1$. Thus,

$$\mathbf{r} = \mathbf{T}_0^1 \mathbf{r}_1 = \begin{bmatrix} \cos\theta_1 & -\sin\theta_1 \\ \sin\theta_1 & \cos\theta_1 \end{bmatrix} \begin{bmatrix} (j-1)l_1 + x \\ y \end{bmatrix} \tag{5.34}$$

Using equation (5.34) with equations (5.23)–(5.33), the inertia and stiffness matrices, and the kinetic and potential energy functions for the 'jth' FE, and thence the corresponding matrices and energy functions for the whole link can be determined. The symmetric generalized inertia matrix $\mathbf{M}_{1j}$ of the jth element is of order 5 and, by simple manipulations, it can be shown that the first-row elements of this matrix are

$$M_{1j}(1, 1) = \frac{m_1 l_1^3}{3}(3j^2 - 3j + 1) + \psi_{1j}^T P_{1j} \psi_{1j}, \quad M_{1j}(1, 2) = \frac{m_1 l_1^2}{20}(10j - 7)$$

$$M_{1j}(1, 3) = \frac{m_1 l_1^3}{60}(5j - 3), \quad M_{1j}(1, 4) = \frac{m_1 l_1^2}{60}(10j - 3)$$

$$M_{1j}(1, 5) = -\frac{m_1 l_1^3}{60}(5j - 2)$$

where

$$\mathbf{P}_{1j} = \frac{m_1 l_1}{420} \mathbf{P}(l_1)$$

The matrix $\mathbf{P}_{1j}$ is also the lower 4×4 sub-matrix in the right-hand corner of the inertia matrix $\mathbf{M}_{1j}$. The total *kinetic* and *potential energies* T_1 and V_1 of link '1' are

$$T_1 = \frac{1}{2} \dot{\mathbf{q}}_1^T \widetilde{\mathbf{M}}_1 \dot{\mathbf{q}}_1$$

$$V_1 = \frac{1}{2} \widetilde{\boldsymbol{\psi}}_1^T \widetilde{\mathbf{K}}_{1\omega\omega} \widetilde{\boldsymbol{\psi}}_1$$

The matrices $\widetilde{\mathbf{M}}_1$ and $\widetilde{\mathbf{K}}_{1\omega\omega}$ are the generalized inertia matrix and the stiffness elastic submatrix of the link, respectively. The matrix elements can be easily computed from equations (5.23) to (5.33) by appropriate manipulations (Usoro *et al.*, 1983, 1984).

5.4.3 A two-link flexible manipulator

In this case, the position of a differential element located on the jth FE of link '2', is given by the vector

$$\mathbf{r} = \mathbf{T}_0^1 \left[\begin{pmatrix} L_1 \\ u_{2n_1+1} \end{pmatrix} + \mathbf{T}_1^2 \begin{pmatrix} (j-1)l_2 + x \\ y \end{pmatrix} \right] \tag{5.35}$$

in inertial coordinates, where $\mathbf{T}_0^1$ is as defined earlier in equation (5.12), and

$$\mathbf{T}_1^2 = \begin{bmatrix} \cos(\theta_2 + u_{2n_1+2}) & -\sin(\theta_2 + u_{2n_1+2}) \\ \sin(\theta_2 + u_{2n_1+2}) & \cos(\theta_2 + u_{2n_1+2}) \end{bmatrix} \tag{5.36}$$

is the transformation from body-fixed system of coordinates $O_2 X_2 Y_2$ to the body-fixed system of coordinates $O_1 X_1 Y_1$. This vector is independent of coordinates u_i, $i = 1, 2, \ldots, 2n_1$. Proceeding exactly as for link '1', and using equations (5.35) and (5.36) with equations (5.23) to (5.33), the total kinetic and potential energies for link '2' are obtained as

$$T_2 = \frac{1}{2} \dot{\mathbf{q}}_2^T \widetilde{\mathbf{M}}_2 \dot{\mathbf{q}}_2$$

$$V_2 = \frac{1}{2}\widetilde{\boldsymbol{\psi}}_2^T \widetilde{\mathbf{K}}_{2\omega\omega} \widetilde{\boldsymbol{\psi}}_2$$

The elements of the inertia matrix $\widetilde{\mathbf{M}}_2$ and the stiffness submatrix $\widetilde{\mathbf{K}}_{2\omega\omega}$ for link '2' can be found in (Usoro *et al.*, 1983).

5.4.3.1   Boundary conditions, payload and damping

When the first joint is constrained to have purely rotational displacement then $u_1(t) = u_2(t) = 0$ for all 't'. Similarly, for a purely rotational second joint, $w_1(t) = w_2(t) = 0$. The appropriate rows and columns are hence eliminated from the matrices $\widetilde{\mathbf{M}}_1$, $\widetilde{\mathbf{M}}_2$, $\widetilde{\mathbf{K}}_1$, $\widetilde{\mathbf{K}}_2$ obtained above. The payload can be incorporated as a point mass at the end-point of the manipulator. For a two-link manipulator, for instance, the vector

$$\mathbf{r} = \mathbf{T}_0^1 \left[\begin{pmatrix} L_1 \\ u_{2n_1+1} \end{pmatrix} + \mathbf{T}_1^2 \begin{pmatrix} L_2 \\ w_{2n_2+1} \end{pmatrix} \right]$$

describes the payload position in inertial coordinates. Using this equation with equations (5.23)–(5.33), the payload's contribution to the inertia matrix can be determined. The dynamic model is obtained by formulating the Lagrangian of the system and using Lagrange's equations (Crandall *et al.*, 1968). Structural damping is usually present in flexible manipulator systems, and can be included by assuming that the manipulator exhibits Rayleigh damping. Experimentally determined damping ratios can then be assigned to the individual modes, and the damping matrix $\mathbf{D}$ formulated. The final dynamic model has the form

$$\mathbf{M}\ddot{\mathbf{q}}(t) + \mathbf{D}\dot{\mathbf{q}}(t) + \mathbf{K}\mathbf{q}(t) = \mathbf{Q}(t) \tag{5.37}$$

where, $\mathbf{q}(t)$ is an $m \times 1$ vector of generalized coordinates, and $\mathbf{Q}(t)$ is the generalized force vector. The other matrices are of compatible dimensions. For a one-link manipulator with 'n' FEs, $m = 2n+1$. It is noted that the flexible manipulator has two types of motion: a rigid-body motion and an elastic motion. There is coupling between the motions.

The model in equation (5.37) is easily converted to the state-space form as

$$\begin{aligned} \dot{\mathbf{v}} &= \mathbf{A}\mathbf{v} + \mathbf{B}\mathbf{u} \\ \mathbf{y} &= \mathbf{C}\mathbf{v} \end{aligned} \tag{5.38}$$

where

$$\mathbf{A} = \begin{bmatrix} \mathbf{0}_m & \mathbf{I}_m \\ -\mathbf{M}^{-1}\mathbf{K} & -\mathbf{M}^{-1}\mathbf{D} \end{bmatrix}, \qquad \mathbf{B} = \begin{bmatrix} \mathbf{0}_{m\times 1} \\ \mathbf{M}^{-1}\mathbf{e}_1 \end{bmatrix}, \qquad \mathbf{C} = \begin{bmatrix} \mathbf{I}_m & \mathbf{0}_m \end{bmatrix}$$

The vector $\mathbf{e}_1$ is the first column of the identity matrix, $\mathbf{0}_{m\times 1}$ is an $m \times 1$ null vector, $\mathbf{v}^T = \begin{bmatrix} \mathbf{q}^T & \dot{\mathbf{q}}^T \end{bmatrix}$ and $\mathbf{u} = \tau$.

5.5   Validation of the FD and FE/Lagrangian methods

To validate the FD and FE/Lagrangian methods, the model of an experimental manipulator rig having a single flexible link is obtained and verified by simulation and actual experimentation.

5.5.1   The experimental manipulator system

The dynamic model of an experimental single-link flexible manipulator is obtained using FE/Lagrangian technique. The flexible arm is made of a thin aluminium alloy with length, $L = 0.9$ m, width $= 19.008$ mm, thickness $= 3.2004$ mm, $E = 7.1 \times 10^{10}$ N/m^2, $I = 5.1924 \times 10^{-11}$ m^4, $\rho = 2710$ kg/m^3 and hub inertia $= 5.8598 \times 10^{-11}$ kgm^2. The rig is equipped with a U9M4AT type printed circuit motor at the hub. It is chosen due to its low inertia and inductance, and due to its physical structure. The printed armature has a smooth torque output even at low speeds. Also, the torque-to-current relationship is linear due to the absence of magnetic material in the armature. A shaft encoder, a tachometer and an accelerometer are used to measure the hub-angle, hub-angular velocity and the end-point acceleration, respectively. The manipulator is constrained to horizontal plane motion. This is described in Chapter 1 and referred to as the Sheffield manipulator. Further information on the design and development of the experimental system is given in Tokhi and Azad (1997).

Substituting the link parameters in the appropriate equations, the matrices **M** and **K** in equation (5.37) are obtained. The first two resonance frequencies of the test-rig are determined experimentally as 12 and 35 Hz, respectively. The damping ratios of the system modes are typically in the range between 0.007 and 0.01 (Hastings and Book, 1987). In this work, these ratios are deduced as $\zeta_1 = 0.007$, and $\zeta_2 = 0.01$ for the first two modes respectively. An approximate damping matrix **D** for these modes can then be determined from (Chapnik *et al.*, 1991)

$$\mathbf{D}_{\omega\omega} = \gamma_1 \mathbf{M}_{\omega\omega} + \gamma_2 \mathbf{K}_{\omega\omega}$$

where

$$\gamma_1 = \frac{2\,\omega_1\omega_2(\zeta_1\omega_2 - \zeta_2\,\omega_1)}{\omega_2^2 - \omega_1^2}, \qquad \gamma_2 = \frac{2\,(\zeta_2\,\omega_2 - \zeta_1\,\omega_1)}{\omega_2^2 - \omega_1^2}$$

The state-space model of the test-rig is then obtained from equation (5.38).

5.5.2   Simulation and experiments

This section compares the results of FE and FD methods along with the responses obtained from the experimental system. All of these are for the single-link flexible manipulator system described above. For the FE method, the number of elements is varied from 1 to 20, while for the FD simulation the number of sections is varied from 3 to 20. For the latter case, with smaller number of sections the accuracy of results deteriorates significantly. Payloads of 0–60 per cent of the manipulator mass are considered. For both the simulation and experimental system, a single-switch bang-bang signal of amplitude ± 0.3 Nm, is used as the input torque applied at the hub so

Table 5.1   Relation between number of elements/sections, execution time and resonance frequencies of the flexible manipulator with FE/FD simulation

Number of elements/ sections	FE simulation				FD simulation			
	Execution time (s)	Resonance frequency (Hz)			Execution time (s)	Resonance frequency (Hz)		
		Mode 1	Mode 2	Mode 3		Mode 1	Mode 2	Mode 3
1	0.38	14.49	47.7	—	—	—	—	—
2	0.44	11.99	35.71	77.17	—	—	—	—
3	0.55	11.99	35.46	65.68	0.54	12.38	36.44	66.86
5	0.67	11.99	35.36	65.43	0.68	11.97	35.31	65.67
10	0.98	11.99	35.22	65.20	0.95	11.89	35.20	65.34
20	2.72	11.99	35.22	65.20	1.91	11.76	35.22	65.32

as to allow the manipulator to initially accelerate, then decelerate and eventually stop at a target location. Two system responses, namely, time domain response of the hub-angle and power spectral densities (PSDs) of end-point acceleration are evaluated. The responses are monitored for duration of 2.5 s. For the experimental system the results are recorded with a sampling time of 2 ms.

The relationship between the resonance frequencies and the number of elements for FE and number of sections for the FD method are shown in Table 5.1. In case of the FE method, increasing the number of elements increases directly the order of the submatrix $M_{\omega\omega}$, the number of elastic modes, and the execution time. While for the FD method, an increase in number of sections would increase the matrix size in equation (5.22) and also increase the execution time. The first three modes characterise the elastic behaviour reasonably well. It is noted that for FE simulation, a three-element model depicts approximately the resonance frequencies of the first three modes; a ten-element model is quite precise. In case of FD simulation, minimum three elements were required to obtain a reasonable result.

The response from the experimental system with various payloads is shown in Figure 5.4. Figure 5.4(a) shows the time response at the hub point of the manipulator, while Figure 5.4(b) shows the PSD of the end-point acceleration. In case of time response of the hub-angle, the total angle reached by the manipulator is less with higher payloads with same amount of torque input. This is to be expected since inertia makes a system sluggish. The payload also affected the resonant frequencies as it is observed at the end-point of the manipulator. The resonance mode frequencies appear to have slightly decreased with higher payloads.

The resonance frequencies obtained from the experimental system are compared with the simulated values in Table 5.2. It is noted that for FE simulation with increasing payloads, the resonance frequencies decrease. While for FD simulation, the change in resonance frequencies is not that prominent. However, the magnitude of the spectral power is slightly decreased with the increase in payload.

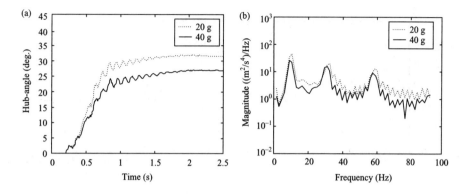

Figure 5.4   *Performance of the experimental system: (a) hub-angle profile, (b) PSD of end-point acceleration*

Table 5.2   *Relationship between payload and resonance frequencies*

Payload (g)	Mode 1 (Hz)			Mode 2 (Hz)			Mode 3 (Hz)		
	FE	FD	Rig	FE	FD	Rig	FE	FD	Rig
0	11.99	11.89	11.72	35.22	35.36	35.15	65.20	65.11	65.60
10	11.65	11.87	10.97	33.22	34.18	33.40	61.19	64.78	61.85
20	11.49	11.57	10.23	32.22	33.53	32.95	59.44	64.56	59.35
30	10.99	11.15	9.97	31.72	33.28	31.42	58.69	64.18	59.35
40	10.74	11.05	9.48	31.22	32.87	30.92	58.19	63.68	58.85
50	10.49	10.68	9.23	30.97	32.12	30.42	57.69	63.16	58.10
60	10.24	10.37	8.97	30.97	32.03	30.17	57.44	62.72	56.61

In case of the FE method the maximum error between experimental and simulation results in modes 2 and 3 is negligibly small, and is less than 2.65 per cent. The error is larger for mode 1, and increases from 6.19 to 14.15 per cent with payloads. Similarly the FD method shows maximum error for modes 2 and 3. In this case, the error increases with an increase of payload. Without any payload, the error is 0.59 and 0.6 per cent for modes 2 and 3 respectively. While for 60 per cent payload the error increases to 6.17 and 10.79 per cent for modes 2 and 3 respectively. Possible reasons may be due to: (a) payload rotary inertia, (b) the gravity effect, and (c) the influence of elastic motion on the hub-angle time constant, which have been ignored in the development of the simulation algorithms. Overall, the models characterise the manipulator system reasonably well. This conclusion is also confirmed by comparing Figure 5.4(b) and Figure 5.5, which show the simulated and experimental PSD of the end-point acceleration for two payloads for FE (Figure 5.5(a)) and FD (Figure 5.5(b)) methods.

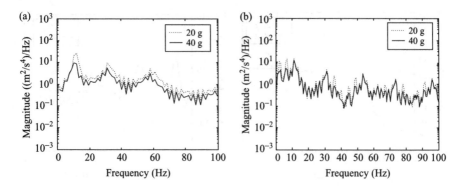

Figure 5.5   PSD of the end-point acceleration with payload: (a) FE simulated system, (b) FD simulated system

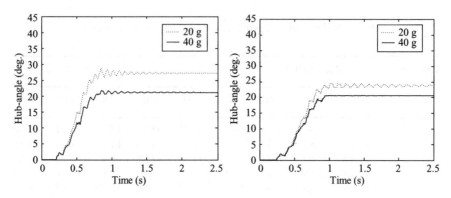

Figure 5.6   Time-domain hub-angle response with various payloads: (a) FE simulated system, (b) FD simulated system

Time-domain responses show the influence of approximations and non-linear terms in the model. These responses shown in Figure 5.4(a) and Figure 5.6 for the hub-angle further confirm the simulation models. The vibrations due to elastic motion are noticeable in the responses. It is noted that increasing the payload from 20 to 40 g reduces the simulated steady-state value of the hub-angle from 27° to 21° for FE simulation (Figure 5.6(a)) and from 24.5° to 21° for FD simulation (Figure 5.6(b)). While the actual experimental values reduce from 31° to 26°, respectively. The differences are within the 2–6° range. This is a small difference, which could be due to the ignored effects mentioned earlier. Overall, the models characterise the manipulator behaviour reasonably well with increasing payloads.

Such characteristic behaviour is important and essential for development of suitable control strategies for flexible manipulator systems. Though not given here, the simulated and experimental responses were observed to be better with no load; the responses were quite close. A steady-state level of 38° was achieved within 1.8 s in both cases. The hub-velocity, the end-point deflection and the end-point acceleration were also observed to converge to zero within the period. Further, the maximum end-point acceleration response was found as $230 \mathrm{m/s^2}$.

5.6 Summary

The modelling of a single-link flexible manipulator system using FD and FE methods has been presented. The results obtained from both methods have been validated by comparing with the response from an experimental system.

The FD method has been used to solve the governing PDE describing the characteristic behaviour of a flexible manipulator system incorporating the effects of hub inertia, payload and damping. It has been demonstrated that incorporating a payload at the end-point will reduce the angular displacement of the flexible manipulator. The level of vibration at the resonance modes varies with the load. Moreover, higher loads result in slightly lower resonance frequencies along with reduced level of vibration.

A dynamic model of an experimental manipulator rig including the effects of structural damping, hub inertia, and payload has been developed using the FE/Lagrangian technique. Experiments have been performed using the experimental rig and used for validation of the FE/Lagrangian model.

For both modelling approaches, a close agreement between simulation and experimental results within reasonable error margins has been observed. Thus, confidence in the accuracy of the model, for utilization in subsequent investigations and development of control strategies for flexible manipulator systems, has been established.

Chapter 6

Dynamic characterisation of flexible manipulators using symbolic manipulation

Z. Mohamed, M.O. Tokhi and H.R. Pota

This chapter presents the dynamic characterisation of flexible manipulators using a symbolic manipulation approach. Simulation algorithms of the system characterising the dynamic behaviour of the system with varying parameters are developed using a symbolic language. Analyses and investigations in terms of stability, time response and vibration frequencies of the system are presented. Moreover, effects of significant parameters on the behaviour of the flexible manipulator are studied. The performance of the approach is verified through numerical and experimental exercises.

6.1 Introduction

Various approaches have previously been devised for the modelling of flexible manipulators. These include the assumed modes, singular perturbation, frequency domain, finite difference (FD) and finite element (FE) methods to solve the partial differential equation (PDE) characterising the dynamic behaviour of a flexible manipulator system. However, in most cases, modelling and analysis of flexible manipulators are numerically based. Dynamic characteristics of the manipulator including stability, time response and vibration frequencies are interpreted on the basis of a single particular case, with no provision for generality. Moreover, numerical systems must operate using numeric approximations whose precision is limited by the computer hardware. Alternatively, exact quantities can be obtained by retaining the computations in a symbolic form. A distinguishing feature of symbolic-based methods is the mathematically comprehensive output they generate, so that the significance of individual terms, or group of terms, may be identified. This brings with it the opportunity to gain insights into the model that would otherwise not be available. A symbolic manipulation will open up the possibility of analysing a system in both new and interesting ways. It can be seen that the trend over time has been away from fully numeric methods of formulation towards those with a strong and total symbolic flavour to them. This is due to the overwhelmingly rapid improvements in computer hardware technology in general and in computer algebra software in particular (Larcombe and Brown, 1997).

Investigations have been carried out on symbolic approaches for modelling and simulation of flexible manipulators. Most of these investigations have developed automated symbolic derivations of dynamic equations of motion of rigid and flexible manipulators utilizing Lagrangian formulation and assumed-mode methods (Cetinkunt and Ittop, 1992; Lin and Lewis, 1994). In Centikunt and Ittop (1992), a method and a script in REDUCE is given to obtain symbolic dynamic models for robotic manipulators with flexible links. The kinematic relationships between frames attached to revolute joints are obtained using modal functions. The kinematic relationship first obtains the position of the end-point of a particular link with respect to its origin (using modal functions) and then the end-point orientation using the sum of differential rotations from the slopes of the link at the end-point in all three directions. The dynamic model is obtained by writing the system Lagrangian and then using the Euler–Lagrange equations to obtain the model. This process can be automated as evidenced by the REDUCE script.

In Pota and Alberts (1995) and Alberts *et al.* (1995), an infinite-dimensional model is developed using symbolic manipulation. The model is obtained for various boundary conditions. From this model, the effect of link parameters and different boundary conditions can be explicitly seen. Moreover, the order of approximation as a result of finite truncation can also be clearly obtained from the complete infinite-dimensional model. Alternative approaches to symbolic modelling include the use of Hamilton's principle and non-linear integro-differential equations (Low and Vidyasagar, 1988) and FD approximations (Tzes *et al.*, 1989). These have demonstrated that the approach has some advantages, such as allowing independent variation of flexure parameters. Moreover, relations between system parameters including payload and hub inertia and the system characteristics can be further investigated. The effect of payload on the manipulator is important for modelling and control purposes, as successful implementation of a flexible manipulator control is contingent upon achieving acceptable uniform performance in the presence of payload variations.

This chapter presents analysis of dynamics of flexible robot manipulators using symbolic manipulation. To reveal the advantages of a symbolic approach, two flexible manipulators with different actuators and sensors are considered. The dynamic models of the systems are developed using the FE method and infinite-dimensional transfer function based on a symbolic approach.

6.2   FE approach to symbolic modelling

This section focuses on the development of a symbolic algorithm in characterising the dynamic behaviour of a flexible manipulator system using the FE method. In this approach, all manipulations are carried out symbolically using Macsyma, a symbolic algebraic manipulation language. Two transfer functions, namely, from torque input to end-point displacement and from torque input to hub-angle of the manipulator are considered.

6.2.1 The flexible manipulator

A schematic diagram of the single-link flexible manipulator system, considered in this chapter, is shown in Figure 6.1, where X_0OY_0 and XOY represent the stationary and moving coordinate frames, respectively. The axis OX coincides with the neutral line of the link in its undeformed configuration, and is tangent to it at the clamped end in a deformed configuration. τ represents the applied torque at the hub. E, L, I, ρ, A, I_h and M_p represent the Young's modulus, length, area moment of inertia, mass density per unit volume, cross-sectional area, hub inertia and payload of the manipulator, respectively. $\theta(t)$ denotes angular displacement (hub-angle) of the manipulator and $w(x,t)$ denotes an elastic deflection (deformation) of a point along the manipulator at a distance x from the hub of the manipulator. In this work, the motion of the manipulator is confined to the X_0OY_0 plane. Since the manipulator is long and slender, the shear deformation and rotary inertia effects are neglected. This allows the use of the Bernoulli–Euler beam theory to model the elastic behaviour of the manipulator. The manipulator is assumed to be stiff in vertical bending and torsion, allowing it to vibrate dominantly in the horizontal direction and thus, gravity effects are neglected. Moreover, the manipulator is considered to have constant cross section and uniform material properties throughout.

6.2.2 Dynamic equation of motion

The total displacement $y(x,t)$ of a point along the manipulator at a distance x from the hub can be obtained as

$$y(x,t) = x\theta(t) + w(x,t)$$

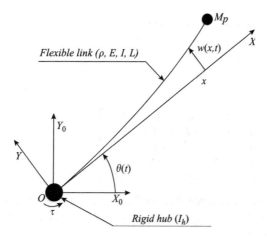

Figure 6.1 Schematic of the flexible manipulator system

Using the FE method to solve dynamic problems leads to the well-known equation

$$w(x,t) = \mathbf{N}_a(x)\,\mathbf{Q}_a(t)$$

where $\mathbf{N}_a(x)$ and $\mathbf{Q}_a(t)$ represent the shape function and nodal displacement, respectively. Hence, the displacement can be obtained as

$$y(x,t) = \mathbf{N}(x)\,\mathbf{Q}_b(t) \tag{6.1}$$

where

$$\mathbf{N}(x) = \begin{bmatrix} x & \mathbf{N}_a(x) \end{bmatrix} \quad \text{and} \quad \mathbf{Q}_b(t) = \begin{bmatrix} \theta(t) & \mathbf{Q}_a(t) \end{bmatrix}^{\mathrm{T}}$$

The shape function $\mathbf{N}(x)$ and nodal displacement vector $\mathbf{Q}_b(t)$ in equation (6.1) incorporate local and global variables. Among these, the angle $\theta(t)$ and the distance x are global variables while $\mathbf{N}_a(x)$ and $\mathbf{Q}_a(t)$ are local variables. Defining $k = x - \sum_{i=1}^{n-1} l_i$ as a local variable of the nth element, where l_i is the length of the ith element and utilizing Macsyma, the shape function can be expressed in symbolic form as

$$\mathbf{N}(k) = \left[k + l(n-1) \quad 1 - \frac{3k^2}{l^2} + \frac{2k^3}{l^3} \quad k - \frac{2k^2}{l} + \frac{k^3}{l^2} \quad \frac{3k^2}{l^2} - \frac{2k^3}{l^3} \quad -\frac{k^2}{l} + \frac{k^3}{l^2} \right]$$

Defining

$$\mathbf{M}_n = \int_0^l \rho A (\mathbf{N}^{\mathrm{T}} \mathbf{N})\, dk = \text{element mass matrix} \tag{6.2}$$

$$\mathbf{K}_n = \int_0^l EI\,(\mathbf{\Phi}^{\mathrm{T}} \mathbf{\Phi})\, dk = \text{element stiffness matrix} \tag{6.3}$$

where $\mathbf{\Phi} = \frac{d^2 \mathbf{N}(k)}{dk^2} = \begin{bmatrix} 0 & \frac{12k}{l^3} - \frac{6}{l^2} & \frac{6k}{l^2} - \frac{4}{l} & \frac{6}{l^2} - \frac{12k}{l^3} & \frac{6k}{l^2} - \frac{2}{l} \end{bmatrix}$, and solving equations (6.2) and (6.3) for n elements, the element mass and stiffness matrices can be obtained as

$$\mathbf{M}_n = \frac{\rho A l}{420} \begin{bmatrix} 140l^2(3n^2 - 3n + 1) & 21l(10n - 7) & 7l^2(5n - 3) & 21l(10n - 3) & -7l^2(5n - 2) \\ 21l(10n - 7) & 156 & 22l & 54 & -13l \\ 7l^2(5n - 3) & 22l & 4l^2 & 13l & -3l^2 \\ 21l(10n - 3) & 54 & 13l & 156 & -22l \\ -7l^2(5n - 2) & -13l & -3l^2 & -22l & 4l^2 \end{bmatrix}$$

$$\mathbf{K}_n = \frac{EI}{l^3} \begin{bmatrix} 0 & 0 & 0 & 0 & 0 \\ 0 & 12 & 6l & -12 & 6l \\ 0 & 6l & 4l^2 & -6l & 2l^2 \\ 0 & -12 & -6l & 12 & -6l \\ 0 & 6l & 2l^2 & -6l & 4l^2 \end{bmatrix}$$

The matrices above are assembled to obtain mass and stiffness matrices of the system, $\mathbf{M}$ and $\mathbf{K}$, and used in the Lagrange equation to obtain the dynamic equation of the flexible manipulator as

$$\mathbf{M}\ddot{\mathbf{Q}}(t) + \mathbf{K}\mathbf{Q}(t) = \mathbf{F}(t) \tag{6.4}$$

where $\mathbf{F}(t)$ is the vector of external torques and $\mathbf{Q}(t) = \begin{bmatrix} \theta & w_0 & \theta_0 & \cdots & w_\alpha & \theta_\alpha \end{bmatrix}^{\mathrm{T}}$, with w_α and θ_α representing the end-point deflection and rotation of the manipulator, respectively. Using a single element, $n = 1$, the dynamic equation of motion of the flexible manipulator can be obtained as in equation (6.4) with

$$\mathbf{M} = \frac{\rho A l}{420} \begin{bmatrix} 140l^2 & 63l & 14l^2 & 147l & -21l^2 \\ 63l & 156 & 22l & 54 & -13l \\ 14l^2 & 22l & 4l^2 & 13l & -3l^2 \\ 147l & 54 & 13l & 156 & -22l \\ -21l^2 & -13l & -3l^2 & -22l & 4l^2 \end{bmatrix}$$

$$\mathbf{K} = \frac{EI}{l^3} \begin{bmatrix} 0 & 0 & 0 & 0 & 0 \\ 0 & 12 & 6l & -12 & 6l \\ 0 & 6l & 4l^2 & -6l & 2l^2 \\ 0 & -12 & -6l & 12 & -6l \\ 0 & 6l & 2l^2 & -6l & 4l^2 \end{bmatrix}$$

$$\mathbf{Q}(t) = \begin{bmatrix} \theta & w_0 & \theta_0 & w_\alpha & \theta_\alpha \end{bmatrix}^{\mathrm{T}} \quad \text{and} \quad \mathbf{F}(t) = \begin{bmatrix} \tau & 0 & 0 & 0 & 0 \end{bmatrix}^{\mathrm{T}}$$

By incorporating the payload and hub inertia into the dynamic model of the system, for a single element, a new system mass matrix that incorporates the hub inertia and payload can be obtained as

$$\mathbf{M} = \frac{\rho A l}{420} \begin{bmatrix} 140l^2 + l^2 M_p + I_h & 63l & 14l^2 & 147l + l M_p & -21l^2 \\ 63l & 156 & 22l & 54 & -13l \\ 14l^2 & 22l & 4l^2 & 13l & -3l^2 \\ 147l + l M_p & 54 & 13l & 156 + M_p & -22l \\ -21l^2 & -13l & -3l^2 & -22l & 4l^2 \end{bmatrix}$$

For the manipulator considered as a clamped-free arm, with the applied torque τ at the hub, the flexural and angular displacements, velocities and accelerations are all zero at the hub at $t = 0$. Moreover, it is assumed here that $\mathbf{Q}(0) = 0$. Incorporating the initial conditions, with flexural and angular displacements at the hub as zero, the second and third rows and columns in $\mathbf{M}$, $\mathbf{K}$, $\mathbf{Q}$ and $\mathbf{F}$ are thus ignored. This yields

$$\mathbf{M} = \frac{\rho A l}{420} \begin{bmatrix} 140l^2 + l^2 M_p + I_h & 147l + l M_p & -21l^2 \\ 147l + l M_p & 156 + M_p & -22l \\ -21l^2 & -22l & 4l^2 \end{bmatrix},$$

$$\mathbf{K} = \frac{EI}{l^3} \begin{bmatrix} 0 & 0 & 0 \\ 0 & 12 & -6l \\ 0 & -6l & 4l^2 \end{bmatrix}$$

$$\mathbf{Q}(t) = \begin{bmatrix} \theta & w_\alpha & \theta_\alpha \end{bmatrix}^{\mathrm{T}} \quad \text{and} \quad \mathbf{F}(t) = \begin{bmatrix} \tau & 0 & 0 \end{bmatrix}^{\mathrm{T}}$$

6.2.3   Transfer functions

For control purposes, the matrix differential equation in equation (6.4) is represented in a state-space form as

$$\dot{v} = \mathbf{A}v + \mathbf{B}u$$
$$y = \mathbf{C}v$$

(6.5)

where

$$\mathbf{A} = \left[\begin{array}{c|c} 0_3 & \mathbf{I}_3 \\ \hline -M^{-1}K & 0_3 \end{array}\right], \qquad \mathbf{B} = \left[\begin{array}{c} 0_{3\times 1} \\ \hline M_1^{-1} \end{array}\right]$$

$u = [\tau]$ and the state vector $v = \begin{bmatrix} \theta & w_\alpha & \theta_\alpha & \dot{\theta} & \dot{w}_\alpha & \dot{\theta}_\alpha \end{bmatrix}^T$ incorporates the angular, end-point flexural and rotational displacements and velocities. 0_3 is a 3×3 null matrix, $\mathbf{I}_3$ is a 3×3 identity matrix, $0_{3 \times 1}$ is a 3×1 null vector and M_1^{-1} is the first column of M^{-1}. The output matrix $\mathbf{C}$ depends on desired transfer functions. For example, for torque input to end-point displacement output, $\mathbf{C} = \begin{bmatrix} L & 1 & 0 & 0 & 0 & 0 \end{bmatrix}$ while for torque input to hub-angle output, $\mathbf{C} = \begin{bmatrix} 1 & 0 & 0 & 0 & 0 & 0 \end{bmatrix}$. Using a single element, the transfer function from torque input to end-point displacement can be obtained as

$$G_1(s) = \cfrac{30\alpha^2 l^7 s^4 - 48\,600\alpha\beta\, l^4 s^2 + 4\,536\,000\beta^2 l}{\begin{array}{l} s^2[((15\alpha^2 l^8 + 3\,600\alpha\, l^6 I_h)M_p + \alpha^3 l^8 + 300\alpha^2\, l^6 I_h)s^4 \\ +((39\,600\alpha\beta\, l^5 + 1\,512\,000\beta\, l^3 I_h)M_p + 5\,220\alpha^2\beta\, l^5 + 367\,200\alpha\beta\, l^3 I_h)s^2 \\ +(4\,53\,6000\beta^2 l^2 M_p + 1\,512\,000\alpha\beta^2 l^2 + 4\,536\,000\beta^2 I_h)]\end{array}}$$

(6.6)

where $\alpha = \rho A l$ represents the weight, $\beta = EI$ represents the flexural rigidity of the manipulator and s is the Laplace variable. Similarly, the transfer function from torque input to hub-angle output of the manipulator can be obtained as

$$G_2(s) = \cfrac{\begin{array}{l}(3\,600\alpha\, l^6 M_p + 300\alpha^2 l^6)s^4 + (1\,512\,000\beta\, l^3 M_p + 367\,200\alpha\beta\, l^3)s^2 \\ +4\,536\,000\beta^2\end{array}}{\begin{array}{l} s^2[((15\alpha^2 l^8 + 3\,600\alpha\, l^6 I_h)M_p + \alpha^3 l^8 + 300\alpha^2\, l^6 I_h)s^4 + \\ ((39\,600\alpha\beta\, l^5 + 1\,512\,000\beta\, l^3 I_h)M_p + 5\,220\alpha^2\beta\, l^5 + 367\,200\alpha\beta\, l^3 I_h)s^2 \\ +(4\,536\,000\beta^2 l^2 M_p + 1\,512\,000\alpha\beta^2 l^2 + 4536\,000\beta^2 I_h)]\end{array}}$$

(6.7)

6.2.4   Analysis

In this section, the transfer functions obtained in the previous section are analysed and assessed in the dynamic characterisation of the flexible manipulator system. First, a system without payload and hub inertia is considered. Then effects of payload on the dynamic behaviour of the system are examined. This involves obtaining and investigating the system characteristics including poles, zeros, stability, vibration frequencies and time response to an input command. Relationships between the physical parameters and the system characteristics are then investigated.

6.2.4.1 System without payload and hub inertia

For a flexible manipulator without payload and hub inertia, the system transfer functions can be obtained by solving equations (6.6) and (6.7) with $M_p = 0$ and $I_h = 0$. Thus, the transfer function from torque input to end-point displacement can be obtained as

$$G_{1a}(s) = \frac{30\alpha^2 l^7 s^4 - 48\,600\alpha\beta\, l^4 s^2 + 4\,536\,000\beta^2 l}{s^2(\alpha^3 l^8 s^4 + 5\,220\alpha^2\beta\, l^5 s^2 + 1\,512\,000\alpha\beta^2 l^2)} \tag{6.8}$$

and the transfer function from torque input to hub-angle output can be obtained as

$$G_{2a}(s) = \frac{300\alpha^2 l^6 s^4 + 367\,200\alpha\beta l^3 s^2 + 4\,536\,000\beta^2}{s^2(\alpha^3 l^8 s^4 + 5\,220\alpha^2\beta\, l^5 s^2 + 1\,512\,000\alpha\beta^2 l^2)} \tag{6.9}$$

Factoring the denominator polynomial of equation (6.8) yields system poles as

$$p = \pm j17.5444\sqrt{\tfrac{\beta}{\alpha\, l^3}}, \quad \pm j70.087\sqrt{\tfrac{\beta}{\alpha\, l^3}}, \quad 0, \ 0. \tag{6.10}$$

The poles on the imaginary axis give the system natural frequencies. These, in turn, determine vibration modes of the system. Evaluating equation (6.10) yields vibration frequencies at modes 1 and 2 as $(17.5444/2\pi)\sqrt{(\beta/\alpha\, l^3)}$ Hz and $(70.087/2\pi)\sqrt{(\beta/\alpha\, l^3)}$ Hz, respectively. These demonstrate that the frequencies depend on α, β and l. Figure 6.2 shows the effect of flexural rigidity, β, of a manipulator on the vibration frequencies of modes 1 and 2 with a constant weight of the manipulator. In this case, $\alpha = 0.15$kg is used. Practically, changing the value of β implies changing the manipulator material. Similarly, the interrelation between α and the vibration frequencies of the system with a constant flexural rigidity can be investigated. The result with $\beta = 3.69$ Nm^2 is shown in Figure 6.3. It is also important to investigate, for the same material, the effect of changing the manipulator length

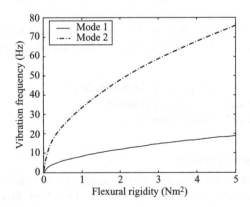

Figure 6.2   Effect of flexural rigidity on vibration frequency ($\alpha = 0.15$ kg)

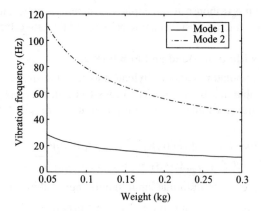

Figure 6.3   Effect of the weight of a manipulator on vibration frequency ($\beta = 3.69 \ Nm^2$)

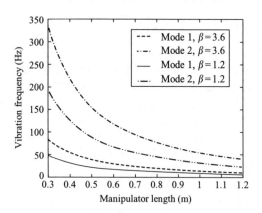

Figure 6.4   Effect of length of manipulator on vibration frequency

on vibration frequencies. Figure 6.4 shows the result with two different materials, $\beta = 1.2$ and $3.6 \ Nm^2$. It follows from these results that system vibration frequencies can be increased by

- increasing the flexural rigidity of the manipulator
- reducing the weight of the manipulator or
- reducing the length of the manipulator.

6.2.4.2   System with payload

It is noted from the transfer function $G_1(s)$ in equation (6.6) that the payload terms are not in the numerator of the transfer function. Therefore, they do not affect the system zeros. The zeros will determine whether the system exhibits minimum phase or non-minimum phase behaviour and will determine the magnitude of response of the system. The system poles, on the other hand, are affected by the payload, see equation (6.6). Conversely, for the transfer function $G_2(s)$ in equation (6.7), it is noted that the zeros are affected by the payload. Furthermore, it is noted that the flexible manipulator is a type two system, which implies that zero steady-state error can only be achieved using step and ramp command inputs to the system.

Equating the numerator of the transfer function $G_1(s)$ to zero and solving yields the zeros as

$$z = \pm 38.9944 \sqrt{\frac{\beta}{\alpha l^3}}, \quad \pm 9.9718 \sqrt{\frac{\beta}{\alpha l^3}}$$

It is noted that, with any α, β and l values, two zeros lie on the right half of s-plane (RHP) and the others on the left half of s-plane (LHP). Thus, the system model is non-minimum phase and undershoot is expected at the start in the end-point displacement response. This agrees, with the result reported earlier in respect of a system incorporating non-collocated sensors and actuators (Tokhi *et al.*, 1997).

To investigate the effect of payload on the dynamic behaviour of the system, the transfer functions $G_1(s)$ and $G_2(s)$ were solved with a system constituting a flexible arm of dimensions $900 \times 3.2004 \times 19.008 \text{mm}^3$, $E = 71 \times 10^9 \text{N/m}^2$, $I = 5.1924 \times 10^{-11} \text{m}^4$, $\rho = 2710 \text{kg/m}^3$ and $I_h = 5.8598 \times 10^{-4} \text{kgm}^2$. These parameters correspond to those of the Sheffield flexible manipulator experimental rig introduced in Chapter 1. Thus, the transfer function from the torque input to end-point displacement can be obtained as

$$G_{1b}(s) = \frac{0.32s^4 - 17669.9s^2 + 5.69 \times 10^7}{(0.3M_p + 0.0035)s^6 + (15370.2M_p + 340.36)s^4 + (5.13 \times 10^7 M_p + 2571860)s^2}$$

$$(6.11)$$

and the transfer function from torque input to hub-angle can be obtained as

$$G_{2b}(s) = \frac{(283.86M_p + 3.51)s^4 + (4116760M_p + 148339.8)s^2 + 6.33 \times 10^7}{(0.3M_p + 0.0035)s^6 + (15370.2M_p + 340.36)s^4 + (5.13 \times 10^7 M_p + 2571860)s^2}$$

$$(6.12)$$

Factorising the denominator of the system transfer functions $G_{1b}(s)$ and $G_{2b}(s)$, the system poles in terms of payload can be obtained as

$$p_{1,2} = \pm 6.09 \sqrt{\frac{6.47h_1 - (351.5M_p + 7.8)}{0.5M_p + 0.006}}$$

$$p_{3,4} = \pm 6.09 \sqrt{\frac{-6.47h_1 - (351.5M_p + 7.8)}{0.5M_p + 0.006}}$$

$$p_5 = 0, p_6 = 0$$

$$(6.13)$$

where $h_1 = \sqrt{2161.6M_p^2 + 82.2M_p + 1}$. Note that for a single element, the system has six poles, two of which are at the origin. Since for $M_p \geq 0$, the terms under the square roots in equation (6.13) are negative, the remaining poles are purely imaginary

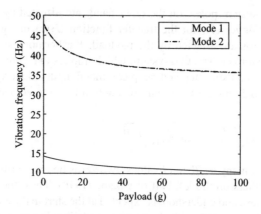

Figure 6.5   Effect of payload on the vibration frequency of the system

and lie on the imaginary axis of the s-plane. These result in, as expected for a system without damping, a marginally stable system.

The system poles give the system vibration frequencies. These, in turn, determine vibration modes of the system, and thus the effect of payload on the vibration frequency can be investigated by solving equation (6.13). Figure 6.5 shows the relation, thus obtained, between payload and system vibration frequencies for modes 1 and 2. It is noted that with increasing payload, the vibration frequencies decrease significantly.

For the transfer function $G_2(s)$, factorising the numerator yields the zeros as

$$z_{1,2} = \frac{\pm 3.5}{l} \sqrt{\frac{3.4\beta h_2 - (210\beta M_p + 51\alpha\beta)}{12\alpha l M_p + \alpha^2 l}}$$

$$z_3 = \frac{-3.8}{l} \sqrt{\frac{12.9 h_3 - (784.3\, M_p + 28.2672)}{1.6 M_p + 0.02}} \tag{6.14}$$

$$z_4 = \frac{-3.8}{l} \sqrt{\frac{-12.9 h_3 - (784.3\, M_p + 28.2672)}{1.6 M_p + 0.02}}$$

where $h_2 = \sqrt{3675 M_p^2 + 1680\alpha\, M_p + 208\alpha^2}$ and $h_3 = \sqrt{3675 M_p^2 + 249.3\alpha\, M_p + 4.6\alpha^2}$.

Similarly, since for $M_p \geq 0$ and $0 < \alpha \leq 1$, the terms under square roots in equation (6.14) are negative, all system zeros lie on the imaginary axis. Thus, as expected of a system with collocated sensors and actuator, the transfer function from torque input to hub-angle response exhibits minimum phase behaviour. Figure 6.6 shows the relation between the payload and the system zeros thus obtained. It is noted that the zeros move towards the origin of the s-plane with increasing payload.

Since control of a non-minimum phase system is rather involved, this aspect is further analysed in this section. For the transfer function from the torque input to end-point displacement that exhibits a non-minimum phase characteristic, it is important

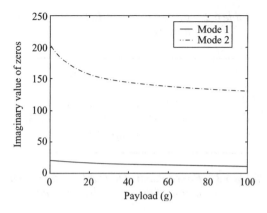

Figure 6.6   Effect of payload on zeros of hub-angle response of the manipulator

Table 6.1   First column of RH table for the numerator of G_1 (s)

s^4	$30\alpha^2 l^7$
s^3	$120\alpha^2 l^7$
s^2	$-24\,300\alpha\,\beta\,l^4$
s^1	$-74\,800\alpha\,\beta\,l^4$
s^0	$4\,536\,000\,\beta^2\,l$

to investigate whether the zeros can be relocated to the LHP by altering the physical parameters of the system. If so, in designing a flexible manipulator, certain parameter values can be considered to make the system minimum phase. To allow this, an analysis using the Routh–Hurtwiz (RH) criterion is carried out. Accordingly, if there is no sign change in the first column of the RH table, then all roots of the polynomial will be on the LHP. Utilizing the RH criterion, the first column of RH table for numerator of $G_1(s)$ can be obtained as shown in Table 6.1. It is noted that there are two sign changes, namely from s^3 to s^2 and s^1 to s^0, indicating that there are two zeros of the transfer function on the RHP. Moreover, the result shows that, since all terms are single, the zeros cannot be relocated by altering any physical parameter value.

For the transfer function $G_2(s)$, the effect of payload on the location of zeros can be investigated in the same manner as above. Table 6.2 shows the first column of RH table for the numerator of $G_2(s)$. Since the numerator is an even polynomial, a row of zeros exists at s^3. Thus, if there is no sign change, all zeros of the transfer function lie on the imaginary axis. It is noted, as all the physical parameters, including the payload are positive, no sign change occurs in the first column of the RH table. This implies that all the system zeros lie on the imaginary axis and the system is minimum phase. Similarly, the effect of payload on the system poles and stability can be studied. Furthermore, a range of payload that ensures system stability could be

Table 6.2   First column of RH table for the numerator of G_2 (s)

s^4	$30\alpha\, l^6 (12 M_p + \alpha)$
s^3	$120\alpha\, l^6 (12 M_p + \alpha)$
s^2	$10\,800\beta\, l^3 (17\alpha + 70 M_p)$
s^1	$57\,600\beta\, l^3 \left(17\alpha + 70 M_p\right)/(17\alpha + 70 M_p)$
s^0	$4\,536\,000\, \beta^2$

Table 6.3   First column of RH table for the denominator of the system transfer function

s^6	$0.3 M_p + 0.0035$
s^5	$1.9 M_p + 0.0208$
s^4	$5\,123.4 M_p + 113.5$
s^3	$\dfrac{7.6\times10^8 M_p^2 + 3.0\times10^7 M_p + 356\,584.8}{15\,370.2 M_p + 340.4}$
s^2	$\dfrac{1.4\times10^{27} M_p^3 + 1.3\times10^{26} M_p^2 + 3.4\times10^{24} M_p + 3.3\times10^{22}}{6.1\times10^{19} M_p^2 + 2.4\times10^{18} M_p + 2.8\times10^{16}}$
s^1	$1.0 \times 10^8 M_p + 5\,143\,720$
s^0	0

determined. Table 6.3 shows the first column of RH table for the denominator of the transfer function, equation (6.14). Again, as the denominator is an even polynomial, a row of zeros exists at s^5. It is noted that, no sign change occurs. This implies that all poles of the system lie on the imaginary axis and the system is marginally stable.

The effect of payload on the dynamic behaviour of the system is further analysed by obtaining the time responses of the end-point displacement and hub-angle of the manipulator. Both transfer functions $G_1 (s)$ and $G_2 (s)$ are considered. For this purpose, a single-switch bang-bang torque input is applied at the hub of the manipulator. Multiplying the system transfer functions in equations (6.6) and (6.7) with the input torque and utilizing the inverse Laplace transform yield the time-domain expressions of the end-point displacement and hub-angle responses of the system.

To demonstrate the performance of the developed symbolic algorithm, simulated exercises with the flexible manipulator system were carried out. In these exercises, a bang-bang input torque, as shown in Figure 6.7, with amplitude of ±0.3 Nm was used. The system responses at the end-point and hub-angle were obtained over a period of 3 s. Figures 6.8 and 6.9 show the simulated responses of the end-point displacement and hub-angle of the manipulator with a payload of 20 g, respectively. It is noted that the responses of the system reached steady-state values of 0.43 m and 27° within 0.9 s with persistent oscillation. The end-point response shows that the system is non-minimum phase, as the response slightly undershoots at start up. This agrees with the symbolic results that were presented and discussed earlier.

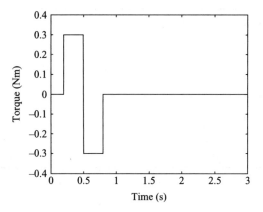

Figure 6.7   Single-switch bang-bang input torque

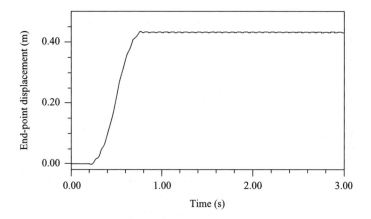

Figure 6.8   Simulated end-point displacement response of the flexible manipulator
($M_p = 20$ g)

To investigate the effect of payload on the system response, steady-state values of the system responses were monitored with various payloads. Based on the assumption that the manipulator reaches the steady state after 4 s, the relations between payload and steady state values of end-point displacement and hub-angle are shown in Figures 6.10 and 6.11, respectively. In both cases, the output levels decrease significantly with increasing payload. The results demonstrate that controllers that are capable of adapting with changing system characteristics have to be developed.

6.2.5   Validation and performance analysis

To validate the developed symbolic model of the flexible manipulator for use in simulation and control, experimental investigations using the Sheffield manipulator

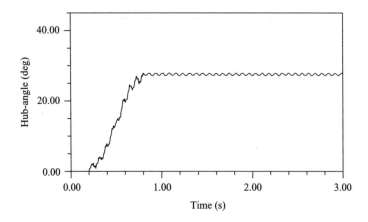

Figure 6.9   Simulated hub-angle response of the flexible manipulator ($M_p = 20$ g)

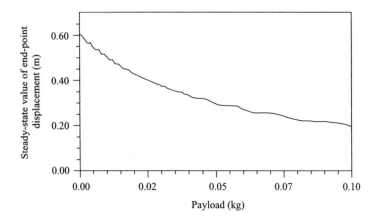

Figure 6.10   Effect of payload on the steady-state value of end-point displacement ($t = 4$ s)

experimental rig are carried out and the results are presented in this section. Accordingly, performance of the symbolic manipulation approach is assessed by comparing the symbolic, simulation and experimental results. The flexible manipulator used is described in Chapter 1. In the experiments, a bang-bang input torque as shown in Figure 6.7 was used. The hub-angle was measured and the corresponding power spectral density was obtained. To investigate the effect of payload on the performance of the manipulator, experiments were performed using various payloads.

   Figure 6.12 shows the experimental hub-angle response of the flexible manipulator without payload and with a payload of 20 g. In the former, it is noted for the hub-angle that the steady-state level of 38° was reached within 1.8 s. The first three modes of vibration were located at 11.72, 35.15 and 65.60 Hz. Furthermore, it is noted that the hub-angle response is minimum phase as proved in the analysis

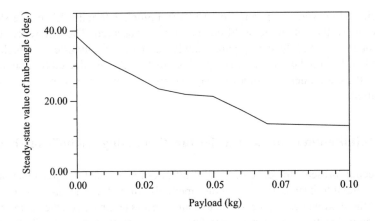

Figure 6.11   Effect of payload on the steady-state value of hub-angle (t = 4 s)

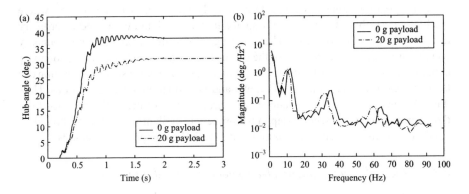

Figure 6.12   Hub-angle response of the flexible manipulator experimental rig: (a) time domain, (b) spectral density

using symbolic manipulation. However, with a single element in the FE method, the symbolic approach gave the first two modes of vibration frequencies as shown in Figure 6.5 as 14 and 47.5 Hz. More accurate results can be obtained with increasing number of elements (Tokhi *et al.*, 1997).

Comparison of Figures 6.9 and 6.12, for the manipulator, with a 20 g payload, shows that a reasonably close agreement between symbolic and experimental results in the time response was reached. It is noted from the experimental results that the steady-state level of hub-angle response decreases with increasing payloads. For a 20 g payload, the steady-state hub-angle level reached was 31°. Analysing the power spectral density, it is also noted that the resonance frequencies of the system decrease with increasing payloads. Using the symbolic approach, similar results were obtained as shown in Figures 6.5 and 6.11. This validates the symbolic model in characterising the dynamic behaviour of a flexible manipulator for development of suitable control strategies. However, comparing the results in Figures 6.9 and 6.12 for the steady-state value of the hub-angle, a difference of 4° is observed. The difference, which is considered negligibly small, could mainly be due to gravity effects, which were ignored in the simulation, whereas a payload that might be affected by gravity was used in the experiments. Moreover, payload rotary inertia was also ignored in the simulation.

6.3   Infinite-dimensional transfer functions using symbolic methods

This section presents the development of infinite-dimensional transfer functions for slewing flexible beams using symbolic methods. In addition to an actuator at the base of the flexible beam, piezo-actuators and sensors laminated on the beam are also considered. The transfer functions are derived from fourth-order linear PDEs, which describe flexible beams. General boundary conditions are assumed in the derivation of the transfer functions.

There are two main steps in this derivation. The first step is a standard one, where the PDEs describing the beam are Laplace transformed and converted to ordinary differential equations (ODEs). The resulting ODEs are linear but with boundary conditions specified at either end of the beam. The need to satisfy two-point boundary conditions makes this a difficult problem. Since this is a linear system, the solution can be analytically written in terms of known and unknown boundary conditions. Here this solution is obtained using state-space representation of the ODEs obtained from the original PDEs. The second step in the process is of obtaining unknown boundary conditions using the analytical solution with symbolic computation methods.

6.3.1   Piezoelectric laminate electromechanical relationships

In this section, a dynamic model is obtained for piezoelectric actuators bonded to flexible structures. These models are used in obtaining transfer functions between piezo-actuators and slewing beam deflections. Figure 6.13 illustrates a typical section of the beam/piezoelectric layer laminate. The subscripts *a*, *b* and *c* correspond to the top piezoelectric layer, the base structure and the lower piezoelectric layer, respectively.

In the present development, the *a* layer will serve as actuator and the *c* layer as sensor. The actuating and sensing layers are assumed to be thin relative to the

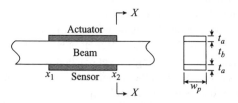

Figure 6.13   Schematic diagram of piezoelectric laminate

base structure, and perfectly bonded to its surface. For thin piezoelectric layers it is reasonable to assume uniform extensional strain in these layers. It is shown in Crawley and Anderson (1990) that this assumption produces good results when $t_b/t_a > 7$. According to the uniform strain model, a voltage V_a applied across the actuating layer a induces a longitudinal stress σ_a given by

$$\sigma_a(x,t) = E_a \frac{d_{31}}{t_a} V_a(x,t) \tag{6.15}$$

where d_{31} is the electric charge constant of the piezoelectric material (m/V), and E_a its Young's modulus (N/m^2). This stress in turn generates a bending moment M_a about the composite system's neutral axis, along with an axial force that does not influence transverse vibrations. The moment is given by (Baz and Poh, 1988)

$$M_a(x,t) = \int_{(t_a/2)}^{(t_b/2)+t_c} \sigma_a w_p y \, dy = \frac{1}{2} E_a d_{31} w_p (t_a + t_b) V_a(x,t) = C_a V_a(x,t) \tag{6.16}$$

where $C_a = 0.5 E_a d_{31} w_p (t_a + t_b)$ is a geometric constant, the form of which is determined by the geometry of the composite beam–actuator–sensor system. The constant changes if, for example, both layers a and c are used as actuators with opposite polarity, in which case $C_a = E_a d_{31} w_p (t_a + t_b)$. It is also worth noting that the expression for C_a is significantly simplified by the fact that the laminated beam's neutral axis coincides with its centre due to the symmetry of the added piezoelectric elements.

Note that the strain ε_c is related to the curvature of the beam as (Baz and Poh, 1988)

$$\varepsilon_c(x,t) = -\left(\frac{t_b}{2} + t_a\right) \frac{\partial^2 y}{\partial x^2} \tag{6.17}$$

Owing to the piezoelectric effect, this strain gives rise to a charge distribution $q(x,t)$, which is given by Alberts and Colvin (1991) as

$$q(x,t) = \left(\frac{k_{31}^2}{g_{31}}\right) \varepsilon_c(x,t) w_p \tag{6.18}$$

where k_{31} is the piezoelectric electromagnetic coupling constant and g_{31} is the piezoelectric stress constant. The total charge developed on the sensing layer is obtained by integrating $q(x,t)$ over the entire length of the piezoelectric element;

$$Q(t) = \int_{x_1}^{x_2} q(x,t) dx = -w_p \left(\frac{t_b}{2} + t_a\right) \frac{k_{31}^2}{g_{31}} \frac{\partial y(x,t)}{\partial x} \bigg|_{x_1}^{x_2} \tag{6.19}$$

The piezoelectric material is similar to an electric capacitor and the voltage across the two layers can thus be written as

$$V_s(t) = \frac{Q(t)}{C w_p (x_2 - x_1)} = C_s \frac{\partial y(x,t)}{\partial x} \bigg|_{x_1}^{x_2} \tag{6.20}$$

where C_s is a constant incorporating various structural and piezoelectric constants in equations (6.19) and (6.20), C is the capacitance per unit area, and $w_p(x_2 - x_1)$ is the surface area of the piezoelectric element. In the next section a dynamic modelling method for piezoelectric laminate beams is presented using the relationships given in this section.

6.3.2   *Dynamic modelling*

Figure 6.14 illustrates the configuration of the system under consideration. It is regarded as generic model representing slewing flexible beams with piezoelectric actuators and sensors. The beam is mounted on rotary base, actuated by a d.c. torque motor, and rotating in the horizontal plane. The piezoelectric patches are attached to both sides of the beam. In this development, one side acts as an actuator and the other acts as a sensor. Thus, the system inputs consist of voltage V_a applied to the actuating layer and the torque τ applied by the torque motor. Outputs include the voltage V_s sensed by the sensing layer, motor hub-angle θ and beam end-point position Y_{tip}.

With the usual assumptions for technical beam theory, the differential equation of motion for the composite beam–actuator–sensor system can be expressed as a Bernoulli–Euler beam equation with an additional term due to the actuating layer (Bailey and Hubbard, 1985)

$$\frac{\partial^2}{\partial x^2}\left[EI\frac{\partial^2 y(x,t)}{\partial x^2} - C_aV_a(x,t)\right] + \rho A\frac{\partial^2 y(x,t)}{\partial t^2} = 0 \tag{6.21}$$

where E and I represent the effective Young's modulus and area moment of inertia for the composite system, respectively. Notice that if the piezoelectric laminates do not cover the entire beam surface then both EI and ρA are functions of x. In the laminated portions of the beam, the product EI is given by $EI = E_aI_a + E_bI_b + E_cI_c$, and the product of the effective mass density ρ with the cross-sectional area A of the composite system is $\rho A = \rho_aA_a + \rho_bA_b + \rho_cA_c$. Since the piezoelectric layers are often thin, relative to the beam structure, the present analysis is simplified by assuming that EI and ρA are uniform over the length of the beam. The boundary conditions are

$$y(0,t) = 0 \tag{6.22}$$

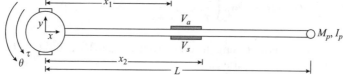

Figure 6.14   Schematic diagram of the slewing beam

$$EI\frac{\partial^2 y\,(0,t)}{\partial x^2} - I_h\frac{\partial^3 y\,(0,t)}{\partial t^2 \partial x} + \tau\,(t) = 0 \tag{6.23}$$

$$EI\frac{\partial^2 y\,(L,t)}{\partial x^2} + I_h\frac{\partial^3 y\,(L,t)}{\partial t^2 \partial x} = 0 \tag{6.24}$$

$$EI\frac{\partial^3 x\,(L,t)}{\partial y^3} - M_p\frac{\partial^2 y\,(L,t)}{\partial t^2} = 0 \tag{6.25}$$

where equation (6.22) represents the inability for the pinned joint to undergo transverse displacement, equation (6.23) includes the torque input $\tau\,(t)$ and hub inertia I_h at the pinned end, equation (6.24) accounts for the end-point inertia I_p and equation (6.25) accounts for the end-point mass M_p (Schmitz, 1985). To allow conversion to transfer functions, the differential equation in equation (6.21) and boundary conditions in equations (6.22)–(6.25) are Laplace transformed. Letting $\varphi^4 = -\rho A s^2/EI$, equation (6.21) becomes

$$Y''''\,(x,s) - \varphi^4 Y\,(x,s) = \frac{C_a V_a''\,(x,s)}{EI} \tag{6.26}$$

with boundary conditions

$$Y\,(0,s) = 0$$

$$EIY''\,(0,s) - I_h s^2 Y'\,(0,s) + \tau\,(s) = 0$$

$$EIY''\,(L,s) - I_p s^2 Y'\,(L,s) = 0 \tag{6.27}$$

$$EIY'''\,(L,s) - M_p s^2 Y\,(L,s) = 0$$

where the primes indicate spatial derivatives. Equations (6.26) and (6.27) together constitute a linear ODE in x with mixed boundary conditions, two at $x = 0$ and two at $x = L$. The differential equation in equation (6.26) is a Laplace transformed version of the classical Bernoulli–Euler beam equation with the second spatial derivative of $V_a\,(x,s)$ as a force input. Note that the actuating voltage V_a is constant in the interval $x_1 < x < x_2$ but undergoes a step change at each of the boundaries of this interval. The second spatial derivative of the actuating voltage function yields

$$V_a''\,(x,s) = \left[\delta'\,(x - x_1) - \delta'\,(x - x_2)\right] V_a\,(s) \tag{6.28}$$

where $\delta'\,(\cdot)$, the spatial derivative of the Dirac delta, represents the unit dipole function. It can be seen that when the spatial distribution of the actuating layer is uniform with respect to x, then the voltage term has a non-zero contribution only at the boundaries of the actuating layers, x_1 and x_2. For a system with P piezoelectric actuator–sensor pairs, a state-space type representation can be formed by putting equations (6.26) and (6.27) together as

$$
\begin{bmatrix} Y'(x,s) \\ Y''(x,s) \\ Y'''(x,s) \\ Y''''(x,s) \end{bmatrix} = \begin{bmatrix} 0 & 1 & 0 & 0 \\ 0 & 0 & 1 & 0 \\ 0 & 0 & 0 & 1 \\ \varphi^4 & 0 & 0 & 0 \end{bmatrix} \begin{bmatrix} Y(x,s) \\ Y'(x,s) \\ Y''(x,s) \\ Y'''(x,s) \end{bmatrix}
$$

$$
+ \begin{bmatrix} 0 \\ 0 \\ 0 \\ 1 \end{bmatrix} \frac{C_a V_a(s)}{EI} \sum_{i=1}^{2P} \delta'(x-x_i)(-1)^{i+1} \tag{6.29}
$$

Defining

$$
Z(x,s) = \begin{bmatrix} Y(x,s) & Y'(x,s) & Y''(x,s) & Y'''(x,s) \end{bmatrix}^{\mathrm{T}},
$$

$$
U(x,s) = \frac{C_a V_a(s)}{EI} \sum_{i=1}^{2P} \delta'(x-x_i)(-1)^{i+1},
$$

and

$$
\mathbf{A} = \begin{bmatrix} 0 & 1 & 0 & 0 \\ 0 & 0 & 1 & 0 \\ 0 & 0 & 0 & 1 \\ \varphi^4 & 0 & 0 & 0 \end{bmatrix}, \quad \mathbf{B} = \begin{bmatrix} 0 \\ 0 \\ 0 \\ 1 \end{bmatrix}
$$

Equation (6.29) can be written in compact form as

$$
Z'(x,s) = \mathbf{A}Z(x,s) + \mathbf{B}U(x,s) \tag{6.30}
$$

In the literature, state-space equations such as equation (6.29) have time t as the independent variable. The system under consideration has time t and linear distance x as its two independent variables. As a first step in getting the solution to the original problem, the differential equation in equation (6.21) and boundary conditions in equations (6.22)–(6.25) are Laplace transformed. The next step is to write the resulting equations (6.26) and (6.27) in a state-space form with the linear distance x as the independent variable. The resulting equation (6.29) does appear in a conventional state-space form, but in principle it is no different from the standard state-space representation (Friedland, 1986). Then using the fact that $\int_{-\infty}^{\infty} \delta^n(x)\phi(x)\,dx = (-1)^n \phi^n(0)$ (Papoulis, 1962), equation (6.30) can be expressed for $P = 1$, (i.e. only one piezoelectric actuator–sensor pair) as

$$
Z(L,s) = e^{\mathbf{A}L}Z(0,s) + [\mathbf{A}e^{\mathbf{A}(L-x_1)}\mathbf{B} - \mathbf{A}e^{\mathbf{A}(L-x_2)}\mathbf{B}]\frac{C_a V_a(s)}{EI} \tag{6.31}
$$

Combining equation (6.31) with the boundary conditions in equation (6.27) yields eight linear algebraic equations in eight unknowns, which can be solved simultaneously for elements of $Z(0,s)$ and $Z(L,s)$. Once the boundary conditions $Z(0,s)$ and

$Z(L, s)$ are known, a general expression for $Z(x, s)$ can be written as

$$Z(x, s) = \begin{cases} e^{Ax}Z(0, s) & 0 \le x \le x_1 \\ e^{Ax}Z(0, s) + Ae^{A(x-x_1)}\mathbf{B}\frac{C_aV_a(s)}{EI} & x_1 \le x \le x_2 \\ e^{Ax}Z(0, s) + A[e^{A(x-x_1)} - e^{A(x-x_2)}]\mathbf{B}\frac{C_aV_a(s)}{EI} & x_2 \le x \le L \end{cases}$$

(6.32)

The matrix exponential function e^{Ax} corresponding to equation (6.29) can be evaluated using inverse Laplace transform of the matrix $(s\mathbf{I} - A)^{-1}$;

$$e^{Ax} = \begin{bmatrix} f'''(x) & f''(x) & f'(x) & f(x) \\ \varphi^4 f(x) & f'''(x) & f''(x) & f'(x) \\ \varphi^4 f'(x) & \varphi^4 f(x) & f'''(x) & f''(x) \\ \varphi^4 f''(x) & \varphi^4 f'(x) & \varphi^4 f(x) & f'''(x) \end{bmatrix}$$

(6.33)

where $f(x) = \frac{1}{2\varphi^3}[\sinh(\varphi x) - \sin(\varphi x)]$.

The solution expressed by equation (6.32) has been derived for a single piezo-electric actuating element, however, because of linearity of the system, superposition allows the same analysis to be applied to systems with any number of piezoelectric elements. In the analysis, which follows, transcendental expressions are developed for the transfer functions of interest. To reduce the complexity of the expressions, the end-point mass M_p and inertia I_p are set to zero.

6.3.3   Transfer functions

The system under consideration has two discrete inputs, V_a and τ, and three outputs, Y_{tip}, θ and V_s. Thus, the transfer matrix can be expressed as

$$\begin{bmatrix} \theta(s) \\ V_s(s) \\ Y_{tip}(s) \end{bmatrix} = \begin{bmatrix} G_{\theta,V_a}(s) & G_{\theta,\tau}(s) \\ G_{V_s,V_a}(s) & G_{V_s,\tau}(s) \\ G_{y,V_a}(s) & G_{y,\tau}(s) \end{bmatrix} \begin{bmatrix} V_a(s) \\ \tau(s) \end{bmatrix}$$

(6.34)

The subscripts p and q of each element $G_{p,q}(s) = N_{p,q}(s)/D(s)$ identify the associated output and input variables, respectively; subscript y is used to denote the absolute end-point position. For example, $G_{y,\tau}(s)$ is the transfer function between the end-point position and the input torque. Note that $Z(x, s)$ obtained in equation (6.32) is a vector with the position $Y(x, s)$ and its spatial derivatives as its elements. The transcendental functions $N_{p,q}(s)$ and $D(s)$ are obtained symbolically to form $Z(x, s)$ and extracting the numerators and denominator according to form the following relations:

$$G_{\theta,V_a}(s) = \frac{N_{\theta,V_a}(s)}{D(s)} = \left. \frac{Y'(0, s)}{V_a(s)} \right|_{\tau(s)=0},$$

$$G_{V_s,V_a}(s) = \frac{N_{V_s,V_a}(s)}{D(s)} = \left. \frac{C_s(Y'(x_2, s) - Y'(x_1, s))}{V_a(s)} \right|_{\tau(s)=0},$$

$$G_{V_s,\tau}(s) = \frac{N_{V_s,T}(s)}{D(s)} = \left.\frac{C_s(Y'(x_2,s) - Y'(x_1,s))}{\tau(s)}\right|_{V_a(s)=0},$$

$$G_{\theta,\tau}(s) = \frac{N_{\theta,\tau}(s)}{D(s)} = \left.\frac{Y'(0,s)}{\tau(s)}\right|_{V_a(s)=0},$$

$$G_{y,V_a}(s) = \frac{N_{y,V_a}(s)}{D(s)} = \left.\frac{Y(L,s)}{V_a(s)}\right|_{\tau(s)=0},$$

$$G_{y,\tau}(s) = \frac{N_{y,\tau}(s)}{D(s)} = \left.\frac{Y(L,s)}{\tau(s)}\right|_{V_a(s)=0}$$

Each of the six transfer functions for this system have the common denominator $D(s)$ as

$$D(s) = 2\varphi^2 EI(\rho A(\cos(\varphi l)\sinh(\varphi l) - \cosh(\varphi l)\sin(\varphi l))$$
$$- \varphi^3 I_h(1 + \cos(\varphi l)\cosh(\varphi l)) \tag{6.35}$$

The transcendental expressions of the individual numerator functions are similar to equation (6.24) but they contain several more terms. A few simple results are

$$N_{V_s,\tau}(s) = -\frac{C_s}{C_a}N_{\theta,V_a}(s), \qquad N_{\theta,\tau}(s) = 2\varphi^2\rho A(1 + \cos(\varphi l)\cosh(\varphi l))$$

$$N_{y,\tau}(s) = 2l\rho A(\sin(\varphi l) + \sinh(\varphi l))$$

and the duality of $N_{\theta,V_a}(s)$ and $N_{V_s,\tau}(s)$ has interesting physical implications.

The transcendental equations in the transfer functions in equation (6.34) may seem too complex to be of practical utility, but it is the finite truncation of these transcendental functions that is of ultimate interest for control purposes. To get an n mode finite truncation the first n roots of these transcendental equations must be obtained numerically, and with the use of numerical methods the complexity of the transcendental equations is not an issue as long as it is practical to obtain the roots of these equations. Section 6.3.4 outlines the procedure for reducing the transcendental equations to rational Laplace domain transfer functions.

The terms $D(s)$ in equation (6.35), $N_{Y,V_a}(s)$ and $N_{V_s,V_a}(s)$ can be reduced to the special case of a cantilever beam ($I_h \rightarrow \infty$) with piezoelectric actuators and sensors spanning its entire surface ($x_1 = 0$, $x_2 = L$). First, it is seen that significant simplification occurs when the piezoelectric laminates span the entire beam surface. Letting $x_1 = 0$ and $x_2 = L$, $N_{V_s,V_a}(s)$ and $N_{Y,V_a}(s)$ reduce to

$$N_{V_s,V_a}(s)\big|_{x_1=0,x_2=L} = 2C_aC_sEI\left[-\rho A + 2\rho A\cos(\varphi L) + 2\rho A\cosh(\varphi L)\right.$$

$$-3\rho A\cos(\varphi L)\cosh(\varphi L) + \varphi^3 I_h\cosh(\varphi L)\sin(\varphi L)$$

$$\left.+\varphi^3 I_h\cos(\varphi L)\sinh(\varphi L)\right] \tag{6.36}$$

Furthermore, letting $I_h \to \infty$ in equations (6.35) and (6.36) and forming the transfer functions results in

$$G_{y,V_a}(s)\Big|_{\substack{x_1 = 0, x_2 = L \\ I_h \to \infty}} = \frac{-C_a \sin(\varphi L) \sinh(\varphi L)}{2\varphi EI\, [1 + \cos(\varphi L) \cosh(\varphi L)]} \tag{6.37}$$

$$G_{V_s,V_a}(s)\Big|_{\substack{x_1 = 0, x_2 = L \\ I_h \to \infty}} = \frac{-2C_a C_s EI\, [\cosh(\varphi L) \sin(\varphi L) + \cos(\varphi L) \sinh(\varphi L)]}{2\varphi EI\, [1 + \cos(\varphi L) \cosh(\varphi L)]}$$

$$\tag{6.38}$$

The transfer functions in equations (6.37) and (6.38) have been derived and experimentally validated (Alberts and Colvin, 1991). In the next section finite truncations of the transcendental expressions in terms of equation (6.34) are given.

6.3.4   Rational Laplace domain transfer functions

The transfer functions presented in the preceding section are exact relationships (for the model used) in terms of transcendental functions of φ. The desired form of the transfer functions in equation (6.34) is in terms of ratios of polynomials in the Laplace operator s. For this purpose, the numerator and denominator transcendental functions can be rationalised as shown in Schmitz (1985) using the Maclaurin series expansion;

$$f(\varphi) = \frac{\varphi^p}{p!} \frac{d^p f(0)}{d\varphi^p} (1 + k_1 \varphi^4 + k_2 \varphi^8 + \cdots) \tag{6.39}$$

or in equivalent product form as

$$f(\varphi) = \frac{\varphi^p}{p!} \frac{d^p f(0)}{d\varphi^p} \prod_{\iota=1}^{\nu} \left[1 - \left(\frac{\varphi}{\varphi_\iota} \right)^4 \right] \tag{6.40}$$

Here $f(\varphi)$ represents a transcendental function to be rationalised, φ_i is the ith root of $f(\varphi)$, and the derivative term in the leading coefficient $d^p f(0)/d\varphi^p$ is obtained by differentiating the function $f(\varphi)$ until a non-zero constant is obtained. p is the index of the first non-zero derivative and n corresponds to the number of modes to be retained in the rationalised representation. The form of equations (6.39) and (6.40), which expands in powers of 4 in φ is possible due to symmetry in φ such that for every root φ_i there are corresponding roots $-\varphi_i$, $j\varphi_i$, and $-j\varphi_i$. (Recall that $\varphi^4 = -\rho A s^2 / EI$.) The expansion leads to the following rational expressions for the elements of equation (6.34) in terms of s:

$$G_{V_s,V_a}(s) = \frac{C_a C_s (x_1 - x_2)}{EI} \prod_{i=1}^{n} \frac{((s/\omega_{v_i})^2 + 1)}{((s/\Omega_\iota)^2 + 1)} \tag{6.41}$$

$$G_{V_s,\tau}(s) = \frac{C_s(x_2 - x_1)(8l^3 - 6l^2 x_1 + x_1^3 - 6l^2 x_2 + x_1^2 x_2 + x_1 x_2^2 + x_2^3)}{24 I_T EI} \prod_{i=1}^{n} \frac{(s/\gamma_i)^2 + 1}{(s/\Omega_i)^2 + 1}$$

(6.42)

$$G_{\theta,V_a}(s) = -\frac{C_a}{C_s} G_{V_s,\tau}(s)$$

(6.43)

$$G_{y,\tau}(s) = \frac{l}{s^2 I_T} \prod_{i=1}^{n} \frac{((s/\lambda_{T_i})^2 - 1)}{((\sigma/\Omega_i)^2 + 1)}$$

(6.44)

$$G_{\theta,\tau}(s) = \frac{1}{s^2 I_T} \prod_{i=1}^{n} \frac{((s/\gamma_i)^2 + 1)}{((\sigma/\Omega_i)^2 + 1)}$$

(6.45)

$$G_{y,V_a}(s) = \frac{-C_a(12 I_h(2l(x_1 - x_2) + x_2^2 - x_1^2) + \rho Al(2l^2(x_1^2 - x_2^2) + x_2^4 - x_1^4))}{24 I_T EI} \prod_{i=1}^{n} \frac{(s/\lambda_{v_i})^2 + 1}{(s/\Omega_i)^2 + 1}$$

(6.46)

where the total rotary inertia I_T is defined as $I_T \triangleq I_h + (l^3 \rho A/3)$. These transfer functions have been experimentally verified in Alberts *et al.* (1995). A brief description of the experimental verification is given in the next section.

6.3.5   *Experimental system*

The transfer functions were verified using a beam experiment at the US Air Force Academy, Frank J. Seiler Research Laboratory. The experiment consists of a 2 m long uniform aluminium beam of rectangular cross section (76.2 mm × 6.35 mm), hanging vertically from a very low friction hinge. The hinge is in a knife-edge arrangement as illustrated in Figure 6.15, and is situated in such a way that the root of the flexible beam is coincident with the hinge axis. This allows PZT (Lead Zirconate Titanate) actuator–sensor pairs to be located with one edge coincident with the hinge axis. The hinge arrangement has effectively zero inertia, however, hub inertia has been added in order to increase pole-zero separation. Although the system has no torque motor, for purposes of modal testing, torque inputs are achieved by applying impulsive force to a rigid moment arm extending from the hinge axis. A PCB Piezotronics force hammer is used to provide the force. Since the angular displacements of the hinged joint are very small during modal testing, it was possible to obtain an adequate measurement of hub-angle by measuring translational displacement of a point on the arm extending from the hinge. A Kaman KD-2300-10CU non-contact displacement measuring system was used for this measurement. This sensor has a rated 3 dB frequency response range of static to 50 kHz, a measurement range of 25.4 mm, and midrange resolution of 0.254 mm. The piezoelectric ceramic elements are Vernitron PZT5A patches mounted in pairs on opposite faces of the beam. A 63.5 mm long by

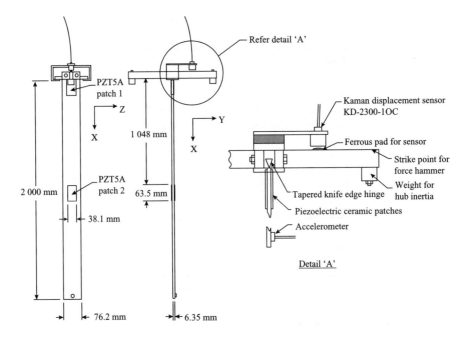

Figure 6.15   Experimental system

38.1 mm wide piezoelectric material patch was attached to each side of the root of the beam, thus $x_1 = 0$, $x_2 = 63.5$ mm and $w_p = 38.1$ mm. The PZT patches used for both actuation and sensing are identical, and hence $t_a = t_c$.

Unlike the idealised sensing relationship given in equation (6.20), a more precise model of the PZT sensing behaviour includes a resistance–capacitance (RC) high-pass filter characteristic, which can be expressed (in the Laplace domain) as

$$V_s(s) = C_s \left(\frac{s R_m C_t}{s R_m C_t + 1} \right) \left. \frac{\partial y(x, s)}{\partial x} \right|_{x_1}^{x_2} \qquad (6.47)$$

where R_m is the resistance of the measurement process (data acquisition system or oscilloscope) and C_t is the effective total capacitance of the PZT patch (as opposed to C in equation (6.20), which is capacitance per unit area). In order to force the sensor response to behave as equation (6.20) in the frequency range of interest (i.e. flat frequency response) it is desirable to introduce a large measurement impedance R_m. For this purpose, the output of the PZT sensor is processed through a high impedance ($10^{12} \Omega$) unity gain buffer amplifier. This moves the high-pass cut-off frequency $1/R_m C_t$ to well below 1 Hz. Beam end-point motion is measured using a PCB Piezotronics Model 336A02 accelerometer. The rated ±5 per cent amplitude sensitivity deviation range for the accelerometer is 1–2000 Hz. A Tektronix model 2630 Fourier Analyser was used for modal testing. The parameters for experimental system are given in Table 6.4. Table 6.5 contains the parameters of the piezoelectric material

Table 6.4   Experimental beam parameters

Parameter	Value
Hub inertia, I_h	0.0338 kgm^2
Cross-sectional area, A	$4.838 \times 10^{-4} \text{ m}^2$
Volumetric mass density	2710 kg/m^3
Young's Modulus, E	$71 \times 10^9 \text{ N/m}^2$
Beam length, L	2 m
Linear mass density, ρ	1.3125 kg/m
End-point mass, M_p	0.0 kg
Area moment of inertia, I	$16.257 \times 10^{-10} \text{ m}^4$

Table 6.5   PZT5A parameters

Parameter	Value
Charge constant, d_{31}	$-171 \times 10^{-12} \text{ m/V}$
Coupling coefficient	-0.340
Surface area	$0.038 \times 0.064 \text{ m}^2$
Voltage constant, g_{31}	$-11.4 \times 10^{-3} \text{ Vm/N}$
Capacitance, C	$68.35 \ \mu\text{F/m}^2$
Thickness, t_a	$0.305 \times 10^{-3} \text{ m}$

Table 6.6   Roots of various transcendental equations in hertz

Mode	$\Delta(s)$ (Ω_i)	(ω_{T_i})	(γ_i)	(λ_{T_i})	(λ_{v_i})	(ω_{v_i})	(γ_i)
1	$\pm j5.477$	$\pm j1.313$	$\pm j5.851$	$\pm j4.176$	$\pm j5.544$	$\pm j5.487$	$\pm j5.851$
2	$\pm j15.33$	$\pm j8.226$	$\pm j18.97$	$\pm j22.57$	$\pm j24.06$	$\pm j15.84$	$\pm j18.97$
3	$\pm j27.55$	$\pm j23.03$	$\pm j39.59$	$\pm j55.73$	$\pm j58.29$	$\pm j30.08$	$\pm j39.59$
4	$\pm j47.31$	$\pm j45.14$	$\pm j67.73$	$\pm j103.6$	$\pm j108.1$	$\pm j51.62$	$\pm j67.73$
5	$\pm j75.84$	$\pm j74.61$	$\pm j103.4$	$\pm j166.3$	$\pm j173.5$	$\pm j81.91$	$\pm j103.4$

PZT5A. A 63.5 mm long by 38.1 mm wide piezoelectric material patch is attached to each side of the base of the beam, thus $x_1 = 0 \text{ mm}$, $x_2 = 63.5 \text{ mm}$ and $w_p = 38.1 \text{ mm}$.

Roots of the various transcendental equations, needed to model the system, are given in Table 6.6. These roots are obtained numerically for the parameter values, given in Table 6.5 for the experimental beam. Note that the system poles and zeros do not depend on the particular piezoelectric element chosen and depend only on the location of the element (in the case of the zeros) and the beam parameters.

6.3.6 Experimental results

The results of the experiments conducted to verify two of the transfer functions are presented as Bode magnitude plots in Figures 6.16 and 6.17. Additional experimental results are presented in Alberts *et al.* (1995). The measured values and the theoretical values, for a particular transfer function, are both plotted on the same graph. The results involving end-point position $G_{y, V_{a1}}(s)$ are presented in terms of the measured output variable, end-point acceleration. From the plots, the accuracy of the theoretical predictions can be easily verified. A damping ratio of 0.2 per cent is added to each

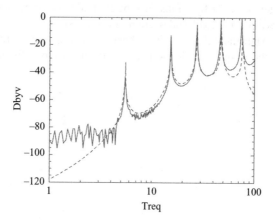

Figure 6.16   Bode magnitude plot for applied voltage to end-point displacement

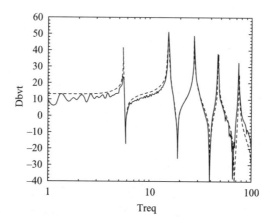

Figure 6.17   Bode magnitude plot for applied torque to voltage sensed

complex conjugate pair of poles and zeros in the theoretical plots to approximate material damping.

The experimental results (frequency response plots) showed that the roots obtained using analytical model are within 5 per cent of the experimental results and the gains of the various transfer functions match almost exactly.

6.4   Summary

The application of a symbolic manipulation approach for modelling and analysis of a flexible manipulator system has been presented. It has been demonstrated that the approach provides several advantages in characterising the dynamic behaviour of the manipulator, and in assessing the stability, response and vibration frequency of the system. The system transfer functions have been obtained in symbolic form and thus interrelations between payload and system characteristics have been investigated. Simulation and experimental results have been presented demonstrating the performance of the symbolic approach in modelling and simulation of flexible manipulator systems.

Chapter 7

Flexible space manipulators: Modelling, simulation, ground validation and space operation

C. Lange, J.-C. Piedboeuf, M. Gu and J. Kövecses

This chapter deals with the development and validation of tools for simulation of flexible space robotic systems using Symofros, a modelling and simulation software program that was developed at the Robotics Section of the Canadian Space Agency. The first part of the chapter describes general modelling issues, Symofros' software architecture, and advanced modelling techniques for flexible arms. The second part presents an experimental validation of the flexible-link model used in Symofros. Additionally, an experimental approach for the end-point detection of flexible links will be discussed. The third part introduces the special purpose dextrous manipulator (SPDM) test verification facility, which is used to emulate the contact dynamic effects arising during payload insertion and extraction tasks of the SPDM on-board the International Space Station (ISS) based on hardware-in-the-loop simulation. Furthermore, an on-board training simulator that is used by astronauts to evaluate their ability to operate the robotic arm on-board the ISS will be introduced.

7.1   Introduction

The International Space Station (ISS) is one of the most ambitious engineering projects. Its construction began in December 1998, when the Russian module Zarya, which means 'sunrise' in Russian, and the American module Unity were successfully joined in space. Canada's contribution to the ISS is the mobile servicing system (MSS) (Doetsch and Middleton, 1987). As shown in Figure 7.1, the MSS is composed of the mobile base system (MBS), the space station remote manipulator system (SSRMS) and the special purpose dextrous manipulator (SPDM).

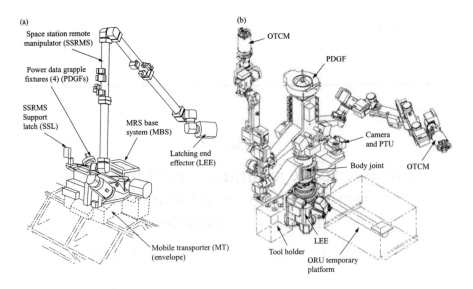

Figure 7.1   Mobile servicing system

The SSRMS is a 17 m long, 7 degrees of freedom (DOF) flexible manipulator. The SPDM consists of two 3.5 m long, 7 DOF arms with an additional DOF through body rotation. Terminating each arm of SPDM is an orbital tool change-out mechanism (OTCM) including a gripper, a camera with two lights, a socket drive mechanism and a force–moment sensor.

While the SSRMS is and will be used to assemble the ISS, the SPDM (to be launched in 2008) will be required to perform maintenance tasks. Essentially, the SPDM will be used to replace the so-called orbital replacement units (ORU), the components of the ISS systems that are replaceable on-orbit. The SPDM will operate directly connected to the ISS or to the tip of the SSRMS. Both the SSRMS and the SPDM are tele-operated by an operator located inside the ISS. Owing to the important flexibility in the SSRMS/SPDM system, all insertion/extraction tasks involving the SPDM will be done using only one arm with the other arm grasping a stabilisation point. As an example, with the SPDM mounted at the end of the SSRMS, five structural modes are present below 2 Hz.

As common to all space applications, all tasks related to the operation of a space manipulator have to be verified on earth prior to the execution in space. These verifications are usually performed using software simulators. It is, in general, impossible to use the space hardware for ground testing since most of the space manipulators cannot support their own weight under 1 *g* environment. As a result, simulations play an important role in the control of space manipulators. The software simulators should reflect the real dynamic behaviour of the robots. However, structural flexibility represents a challenge in the model validations since it is difficult to verify the correctness of the model from the analysis of simulation results. Therefore, the validation should be done based on both theoretical analysis and experimental verifications.

This chapter deals with the development of tools for simulation of flexible space robotic systems using Symofros, a modelling and simulation environment that was developed at the Canadian Space Agency (CSA) to support the research in space robotics. In particular, focus is given on the validations of models for link flexibility and their applications.

The first part of this chapter, Section 7.2, will give an overview of Symofros and describe its software architecture, which is based on the Maple symbolic modelling engine and the Matlab–Simulink environment. In addition to the model approaches introduced in previous chapters, a combined finite element (FE) and assumed-mode modelling approach for flexible arms subjected to both bending and torsional deformation is discussed.

Section 7.3 describes experimental validations of the modelling approach detailed in the previous Section using models of order 1 and 2. In particular, experimental results underlining the foreshortening effect will be presented. Furthermore, this experimental set-up also allows identifying model parameters, such as moment of inertia and bending stiffness parameters. Moreover, a practical technique to obtain the end-point position and force of a flexible manipulator using strain gauges in conjunction with the joint sensing will be developed.

Section 7.4 introduces the utilisation of hardware-in-the-loop simulation (HLS) to solve the contact dynamics problem that arises during payload insertion and extraction tasks of the SPDM. The SPDM test verification facility (STVF) designed in support of the HLS will be discussed in detail. Furthermore, as an alternative to this HLS approach, a contact dynamics software simulator is introduced. In particular, an experimental approach to determine the contact parameters for the contact dynamics toolkit (CDT) of MacDonald Dettwiler Space and Advanced Robotics Ltd for an arm computer unit (ACU) insertion into its berth using STVF and a simulator thereof is presented.

Section 7.5 introduces the on-orbit MSS training simulator SMP. This system is used on-orbit to keep the skills of the astronauts at the required level in order to perform complicated tasks, such as a free-flyer capture.

7.2 Symofros

Multi-body dynamics is of central importance in design and analysis of mechanical systems and their controllers. In space systems, multi-body modelling and analysis is fundamental in developing operating systems and technologies. For space robotics and space systems, both non-real-time and real-time simulations are required in general. The CSA's in-house multi-body dynamics software package Symofros has been developed since 1994. This environment allows the user to efficiently model and simulate in non-realtime as well as in real time, and then to perform the implementation on the real hardware. The software architecture of Symofros is based on the Maple symbolic modelling engine and the Matlab–Simulink environment. Symofros is used for various projects in robotics both inside and outside CSA. This section describes the integrated virtual environment provided by Symofros.

7.2.1   Overview

Symofros' multi-body dynamics engine is based on a formulation relying on Jourdain's principle. Jourdain's principle provides a physically clear framework for multi-body analysis of both holonomic and non-holonomic systems, as already detailed in Chapter 3. Open-loop systems and subsystems are modelled in Symofros using a generic recursive formulation, which can consider both rigid and flexible elements in the system (Piedboeuf, 1998). Complex systems, for example, closed-loop multi-body systems, parallel robots, are currently modelled by splitting them into subsystems. The system model can then be assembled by employing constraints between the various subsystems. In general, consideration of system constraints is a key issue in multi-body dynamics. Symofros is able to handle both holonomic and non-holonomic constraints based on the Lagrangian multiplier technique with Baumgarte stabilisation and the use of projection and decomposition techniques. Various parts of Symofros' modelling engine have been extensively validated by experiments, analytical examples and simulations.

Flexible beams are implemented for flexible-body modelling with various choices of shape functions. Besides the traditional assumed-modes approximations, a characteristic modelling approach employed is the advanced use of the assumed-modes method (AMM), where discretisation is carried out in a way similar to the FE method, that is, interpolation functions are generated locally for an element, but the shape functions are represented globally as in the traditional (AMM), as detailed in section 7.2.3.

Besides body flexibility, finite stiffness of mechanical structure of the connecting joints is also a dominant effect in multi-body systems. Symofros is capable of modelling joint flexibility using discrete stiffness models.

For contact mechanics modelling, Symofros seamlessly interfaces the CDT developed by MacDonald Dettwiler Space and Advanced Robotics Ltd (Ma, 2000). Furthermore, work is in progress to extend the contact-impact modelling capabilities of the Symofros environment with special attention to the real-time aspects (Gonthier *et al.*, 2004). Symofros also provides an interface to the contact parameter identification tools developed by Weber *et al.* (2002). Furthermore, it provides the analytical framework for dynamic parameter identification algorithms, which is an important area in multi-body system simulation, analysis and control. The two main purposes of the identification toolbox are to facilitate the optimum generation of experimental data for identification, and to process the measured data to determine the required parameters. This work involves (a) the formulation and analysis of the dynamic equations in a form suitable for identification and (b) the solution techniques of these equations for the parameters, details are given in Moore *et al.* (2003).

Symofros also includes a control system toolbox comprising a library of Simulink blocks of various control algorithms (e.g. model-based control with PD compensation). These can be easily linked and tested with the dynamic model of a multi-body system to form the model of a controlled system. Also, new control algorithms can be readily built from the existing primitives.

7.2.2   *Software architecture*

Symofros' modelling and simulation environment is composed of three main modules, see Figure 7.2. The first module is a graphical Java user interface that is provided to the user to describe the topology of the system (links, joints, etc.). This topology is then passed as XML or special Maple-input file to the second module, the Maple-based symbolic model generator (SMG). The SMG comprises modules written in Maple language to perform the symbolic modelling, that is, to compute the kinematic and dynamic quantities of the bodies and the joints. To do so, first, the topology of the mechanical system is analysed to generate a graph model. Using the topology with the body and joint data, the SMG develops the kinematic equations. Using the kinematic formulation, the SMG builds the dynamic equations in various forms, for simulation (forward dynamics) control (inverse dynamics) and parameter identification (e.g., regression matrix). The SMG is normally used as an automatic model generator, but it is also a powerful tool to analyse the dynamic equations and to develop models online. More details on the symbolic modelling part of Symofros can be found in Piedboeuf (1996) and Moore *et al.* (2002).

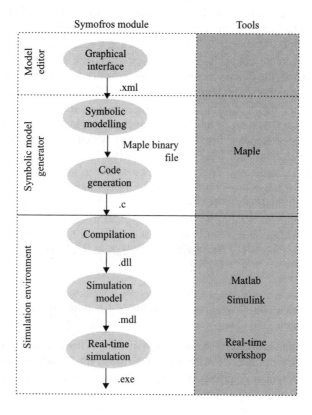

Figure 7.2   Overview of Symofros modules

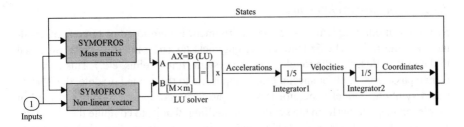

Figure 7.3   Example of a simple simulation model within Simulink

For simulation and real-time implementation, the SMG generates C-code to represent the multi-body system. The code generation uses optimisation tools to break the
complex expressions down to smaller expressions and to avoid redundant computation. The generated C-functions are the links between the modelling part of Symofros
and the simulation/real-time implementation parts.

To allow an efficient and convenient use of the derived mathematical model, and to
enable the numerical simulation, Symofros is directly linked to the Matlab–Simulink
environment. The Simulink environment allows users to create complex models and
to generate complex simulation systems in only a few simple steps without the need
of advanced programming skills. Within this environment, special Symofros blocks
are available to call the functions generated symbolically and written in the C-file. As
an example, Figure 7.3 shows how the forward dynamics can be computed. Here, the
dark blocks (mass matrix, non-linear vector) are used to call the functions to compute
the mass matrix and the non-linear force vector contained in the previously generated
C-file. Then, using standard Simulink blocks, the system of equations is solved to
obtain the accelerations, which are then integrated to obtain the generalized velocities
and generalized coordinates. This complete block describing the forward dynamics
can be found in the Symofros library and reused with other models.

Real-time simulation and HLS can be achieved by using complementary tools
like the Real-Time Workshop and RT-Lab[1] for generating real-time simulation code
and distributing the computations on several computers. More details on this topic
can be found in Lambert *et al.* (2001), Piedboeuf *et al.* (2001) and L'Archeveque
et al. (2000).

7.2.3   Flexible beam modelling: a combined FE and assumed-modes approach

As already detailed in previous chapters, the dynamics of flexible manipulators is governed by an infinite-dimensional hybrid mathematical model, which consists of a set
of ordinary differential equations (ODEs) and partial differential equations (PDEs)

---

[1] www.opal-rt.com

associated with constraints known as boundary conditions (BCs). On the basis of modal expansion, the solution of the PDEs, when possible, leads to ODEs, which generate an infinite-dimensional eigenvalue problem (Meirovitch, 1967). In theory, flexible manipulators require an infinite number of elastic modes to completely describe their behaviour. In practice, however, the dynamic model of a flexible manipulator is usually developed based on a finite number of elastic modes.

Since exact solutions of PDEs are only possible for simple cases, approximate models are required. A very popular class of methods is based on the Rayleigh–Ritz expansion (Meirovitch, 1967). The deflection is represented by a finite series of space-dependent trial functions that are multiplied by time-dependent amplitude functions. A first option is to select the space-dependent functions, known as shape functions, from a set of complete functions defined over the entire length of the flexible link. This is the classical Rayleigh–Ritz or the assumed-modes method (AMM) (Meirovitch, 1967). A second option is to define the space-dependent functions over subintervals of the length of the flexible link. These functions are known as interpolation functions and the resulting method is known as the FE method (Reddy 1993; Strang and Fix, 1973).

In the AMM, the shape functions are selected from a set of comparison functions, that is, any function, which is $2p$ times differentiable and satisfies all the BCs, where $2p$ is the order of the PDE characterising the flexible manipulator. The shape functions may be selected from the set of admissible functions, namely, the set of p times differentiable functions that satisfy only the geometric BCs (Meirovitch, 1967). Obviously, the set of comparison functions is a subset of the set of admissible functions. Typically, the selected shape functions are eigenfunctions of a closely related simpler problem with standard BCs, for example, the Euler–Bernoulli beam in one of the following configurations: clamped-free (Book *et al.*, 1975; Buffinton and Lam, 1992; Siciliano, 1990; Yim, 1993; Yuan *et al.*, 1989), pinned-free (Cetinkunt and Yu, 1991; Chen and Menq, 1990), clamped-mass (de Luca and Siciliano, 1991; Hastings and Book, 1987; Theodore and Ghosal, 1995; Yigit, 1994*a*) or pinned-mass (Cannon and Schmitz, 1984; Chiang *et al.*, 1991; Hastings and Book, 1987).

In the FE method, the interpolation functions are simple polynomials of degree p, which verify the continuity conditions between two adjacent elements or nodes (Reddy, 1993). These polynomials need to be at least $p - 1$ times differentiable at the interior nodes. The interpolation functions usually chosen in the literature are Hermite cubic functions (Bayo, 1987; Chen and Menq, 1990; Meirovitch, 1986), cubic splines (Cho *et al.*, 1991; Dancose and Angeles, 1989; Dancose *et al.*, 1989) or cubic B-splines (Truckenbrodt, 1980; Yang and Gibson, 1989).

In this section, a combined AMM and FE approach is proposed combining the advantages of both methods. This approach, which will be for the sake of brevity detailed below for a cubic spline in one dimension only, has been implemented in Symofros for the spatial case and forms a critical part of symbolic modelling of flexible beams. A more detailed work has been reported in Saad *et al.* (2000*a* and *b*).

Let the discrete model be given by approximating the continuous deflection $y(x, t)$ by a finite series of space-dependent functions $\phi_i(x)$ multiplied by time-dependent amplitude functions $\eta_i(t)$, that is,

$$y(x,t) = \sum_{i=1}^{n} \phi_i(x)\eta_i(t) = \mathbf{\Phi}^{\mathrm{T}}(x)\mathbf{\eta}(t),$$

where n is the number of modes, vector $\mathbf{\Phi}(x) = [\phi_1(x), \ldots, \phi_n(x)]^{\mathrm{T}}$ and vector $\mathbf{\eta}(t) = [\eta_1(t), \ldots, \eta_n(t)]^{\mathrm{T}}$. This can now be rewritten as the summation over n beam segments as

$$y(x,t) = \sum_{i=1}^{n} y_i(x,t)[\mu_i - \mu_{i+1}] \tag{7.1}$$

using the Heaviside function μ_i defined as

$$\mu_i = \begin{cases} 0 & : & x < x_i \\ 1 & : & x \geq x_i \end{cases}$$

the approximation of the deformation by a cubic spline function (Gerald and Wheatley, 1985) can be written on the interval $x_i \leq x \leq x_{i+1}$ as

$$y_i(x,t) = a_i(x - x_i)^3 + b_i(x - x_i)^2 + c_i(x - x_i) + d_i$$

The coefficients a_i, b_i, c_i and d_i can easily be obtained, as shown in many text books, as

$$a_i = \frac{y''_{i+1} - y''_i}{6h_i}, \quad b_i = \frac{1}{2}y''_i, \quad c_i = \frac{y_{i+1} - y_i}{h_i} - \frac{1}{6}h_i\left(2y''_i + y''_{i+1}\right), \quad d_i = y_i$$

$$\tag{7.2}$$

with $i = 1, \ldots, n$ and $h_i = x_{i+1} - x_i$. A linear relation between y_i and y''_i can be obtained for $i = 2, \ldots, n$ as

$$6\left[\frac{1}{h_{i-1}}y_{i-1} - \left(\frac{1}{h_{i-1}} + \frac{1}{h_i}\right)y_i + \frac{1}{h_i}y_{i+1}\right] = h_{i-1}y''_{i-1} + 2(h_{i-1} + h_i)y''_i + h_iy''_{i+1}$$

$$\tag{7.3}$$

Now, let $\mathbf{Y}$ be the vector of node displacements whose ith component is y_i, and with vector $\mathbf{Y}''$ of node curvatures defined likewise, equation (7.3) can be rewritten in a more compact matrix form as

$$6\mathbf{CY} = \mathbf{AY}'' \tag{7.4}$$

with the $(n+1) \times (n+1)$ matrices $\mathbf{A}$ and $\mathbf{C}$ defined as

$$\mathbf{A} = \begin{bmatrix} A_1 & B_1 & C_1 & 0 & \cdots & 0 \\ h_1 & 2(h_1+h_2) & h_2 & 0 & \cdots & 0 \\ 0 & h_2 & 2(h_2+h_3) & h_3 & \ddots & \vdots \\ \vdots & \ddots & \ddots & \ddots & \ddots & \vdots \\ 0 & \cdots & 0 & h_{n-1} & 2(h_{n-1}+h_n) & h_n \\ 0 & \cdots & 0 & A_n & B_n & C_n \end{bmatrix}$$

$$\mathbf{C} = \begin{bmatrix} \alpha_1 & \beta_1 & \gamma_1 & 0 & \cdots & 0 \\ \frac{1}{h_1} & -\left(\frac{1}{h_1}+\frac{1}{h_2}\right) & \frac{1}{h_2} & 0 & \cdots & 0 \\ 0 & \frac{1}{h_2} & -\left(\frac{1}{h_2}+\frac{1}{h_3}\right) & \frac{1}{h_3} & \ddots & \vdots \\ \vdots & \ddots & \ddots & \ddots & \ddots & \vdots \\ 0 & \cdots & 0 & \frac{1}{h_{n-1}} & -\left(\frac{1}{h_{n-1}}+\frac{1}{h_n}\right) & \frac{1}{h_n} \\ 0 & \cdots & 0 & \alpha_n & \beta_n & \gamma_n \end{bmatrix}.$$

The coefficients A_1, B_1, C_1 and $\alpha_1, \beta_1, \gamma_1$ depend on the boundary condition of the base point of the beam, while A_n, B_n, C_n and $\alpha_n, \beta_n, \gamma_n$ depend on the boundary condition of the end-point of the beam, which are discussed in the sequel.

In the case of a clamp boundary condition at the base point (node 1), the deformation y and the slope y' at the beginning of the beam are zero, that is,

$$y_1 = y(0,t) = 0$$
$$y_1' = y'(0,t) = 0$$

Hence, the first column of $\mathbf{C}$ disappears and for the coefficients depending on the boundary conditions it follows that

$$\alpha_1 = 0, \quad \gamma_1 = 0, \quad \beta_1 = \frac{1}{h_1}, \quad A_1 = 2h_1, \quad B_1 = h_1, \quad C_1 = 0$$

In the case of a free boundary condition (no bending moment) at the end-point (node $n+1$), one obtains

$$y_{n+1}'' = 0$$

and hence, the last column of matrix $\mathbf{A}$ disappears.

In the case of a constant bending moment on the last element, the boundary condition can be obtained as

$$y_{n+1}'' = y_n''$$

from which the coefficients follow as

$$\alpha_{n+1} = 0, \quad \gamma_{n+1} = 0, \quad \beta_{n+1} = 0, \quad A_{n+1} = 0, \quad B_{n+1} = 1, \quad C_{n+1} = -1$$

For a clamp-free boundary condition the relation of node displacements and curvatures, equation (7.4), follows with the now n-dimensional vectors $\mathbf{Y}$ and $\mathbf{Y}''$ and the $n \times n$ matrices $\mathbf{A}$ and $\mathbf{C}$ as

$$6\mathbf{CY} = \mathbf{AY}'' \tag{7.5}$$

where

$$\mathbf{Y} = \left[\, y_2 \quad \cdots \quad y_{n+1} \,\right]^{\mathrm{T}}, \qquad \mathbf{Y}'' = \left[\, y_1'' \quad \cdots \quad y_n'' \,\right]^{\mathrm{T}}$$

$$\mathbf{A} = \begin{bmatrix} 2h_1 & h_1 & 0 & \cdots & & 0 \\ h_1 & 2(h_1+h_2) & h_2 & & & \vdots \\ 0 & \ddots & \ddots & \ddots & & 0 \\ \vdots & \ddots & h_{n-2} & 2(h_{n-2}+h_{n-1}) & h_{n-1} & \\ 0 & \cdots & 0 & h_{n-1} & 2(h_{n-1}+h_n) \end{bmatrix}$$

$$\mathbf{C} = \begin{bmatrix} \frac{1}{h_1} & 0 & \cdots & \cdots & 0 \\ -\left(\frac{1}{h_1}+\frac{1}{h_2}\right) & \frac{1}{h_2} & 0 & \cdots & 0 \\ \frac{1}{h_2} & -\left(\frac{1}{h_2}+\frac{1}{h_3}\right) & \frac{1}{h_3} & \ddots & \vdots \\ \vdots & \ddots & \ddots & \ddots & 0 \\ 0 & \cdots & \frac{1}{h_{n-1}} & -\left(\frac{1}{h_{n-1}}+\frac{1}{h_n}\right) & \frac{1}{h_n} \end{bmatrix}$$

In order to determine the modal shape function, one can resort to equations (7.1) and (7.2), leading to

$$y(x,t) = \sum_{i=1}^{n} y_i(x,t)[\mu_i - \mu_{i+1}] = \mathbf{U}^{\mathrm{T}}(x)\mathbf{Y} + \mathbf{V}^{\mathrm{T}}(x)\mathbf{Y}'' \tag{7.6}$$

where the vectors $\mathbf{U}(x)$ and $\mathbf{V}(x)$ are given as

$$\mathbf{U}(x) = \begin{bmatrix} \frac{1}{h_1}(x-x_1)(\mu_1-\mu_2) + \left[1 - \frac{1}{h_2}(x-x_2)\right](\mu_2-\mu_3) \\ \frac{1}{h_2}(x-x_2)(\mu_2-\mu_3) + \left[1 - \frac{1}{h_3}(x-x_3)\right](\mu_3-\mu_4) \\ \vdots \\ \frac{1}{h_{n-1}}(x-x_{n-1})(\mu_{n-1}-\mu_n) + \left[1 - \frac{1}{h_n}(x-x_n)\right](\mu_n-\mu_{n+1}) \\ \frac{1}{h_n}(x-x_n)(\mu_n-\mu_{n+1}) \end{bmatrix}$$

$$
\mathbf{V}(x) =
\begin{bmatrix}
\left[-\frac{h_1}{3}(x-x_1)+\frac{1}{2}(x-x_1)^2-\frac{1}{6h_1}(x-x_1)^3\right](\mu_1-\mu_2) \\[4pt]
\left[-\frac{h_1}{6}(x-x_1)+\frac{1}{6h_1}(x-x_1)^3\right](\mu_1-\mu_2)+ \\[4pt]
\left[-\frac{h_2}{3}(x-x_2)+\frac{1}{2}(x-x_2)^2-\frac{1}{6h_2}(x-x_2)^3\right](\mu_2-\mu_3) \\[4pt]
\vdots \\[4pt]
\left[-\frac{h_{n-1}}{6}(x-x_{n-1})+\frac{1}{6h_{n-1}}(x-x_{n-1})^3\right](\mu_{n-1}-\mu_n)+ \\[4pt]
\left[-\frac{h_n}{3}(x-x_n)+\frac{1}{2}(x-x_n)^2-\frac{1}{6h_n}(x-x_n)^3\right](\mu_n-\mu_{n+1}) \\[4pt]
\left[-\frac{h_n}{6}(x-x_n)+\frac{1}{6h_n}(x-x_n)^3\right](\mu_n-\mu_{n+1})
\end{bmatrix}
$$

Solving equation (7.5) for $\mathbf{Y}''$, that is,

$$\mathbf{Y}'' = 6\mathbf{A}^{-1}\mathbf{C}\mathbf{Y}$$

equation (7.6) results as

$$y(x,t) = \left[\mathbf{U}^{\mathrm{T}}(x) + 6\mathbf{V}^{\mathrm{T}}(x)\,\mathbf{A}^{-1}\mathbf{C}\right]\mathbf{Y}$$

from which $\boldsymbol{\Phi}(x)$ and $\boldsymbol{\eta}(t)$ follow as

$$\boldsymbol{\Phi}^{\mathrm{T}}(x) = \mathbf{U}^{\mathrm{T}}(x) + 6\mathbf{V}^{\mathrm{T}}(x)\,\mathbf{A}^{-1}\mathbf{C}, \qquad \boldsymbol{\eta}(t) = \mathbf{Y} = \begin{bmatrix} y_2 & \cdots & y_{n+1} \end{bmatrix}^{\mathrm{T}}$$

Alternatively, when solving equation (7.5) for $\mathbf{Y}$, that is,

$$\mathbf{Y} = \frac{1}{6}\mathbf{C}^{-1}\mathbf{A}\mathbf{Y}''$$

equation (7.6) can be obtained as

$$y(x,t) = \left[\frac{1}{6}\mathbf{U}^{\mathrm{T}}(x)\,\mathbf{C}^{-1}\mathbf{A} + \mathbf{V}^{\mathrm{T}}(x)\right]\mathbf{Y}''$$

with

$$\boldsymbol{\Phi}^{\mathrm{T}}(x) = \frac{1}{6}\mathbf{U}^{\mathrm{T}}(x)\,\mathbf{C}^{-1}\mathbf{A} + \mathbf{V}^{\mathrm{T}}(x), \qquad \boldsymbol{\eta}(t) = \mathbf{Y}'' = \begin{bmatrix} y_1'' & \cdots & y_n'' \end{bmatrix}^{\mathrm{T}}$$

As shown in Saad *et al.* (2000*a* and *b*) the cubic splines with curvatures as coordinates provide a fast convergence rate. The modelling practice based on Symofros also illustrates the advantages of this approach in auto symbolic model generation. Furthermore, the curvatures, so-called flexible coordinates, can be directly detected using strain gauges and greatly benefit the control of flexible manipulators. As an example given in section 7.3.2, a self-sensory flexible link is developed based on curvature measurements and used for the detection of end-point pose and force of a flexible manipulator.

7.3   Experimental validation

This section discusses the experimental model validation and exploits the derived models to enhance control capabilities of flexible links.

Section 7.3.1 describes an experimental validation of the model approach detailed in the previous section using models of order 1 and 2. In particular, experimental results for the foreshortening effect (Piedboeuf and Moore, 2002) will be presented. Furthermore, the set-up allows us to identify model parameters, such as the moment of inertia and bending stiffness parameters.

In section 7.3.2, a practical technique to obtain the end-point position and force of a flexible manipulator using strain gauges in conjunction with joint sensing will be developed.

7.3.1   *Experimental model validation using a single flexible link*

On the basis of Symofros modelling and simulation environment described above, this section is concerned with its model validation. In particular, a comparison is given for a Symofros model of a single flexible link using order 1 and order 2 models with experimental data. Since the curvature is used as generalized coordinates in the simulations, it can be directly compared with strain gauge measurements obtained from the physical test-bed.

7.3.1.1   Experimental set-up

The flexible link used in this validation is shown in Figure 7.4, with the list of corresponding model parameters given in Table 7.1. Hereby, ρ defines the density of the beam material, A is the area of cross section of the beam, E and G are the Young's modulus and shear modulus of the beam material, respectively, D_o and D_i are the outside and inside diameters of the beam, respectively. The value of the parameter Idmx shown in the table includes the hub inertia and the inertia caused by cables for the strain gauges. Therefore, the value is larger than that from the calculation. The damping coefficient Ke is determined experimentally.

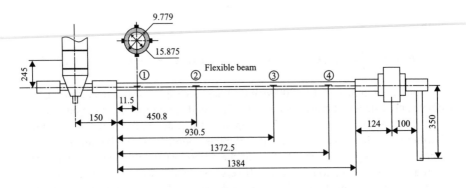

Figure 7.4   Experimental set-up for flexible link validation

Table 7.1 Physical parameters used for building the Symofros model

Parameter	Value	Calculation	Meaning
EIy	533.747 Nm^2	$EI_y, I_y = \pi(D_0^4 - D_i^4)/64$	Bending stiffness in y
EIz	533.747 Nm^2	$EI_z, I_z = I_y$	Bending stiffness in z
GIx	513.654 Nm^2	$GI_x, I_z = \pi(D_0^4 - D_i^4)/32$	Torsion stiffness in x
Idmx	0.201 kgm^2	$\rho L I_x$	Moment of inertia of mass in x
Idmy	0.488 kgm^2	$\rho L I_y + \frac{1}{3}\rho A L^3$	Moment of inertia of mass in y
Idmz	0.488 kgm^2	$\rho L I_z + \frac{1}{3}\rho A L^3$	Moment of inertia of mass in z
Ke	0.045	By experiments	Damping coefficient
L	1.384 m	measured	Length of link
Rho	0.9642 kg/m	$\rho A, A = \pi(D_0^2 - D_i^2)/4$	Mass density per unit length

To determine the moment of inertia and bending stiffness parameters, a frame system is set up as shown in Figure 7.5, with the *x*-axis being the beam axis. The strain gauges are located at four positions on the beam and are used to detect the curvatures of the beam. These curvatures are comparable to the simulation outputs, and therefore, comparison between the measured curvatures and simulation outputs will verify the correctness of the model developed using Symofros.

Ten strain bridges are used for detecting the curvatures of the beam, among them eight are used for detecting bending strain and two are used for detecting torsion strain. The torsion strain gauge is a four-element rosette, which provides a full-bridge connection. The two torsion strain bridges are located at the position 1 and position 4 as shown in Figure 7.4. They detect the torsion strain about the beam axis. For bending strain, 16 strain gauges are used and they are located at the four positions as shown in Figure 7.5. Each position has four strain gauges. The top and bottom strain gauges compose a strain bridge that detects the bending strain (curvature) about the *y*-axis, and the left and right strain gauges compose a strain bridge that detects the bending strain (curvature) about the *z*-axis. Therefore, 16 strain gauges form 8 bridges and detect 4 bending strains about the *y*-axis and 4 bending strains about the *z*-axis.

7.3.1.2 Simulation results

The Symofros model for the three-dimensional flexible link includes two rigid bodies and one flexible beam. In order to validate the Symofros model, simulations are performed and the results are compared with those from the experimental tests. The inputs to the system are the end-point force and moment that come directly from the measurements of the JR3 force sensor at the end of the beam during the tests. The outputs are the states of the system, which have been chosen as the curvatures when modelling the flexible beam in Symofros. Since the same end-point forces were used in the tests and in the simulations, the simulation should have the same responses as the experiments. If this is true, the simulation model matches the real flexible system.

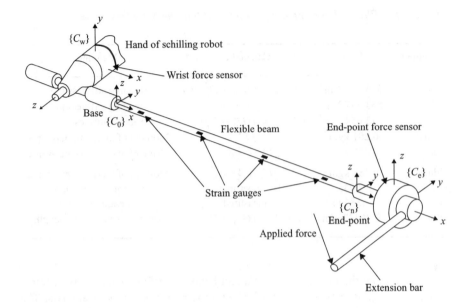

Figure 7.5   Experimental set-up with frame system

To illustrate the importance of the model order, two different models are generated for the simulations. One is the order 1 model and the other is the order 2 model. The two models have exactly the same parameters except for setting different order for strain–displacement relation when generating the Symofros model.

Figures 7.6–7.8 show the curvature responses of the flexible beam under a static load of 1.357 kg, where Figure 7.6 shows the torsion curvature about the x-axis, Figure 7.7 shows the bending curvature about the y-axis and Figure 7.8 shows the bending curvature about the z-axis. In these figures, the solid lines indicate the results obtained from experimental tests, dotted lines are the simulation results using the order 1 model and dashed lines show the simulation results using the order 2 model. Figures 7.9–7.11 show the curvature responses under a dynamic load.

In conclusion, the Symofros order 2 model provides good simulation outputs that match very well with the results from experimental tests. The Symofros order 1 model produces some error in curvature responses, especially for larger deformation cases. Torsion curvature remains unchanged along the flexible beam. There is no difference for the torsion curvature in gauge #1 and gauge #4. There are some errors in dynamic responses (the simulation responses appear smaller than those from the experimental tests). This is because the dynamic force generated by the force sensor and fixture between force sensor and end-point of the beam are not included in the force measurement. This dynamic force depends on the acceleration of the end-point. In static and quasi-static cases, the end-point acceleration is very small and the dynamic force vanishes. Therefore, there is no difference seen between simulations and experiments.

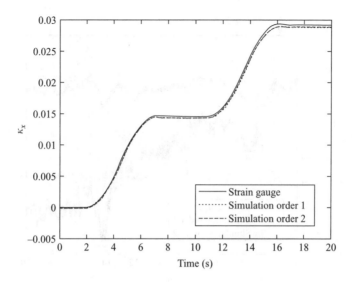

Figure 7.6   Torsion curvatures about x-axis under static load 1.357 kg

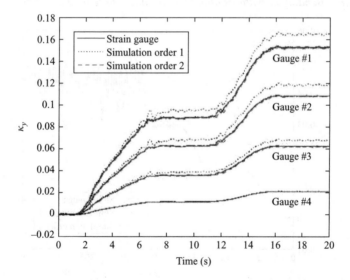

Figure 7.7   Bending curvatures about y-axis under static load 1.357 kg

7.3.2   Flexible manipulator end-point detection and validation

Robot manipulators are typically designed to perform tasks at their end-point. In order to achieve precise end-point pose and force control, accurate detection of the end-point position and force is usually required. This section develops an approach to detect the end-point of a flexible link based on strain measurements. The analytical

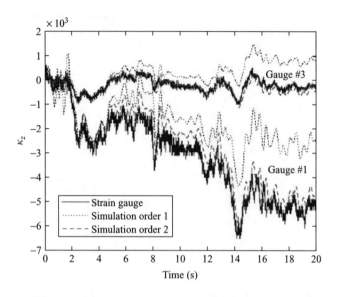

Figure 7.8   Bending curvatures about z-axis under static load 1.357 kg

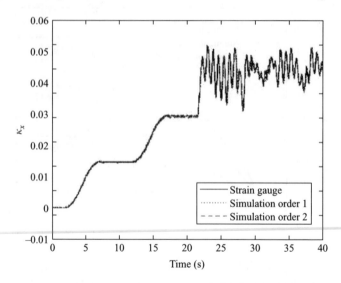

Figure 7.9   Torsion curvatures about x-axis under dynamic load

formulation is compared to experiments using strain gauge measurements and a vision system.

7.3.2.1   Flexible manipulator kinematics

In the sequel, a general n-link flexible manipulator shown in Figure 7.12 is considered.

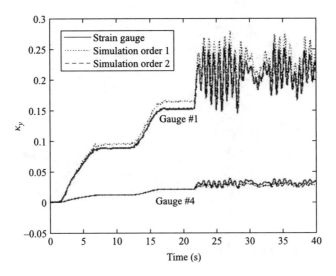

Figure 7.10   Bending curvatures about y-axis under dynamic load

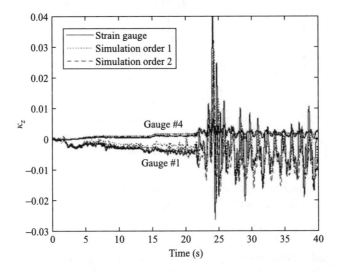

Figure 7.11   Bending curvatures about z-axis under dynamic load

The end-point of the manipulator, that is, the position and orientation of the last link relative to the base, $\{R_0\}$, can be determined using a homogeneous transformation

$$_n^0\mathbf{T} =\,_1^0\mathbf{T}\,_2^1\mathbf{T} \cdots\,_{j+1}^{j}\mathbf{T} \cdots\,_n^{n-1}\mathbf{T}$$

The inclusion of flexible arms into this scheme requires the insertion of appropriate transformation matrices, which take into account variations of the position and orientation caused by arm deformations. Assume that the jth link is a flexible arm

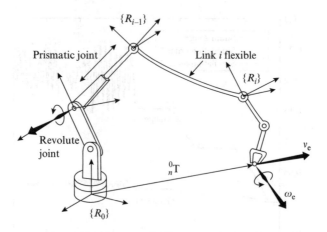

Figure 7.12   A general n-link robot manipulator

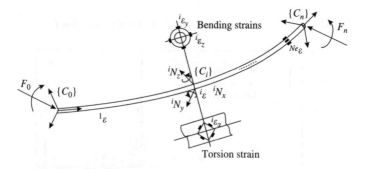

Figure 7.13   Flexible link unit

and the transformation of jth link can be described as a combination of a rigid-body transform and a flexible-body transform, namely,

$$^{j}_{j+1}\mathbf{T} = \mathbf{T}_{\mathrm{r}}(\theta_j, p_j)\, \mathbf{T}_{\mathrm{f}}(\theta_f, p_f)$$

The rigid-body transform $\mathbf{T}_{\mathrm{r}}(\theta_j, p_j)$ is a function of the joint variable θ_j or position variable p_j. These two variables are usually obtainable from joint sensing using conventional sensors such as encoder or LVDT. The flexible-body transform $\mathbf{T}_{\mathrm{f}}(\theta_f, \mathbf{p}_f)$, however, depends on the rotation vector θ_f and position vector $\mathbf{p}_f$. These two vectors represent the orientation and position variations caused by the arm deformations. In the general cases, θ_f and $\mathbf{p}_f$ are not constants, and detection of θ_f and $\mathbf{p}_f$ is a key issue in determination of the end-point position and orientation of a flexible manipulator. A detailed discussion on the issue is given in Gu and Piedboeuf (2002).

   In order to detect the two vectors θ_f and $\mathbf{p}_f$, a self-sensory flexible arm with distributed strain gauges had been built as shown in Figure 7.13. There are N_e positions where the strain gauges are placed. In each position, the strain gauges detect two bending strains $(\varepsilon_y, \varepsilon_z)$ and one torsion strain (ε_α). These bending and torsion strains can be expressed as N_e local strain vectors:

$$^i\varepsilon = \begin{bmatrix} ^i\varepsilon_\alpha & ^i\varepsilon_y & ^i\varepsilon_z \end{bmatrix}^T, \quad i = 1, \ldots, N_e$$

These strain vectors provide the basic information of the link deformations, which can be used to determine the end-point position and orientation as well as the end-point force and moment of a flexible manipulator. To do so, the relation between local strains and curvature of the arm as well as the relation between strains and internal forces of link sections are determined. According to the beam bending and torsion theory (Timoshenko and MacCullough, 1949), the curvature of a flexible arm is, within the elastic limits of material, proportional to the strain vector, that is,

$$^i\kappa = \mathbf{C}_a{}^i\varepsilon, \quad i = 1, \cdots, N_e$$

where

$$^i\kappa = \begin{bmatrix} ^i\kappa_x \\ ^i\kappa_y \\ ^i\kappa_z \end{bmatrix}, \quad \mathbf{C}_a = \begin{bmatrix} 1/c_\alpha & 0 & 0 \\ 0 & 1/c_y & 0 \\ 0 & 0 & 1/c_z \end{bmatrix}$$

The constants c_α, c_y and c_z represent the active radius, which depends on the distance from the neutral axis of the link to the locations of the strain gauges. For a cylindric flexible beam with outside diameter of D_o, this radius is $D_o/2$.

The internal forces associated with the strain measurement include two bending moments and a torsion moment, and can be expressed as

$$^i\mathbf{N} = \begin{bmatrix} ^iN_x & ^iN_y & ^iN_z \end{bmatrix}^T, \quad i = 1, \ldots, N_e$$

A proportional relation between the moment vector $^i\mathbf{N}$ and the local strain vector $^i\varepsilon$ can be found as

$$^i\mathbf{N} = \mathbf{C}_e{}^i\varepsilon, \quad i = 1, \ldots, N_e \tag{7.7}$$

where the strain-to-force gains are given by

$$\mathbf{C}_e = \begin{bmatrix} \frac{GI_x}{c_\alpha} & 0 & 0 \\ 0 & \frac{EI_z}{c_y} & 0 \\ 0 & 0 & \frac{EI_y}{c_z} \end{bmatrix}$$

The terms EI_z, EI_y and GI_x represent the bending and torsion stiffness of the cross Section of the link.

7.3.2.2 Statics

When external forces are applied to both ends of the flexible arm, an internal force $^i\mathbf{F}$ is generated in the arm section $\{C_i\}$. As shown in Figure 7.13, $^n\mathbf{F}_n$ denotes the force applied at the end-point of the arm and $^0\mathbf{F}_0$ represents the reaction force at the base point of the arm. For convenience the forces are expressed in the local frames and $\{C_i\}$ is located at the ith strain gauge position.

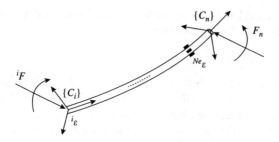

Figure 7.14   Force balance for a flexible arm section

To determine the end-point force and moment from strain measurement, the flexible arm is partitioned into two sections at the frame $\{C_i\}$ with one of the two shown in Figure 7.14. The force balance of this section gives the relation of the internal force $^i\mathbf{F}$ and the end-point force $^n\mathbf{F}_n$ as

$$^i\mathbf{F} = {}_n^i\mathbf{T}_f\, {}^n\mathbf{F}_n \tag{7.8}$$

with

$$^i\mathbf{F} = \begin{bmatrix} ^i\mathbf{V} \\ ^i\mathbf{N} \end{bmatrix}, \quad ^n\mathbf{F}_n = \begin{bmatrix} ^n\mathbf{f}_n \\ ^n\mathbf{n}_n \end{bmatrix}, \quad ^i_n\mathbf{T}_F = \begin{bmatrix} ^i_n\mathbf{R} & \mathbf{0} \\ ^i\mathbf{p}_n \times ^i_n\mathbf{R} & ^i_n\mathbf{R} \end{bmatrix}$$

where $^i\mathbf{p}_n$ and $^i_n\mathbf{R}$ represent the position and orientation of the end-point frame $\{C_n\}$ relative to frame $\{C_i\}$. For simplicity, the weight of the flexible arm is not considered in the above formulation. A more detailed formulation that considers the weight of the arm is given by Gu (2002). The internal force $^i\mathbf{F}$ consists of a force vector $^i\mathbf{V}$ and a moment vector $^i\mathbf{N}$. The internal moment vector is the only portion associated with the strain measurements. Hence, one can rewrite the respective part of equation (7.8) as

$$^i\mathbf{N} = \begin{bmatrix} ^i\mathbf{p}_n \times ^i_n\mathbf{R} & ^i_n\mathbf{R} \end{bmatrix} ^n\mathbf{F}_n, \quad i = 1, \ldots, N_e$$

Substituting the proportional relation between the moment vector $^i\mathbf{N}$ and the strain vector $^i\varepsilon$ given in equation (7.7), a linear relation between the end-point force and the measured strain is obtained as

$$\varepsilon = \mathbf{C}_n\, {}^n\mathbf{F}_n \tag{7.9}$$

where

$$\varepsilon = \begin{bmatrix} ^1\varepsilon & ^2\varepsilon & \cdots & ^{N_e}\varepsilon \end{bmatrix}^{\mathrm{T}}$$

and

$$\mathbf{C}_n = \begin{bmatrix} \mathbf{C}_e^{-1}\, ^1\mathbf{p}_n \times ^1_n\mathbf{R} & \mathbf{C}_e^{-1}\, ^1_n\mathbf{R} \\ \mathbf{C}_e^{-1}\, ^2\mathbf{p}_n \times ^2_n\mathbf{R} & \mathbf{C}_e^{-1}\, ^2_n\mathbf{R} \\ \vdots & \vdots \\ \mathbf{C}_e^{-1}\, ^{N_e}\mathbf{p}_n \times ^{N_e}_n\mathbf{R} & \mathbf{C}_e^{-1}\, ^{N_e}_n\mathbf{R} \end{bmatrix}$$

The expressions for ${}^i\mathbf{p}_n$ and ${}^i_n\mathbf{R}$ can be determined using

$${}^i\mathbf{p}_n = {}^0_n\mathbf{R}^{\mathrm{T}}\left({}^0\mathbf{p}_n - {}^0\mathbf{p}_i\right), \qquad {}^i_n\mathbf{R} = {}^0_i\mathbf{R}^{\mathrm{T}}\, {}^0_n\mathbf{R}$$

From equation (7.9), the end-point force can expressed in terms of the measurement strains as

$${}^n\mathbf{F}_n = {}^n\mathbf{C}_{\mathrm{F}}\,\varepsilon \tag{7.10}$$

where the force measurement matrix ${}^n\mathbf{C}_{\mathrm{F}}$ is given by

$${}^n\mathbf{C}_{\mathrm{F}} = \left(\mathbf{C}_n^{\mathrm{T}}\mathbf{C}_n\right)^{-1}\mathbf{C}_n^{\mathrm{T}}$$

Equation (7.10) gives the end-point force and moment expressed in terms of the end-point frame $\{C_n\}$. In some cases, it is desirable to know the end-point force expressed in terms of the base frame $\{C_0\}$ of the flexible arm. This end-point force may be denoted by ${}^0\mathbf{F}_n$. Using rotation matrix ${}^0_n\mathbf{R}$, the end-point force ${}^n\mathbf{F}_n$ can be transformed from the end-point frame to the base frame and the end-point force ${}^0\mathbf{F}_n$ can be expressed as

$${}^0\mathbf{F}_n = {}^0_n\mathbf{R}\, {}^n\mathbf{F}_n$$

It is evident from the above analysis that determination of the end-point force and moment of a flexible arm requires information of the positions and orientations of the end-point as well as those points where the strain gauges are located.

To determine the force and moment acting at the end-effector of the manipulator, the force–moment propagation from the jth flexible link to the end-effector link n is considered. Figure 7.15 shows the forces and moments acting on link i, where the wrench ${}^{i+1}\mathbf{W}_i$ represents the force and torque exerted on the ith link by the $(i-1)$th link and is expressed in terms of frame $\{C_{i+1}\}$, ${}^{i+1}\mathbf{W}_{i+1}$ represents the force and torque exerted on the ith link by the $(i+1)$th link, and ${}^{i+1}\mathbf{G}_i$ denotes the contribution

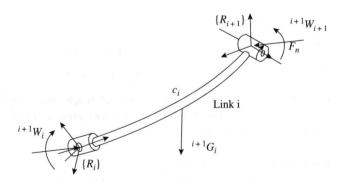

Figure 7.15   Static force-moment balance for a single link

due to gravity of the ith link. Summing up the forces and moments and setting them equal to zero result in an equation for the force balance for the ith link

$$^{i+1}\mathbf{W}_{i+1} = {}^{i+1}_i\mathbf{T}_f \, {}^i\mathbf{W}_i - {}^{i+1}_i\mathbf{T}_g \, {}^i\mathbf{G}_i, \quad i = j, j+1, \ldots, n \tag{7.11}$$

with

$$^{i+1}_i\mathbf{T}_f = \begin{bmatrix} {}^{i+1}_i\mathbf{A} & \mathbf{0} \\ {}^{i+1}\mathbf{r}_i \times {}^{i+1}_i\mathbf{A} & {}^{i+1}_i\mathbf{A} \end{bmatrix}, \quad {}^{i+1}_i\mathbf{T}_g = \begin{bmatrix} {}^{i+1}_i\mathbf{A} \\ {}^{i+1}\mathbf{r}_{ic} \times {}^{i+1}_i\mathbf{A} \end{bmatrix},$$

where $^{i+1}_i\mathbf{A}$ denotes the rotation matrix of link i mapping frame $\{R_{i+1}\}$ to frame $\{R_i\}$, $^{i+1}\mathbf{r}_i$ denotes the position vector from the origin of $\{R_{i+1}\}$ to the origin of $\{R_i\}$ and $^{i+1}\mathbf{r}_{ic}$ denotes the position vector from the origin of $\{R_{i+1}\}$ to the centre of mass of link i (point c_i in Figure 7.15). The weight of link i is determined using

$$^i\mathbf{G}_i = m_i \, {}^i_0\mathbf{A}\,{}^0\mathbf{g}$$

where m_i denotes the mass of ith link, $^i_0\mathbf{A}$ denotes the rotation matrix mapping frame $\{R_i\}$ to the base frame of the manipulator $\{R_0\}$ and $^0\mathbf{g}$ represents the gravitational acceleration vector and is defined by:

$$^0\mathbf{g} = \begin{bmatrix} 0 & 0 & -9.8 \end{bmatrix}^{\mathrm{T}}$$

Summing up the equation set given in equation (7.11), the end-point force and moment can be written in terms of the wrench of the jth link as

$$^n\mathbf{W}_n = {}^n_j\mathbf{T}_f \, {}^j\mathbf{W}_j - \sum_{i=j}^n {}^n_i\mathbf{T}_g \, {}^i\mathbf{G}_i \tag{7.12}$$

with

$$^n_j\mathbf{T}_f = \begin{bmatrix} {}^n_j\mathbf{A} & \mathbf{0} \\ {}^n\mathbf{r}_j \times {}^n_j\mathbf{A} & {}^n_j\mathbf{A} \end{bmatrix}, \quad {}^n_i\mathbf{T}_g = \begin{bmatrix} {}^n_i\mathbf{A} \\ {}^n\mathbf{r}_{ic} \times {}^n_i\mathbf{A} \end{bmatrix}$$

where the wrench $^j\mathbf{W}_j = {}^n\mathbf{F}_n$ is the end-point force–moment vector of the flexible arm unit.

The rotation matrix $^n_j\mathbf{A}$ and position vectors $^n\mathbf{r}_j$ and $^n\mathbf{r}_{ic}$ can be obtained as

$$^n_j\mathbf{A} = \left({}^j_{j+1}\mathbf{A}\,{}^{j+1}_{j+2}\mathbf{A} \cdots {}^{n-1}_n\mathbf{A} \right)^{\mathrm{T}}$$

$$^n\mathbf{r}_j = {}^n_{j+1}\mathbf{A}^{j+1}\mathbf{r}_j + {}^n_{j+2}\mathbf{A}^{j+2}\mathbf{r}_{j+1} + \cdots + {}^n_{n-1}\mathbf{A}^{n-1}\mathbf{r}_{n-2} + {}^n\mathbf{r}_{n-1}$$

$$^n\mathbf{r}_{ic} = {}^n_{i+1}\mathbf{A}^{i+1}\mathbf{r}_{ic} + {}^n_{i+2}\mathbf{A}^{i+2}\mathbf{r}_{i+1} + \cdots + {}^n_{n-1}\mathbf{A}^{n-1}\mathbf{r}_{n-2} + {}^n\mathbf{r}_{n-1}$$

It is evident from equation (7.12) that one can determine the wrench $^n\mathbf{W}_n$ at the end-effector of the manipulator as long as the end-point force and moment of the flexible arm j are available from strain measurement.

7.3.2.3   End-point detection using strain gauges

To verify the proposed strain measurement approach for building self-sensing robot arms, substantial tests have been performed using the experimental set-up shown in Figure 7.16.

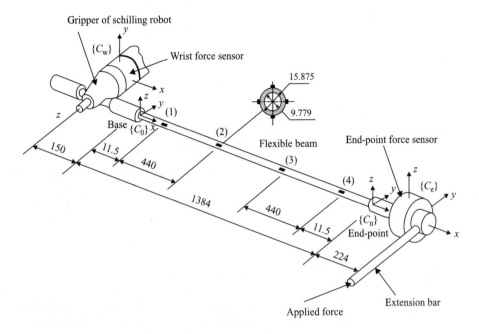

Figure 7.16   Experimental set-up for self-sensing robot arm

In this experiment, a Schilling Titan II robot holds a flexible arm with the gripper of the robot. The flexible arm is a cylindrical beam of 1.384 m with inside and outside diameters of 9.779 and 15.875 mm, respectively. Strain gauges were placed at four positions on the arm to detect both bending and torsional strains. For bending strain measurement, each position contains four CEA-06-125UN-120 strain gauges with a nominal resistance of 120 $\Omega \pm 0.3$ per cent and a gage factor of 2.07 ± 0.5 per cent. The four strain gauges were distributed around the outside surface of the arm in an angle of 90° from each other. The top and bottom strain gauge pairs were connected in a half-bridge configuration and detect the bending strains in the vertical plane. The left and right strain gauge pairs were also connected in half-bridge configuration and detect the bending strains in the horizontal plane. For torsional strain measurement, two CEA-06-062UV-350 strain gauges with a nominal resistance of 350 $\Omega \pm 0.4$ per cent and a gage factor of 2.125 $\pm$ 0.5 per cent were used. They were connected in a full-bridge configuration. A multi-channel transducer-conditioning amplifier (Ectron Model 563H) was used to process the strain signals and the cut-off frequency was set to 100 Hz. The sampling rate for strain data collection was 250 Hz. Using these four sets of strain measurements, the end-point position and force were determined according to the algorithms developed in the above sections.

To validate the results obtained from the strain measurement, two additional measurement systems detected the end-point position and force simultaneously. For end-point position measurement, the vision system Optotrak was used. The resolution of Optotrak is 0.01 mm and the rated RMS accuracy is 0.1 mm. The set-up for the Optotrak vision system is shown in Figure 7.17. In this set-up, two rigid

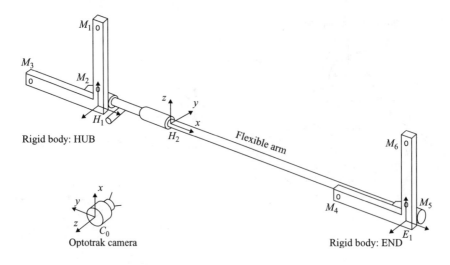

Figure 7.17   End-point detection using Optotrak vision system

bodies: the rigid-body HUB and the rigid-body END were built. Each rigid body was equipped with three infrared markers, which were sensed by an Optotrak camera during the experiment. The positions and orientations of the rigid bodies were determined according to the positions of the markers on the rigid bodies. The rigid-body HUB was attached to the base of the arm while the rigid-body END was attached to the end-point of the arm. The relative motion between the two rigid bodies directly reflects the end-point position and orientation relative to the base frame of the flexible arm.

For end-point force measurements, two force sensors were used. Their locations are shown in Figure 7.16. The end-point force sensor was a JR3-40E15A with a rated output non-linearity of 0.2 per cent and a natural frequency larger than 300 Hz. Another force sensor was the built-in wrist force sensor of the Titan II robot, which has a load rating of 508 Nm and rated accuracy of 0.5 per cent. In order for the results to be comparable to each other, the end-point forces of flexible arm obtained from strain measurement are transformed into the frames where the force sensors are located. Figure 7.16 shows the four frames of interest. For example, if $^{n}\mathbf{F}_{n}$ represents the end-point force from strain measurement, after transforming, this force from the end-point frame $\{C_{n}\}$ to the coordinate frame $\{C_{e}\}$ it will be denoted by $^{e}\mathbf{F}_{n-e}$ and can be determined using equation (7.12)

$$^{e}\mathbf{F}_{n-e} =_{n}^{e} \mathbf{T}_{F}{}^{n}\mathbf{F}_{n}$$

with

$$_{n}^{e}\mathbf{T}_{F} = \begin{bmatrix} _{n}^{e}\mathbf{R} & \mathbf{0} \\ ^{e}\mathbf{p}_{e-n} \times _{n}^{e}\mathbf{R} & _{n}^{e}\mathbf{R} \end{bmatrix},$$

where $_{n}^{e}\mathbf{R}$ and $^{e}\mathbf{p}_{e-n}$ represent the rotation matrix and position vector mapping frame $\{C_{n}\}$ to frame $\{C_{e}\}$, respectively. The end-point force $^{e}\mathbf{F}_{e-n}$ is now comparable to the force $^{e}\mathbf{F}_{JR3}$ of the end-point JR3 force sensor. For simplicity, the weight of end-point fixture has been neglected. To compare the results with the wrist force sensor, the end-point force $^n\mathbf{F}_n$ is transformed to frame $\{C_w\}$ using the same method (Gu, 2002).

Experimentally, the end-point position and orientation of the flexible arm obtained using the results of the strain measurement have been compared with those obtained using the vision system. Also, the end-point force and moment and the torque at the wrist of the robot have been evaluated. The results from strain measurement were compared with those obtained using the force sensors. During the tests, an external force was applied to the end of an extension bar as shown in Figure 7.16. Using an extension bar allows loading that produces a combined bending and torsional effects on the flexible arm. Three kinds of loads, namely, static, quasi-static and dynamic, have been applied in the tests. The static loading was made by hanging a weight block at the end of the extension bar. Since the mass of the weight block is known beforehand, the applied force can be predicted and the sensor output calibrated. The quasi-static loading was made by a hand-push at the extension bar. One can control the push force with slowly varying direction and amplitude so that a quasi-static load is applied to the flexible arm. A quasi-static load is quite similar to the case of slow manipulation such as inserting or assembling a mechanical part using a robot, in which the end-effector usually experiences a slow variation of force and moment. The dynamic load is generated by pushing a weight block that is attached at the end of the extension bar as shown in Figure 7.5. The pushing stroke causes both bending and torsional vibrations of the beam and the dynamic responses of the end-point pose and force can be verified.

Figures 7.18–7.23 show the experimental results of static, quasi-static and dynamic tests. In all these plots, the solid lines indicate the results obtained from strain measurement and the circular markers display the results obtained using Optotrak vision system or JR3 force sensors.

Two static loads, namely, a weight block of 1.357 kg and a weight block of 2.273 kg, have been applied in the static tests. Figure 7.18 shows the responses of the end-point position and orientation under the static load 1.357 kg. It is evident that the results from strain measurement indicated by the solid lines match those from Optotrak vision system illustrated by circular markers. When examining the results, it is noticeable that the static load was applied in vertical direction. Larger end-point deflection appears in the z-axis and the variation of p_z is about 25 mm. The corresponding variation of end-point orientation θ_y is about 1.5°. However, due to the coupling of torsion and bending, there is a small deflection in horizontal direction (y-axis) and a small rotation about z-axis. This can be explained from the definition of $\mathbf{p}_f$ and $\boldsymbol{\theta}_f$, in which the coupling second-order terms in p_y and θ_z depend on the torsion about the x-axis and the bending about the z-axis (Gu and Piedboeuf, 2002). The plot of p_x shows only the foreshortening of the beam, which is about 1.4 mm. The actual value of x-component for the end-point $\{C_n\}$ of the flexible arm relative to the base point $\{C_0\}$ should include the original length (1.384 m) of the arm. The end-point orientation, θ_x is mainly caused by the torsion of the arm, and has a value of 1.25°. Because the flexible arm and end-point fixture are exposed to gravity, there is an initial deformation before the static load was applied to the arm. This deformation yields non-zero values for the end-point position and orientation as shown in Figure 7.18.

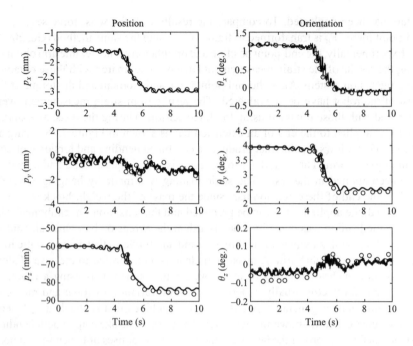

Figure 7.18   Position responses with static loading 1.357 kg (∘∘∘ Optotrak, – strain gauge)

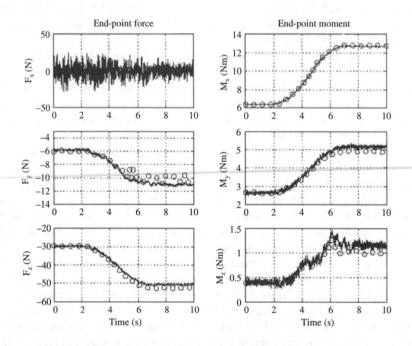

Figure 7.19   End-point force responses with static loading 2.273 kg (∘ ∘ ∘ JR3, – strain gauge)

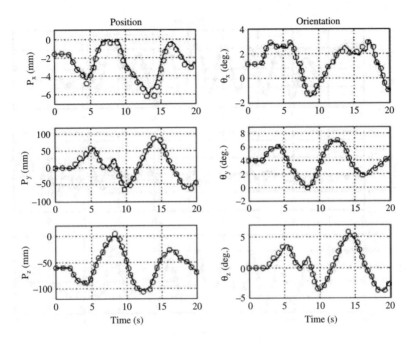

Figure 7.20 Position responses with quasi-static loading (o o o Optotrak, – strain gauge)

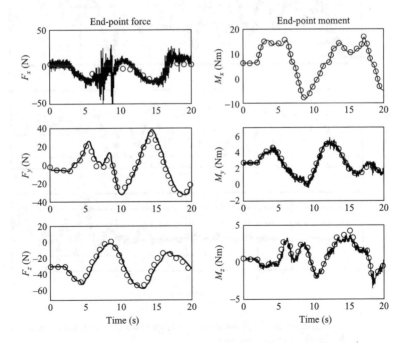

Figure 7.21 End-point force responses with quasi-static loading (o o o JR3, – strain gauge)

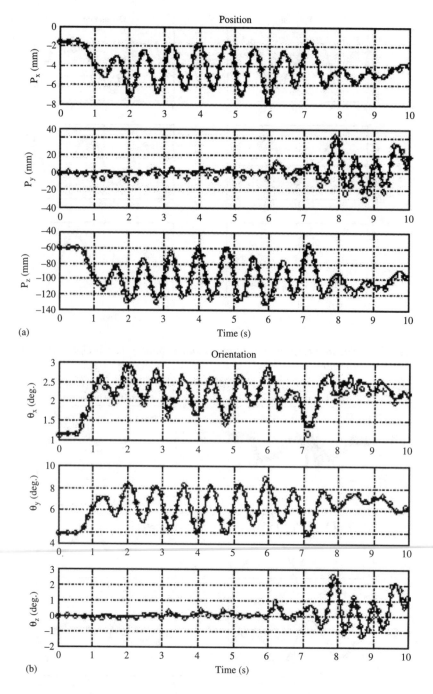

*Figure 7.22   (a) Position and (b) orientation responses for dynamic load (o o o
Optotrak, – strain gauge)*

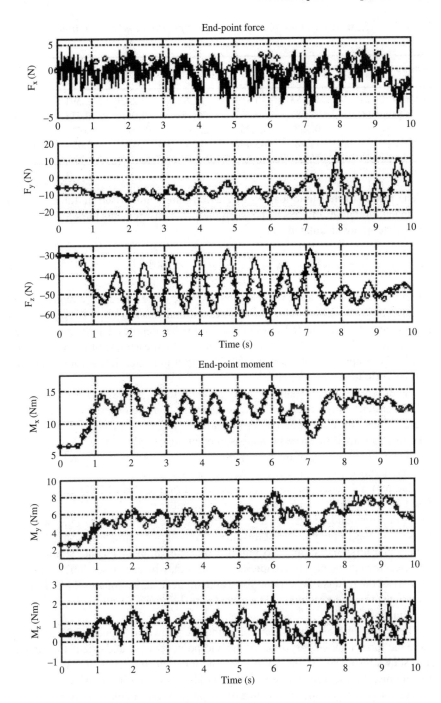

Figure 7.23   Dynamic responses of the end-point force and moment (o o o JR3, – strain gauge)

Table 7.2   *Results for static loading 2.273 kg: position and orientation*

Sensors	p_x(mm)	p_y(mm)	p_z(mm)	θ_x(deg.)	θ_y(deg.)	θ_z(deg.)
Optotrak ($\Delta \bar{y}_v$)	−3.153	−0.519	−42.62	−2.222	−2.591	0.114
Strain gauge ($\Delta \bar{y}_s$)	−2.585	−0.293	−39.783	−2.065	−2.513	0.095
e_a	0.567	0.225	2.837	0.156	0.078	0.019
e_r(%)	17.98	43.44	6.66	7.03	3.02	16.5

The results of the static loading of 2.273 kg are summarised in Table 7.2. The absolute and relative differences between the two measurements (strain gauges and vision system) are calculated using

$$e_a = |\Delta \bar{y}_v - \Delta \bar{y}_s|, \qquad e_r = \frac{|\Delta \bar{y}_v - \Delta \bar{y}_s|}{|\Delta \bar{y}_v|} \times 100\% \qquad (7.13)$$

with $\Delta \bar{y}_v = \bar{y}_{v1} - \bar{y}_{v0}$ and $\Delta \bar{y}_s = \bar{y}_{s1} - \bar{y}_{s0}$, where the subscript v represents the measurement that was made by vision system and the subscript s represents the measurement that was made by strain gauges, the bar denotes that the measurement is an average value, also the subscript 0 indicates the measurement made before the static load was applied to the arm and the subscript 1 indicates the measurement made after the static load was applied.

The difference for p_z results as 6.66 per cent, for θ_y as 3.02 per cent and for θ_x as 7.03 per cent. Large relative differences appear in p_y and θ_z because the loading situation does not produce large deformation in horizontal plane, the resulting p_y and θ_z are too small to exceed the measurement noise and the errors caused by the coordinate inconsistency of the two measurement systems. While the relative errors are large in this direction, the absolute differences are small.

The end-point force and moment responses with a static load 2.273 kg are shown in Figure 7.19. A good agreement can be seen, especially for F_z, M_x and M_y, which have larger variations under static loading. The large noise in F_x for strain measurement is basically due to poor observability in the axial force estimation using only the bending and torsional strains. To improve the estimation of F_x, the strain gauges that detect the extension and compression strains may be used (Gu, 2002). Table 7.3 summarises the results of end-point force and moment, and compares the differences between the two measurements. The e_a and e_r in Table 7.3 take the same definitions as in equation (7.13).

The test results of a quasi-static loading are shown in Figures 7.20 and 7.21, where the former shows responses of the end-point position and orientation and the latter shows responses of end-point force and moment. It is evident that the results of strain measurement match very well with those obtained from vision system and force sensors. A very important feature of the proposed strain measurement approach is the redundant characteristics. In the experimental set-up, four strain gauges have been used. If one strain gauge fails, one can get the results using the remaining

Table 7.3   Results for static loading 2.273 kg: force and moment

Sensor	F_x(N)	F_y(N)	F_z(N)	M_x(Nm)	M_y(Nm)	M_z(Nm)
JR3	2.12	−4.65	−22.72	6.42	2.26	0.61
e_a	0.41	−5.26	−20.91	6.36	2.52	0.77
e_r(%)	80.78	13.09	7.99	0.97	11.37	26.05

three strain gauges. If two strain gauges fail, one can still get reasonable results using the remaining two strain gauges. This redundancy feature is favourable for space applications, as the cost for repairing or replacing failed equipment in space is extremely high and sometimes it is impossible to do so.

The dynamic responses of the end-point position and orientation are shown in Figure 7.22. The dynamic loading was generated by hanging a weight block at the end of the extension bar and making a sudden push on the weight block. Because the push can be in any direction, the vibration appears in bending as well as in torsion and takes the system natural frequencies. The results from strain measurement match very well with those from the vision system. The differences appearing in p_y and θ_z are mainly caused by the coordinate inconsistency of the two measurement systems. The dynamic responses of the end-point forces and moments are given in Figure 7.23. It is evident from these figures that the results from strain measurement displays larger amplitudes than those obtained from the force sensor. Because the JR3 force sensor has a mass, the strain gauges on flexible arm sense the dynamic forces that were generated by both the weight block and the JR3 force sensor, while the JR3 force sensor only detects the dynamic force of the weight block. M_x is exceptional because the inertia about x-axis mainly comes from the weight block attached at the end of the extension bar, as shown in Figure 7.16. The strain gauges and the end-point JR3 force sensor experience the same dynamic torque that is generated by weight block and display the same responses for M_x. When examining the dynamic responses of the wrist torque at the base of the flexible arm, the amplitude differences between strain measurement and torque sensor measurement disappear. However, there is a phase difference between the two measurements. As the flexible arm acts as a second-order mechanical system with respect to the first natural frequency, the force responses at the base of the arm will experience a phase delay compared to the force responses at the end-point of the arm.

The above discussion indicates that the strain gauge approach detects the forces applied at the end-point of the flexible arm. These forces include the force acting at the end-effector and dynamic forces generated by the links attached to the end of the flexible arm. However, the manipulator usually moves slowly when its end-effector contacts the environment. In this case, the dynamic forces generated by the outer links can be ignored and one can obtain the contact force of the end-effector directly from the strain measurement. On the other hand, when the manipulator holds a payload and moves it from one place to another place, there is no contact force acting at the end-effector. In this case, the dynamic force of payload and the outer links obtained from the strain measurement can be used as a feedback to control the motion of manipulator and stabilise the vibrations of the flexible links. Therefore, the strain gauge approach is a promising technique, which can enhance the control capability of a flexible manipulator.

7.4   SPDM task verification facility

In this section, the concept of CSA's STVF is described. The verification process involves three complementary stages. First, a real-time software simulator is used to verify the complete nominal and malfunction procedures. Second, the feasibility of a task involving contact is verified with the help of a hardware-in-the-loop simulator. And finally, a non-real-time simulator is used to perform detailed parametric studies, given the known tolerances on the components.

7.4.1   Background

The cost and risks associated with the execution of robotic tasks in space require that all procedures be verified on earth prior to their execution in space. Canada is responsible for the verification of all the tasks involving the SPDM. The CSA has developed the SPDM task verification facility, a series of simulation and analysis tools to be used for verifying the kinematics (clearance, interface reach, DOF), dynamics (insertion forces, flexibility), visual accessibility (ability to see the worksite) and resource allocations (power, crew time).

One of the main technical challenges with the STVF is verification of the feasibility of insertion/extraction tasks. The forces involved are mainly the result of complex frictional contact between the payload and the worksite. Accurate parameters for contact models are difficult to obtain, especially since friction parameters are inherently different in laboratory and in space. Another important requirement for the STVF is the capability to verify if the selected procedure provides sufficient visual cues on the payload and worksite. The SPDM operator will rely exclusively on camera views for executing the task, and must be able to see properly the edges of the payload and worksite. Although simulating the right surface finish, lighting conditions and camera characteristics in software is feasible, hardware validation is still required.

7.4.2   SPDM task verification facility concept

The concept of STVF is shown schematically in Figure 7.24. It can be split into three main components: a real-time simulator called the MSS operation and training simulator (MOTS), a hardware-in-the-loop simulator called the STVF test-bed (SMT) and a non real-time (NRT) simulator called the manipulator development and simulation facility (MDSF-NRT) (Ma *et al.*, 1997). Each component of the STVF is used to fulfil a particular objective in the verification process. A session on the

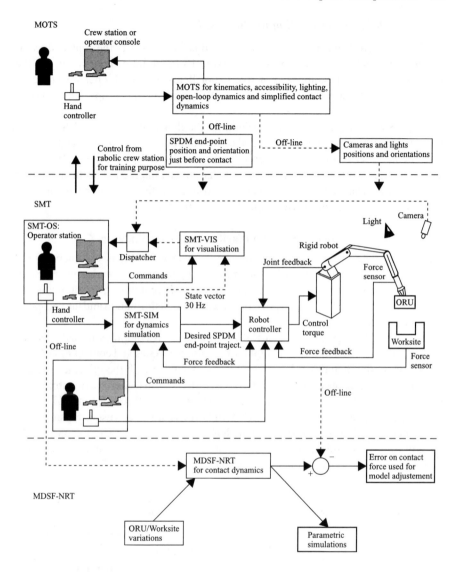

Figure 7.24   STVF concept

MOTS real-time simulator essentially verifies the task execution in its entire scope, including malfunctions and operator interface. For the verification of criteria relying on the less reliable real-time simulation model, such as contact dynamics and visual environment, the real-time hardware-in-the-loop simulator is used. Finally, detailed verification of the effect of parameter tolerances on the contact behaviour is performed using the MDSF-NRT. In the following discussion the focus is given only to the dynamic simulation used in the hardware-in-the-loop simulator.

7.4.3   SPDM task verification facility test-bed

The SMT is a hardware-in-the-loop simulator. It is used to verify the contact dynamics behaviour and the visual aspect of SPDM tasks in the immediate vicinity of the worksite. The SMT is composed of five major items: the SMT operator and controller stations, the simulator, the robot, the ORU/worksite mock-ups and the visual environment.

In the SMT concept, the simulator mimics the dynamic response of the SPDM, submitted to end-point contact forces and operator commands. Its response drives the SMT-robot end-effector motion to replicate the simulated SPDM motion. Mock-ups of the ORU and worksite are used to emulate the contact interaction occurring during the task. The measured contact forces are fed back to the simulator that, in turn, simulates the response of the SPDM to the measured force.

7.4.3.1   The SPDM task verification facility test-bed simulator

The primary role of the SMT dynamic simulator (SMT-SIM) is to generate real-time end-point motion of the SPDM submitted to operator commands and end-point contact forces. As shown in Figure 7.25, the SMT-SIM is composed of four items: the dynamic engine, the SPDM controller, the kinematic transformation and the end-point force transformation.

The dynamic engine is the heart of the simulator. It computes in real time the motion of the SPDM submitted to joint torques, brake torques and contact forces. The simulation model includes at least the SSRMS dynamics with the joints locked in a given configuration, the dynamics of the passive SPDM arm (secured to a stabilisation point or free) and the dynamics of the active SPDM arm with flexible joints. Whenever possible, complete models of the SSRMS/SPDM dynamics are used, replacing the lock joints constraints by proper brake models. Since in normal operation one SPDM arm is grasping a stabilisation point, the complete dynamic model contains a kinematic loop. The complete dynamic model has around 80–90 states. A possible simplification

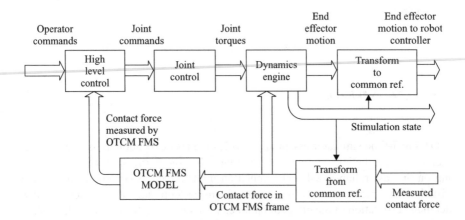

Figure 7.25   SMT dynamic simulator

to speed up the computation is to develop a smaller dynamic model by considering that the active SPDM arm is mounted on an elastic base with a few DOF. The flexible base is then represented by a reduced-order dynamic system (in the modal sense). In any case, the dynamic model of the simulator can be expressed as

$$\mathbf{M}\,(\mathbf{q})\,\ddot{\mathbf{q}} + \mathbf{N}\,(\mathbf{q}, \dot{\mathbf{q}}) + \mathbf{G}\,(\mathbf{q}) = \mathbf{Q}_u\,(\tau) - \mathbf{Q}_f\,(\tau_f) + \mathbf{J}^\mathrm{T}\,(\mathbf{q})\,\mathbf{F}_e$$

where $\mathbf{q}$ is the generalized coordinate vector including rigid body, elastic and flexible coordinates, $\mathbf{M}$ is the inertia matrix, $\mathbf{N}$ is a non-linear vector containing Coriolis and centrifugal terms, $\mathbf{G}$ is the gravity vector, $\mathbf{Q}_u\,(\tau)$ is the generalized commanded joint torque vector, $\mathbf{Q}_f\,(\tau_f)$ is the generalized joint friction torque vector, $\mathbf{J}$ is the manipulator Jacobian and $\mathbf{F}_e$ is the wrench applied to the manipulator end-effector.

A kinematic transformation is required to express the motion of the end-point into a frame of reference that is common to both the SMT simulator and the SMT-robot. The kinematic relationship is given by

$$\mathbf{x}_e = \boldsymbol{\Gamma}\,(\mathbf{q})$$

$$\dot{\mathbf{x}}_e = \mathbf{J}\,(\mathbf{q})\,\dot{\mathbf{q}}$$

where $\mathbf{x}_e$ represents the end-effector position and orientation and $\dot{\mathbf{x}}_e$ represents the end-effector end-point velocity, both expressed in the common reference frame. The force transformation function shown in Figure 7.25 transforms the measured contact force obtained from the hardware and expressed in the common reference frame, into an equivalent force applied at the end-point of the SPDM.

The SSRMS/SPDM high-level and joint controllers are copies of the flight control algorithm. The outputs of the controllers are the joint control torques and the brake commands. The SPDM control scheme includes a force–moment accommodation (FMA), which is used for insertion tasks. The input to the FMA is the real force/torque sensor measurements transformed into the SPDM arm base frame and sampled at the same frequency as the real force/torque sensor. A simplified representation of the operational space rate controller with force/moment accommodation is given by

$$\dot{\mathbf{q}}_d = \mathbf{J}_r^{\#}\left(\dot{\mathbf{x}}_d - \mathbf{K}_f\,(\cdot)\,\mathbf{F}_{e,m}\right) + \dot{\mathbf{q}}_N$$

where $\dot{\mathbf{q}}_d$ is the desired joint rate vector, $\mathbf{J}_r^{\#}$ is the pseudo-inverse of the rigid-body Jacobian relating the end-effector motion to the rigid-body coordinate, $\dot{\mathbf{x}}_d$ is the desired end-effector speed (linear and angular), $\mathbf{F}_{e,m}$ is the measured end-point force, $\mathbf{K}_f\,(\cdot)$ is a linear operator representing the force feedback transfer function and $\dot{\mathbf{q}}_N$ represents a set of joint velocities in null space of $\mathbf{J}_r$.

7.4.3.2 The SPDM task verification facility test-bed robot and robot controller

The SMT-robot reproduces the SPDM end-point trajectories resulting from the SMT simulator. Since the SMT-robot is used to verify the contact portion of the task, the SPDM vernier (slow velocity) control mode is in effect. However, since the dynamic model includes flexibility, the resulting end-point velocities may likely surpass the SPDM's rated vernier velocity.

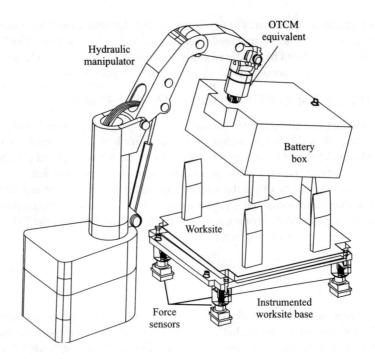

Figure 7.26   SMT robot manipulator

A kinematic equivalence between the SMT-robot and SPDM is not required as the SMT only simulates contact operations in close proximity of the worksite. The robot has 6 DOF and its workspace is large enough to insert or extract ORUs and payloads. Currently, the maximum extraction requirement is around 1.32 m. To reduce the gravity induced moment on the robot wrist, most insertion tasks will be executed vertically as illustrated in Figure 7.26.

The stiffness of the SMT-robot is driven by the need to reproduce SPDM flexible motion. To avoid exciting its own natural modes, the robot natural frequency has to be 5–10 times higher than those of the simulated system. Typically, for SSRMS/SPDM simulation models include flexibility up to 2 Hz. The maximum SPDM payload mass is 600 kg. Such heavy payloads are difficult to handle under 1 g environment, especially for the required robot stiffness. However, the mass of a representative mock-up of payloads and ORU can be limited to 100 kg. The robot includes as well a functional equivalent of the OTCM that includes the gripper capability, the camera and lights, and the socket mechanism used to attach and detach the ORU.

A rigid-body model of the manipulator is given by

$$\mathbf{M}_h\,(\mathbf{q}_h)\,\ddot{\mathbf{q}}_h + \mathbf{N}_h\,(\mathbf{q}_h,\dot{\mathbf{q}}_h) + \mathbf{G}_h\,(\mathbf{q}_h) = \boldsymbol{\tau}_h - \boldsymbol{\tau}_{f,h} + \mathbf{J}_h^{\mathrm{T}}\,(\mathbf{q}_h)\,\mathbf{F}_e$$

where $\mathbf{q}_h$ is the rigid-body generalized coordinate vector, $\mathbf{M}_h$ is the inertia matrix, $\mathbf{N}_h$ is a non-linear vector containing Coriolis and centrifugal terms, $\mathbf{G}_h$ is the gravity vector, $\boldsymbol{\tau}_h$ is the commanded joint torque vector, $\boldsymbol{\tau}_{f,h}$ is the joint friction torque vector, $\mathbf{J_h}$ is the manipulator Jacobian and $\mathbf{F_h}$ is the wrench applied to the manipulator end-effector.

The primary role of the robot controller is to ensure that the robot end-point motion is representative of the simulated response. In essence, the control problem can be approached from two different angles. From the standard control system methodology, the problem appears as a typical impedance matching problem (Hogan, 1987) in which the impedance of the SMT-robot with the simulator in the loop shall match the impedance of the real system. Therefore, shaping the closed-loop impedance of the SMT-robot is one approach to consider. In the second approach, the hardware represents essentially a subsystem of the simulated one. The idea is to use the hardware to physically integrate a portion of the state vector and to numerically integrate the remaining part of the state vector. The robot control problem is then to define a control law that provides decoupled integration of the states associated with the hardware. The two approaches basically result in the same implementation scheme. However, in the latter case, direct update of the simulated state based on the hardware response provides a means to alleviate the limitations encountered with the first one.

In the case of the integrated approach, the idea is to set directly

$$v = \ddot{\mathbf{x}}_e$$

and let the robot integrate the end-effector motion. The coordinate vector is partitioned into

$$\mathbf{q} = \left(\mathbf{q}_1^T, \mathbf{q}_2^T \right)$$

where $\mathbf{q}_1$ is integrated from the hardware response and $\mathbf{q}_2$ is integrated numerically. It is then possible to write the kinematic equations as

$$\mathbf{q}_1 = \Gamma_1^{-1} \left(\mathbf{x}_h, \mathbf{q}_2 \right)$$

$$\dot{\mathbf{q}}_1 = \mathbf{J}_1^{-1} \left(\mathbf{q}_1, \mathbf{q}_2 \right) \left(\dot{\mathbf{x}}_h - \mathbf{J}_2 \left(\mathbf{q}_1, \mathbf{q}_2 \right) \dot{\mathbf{q}}_2 \right)$$

where Γ_1 represents the local inverse function mapping $\mathbf{x}_h$ onto $\mathbf{q}_1$ for a given $\mathbf{q}_2$, and $\mathbf{J}_1$ and $\mathbf{J}_2$ result from the partitioning of $\mathbf{J}$ according to $\mathbf{q}_1$ and $\mathbf{q}_2$. Further details can be found in de Carufel *et al.* (2000) and Doyon *et al.* (2003).

This approach has the advantage of constraining the simulated and real hardware motions to follow a common Cartesian space trajectory. In the impedance matching approach, this equality has to be obtained through high tracking gain on the SMT manipulator control.

7.4.3.3   Computer architecture

The SMT computer architecture is shown schematically in Figure 7.27. The SMT-SIM software is hosted by an SGI Origin computer with four R10000 CPUs, while the visualisation engine runs on an SGI Onyx desktop computer with four R4400 CPUs. The SMT-robot control software runs on a cluster of Pentium processors interconnected through FireWire, a high-speed serial bus, and running under the QNX operating system. The communication between the operator station, the controller, the SMT-SIM host and the SMT-VIS host is through Ethernet. The time-critical real-time

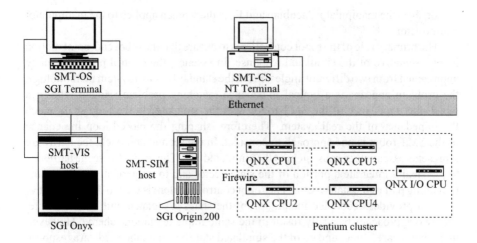

Figure 7.27   SMT computer architecture

communication between the SMT-SIM host and the Pentium cluster is implemented through Firewire, while the non-time-critical communications are transmitted through Ethernet.

7.4.3.4   ORUs and worksite

In the SMT, ORU and worksite mock-ups are designed to emulate the contact forces with high fidelity, provide a representative visual environment and display photogrammetric targets to the vision system. The representative contact forces are obtained by building geometrically equivalent mock-ups with equivalent stiffness characteristics. The surfaces are developed with friction characteristics representative of those that would prevail in space. Since the robot is limited to 100 kg payload, the mass and the centre of mass of the mock-ups are possibly different from those of the real ORUs. However, the SMT simulator model can be adjusted to represent the real mass and inertia characteristics of the ORU or payload.

The ORU mock-ups are also visually representative, having geometric and surface finish properties similar to those of the flight units. The visual targets used are identical to the real targets.

To improve the accuracy of the force–torque measurement, the contact force used by the SMT simulator is measured on the worksite side instead of the ORU side as on SPDM (Figure 7.26). The worksite is designed to be mounted on a platform instrumented with four force–torque sensors. A geometrical transformation brings the measured signal in the common reference frame.

7.4.4   *Experimental contact parameter estimation using STVF*

Using a HLS technique, the SMT-robot emulates the dynamics of the end-effector of the SPDM. To ensure smooth operation, the impact of variations in size and shape as well as misalignments on the successful execution of the tasks must be determined.

Figure 7.28 Arm Computer Unit attached to the OTCME

However, the development of an additional ORU mock-up for each configuration of parameters would be too expensive. Therefore, a pure simulation environment of the STVF using currently available contact models was additionally developed.

In this section, the determination of a set of suitable configuration and model parameters for the CDT of MacDonald Dettwiler Space and Advanced Robotics Ltd (MDR) for an ACU insertion into its berth using STVF is presented. The ACU payload is one typical ORU. Figure 7.28 shows the ACU being attached to the ORU tool change-out mechanism emulator (OTCME) of the SMT-robot.

Contact dynamics toolkit is a general contact dynamics modelling software and its contact models are based on elastic theory with consideration of detailed geometric and material properties. The toolkit is capable of simulating impact, bouncing, sliding, rolling, spinning, sticking and jamming for arbitrary geometries and simultaneous multiple contacts. Details of CDT can be found in (Ma, 1995, 2000, 2002).

In the presented approach, the basic idea for the verification and identification of the contact model and its underlying set of parameters is to match both motion and force signals of the SMT-robot and the SMT-robot simulator while performing an identical spatial contact operation. For this purpose, both robots are driven along the same pre-defined Cartesian velocity trajectory, whereby an FMA control scheme (Aghili *et al.*, 2001) is enabled. During the operation, all signals of interest are recorded, that is, Cartesian and joint motions as well as joint torques, worksite and end-effector forces. The free contact parameters in the simulation model are empirically adjusted based on the analyses of the recorded signals. Owing to the complexity of the occurring contact phenomena, a series of different contact operations with increasing complexity are performed.

7.4.4.1 Description of the simulation environment
The SMT-robot simulator is almost identical with the hardware-in-the-loop architecture used for STVF, except that the hardware components, that is, the SMT-robot

itself, the contact pair (payload) interaction and the force-plate, are modelled as well. The space robot emulation part, that is, the SPDM simulation, is disabled since it is not necessary.

A serial chain of rigid links connected by ideal joints and with lumped masses represents the SMT-robot. The multi-body model is generated using Symofros (Section 7.2). To reproduce the friction effects of the joints a friction model as described by Gonthier *et al.* (2004) was used. The contact pair interaction was modelled using a rigid-body CDT model with local compliance (Ma, 1995, 2000, 2002)[2], which also provides the force-plate signal. The 42 CDT input parameters can be grouped into three categories: (a) sixteen contact model configuration parameters, such as selectors for linear and non-linear force and damping models, different friction models, fixed or variable step size, and so on; (b) ten numerical parameters, which are mainly control parameters for the distance computation algorithm; (c) sixteen 'direct' contact parameters, for example, normal, tangential and rotational stiffness and damping parameters for the contact model, body masses, and so on.

It is important to note that the 'direct' contact parameters do not necessarily relate to true physical quantities, hence, they are difficult to estimate; especially when having in mind that the model does not describe the local contact phenomena but has to represent the test-bed compliance.

While the high-level model-based motion control is identical with the one of the SMT-robot, the low-level torque control does not exist. Assuming the latter to be perfect, the applied torque is set identical to the desired one computed by the high-level motion controller. On the basis of this modelling approach, the following simplifications can be identified:

(a)   link flexibility is not included;
(b)   joint flexibility is not included;
(c)   actuator dynamics (hydraulic fluid, valves, pipes) are not modelled;
(d)   no accurate identification of joint friction characteristics; simulation model parameters for the joint damping and joint friction were based on empirical values and subsequently tuned such that a stable simulation was obtained, while matching the free-space motion characteristics;
(e)   uncertainties in robot kinematics and dynamics parameter identification; the kinematics and inertia parameters of the robot are initially derived from CAD drawings and then adjusted using kinematic and dynamic identification;
(f)   ACU structural compliance could not be modelled, CDT represents contact phenomena by several point contact models of one rigid body;
(g)   force-plate compliance is not modelled;
(h)   joint load cells are not modelled.

Applied joint torque of the SMT-robot is determined based on axial load sensors, then converted into the corresponding joint torque, if the load-cell offsets and gains are not accurately determined and the non-linear conversion function leads to non-constant torque difference.

---

[2] The CDT is a proprietary technology to MDR, and we are not allowed to disclose confidential information regarding its implementation, or the parameters for the contract model it uses.

The use of rigid-body models for the robot, the contact pair and the force-plate makes it necessary to lump all compliance into the local compliance of the multi-point contact model of CDT, that is, the guidance mechanisms between the ACU and its berth. This fact indicates that the contact model with its parameter set represents not only the physics of the local contact phenomena but is bound to comprise the compliance and dynamics of the overall structure of the test-bed. Therefore, the identified contact model and its parameters are inherently related to the facility used to identify them. Furthermore, the model is used (abused) to represent something it is not made for. Since the payload can be regarded as a compliance element (spring) connected in series to the test-bed compliance, this effect becomes significantly important if the stiffness of the payload is close to the one of the test-bed. As a result, the model cannot be expected to represent the transition phases (no contact to contact) with a high fidelity due to this lumped-compliance approach. Furthermore, it is important to note that the end-effector position of the robot, and therefore the payload position, is based on the joint encoder reading and the rigid-body forward kinematics. Hence, even if the robot end-effector does not move, since it is in rigid contact with a surface, the computed end-effector position will indicate a motion as a function of the link and joint flexibility. However, it was experimentally verified that this approach still leads to more accurate results than the available external camera system, that is, OptoTrack made by Northern Digital Inc. (www.northerndigital.ca) with calibration software made by Krypton (www.krypton.be).

7.4.4.2 Experiments, simulations and results

In this section, the experiments and simulations with gradually increasing complexity are described. They were carried out to eventually accomplish a representative simulation of a complex spatial insertion task of an ACU. To this end, free motion experiments, studies of the static stiffness of the test-bed and the payloads, single-point normal as well as single-point normal and tangential contact experiments were conducted before complex spatial contact scenarios were investigated. For each step, the objective, the approach taken as well as the results are presented below.

7.4.4.2.1 *Unconstrained motion in joint position control mode*

Objective: For the envisioned parameter estimation approach, an identical – or more realistically, a similar – kinematics and dynamics response of the SMT-robot and the SMT-robot simulator is crucial. Therefore, this task is defined in order to verify the kinematics and dynamics response of the SMT-robot simulator in free motion. Benchmark criteria are (a) primarily, agreement of motion (position, velocity) of the end-effector and (b) secondarily, agreement of the joint torques of the simulation versus the experimental data.

Approach: In order to obtain experimental results, the SMT-robot is driven along a pre-defined joint position trajectory, whereby each joint position is commanded to move starting from the initial orientation by 10° and then back to the initial configuration. Beginning with the first joint, each joint was rotated sequentially.

The same trajectory is then used to drive the SMT-robot simulator. Even though the SMT-robot simulator (including its controller) contains approximately

200 parameters, 'only' an adjustment of the $6 \times 6 = 36$ joint friction parameters was considered to get a good match between the motion and torque signals of the SMT-robot and its simulator. The reason for this is the fact that all other model parameters are also used by the model-based controller of the SMT-robot itself, while the joint friction parameters are not. Hence, the controller related parameters were kept unchanged in order not to deteriorate the behaviour of the robot optimised for hardware-in-the-loop operations. However, one has to note that this set of controller related parameters is not necessarily the best possible choice in terms of model equivalence between SMT-robot and its simulator.

Results: The chosen joint friction parameters result in an SMT-robot simulator response very close to the one of the SMT-robot, as given in Figure 7.29, where *fb* and *sim* indicate feedback signals from the SMT-robot and the SMT-robot simulator, respectively. Although it was noticed that in order to generate the same motion pattern with the SMT-robot simulator as with the SMT-robot slightly different joint torques have to be applied. The simplifications of the robot simulator, see Section 7.4.4.1, seem to be the reason for that, especially the simplified joint friction model.

7.4.4.2.2   *Measurements of test-bed and payloads stiffness*

Objective: A critical part for contact dynamics simulations is identification of parameters for the normal force model. Since for the SMT-robot simulator model, all compliances of the SMT-robot and the payload are lumped into the contact model not only the stiffness of the payload has to be measured but also of the SMT-robot, which itself is mounted on two force-plates (so-called robot force-plates), and the force-plate to which the stationary contact body is attached (so-called worksite force-plate or force-plate for simplicity). Therefore, this task is defined in order to identify

1.   The test-bed stiffness, that is, SMT-robot with (worksite) force-plate.
2.   Test-bed stiffness with half-sphere payload.
3.   Test-bed stiffness with ACU payload.
4.   Stiffness of the half-sphere alone.

Approach: In test case (1) the SMT-robot directly grasps a micro-fixture, which, in turn, is bolted onto an aluminium plate (thickness 10 mm) directly on top of an aluminium supporting stilt (approximately 100 mm length), which, in turn, is mounted onto the force-plate. Using the FMA control scheme based on the force-plate signal, the desired downward velocity $(-z)$ is stepwise increased up to a resulting reaction force of about 250 N. In test case (2) the robot end-effector grasps the half-sphere, which is interfaced by a micro-fixture. The half-sphere is then brought into contact with the aluminium plate, which is, in turn, mounted onto the force-plate. A position in proximity of one of the four supporting stilts was chosen, similar to test case (1). Then, again, the desired z-velocity is increased stepwise. Test case (3) is similar to (2) but now the ACU is grasped and contacted with its berth mounted onto the aluminium plate. Test case (4) has been carried out at McGill University. To obtain a force–displacement characteristic, the half-sphere was loaded by a steel compression

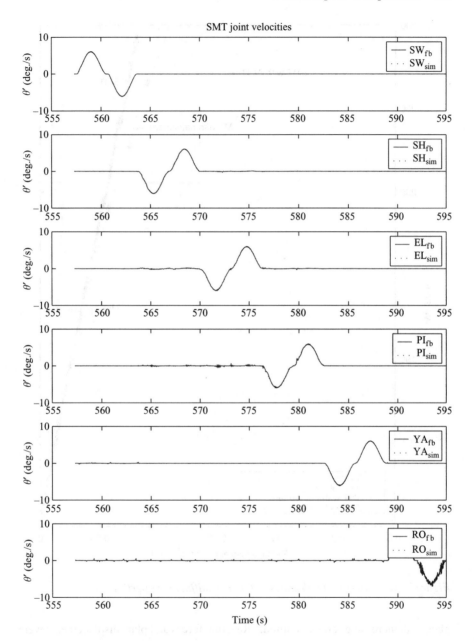

Figure 7.29   SMT-robot validation – joint velocities

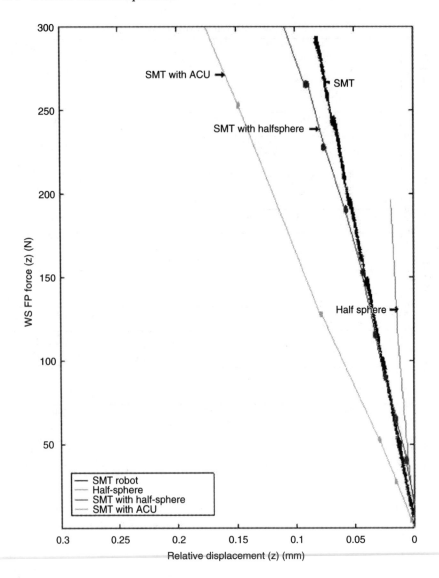

Figure 7.29　SMT-robot validation – test-bed stiffness properties

plate with increasing force magnitude. Reaction forces and plate displacements were recorded.

Results: The results for this test case are given in Figure 7.30. For test cases (1)–(3) an almost linear force–displacement characteristic has been obtained. The highest stiffness value of $k_{SMT} = 3.5e6$ N/m was measured for case (1), the SMT-robot together with the force-plate. This compliance is expected to come from the robot compliance itself and the load cells of the force-plates.

Within a force range below 200 N a slightly reduced stiffness was obtained, that is, $k_{SMT+hsphere} = 2.67e6$ N/m, for test case (2) with the half-sphere. The combined stiffness of test-bed and ACU, which also shows a rather linear behaviour, was determined to be about $k_{SMT+ACU} = 1.7e6$ N/m (test case (3)). The measurements of the half-sphere force–displacement characteristics (case (4)) resemble the relationship $F = k^*_{hsphere} x^{3/2}$, with F being the contact force, x the deformation in normal direction and $k^*_{hsphere} = 1.36e9$ N/m$^{3/2}$. It is interesting to note that if one approximates this with a linear stiffness of about $k_{hsphere} = 9.7e6$ N/m, the serial combination of stiffness of the SMT, k_{SMT}, and the stiffness of the half-sphere, $k_{hsphere}$, are resulting in a combined stiffness of approximately 2.6e6 N/m, which is very close to the experimentally determined value $k_{SMT+hsphere}$. Hence, this approach allows estimation of the stiffness of a payload by measuring the combined stiffness of payload and SMT as well as SMT alone.

7.4.4.2.3 Single-point contact with only normal commanded motion

Objective: The objective of this (and the next) task is to get more knowledge about the SMT-robot simulator with its contact model and its sensitivity to a change in contact model parameters. To make this as easy as possible, the single-point contact with only normal commanded motion is investigated in a first step.

For the numerical simulation of this contact scenario, the determined stiffness value of Section 7.4.4.2.2 could have been directly used as input parameter for the linear CDT contact model. However, the linear contact model was not applicable due to numerical problems of the simulator. The linear stiffness value could, nevertheless, be used as an initial guess for the estimation of the Young's modulus, which is utilized within the non-linear contact model. A direct conversion is, however, not possible, since the force–displacement characteristics of the non-linear model are not only a function of the Young's modulus, the Poisson's ratio and the penetration depth, but also of factors depending on geometric properties of the contact pair. The latter being numerically computed at each time step.

Approach: To obtain the experimental data, the SMT-robot was driven along a pre-defined Cartesian velocity trajectory under FMA control. The SMT-robot was put into a configuration such that the half-sphere was placed directly above the aluminium plate (same location as for Section 7.4.4.2.2, task (2)). Then, it was repeatedly moved down with a velocity of $v_z = -2$ mm/s and up with $v_z = 1$ mm/s. On the simulation side, three different approaches were taken:

1. A kinematics simulation, that is, the motion signals were directly fed into a CDT contact model without any robot in the loop.
2. The pre-defined trajectory was run with the SMT-robot simulator.
3. To gain additional insight into the numerical behaviour of the contact model, a separate simulation of a free falling half-sphere with only 1 DOF (in z-direction) was performed.

Results: In case (1) it was possible to adjust the stiffness parameters of the non-linear force model to get a match of the normal contact forces (z-direction). Although the commanded motion was only in the normal direction, tangential forces appear. This is caused by the fact that the performed motion is not perfectly normal and also that the robot end-effector and the force-plate orientation is not perfectly known in order to compensate for the tilting. After adjusting the z-position of the force-plate (and therewith the position of the fixed contact body) based on the contact period, a Young's modulus of $E = 10.127e7\ N/m^2$ was identified. Here one has to keep in mind that this number does not correspond to the physical value of the half-sphere but of the overall test-bed with the half-sphere. A Poisson's ratio of 0.3 was assumed. Using these values, case (2) was simulated, whereby the damping parameters were adjusted to get similar results compared to the experiment and a numerically stable simulation. The resulting motion, torque and force-plate signals[3] are given in Figures 7.31 and 7.32. A very good match can be observed, even for the small contact forces in the contact plane, that is, F_x and F_y and the moments. However, it is not possible to simulate the transition phase exactly. It was noted that for a simulation of a half-sphere contact with a plane (only 1 DOF, task (3)) the sampling time t_s has to be reduced to $t_s \leq$ 0.001 ms using an ODE5 integration scheme (Dormand–Prince formula) to ensure a physically meaningful simulation of the contact phase if the damping is completely turned off (2 mm/s impact velocity). Physically meaningful means in this regard that there is no 'energy gain' for the fully elastic impact due to the numerical discretisation during the simulation. However, this small sampling time is far too time consuming for the simulation with the SMT-robot simulator, which was finally run with ODE1 (Euler's method) and $t_s = 1$ ms. Hence, the sample time is another source of error.

7.4.4.2.4   *Single-point contact with normal and tangential commanded motion*

Objective: This task is defined in order to identify the static and kinetic friction coefficient to be used for the CDT contact model for the half-sphere–plane contact.

Approach: For this task, a similar approach was taken as in Section 7.4.4.2.3 but this time two horizontal motion components (y and x direction) were additionally commanded. On the simulation side, an approach similar to case 1 of Section 7.4.4.2.3 was tried, but failed since it turned out to be too complicated to find a proper orientation matrix for the force-plate with respect to the robot. In particular, it was found that this matrix as well as the force-plate position were time dependent. Hence, simulations similar to case 2 of Section 7.4.4.2.3 were carried out. This approach (Cartesian velocity controlled motion with FMA) is a lot less stringent in terms of payload and worksite geometric uncertainties due to its compensating effects. The controller always tries to keep the reaction force constant and not the end-effector position. Therefore, considering the FMA functionality, the slope of the contact force at impact is the critical matching criterion and not the contact force at steady state. A steep slope corresponds to a large stiffness and a gentle slope to a small stiffness. The maximal force amplitude is always given by an FMA gain relating the desired velocity and the maximal force, that is, in this case approximately 10 N per 1 mm/s, plus a 10 N deadband.

---

[3] Note: Motion signals are given in the 'SMT world frame' $\mathcal{K}_\mathcal{M}$ and forces are given in the 'work-site force-plate frame' $\mathcal{K}_\mathcal{F}$, with $R : \mathcal{K}_\mathcal{M} \rightarrow \mathcal{K}_\mathcal{F}, R = R_z(\pi/2)R_x(\pi)$.

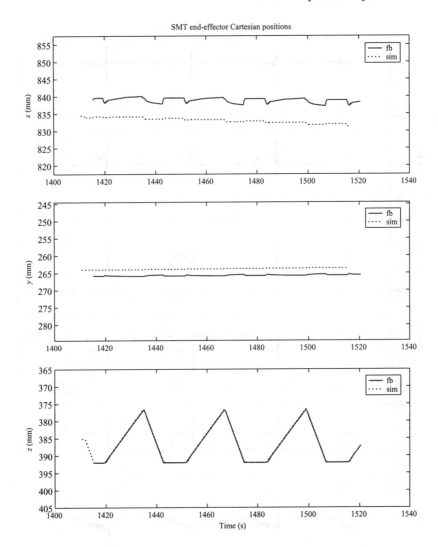

Figure 7.31   Normal contact: half sphere – Cartesian position of EE

Results: Figures 7.33 and 7.34 shows the results of single-point contact with normal and tangential commanded motion. Similar to the results obtained previously, a good match between simulated and experimental results can be observed for the motion signal. The joint torques are somewhat different again.

7.4.4.2.5   General spatial contact under free-fall conditions (ACU–berth without SMT-robot)

Objective: This task is carried out to get acquainted with the numerical properties of the ACU–berth contact and to estimate a set of contact stiffness and

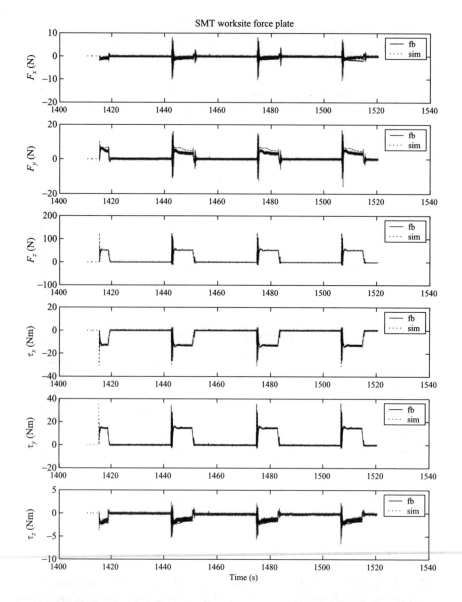

Figure 7.32   Normal contact: half sphere – worksite force plate measurements

damping parameters based on the results of Section 7.4.4.2.2 and engineering assessment.

Approach: Owing to lack of a proper release mechanism and fragility and price of the ACU mock-up, this task was only carried out in simulation. For this purpose, an unconstrained rigid body (ACU) under the influence of gravity and the contact forces computed by the ACU model of CDT was simulated in Simulink. The ACU was

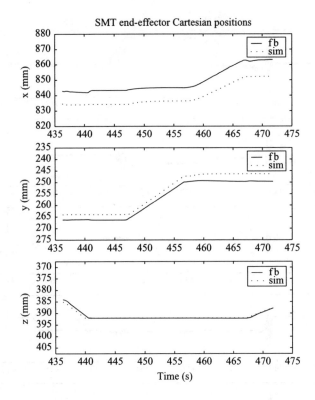

Figure 7.33 *Normal, tangential contact: half sphere – Cartesian position of EE*

dropped from a position which had a minor offset in vertical and horizontal direction compared to the fully inserted one.

    Results: The major work to establish a suitable input parameter file was the adjustment of the damping parameters. The stiffness parameters were estimated based on the results of Section 7.4.4.2.2. Furthermore, it has to be noted that the friction parameters were already identified within another project conducted by MDR. But even if the conditions for these particular experiments might have been slightly different, using the SMT-robot to estimate also the friction parameters would have led to larger errors.

7.4.4.2.6 *General spatial contact (ACU–berth with SMT-robot)*

Objective: This task represents the main objective of the work carried out, that is, to establish the final parameter set suitable for the simulation of the complex spatial ACU insertion into its berth at low speed (2 mm/s). To verify the parameter set, two different insertion scenarios are investigated.

    Approach: To obtain the experimental data, the SMT-robot was driven along a predefined trajectory starting just slightly above the fully inserted position. To establish an initial reference position and orientation, which can be approximately repeated in

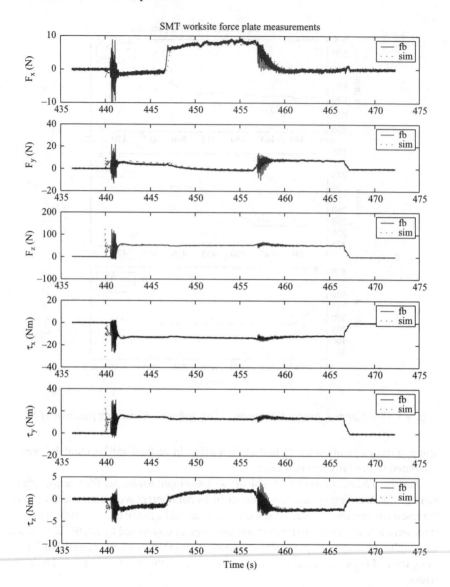

Figure 7.34   Normal, Tangential contact: half sphere – worksite force plate measurements

simulation (in spite of the present absolute pose error), the ACU was first fully inserted and kept inserted for a few seconds. Then, a relative motion profile was executed. The first trajectory prescribes first an upward motion (+z), then a sidewards motion (+x) and finally a downward motion again (−z). The second trajectory uses instead of the sidewards motion a rotation about the vertical axis (+z). The same trajectories are run in order to obtain the simulation results.

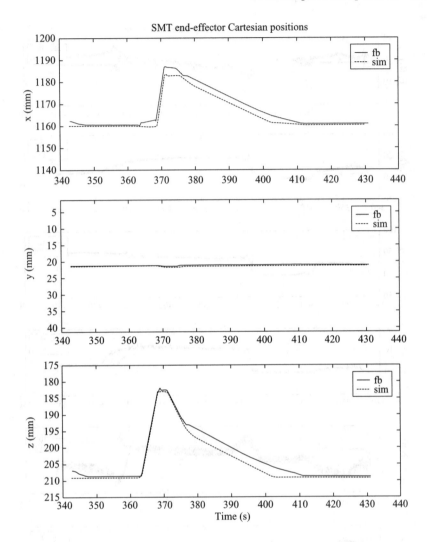

Figure 7.35   3D ACU (case 1 trajectory) – Cartesian position of EE

Results: The comparison of experimental versus simulation results is shown in Figures 7.35 and 7.36 for the first trajectory and in Figures 7.37 and 7.38 for the second trajectory. Having in mind the complexity of the contact task, the complexity of the driving robot and the simplifications of the SMT-robot simulator and the contact model, a good agreement between the signals can be stated. One has to note also that only data after the establishment of the initial contact, that is, before the start of the relative motion (up, offset, down), should be compared ($t > 370$ s for case 1; $t > 190$ s for case 2). It is also important to mention that it is quite difficult to perform contact parameter studies with this final and complex model, that is, to correlate for instance changes in the time history of the contact forces to a specific contact parameter.

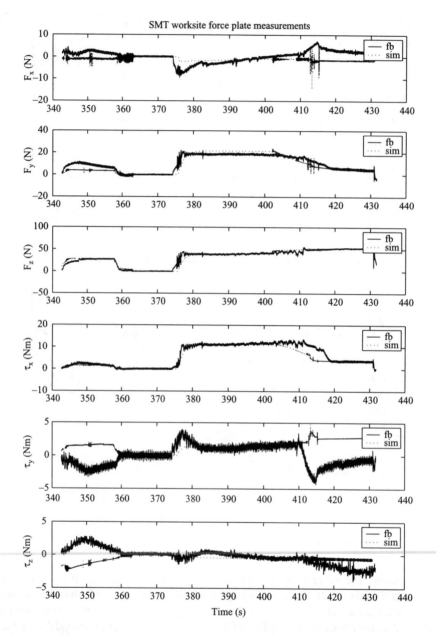

Figure 7.36   3D ACU (case 1 trajectory) – worksite force plate measurements

While it was possible to simulate a free-falling ACU (Section 7.4.4.2.5) using an implicit integration scheme, it was not possible, with the current configuration and environment to run the full SMT-robot simulation with the CDT contact model of the ACU with an implicit integrator. The reason for that seems to be an increased stiffness

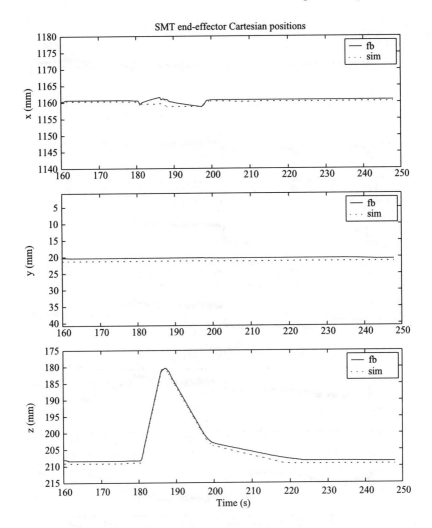

Figure 7.37   3D ACU (case 2 trajectory) – Cartesian position of EE

of the resulting equations due to the interaction of the contact model with the robot model including its controller. Furthermore, it has to be noted that due to stochastic phenomena and uncertainties, the 'same' experiment (same driving trajectory) leads in some cases to (slightly) different contact forces. Hence, it is difficult to establish the true set of data to which to compare to.

7.5   On-orbit MSS training simulator

Experimental tests and analyses have shown that the capture of free-flyers is a very complicated task to be performed by a robotic operator on-board the ISS. The understanding of MSS and free-flyer dynamics require highly qualified and well-trained

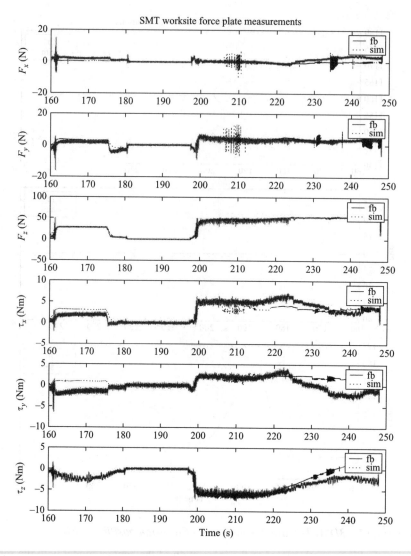

Figure 7.38   3D ACU (case 2 trajectory) – worksite force plate measurements

operators. The dexterity and accuracy of the astronauts may decrease over time if they are not trained on-board. It was an obvious choice to have a simulator on-orbit to keep the skills of the astronauts at the required level. In order to support the training scenarios required by the on-orbit training, the SMP[4] simulator has been developed. This section presents the goals and the architecture of SMP and provides some results obtained with the system.

[4] System for maintaining, monitoring MRP performance on-board the ISS.

7.5.1   On-orbit training and simulation

The safety and success of MSS operations will depend on reliable performance by the MSS robotics operator (MRO). This is particularly important for the human-in-loop modes involved in such critical tasks as free-flyer capture, payload berthing/unberthing, capture/release, extravehicular activity (EVA) support and SPDM positioning. Crew time is an expensive resource on the ISS, and training needs to be adapted to each individual, performed only when needed, and focused on the skills that degrade the most on-orbit. As a result of analysis of operators' pre-flight performance during training using the MOTS and on-orbit operation using the MSS on the ISS, critical MRO skills have been identified and prioritised. At the top of the list are the skills required to perform a free-flyer capture. This involves significant target velocity and strict execution time constraint that requires excellent hand-controller skills and complex three-dimensional transformations from the operator. SMP uses a new concept of on-board training developed at CSA. It includes the monitoring of pre-selected MRO critical skills using simulator exercises to identify MRO weaknesses based on analysis of on-orbit performance against pre-flight baseline. On the basis of such analysis, the training is adjusted and repeated until the operator performance recovers to the baseline level. In order to identify skill degradation, SMP implements the following elements: (a) use of a skill-centred task (free-flyer capture); (b) objective performance index; (c) immediate user performance feedback on a task by task basis; (d) a trend analysis of performance over time.

Implementing these elements requires a realistic on-board simulation of the free-flyer capture, the capability to record performance data on both pre-flight and on-orbit exercises, and a simple-to-use analysis tools to provide feedback to the crew as well as detailed analysis to the training community on the ground.

7.5.2   Hardware architecture

As shown in Figure 7.39(a), SMP comprises an IBM Thinkpad Laptop (P-III 800 MHz), hand controllers (similar to the ones used to operate the MSS) and an electronic interface module acting as a Universal Serial Bus (USB) interface on the computer side and as a data acquisition interface on the hand-controller side. The use of USB has numerous advantages, one of the most interesting being that connected devices can draw power from the host, thus not requiring separate power supply. The standard is widely in use in industry and thus has widespread, extensive software support. The controlling entity in the interface is the Q-Card. It supplies power to the controllers, sampled the analogue data as well as their digital button outputs and controls a USB link layer that handles the low-level protocol issues. The Q4-SMP USB interface unit consists of two circuit boards: a Q4 and a purpose-built daughterboard (Figure 7.39(b)).

7.5.3   Software architecture

The simulator includes four modules, the graphical user interface (GUI), the analysis module, the visual renderer (VR) and the dynamic simulator (SIM) as shown in Figure 7.40(b).

(a) (b)

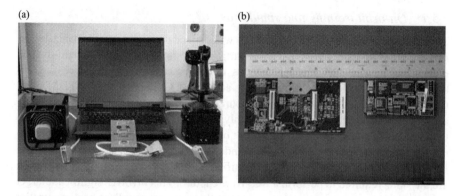

Figure 7.39   SMP hardware architecture

(a) (b)

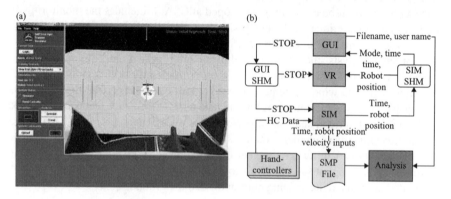

Figure 7.40   SMP graphical user interface and software architecture

The SMP GUI has been developed with Java and runs on a Windows operating system (Figure 7.40(a)). It manages user logging, user session files and user profiles. It also spawns other SMP modules and displays data during simulation or analysis sessions. It also controls the SMP modules execution using shared memory (SHM). The VR provides a virtual environment that models a free-flyer, the SSRMS end-effector camera view and the SSRMS capture overlays (Figure 7.40(a)). The engine is driven by the data generated in real time by the simulator during simulation sessions.

7.5.4   Simulation validation

The SSRMS model developed with Symofros has been validated using the MDSF. MDSF includes the model of the SSRMS, which has been identified and validated against flight data and considered as a truth model of the system. This model includes all the relevant physical characteristics of the SSRMS: detailed joint models that account for friction, non-linear stiffness, backlash and damping; flexibility of the two long booms, validated mass and inertial parameters; models of the controllers and algorithms. The simplified SSRMS model used in the SMP simulator has been compared to the MDSF model in terms of the fundamental structural frequencies, and based on simulating the motion for end-effector trajectories. It was found that there is a good agreement between the first few fundamental frequencies. These govern the response of the robot for the majority of robotics operations. The agreement in performance was also confirmed based on comparing the simulated responses for end-effector trajectories. Although, it is important to note that the validation was performed only for one configuration of the SSRMS. Since the simulator always starts in this configuration and this is the only training scenario for SMP, this configuration was deemed sufficient for this validation of the simulator. However, results as good are not expected for other configurations of the manipulator.

7.5.5   Symofros simulator engine

This module has been developed with Symofros and simulates the motion of the SSRMS end-effector and the free-flyer. For the purpose of the free-flyer scenario, astronauts will always use the manual augmented mode (MAM). This is a Cartesian velocity mode where the desired velocity signals come from the hand-controller inputs. The desired Cartesian velocity is mapped in joint space using the Jacobian matrix of the manipulator. The arm has 7 DOF, thus the Jacobian matrix is not square. In the first rate resolver mode, this matrix is inverted using the pseudo-inverse. In the second mode, the first joint is locked assuming a zero velocity command and the first column of the Jacobian is eliminated accordingly to obtain a square matrix, easily invertible. The third rate resolver mode consists in locking the second joint, again assuming a zero velocity command for that joint, and by inverting the Jacobian matrix with its second column removed. The output of the rate resolver, the desired joint velocity command, is tracked using a PI controller with collocated feedback of the actual joint velocity.

7.5.6   Analysis module

The multi-level automated feedback informs the MRO about his level of skills. This allows the MRO to find out if more training is needed or not, and also to find out the cause of a particular success or mistake. The session analysis provides information such as the hand-controller rates, the relative position and velocity between the end-effector and the free-flyer, and the capture status. Operational criteria and heuristics are used to provide a score, which allows the astronaut to have a good picture of the personal progress over time using a trend analysis. The astronaut can then determine if more training is needed or not.

7.5.7   Ground and on-orbit results

The skills required for free-flyer capture have been found to be very close to those required for docking a Soyuz spacecraft (precise sensory-motor coordination and complicated cognitive spatial–logical transformations). Research conducted on docking skill degradation has been used as the basis for the study of MRO performance.

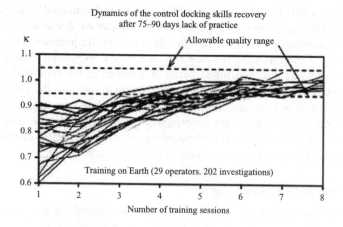

Figure 7.41   Soyuz docking skill degradation

These studies show that a period of 25–30 days without docking practice results in 20–25 per cent decrease in performance and that operators require 3–4 training sessions to recover. Periods of 75–90 days typically cause performance degradation of up to 40 per cent, as shown in Figure 7.41. The same trend, even more pronounced according to preliminary results, is expected for free-flyer capture skills. Figure 7.41 shows the significant differences between subjects (each line represents one subject performance over training sessions) as for skill degradation level and also for their recovery rate. The evaluation of free-flyer capture performance uses the following measurements:

- Operation effectiveness parameters: final positioning accuracy, trajectory deviation (relative to the ideal), oscillation amplitudes, rate, acceleration, time of operation.
- Human motor output parameters: hand-controller's deflection direction, proportions of outputs, smoothness of deflection, concurrent and multi-axes deflections.

A few cosmonauts, already expert in Soyuz docking, were trained for free-flyer capture on SMP on the ground. They considered the free-flyer task more complicated than docking, which is confirmed by experimental results shown in Figure 7.42. It shows that the subject (a professional cosmonaut) got to a certain level of performance stability (successful captures) after about 60 trials. While the average motion roughness decreased and success rate increased in the last few sessions, a few failed captures still occurred, showing that after 60 trials, the subject did not reach yet a plateau in performance.

Russian cosmonauts have used SMP officially and the first trainees have executed more than 75 simulation sessions. SMP was launched on 2 February 2003 on-board the ISS.

Currently, a new experiment (SMP2) designed to study the three-dimensional mental image reconstruction is being developed while performing tasks not only

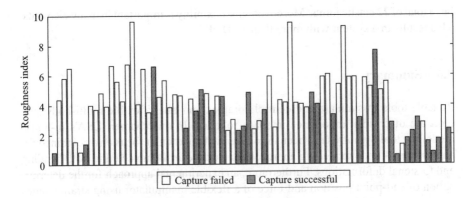

Figure 7.42   *Free-flyer capture skill acquisition dynamics*

Figure 7.43   *Visual rendering of ISS for SMP2*

involving SSRMS but the complete MSS with all its components, that is, MBS, SSRMS and SPDM, including the two arms, the SPDM body roll and the two grippers at their end. The long-term goal of this research project is to gather data to determine the frequency of on-orbit training necessary to ensure that the MSS and similar complex systems are operated safely. In order to study performance degradation and skill recovery, a highly efficient simulator is required in order to ensure on-orbit real-time simulation and fast feedback to the operator. This simulator, based on the dynamics engine of Symofros, visualises the overall ISS and the operator can control the complete MSS, Figure 7.43. The challenge is the real-time simulation of the Canadarm2 and Dextre dynamics while performing graphics rendering of the worksite environment using the same computer, a P4 1.8 Ghz IBM ThinkPad. The combined system has a total of 22 elastic joints. Moreover, link flexibility is important in many elements. This results in a system with more than 50 DOF.

7.6   Summary

Reliable tools to simulate and control complex and often flexible multi-body systems are of critical importance in space robotics. In this chapter, CSA's in-house modelling tool Symofros has been introduced with particular focus on a combined FE and assumed-mode modelling approach for flexible arms subjected to both bending and torsional deformations. Furthermore, an experimental approach for the determination of end-point position and force of a flexible manipulator using strain gauges in conjunction with joint sensing has been presented. The SPDM test verification facility has been discussed as an example using HLS to solve the contact dynamics problem that arises during payload insertion and extraction tasks of flexible space robots. Furthermore, the on-board training simulator SMP was detailed as an example for an application of the modelling of flexible space robots as training simulator on-board the ISS.

7.7   Acknowledgements

The work presented in this chapter has resulted from the work of several engineers and researchers of the Robotics Section, Space Technologies of the CSA over last several years, which the authors would herewith like to acknowledge.

Chapter 8

Open-loop control of flexible manipulators using command-generation techniques

*A.K.M. Azad, M.H. Shaheed, Z. Mohamed, M.O. Tokhi
and H. Poerwanto*

This chapter presents the development of open-loop command-generation techniques for control of flexible manipulators based on filtered input, Gaussian shaped input and input shaping and provides a comparative assessment of the performance of these techniques. It is assumed that the motion itself is the main source of system vibration. Thus, input torque profiles that do not contain energy at system natural frequencies will not excite system resonances and hence will not result in structural vibration. Accordingly, shaped torque inputs, including Gaussian shaped, low-pass and band-stop filtered torque input functions and input shaping profiles are developed on the basis of identified resonance modes of the system using parametric and non-parametric modelling methods. Experimental results verifying the performance of the developed control strategies are presented and discussed. Performances of the techniques are assessed in terms of level of vibration reduction at the natural frequencies, time response specifications and robustness to natural frequency variation. The effects of various loading conditions on the performance of the system are also studied.

8.1   Introduction

The control strategies for flexible manipulator systems can be classified as open-loop (feedforward) and closed-loop (feedback) control schemes. Open-loop techniques for vibration suppression involve developing the control input through consideration of the physical and vibration properties of the system, so that system vibrations at dominant response modes are reduced. This method does not require additional sensors or actuators and does not account for changes in the system once the input is developed.

On the other hand, feedback control techniques use measurement and estimations of the system states to reduce vibration. Feedback controllers can be designed to be robust to parameter uncertainty. For flexible manipulators, feedforward and feedback control techniques are used for vibration suppression and end-point position control, respectively (Moulin and Bayo, 1991; Wang, 1986). An acceptable system performance without vibration that accounts for system changes can be achieved by developing a hybrid controller consisting of both control techniques. Thus, with a properly designed feedforward controller, the complexity of the required feedback controller can be reduced. Various feedforward control strategies have been investigated for control of vibration. These include the use of Fourier expansion as a forcing function (Aspinwall, 1980), the application of computed torque control (Moulin and Bayo, 1991), utilization of single and multiple-switch bang-bang control function (Onsay and Akay, 1991) and construction of input functions from ramped sinusoids and versine functions (Meckl and Seering, 1990).

   This chapter presents the development of open-loop command-shaping techniques for control of a flexible manipulator. Moreover, the chapter provides a comparative assessment of the performance of these techniques. The results presented in this chapter are all based on the Sheffield manipulator described in Chapter 1. For input shaping with a four-impulse sequence and sixth-order low-pass and band-stop filters are considered. Initially, to identify the characteristic parameters of the system, the flexible manipulator is excited with a single-switch bang-bang torque input and its vibration behaviour is monitored. Then the filters and input shapers are designed and used for pre-processing the input, so that no energy is fed into the system at the natural frequencies. Experiments are performed on a laboratory scale single-link flexible manipulator and results are presented to verify the performance of the control strategies. Performances of the techniques are assessed in terms of level of vibration reduction at the natural frequencies, time response specifications and robustness to natural frequency variation. These are accomplished by comparing the system response with the unshaped bang-bang input. The robustness of the control schemes is assessed with up to 30 per cent tolerance in vibration frequencies. As the dynamic behaviour and vibration of flexible manipulators is significantly affected by payload variations, the performance of the control strategies is also assessed with a flexible manipulator incorporating a payload. Finally, a comparative assessment of the performances of the control strategies for open-loop control of a flexible manipulator is presented.

8.2   Identification of natural frequencies

The open-loop control strategies were designed on the basis of identified natural frequencies and damping ratios of the flexible manipulator system. So, the identification of natural frequencies is an important factor for efficient design of an open-loop controller. The natural frequencies are identified using four distinct approaches. These are

- analytical
- experimental

- genetic modelling
- neural network (NN) modelling.

Through the solution of the system's equation, one can find the natural frequencies of a flexible manipulator system, as described in Chapter 2. There are a number of experimental procedures, which can be used to find the natural frequencies of the system (Tokhi and Azad, 1997). In this chapter, the natural frequencies of a flexible manipulator are obtained experimentally by exciting the manipulator with the unshaped bang-bang torque input in an open-loop configuration. Genetic algorithm (GA) and NN-based system identification techniques are used to find the dominant modes of vibration of the system.

8.2.1 Analytical approach

The natural frequencies for a flexible manipulator system can be obtained through analytical process by solving the system's dynamic equation of motion. The details of the process are given in Chapter 2 of this book. The analytical values of natural frequencies ω_i can be obtained using

$$\lambda_i = \beta_i l; \qquad \varepsilon = \frac{I_h}{Ml^2} = \frac{3I_h}{I_b} \qquad (8.1)$$

The parameter ε determines the vibration frequencies of the manipulator; a small ε implies lower vibration frequencies. For a very large ε the vibration frequencies correspond to those of a cantilever beam. The effect of a payload mass, on the other hand, is significant on the vibration frequencies. The natural frequencies for the first three modes of the (Sheffield manipulator) system obtained through this analytical method are 12.499, 36.36 and 64.56 Hz, respectively.

8.2.2 Experimental approach

The natural frequencies can also be obtained through experimental process. In one approach the manipulator system can be excited by a random signal, which covers all the vibration modes of interest. The responses at various points are measured for further analysis. There are two distinct approaches to analyse the collected data for natural frequencies. The first approach is based on the measurement of the autopower spectral density of the response of the system. This is referred to as the spectral density method. The second method is based on the measurements of the frequency response function (FRF) and coherency function of the system. This is referred to as the FRF method. To obtain a better accuracy around the resonance, the manipulator system can be excited by a stepped sine wave from a spectrum analyser and the response around the natural frequencies obtained.

The autopower spectral density $S_{xx}(\omega)$ of a signal x is defined as

$$S_{xx}(\omega) = S_x(j\omega) S_x^*(j\omega) \qquad (8.2)$$

where $S_{xx}(\omega)$ is a real-valued function containing the magnitude information only, $S_x(j\omega)$ is the linear spectrum of x given by the Fourier transform of the time signal $x(t)$. $S_x^*(j\omega)$ is the complex conjugate of $S_x(j\omega)$.

The response of the system can alternatively be described by the FRF. The equations relating the response of a system in random vibrations to the excitation are given as (Newland, 1996)

$$S_{yy}(\omega) = |H(j\omega)|^2 S_{xx}(\omega)$$
$$S_{xy}(j\omega) = H(j\omega) S_{xx}(\omega)$$
$$S_{yy}(\omega) = H(j\omega) S_{yx}(j\omega)$$

(8.3)

where $S_{xx}(\omega)$ and $S_{yy}(\omega)$ are the autopower spectral densities of the excitation signal x and the response signal y, respectively. $S_{xy}(j\omega)$ and $S_{yx}(j\omega)$ are the cross-spectral densities between these two signals and $H(j\omega)$ is the FRF of the system. Let $H_1(j\omega)$ and $H_2(j\omega)$ denote two estimates of the FRF obtained according to the relations in equation (8.3) as

$$H_1(j\omega) = \frac{S_{xy}(j\omega)}{S_{xx}(\omega)}$$

$$H_2(j\omega) = \frac{S_{yy}(\omega)}{S_{yx}(j\omega)}$$

(8.4)

The error between the two functions is given by the coherency function defined as

$$\gamma^2(\omega) = \frac{H_1(j\omega)}{H_2(j\omega)}$$

(8.5)

In this manner, the coherency function gives a measure of the estimation error and indicates the level of coherence between the input and the output. If γ^2 is unity at some frequency ω, this means that the output is entirely due to the input at that frequency. However, a value of γ^2 less than unity means that either the output is due to the input as well as other inputs or the output is corrupted with noise.

The time responses of the system were measured at the hub, at the four strain gauge locations (along with the length of the manipulator) and at the end-point. These were used to obtain autopower spectral densities of the signals using equation (8.2) and the FRFs and coherency functions between the input torque and the responses using equations (8.3), (8.4) and (8.5). The natural frequencies for the first three modes were obtained by identifying the peaks (maximum amplitudes) in the autopower spectral density functions and the FRFs. The average values of these natural frequencies obtained through this process are shown in Table 8.1, along with the corresponding analytical values. The details of the measurement process are given in Chapter 2.

Table 8.1   Comparison of the analytical and experimental values for the first three natural frequencies

Methods	First mode (Hz)	Second mode (Hz)	Third mode (Hz)
Analytical	12.499	36.36	64.56
Spectral density	12.068	35.993	63.33
FRF	12.137	36.132	63.01

8.2.3   Genetic modelling

In this approach, a GA is used for parametric identification of the flexible manipulator, and is used for developing shaped torque on the basis of the detected vibration modes. The operating mechanism of a GA was described in Chapter 4.

For parametric identification of the manipulator with GA, randomly selected parameters are optimised for different, arbitrarily chosen, order to fit to the system by applying the working mechanism of GA. The fitness function utilized is the sum-squared error between the actual output, $y(n)$, of the system and the predicted output, $\hat{y}(n)$, produced from the input to the system and the optimised parameters:

$$f(e) = \sum_{i=1}^{n} (|y(n) - \hat{y}(n)|)^2 \tag{8.6}$$

where n represents the number of input/output samples. With the fitness function given above, the global search technique of the GA is utilized to obtain the best set of parameters among all the attempted orders for the system. The output of the system is thus simulated using the best sets of parameters and the system input.

The manipulator is modelled from the input torque to hub-angle, hub-velocity and end-point acceleration. These are referred to as the hub-angle model, hub-velocity model and end-point acceleration model respectively. In all the three cases, it was attempted to simulate the GA with different initial values and operator rates. Satisfactory results were achieved with the following set of parameters:

Generation gap:	0.9
Crossover rate:	0.7
Mutation rate (hub-angle model):	0.00313
Mutation rate (hub-velocity model):	0.00357
Mutation rate (end-point acceleration model):	0.00417

8.2.4   Neural modelling

Various modelling techniques can be used with NNs to identify non-linear dynamic systems. These include state-output model, recurrent state model and non-linear autoregressive moving average process with exogeneous input (NARMAX) model. However, it is evident from the literature that if the plant's input and output data are available, the NARMAX model is a suitable choice, for modelling systems having non-linearities, with standard backpropagation and radial basis function (RBF) learning algorithms. Mathematically, the model is given by (Luo and Unbehauen, 1997)

$$\hat{y}(t) = f[(y(t-1), y(t-2), \ldots, y(t-n_y),$$
$$u(t-1), u(t-2), \ldots, u(t-n_u),$$
$$e(t-1), e(t-2), \ldots, e(t-n_e)] + e(t) \tag{8.7}$$

where $\hat{y}(t)$ is the output vector determined by the past values of the system input vector, output vector and noise with maximum lags n_y, n_u and n_e, respectively, $f(\cdot)$

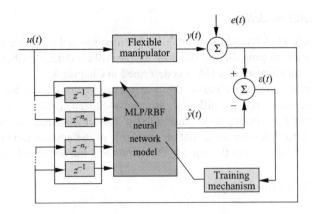

Figure 8.1   NARX model identification with MLP/RBF neural networks

is the system mapping constructed through multi-player perception (MLP) method or RBF NNs with an appropriate learning algorithm. The NN modelling structure in block diagram form is shown in Figure 8.1.

The model is also known as NARMAX equation error model. However, if the model is good enough to identify the system without incorporating the noise term or considering the noise as additive at the output, the model can be represented in a NARX form (Luo and Unbehauen, 1997; Sze, 1995) as

$$\hat{y}(t) = f[(y(t-1), y(t-2), \ldots, y(t-n_y)),$$
$$u(t-1), u(t-2), \ldots, u(t-n_u)] + e(t) \qquad (8.8)$$

8.2.5   Natural frequencies from the genetic and neural modelling

It is highly required to determine the dominant modes of vibration of the system as precisely as possible for achieving the best possible results from the resonance mode-based open-loop control process. The first three modes of vibration of the system as detected from modelling the manipulator with GAs and artificial neural networks (ANNs) are presented in Tables 8.2–8.4.

It is noted that out of the 23 cases, the first mode appeared at 11.112 Hz in 19 cases and hence this can be used as the first vibration mode of the system. The second

Table 8.2   Modes detected from GA-based modelling with composite-PRBS input

Modelling domains	First mode (Hz)	Second mode (Hz)	Third mode (Hz)
GA-based hub-angle model	11.112	34.26	63.89
GA-based hub-velocity model	11.112	36.11	63.89
GA-based end-point acceleration model	11.112	36.11	63.89

Table 8.3   Modes detected from NN-based modelling with Composite-PRBS input

Modelling domains	First mode (Hz)	Second mode (Hz)	Third mode (Hz)
MLP-NN-based hub-angle model (OSAP)	11.112	34.26	63.89
MLP-NN-based hub-velocity model (OSAP)	11.112	36.11	63.89
MLP-NN-based hub-velocity model (MPO)	11.112	36.11	65.74
MLP-NN-based end-point acc. model (OSAP)	11.112	35.185	64.815
MLP-NN-based end-point acc. model (MPO)	12.96	34.26	63.89
RBF-NN-based hub-angle model (OSAP)	11.112	34.26	63.89
RBF-NN-based hub-velocity model (OSAP)	11.112	36.11	63.89
RBF-NN-based hub-velocity model (MPO)	—	36.11	65.74
RBF-NN-based end-point acceleration model (OSAP)	12.96	35.185	64.815
RBF-NN-based end-point acceleration model (MPO)	12.96	36.11	63.89

Table 8.4   Modes detected from NN-based modelling with white noise input

Modelling domains	First mode (Hz)	Second mode (Hz)	Third mode (Hz)
MLP-NN-based hub-angle model (OSAP)	11.112	34.26	62.00
MLP-NN-based hub-velocity model (OSAP)	11.112	35.185	64.815
MLP-NN-based hub-velocity model (MPO)	11.112	35.185	64.815
MLP-NN-based end-point acc. model (OSAP)	11.112	35.185	64.815
MLP-NN-based end-point acc. model (MPO)	11.112	35.185	64.815
RBF-NN-based hub-angle model (OSAP)	11.112	34.26	64.815
RBF-NN-based hub-velocity model (OSAP)	11.112	35.185	64.815
RBF-NN-based hub-velocity model (MPO)	11.112	35.185	64.815
RBF-NN-based end-point acceleration model (OSAP)	11.112	35.185	64.815
RBF-NN-based end-point acceleration model (MPO)	11.112	35.185	64.815

mode appeared at 35.185 Hz in 10 cases, 34.26 Hz in 6 cases and 36.11 Hz in 7 cases. However, the average of 34.26 Hz and 36.11 Hz is also 35.185 Hz. Hence, it is very likely that the second mode of the system is located at 35.185 Hz. The third mode appeared at 64.815 Hz in 11 cases, 63.89 Hz in 9 cases, 65.74 Hz in two cases and 62 Hz in 1 case. Again, the average of 63.89 Hz and 65.74 Hz is also 64.815 Hz. Thus, the probability of the third vibration mode of the system to be located at 64.815 Hz is quite high.

8.3   Gaussian shaped torque input

A Gaussian shaped torque input, that is, the first derivative of the Gaussian distribution function, is examined in this section. The application of this function in the form of an acceleration profile, to develop input torque profile through inverse dynamics of the system, has previously been shown (Bayo, 1988). Here the behaviour of the function as an input torque profile for the system is investigated by adopting a much simpler method of developing an input torque profile for a flexible manipulator system. Variation of frequency distribution, duty cycle and amplitude of the Gaussian shaped torque input with various parameters are studied. This enables the generation of an appropriate input trajectory to move the flexible manipulator for a given position with negligible vibration. The Gaussian distribution function can be written as

$$P(x) = \frac{1}{\sqrt{2\pi}\sigma} e^{[-(x-\mu)^2/2\sigma^2]} \tag{8.9}$$

where σ represents the standard deviation and μ the mean of the variable x.

This function is shown in Figure 8.2(a). Taking the first derivative of this function yields

$$\frac{dp(x)}{dx} = \frac{(x-\mu)}{\sqrt{2\pi}\sigma^3} e^{[-(x-\mu)^2/2\sigma^2]} \tag{8.10}$$

Considering the left-hand side of equation (8.10) as a system torque input with x representing time and μ and σ as constants for a given torque input, yields

$$\tau(t) = \frac{(t-\mu)}{\sqrt{2\pi}\sigma^3} e^{[-(t-\mu)^2/2\sigma^2]} \tag{8.11}$$

where, τ is the system input torque and t is time as the independent variable. The Gaussian shaped torque input thus obtained is shown in Figure 8.2(b).

To study the effects of μ and σ on various properties of the driving torque, these are varied and the corresponding torque obtained. Figure 8.3 shows the cut-off

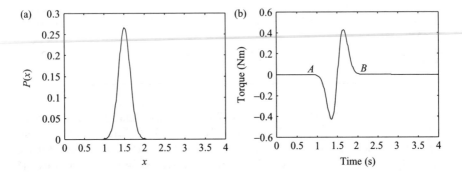

Figure 8.2   Gaussian distribution function and developed torque input: (a) Gaussion distribution function, (b) derivative of Gaussion distribution function

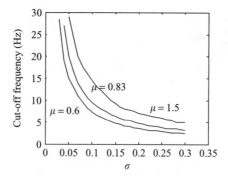

Figure 8.3 Variation of cut-off frequency as a function of σ

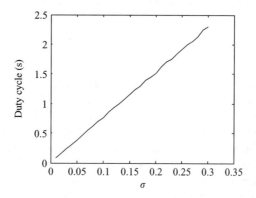

Figure 8.4 Relation between the duty cycle and the σ

frequencies of the Gaussian input torque as a function of σ, with μ as a parameter. The cut-off frequency of this input is obtained from the autopower spectrum of the developed torque profile, where at the cut-off frequency the power level of the input is reduced to two-thirds of its peak value. It is noted that for a given value of μ, the cut-off frequency increases with a decrease in the value of σ. For a given value of σ, on the other hand, the cut-off frequency increases with an increase in the value of μ. Figure 8.4 shows the variation of the duty cycle of the Gaussian torque input as a function of σ. The duty cycle corresponds to the time, between points A and B in Figure 8.2. This is useful in the selection of the value of σ for an allowed period of movement. Figure 8.5 shows the variation of the amplitude of the Gaussian torque input with the value of σ. This is useful that the maximum amplitude of the developed torque profile can be kept within a particular range so as to avoid actuator saturation and structural damage. Note that the last two properties, namely, the duty cycle and amplitude shown in Figures 8.3 and 8.4 are independent of μ. With the set of information given in Figures 8.3–8.5, it is possible to select suitable parameters for the Gaussian torque input and generate the torque profile accordingly.

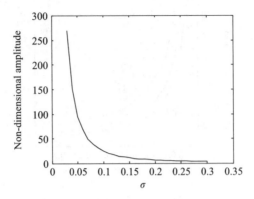

Figure 8.5   Amplitude of the developed torque as a function of σ

8.4   Shaped torque input

In this section, input shaping is introduced for control of flexible manipulators. Further detailed coverage of the approach with applications is provided in Chapter 9. The input shaping method involves convolving a desired command with a sequence of impulses known as an input shaper (Singer and Seering, 1990). The shaped command that results from the convolution is then used to drive the system. The design objectives are to determine the amplitude and time location of the impulses from the natural frequencies and damping ratios of the system, so that the shaped command reduces the system vibration. The method is briefly described in this section. A vibratory system can be modelled as a superposition of second-order systems each with a transfer function

$$G(s) = \frac{\omega_n^2}{s^2 + 2\zeta\omega_n s + \omega_n^2} \tag{8.12}$$

where ω_n is the natural frequency and ζ is the damping ratio of the system. Thus, the impulse response of the system at time t is

$$y_\delta(t) = \frac{J\omega_n}{\sqrt{1-\zeta^2}} e^{-\zeta\omega_n(t-t_0)} \sin(\omega_n\sqrt{1-\zeta^2}(t-t_0)) \tag{8.13}$$

where J and t_0 are the magnitude and time of the impulse, respectively. Further, the response to a sequence of impulses can be obtained using the superposition principle. Thus, for q impulses, with damped frequency, $\omega_d = \omega_n\sqrt{1-\zeta^2}$, the impulse response can be expressed as

$$y_\delta(t) = W \sin(\omega_d t + \chi) \tag{8.14}$$

where

$$W = \sqrt{\left(\sum_{i=1}^{q} \lambda_i \cos \sigma_i\right)^2 + \left(\sum_{i=1}^{q} \lambda_i \sin \sigma_i\right)^2}$$

$$\lambda_i = \frac{J_i \omega_n}{\sqrt{1-\zeta^2}} e^{-\zeta \omega (t-t_i)}, \quad \sigma_i = \omega_d t_i \quad \text{and} \quad \chi = \tan^{-1}\left(\sum_{i=1}^{q} \frac{\lambda_i \cos \sigma_i}{\lambda_i \sin \sigma_i}\right).$$

J_i and t_i are the magnitudes and times at which the impulses occur.

The residual single mode vibration amplitude of the impulse response is obtained at the time of the last impulse, t_N as

$$V = \sqrt{V_1^2 + V_2^2} \tag{8.15}$$

where

$$V_1 = \sum_{i=1}^{q} \frac{J_i \omega_n}{\sqrt{1-\zeta^2}} e^{-\zeta \omega_n (t_N - t_i)} \cos(\sigma_i), \quad V_2 = \sum_{i=1}^{q} \frac{J_i \omega_n}{\sqrt{1-\zeta^2}} e^{-\zeta \omega_n (t_N - t_i)} \sin(\sigma_i)$$

To achieve zero vibration after the last impulse, it is required that both V_1 and V_2 in equation (8.15) are independently zero. Furthermore, to ensure that the shaped command input produces the same rigid body motion as the unshaped command, it is required that the sum of amplitudes of the impulses is unity. To avoid response delay, the first impulse is selected at time $t_1 = 0$. Hence by setting V_1 and V_2 in equation (8.15) to zero, $\sum_{i=1}^{q} J_i = 1$ and solving yields a two-impulse sequence as

$$
\begin{array}{cc}
t_1 = 0, & t_2 = \frac{\pi}{\omega_d} \\
J_1 = \frac{1}{1+H}, & J_2 = \frac{H}{1+H}
\end{array}
\tag{8.16}
$$

where $H = e^{\zeta \pi / \sqrt{1-\zeta^2}}$.

The robustness of the input shaper to errors in natural frequencies of the system can be increased by setting $dV/d\omega_n = 0$. Setting the derivative to zero is equivalent of producing small changes in vibration corresponding to natural frequency changes. By obtaining the first derivatives of V_1 and V_2 in equation (8.15) and simplifying yields (Singer and Seering, 1990)

$$
\begin{aligned}
\frac{dV_1}{d\omega_n} &= \sum_{i=1}^{q} J_i t_i e^{-\zeta \omega_n (t_N - t_i)} \sin(\sigma_i) \\
\frac{dV_2}{d\omega_n} &= \sum_{i=1}^{q} J_i t_i e^{-\zeta \omega_n (t_N - t_i)} \cos(\sigma_i)
\end{aligned}
\tag{8.17}
$$

Hence, by setting equations (8.15) and (8.17) to zero and solving yields a three-impulse sequence as

$$
\begin{array}{ccc}
t_1 = 0, & t_2 = \frac{\pi}{\omega_d}, & t_3 = \frac{2\pi}{\omega_d} \\
J_1 = \frac{1}{1+2H+H^2}, & J_2 = \frac{2H}{1+2H+H^2}, & J_3 = \frac{H^2}{1+2H+H^2}
\end{array}
\tag{8.18}
$$

where H is as in equation (8.16). The robustness of the input shaper can further be increased by taking and solving the second derivative of the vibration in equation (8.15). Similarly, this yields a four-impulse sequence as

$$
\begin{aligned}
&t_1 = 0, \qquad t_2 = \frac{\pi}{\omega_d}, \qquad t_3 = \frac{2\pi}{\omega_d}, \qquad t_4 = \frac{3\pi}{\omega_d} \\
&J_1 = \frac{1}{1+3H+3H^2+H^3}, \qquad J_2 = \frac{3H}{1+3H+3H^2+H^3} \\
&J_3 = \frac{3H^2}{1+3H+3H^2+H^3}, \qquad J_3 = \frac{H^3}{1+3H+3H^2+H^3}
\end{aligned}
\tag{8.19}
$$

where H is as in equation (8.16).

To handle higher vibration modes, an impulse sequence for each vibration mode can be designed independently. Then the impulse sequences can be convoluted together to form a sequence of impulses that attenuate vibration at higher modes. In this manner, for a vibratory system, the vibration reduction can be accomplished by convolving a desired system input with the impulse sequence. This yields a shaped input that drives the system to a desired location with reduced vibration.

8.5   Filtered torque input

Input torque using filtering techniques is developed on the basis of extracting input energies around natural frequencies of the system. The filters are thus used for pre-processing the input signal so that no energy is fed into the system at the natural frequencies. In this manner, the flexural modes of the system are not excited, leading to a vibration-free motion. This can be realised by employing either low-pass or band-stop filters. In the former, the filter is designed with a cut-off frequency lower than the first natural frequency of the system. In the latter case, band-stop filters with centre frequencies at the natural frequencies of the system are designed. This will require one filter for each mode of the system. The band-stop filters thus designed are then implemented in cascade to pre-process the input signal.

There are various filter types such as Butterworth, Chebyshev and Elliptic that can be designed and employed. These filters have the desired frequency response in magnitude, allow for any desired cut-off rate and are physically realisable. The magnitude of the frequency response of a low-pass Butterworth filter is given by (Jackson, 1989)

$$
|H(j\omega)|^2 = \frac{1}{1+[\omega/\omega_C]^{2n}} = \frac{1}{1+\varepsilon[\omega/\omega_p]^{2n}}
\tag{8.20}
$$

where, n is a positive integer signifying the order of the filter, ω_C is the filter cut-off frequency (-3 dB frequency), ω_P is the pass-band edge frequency and $(1+\varepsilon^2)^{-1}$ is the band edge value of $|H(j\omega)|^2$. Note that $|H(j\omega)|^2$ is monotomic in both the pass-band and stop-band. The order of the filter required to yield attenuation δ_2 at a specified frequency ω_S (stop-band edge frequency) is easily determined from equation (8.20) as

$$
n = \frac{\log\left\{(1/\delta_2^2)-1\right\}}{2\log\left\{\omega_S/\omega_P\right\}} = \frac{\log\left\{\delta_1/\varepsilon\right\}}{\log\left\{\omega_S/\omega_P\right\}}
\tag{8.21}
$$

where, by definition, $\delta_2 = (1+\delta_1^2)^{-0.5}$. Thus, the Butterworth filter is completely characterised by the parameters n, δ_2, ε and the ratio ω_S/ω_P.

Equation (8.21) can be employed with arbitrary δ_1, δ_2, ω_C and ω_S to yield the required filter order n from which the filter design is readily obtained. The Butterworth approximation results from the requirement that the magnitude response be maximally flat in both the pass-band and the stop-band. That is, the first $2n - 1$ derivatives of $|H(j\omega)|^2$ are specified to be equal to zero at $\omega = 0$ and at $\omega = \infty$.

The sharpest transition from pass-band to stop-band for a given set of filter specification is achieved by an elliptic filter design. Thus, the elliptic design is optimum in this sense. The magnitude response of an elliptic filter is equi-ripple in both the pass-band and stop-band. The squared magnitude response of a low-pass elliptic filter is of the form (Zverev, 1967)

$$|H(j\omega)|^2 = \frac{1}{1+\eta^2 U_n^2 \left[\omega/\omega_C\right]} \tag{8.22}$$

where, $U_n\{\omega\}$ is a Jacobian elliptic function of order n and η is a parameter related to the pass-band ripple. It is known that most efficient designs occur when the approximation error is equally spread over the pass-band and stop-band. Elliptic filters allow this objective to be achieved easily, thus, being most efficient from the viewpoint of yielding the smallest-order filter for a given set of specifications. Equivalently, for a given order and a given set of specifications, an elliptic filter has the smallest transition bandwidth.

The filter order required for a pass-band ripple γ_1, stop-band ripple γ_2, and transition ratio ω_P/ω_S is given as

$$n = \frac{K\{\omega_S/\omega_C\} K\left\{\left(1-\eta^2/\gamma_1^2\right)^{0.5}\right\}}{K\{\eta/\gamma_1\} K\left\{\left(1-\omega_P^2/\omega_S^2\right)^{0.5}\right\}} \tag{8.23}$$

where, $K\{v\}$ is the complete elliptic integral of the first kind, defined as

$$K\{v\} = \int_0^{\pi/2} \left(1 - v^2\sin^2\varphi\right)^{-0.5} d\varphi$$

$$\gamma_2 = \left(1+\gamma_1^2\right)^{-0.5} \quad \text{and} \quad \gamma_1 = 10\log(1+\eta^2)$$

Values of the above integral are given in tabulated form in a number of textbooks (Dwight, 1957). The phase response of an elliptic filter is more non-linear in the pass-band than a comparable Butterworth filter, especially near the band edge.

The design relations for the low-pass filters given above can be utilized in normalized form to design the corresponding band-stop filters. This involves a transformation from low-pass to band-stop filter (Banks, 1990).

8.6   Experimentation and results

The open-loop control strategies were designed on the basis of identified natural frequencies and damping ratios of the flexible manipulator system. The damping ratios of the system have previously been deduced as 0.026, 0.038 and 0.04 for the first, second and third modes, respectively (Tokhi and Azad, 1997). The natural frequencies were obtained by using various methods as described in Section 8.2. Based on these natural frequencies and damping ratios, the input shapers and filters are designed and used for pre-processing the bang-bang torque input. The developed torque profile is then applied to the system in an open-loop configuration (Figure 8.6) to reduce the vibration of the manipulator. In this process, the unshaped, shaped and filtered inputs were designed with a sampling frequency of 500 Hz.

Experimental results of the response of the flexible manipulator to the unshaped, shaped and filtered inputs are presented in this section in the time and frequency domains. To verify the performance of the control techniques, the results are examined in comparison to the unshaped bang-bang torque input for a similar input level in each case. Three system responses, namely, the hub-angle, hub-angular velocity and end-point acceleration are measured experimentally, and the power spectral density (PSD) of the end-point acceleration is evaluated. Four criteria are used to evaluate the performances of the control schemes:

1. *Level of vibration reduction at the natural frequencies.* This is accomplished by comparing the responses to the shaped and filtered inputs with the response to the unshaped input. The results are presented in dB.
2. *The time response specifications.* Parameters that are evaluated are settling time and overshoot of the hub-angle response. The settling time is calculated on the basis of ±2 per cent of the steady-state value. Moreover, the magnitude of oscillation of the system response is observed.
3. *Robustness to parameter uncertainty.* To examine the robustness of the techniques, the system performance is assessed with 30 per cent error tolerance in natural frequencies. This is incorporated in the design of the input shapers and filters.
4. *Performance with variation in payload.* This is accomplished by testing and evaluating the performance of the system with the control techniques using different payloads at the end-point of the manipulator.

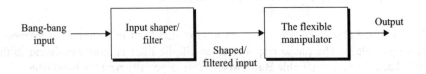

Figure 8.6   Block diagram of the open-loop control configuration

8.6.1 Unshaped bang-bang torque input

Figure 8.7 shows a single-switch bang-bang torque input of amplitude ± 0.3 Nm and duty cycle of 0.6 s applied at the hub of the manipulator. A bang-bang torque has a positive (acceleration) and negative (deceleration) period allowing the manipulator to, initially, accelerate and then decelerate and eventually stop at a target location. In this work, the first three modes of vibration of the system are considered, as these dominate the dynamics of the system.

Figure 8.8 shows the response of the flexible manipulator with an unshaped bang-bang input. Two conditions, namely without payload and with a 40 g payload, are presented. These results were considered as the system response to the unshaped input and subsequently will be used to design and evaluate the performance of the command-shaping techniques. It is noted that without payload a steady-state hub-angle level of 38° was achieved with a settling time and overshoot of 1.555 s and 2.6 per cent respectively. Note that vibration occurs at the hub and end-point during movement of the manipulator, as evidenced in the hub-angle, hub-velocity and end-point acceleration responses. As demonstrated, the dynamic and vibration behaviour of the flexible manipulator is significantly affected with the presence of payload. With a 40 g payload, the hub-angle reached 26° with a settling time of 1.756 s. The residual motion of the system is dominated by the first three modes of vibration. It is noted that the resonance frequencies decrease with payload. As demonstrated in Figure 8.8, the natural frequencies of the system without payload are at 12, 35 and 65 Hz and with payload (40 g) at 10, 31 and 58 Hz. The magnitudes of PSD of the end-point acceleration of the system without payload at the natural frequencies are obtained as 80, 50 and 30 $m^2/s^4/Hz$ for the first three modes, respectively. Similarly, for the system with a payload (40 g), the magnitudes of PSD are obtained as 30, 15 and 9 $m^2/s^4/Hz$ for the first three modes, respectively.

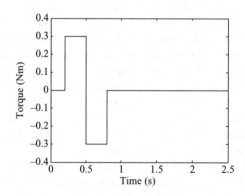

Figure 8.7   The bang-bang input torque

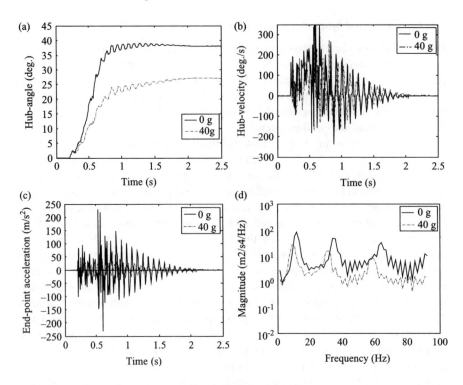

Figure 8.8   *Response to an the unshaped bang-bang torque input: (a) hub-angle,*
(b) hub-velocity, (c) end-point acceleration, (d) PSD of end-point
acceleration

8.6.2   *Shaped torque input*

Using the parameters of the system, an input shaper with a four-impulse sequence
for the first three modes of vibration was designed. As demonstrated in the previous
section, the natural frequencies were 12, 35 and 65 Hz and the damping ratios were
0.026, 0.038 and 0.04 for the first, second and third modes respectively. The magnitude and time location of the impulses were obtained by solving equation (8.19).
Similarly, for evaluation of robustness, input shapers with error in natural frequencies were also evaluated. As a consequence of using 30 per cent error the natural
frequencies of the system were assumed at 15.6, 45.5 and 84.5 Hz. To avoid response
delay, the first impulse is selected at time, $t = 0$. For digital implementation of the
input shapers, locations of the impulses were selected at the nearest sampling time.
Figure 8.9 shows the shaped input using input shaping.

Figure 8.10 shows the response of the flexible manipulator without payload with
the PSD of the end-point acceleration to the shaped inputs with exact and erroneous
natural frequencies. It is noted that the magnitudes of vibration of the system, with
the hub-angle, hub-velocity and end-point acceleration responses, have significantly
been reduced at the natural frequencies. These can be observed by comparing the

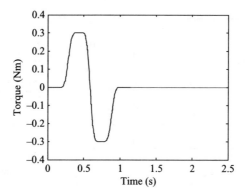

Figure 8.9   A shaped torque input using four-impulse sequences

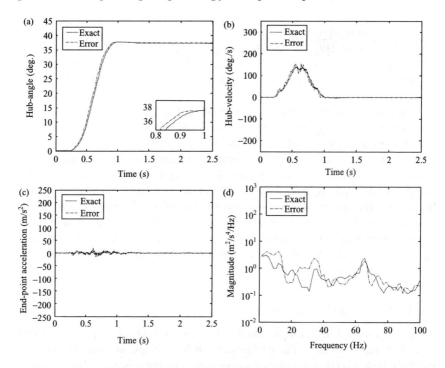

Figure 8.10   Response of the flexible manipulator to shaped inputs with exact and erroneous natural frequencies: (a) hub-angle, (b) hub-velocity, (c) end-point acceleration, (d) PSD of end-point acceleration

system responses to the unshaped input (Figure 8.8). With exact frequencies, the oscillations in the hub-angle, hub-velocity, and end-point acceleration response were found to have almost reduced to zero. Hence, a smoother response was achieved. The level of vibration reduction with the end-point acceleration in comparison to the

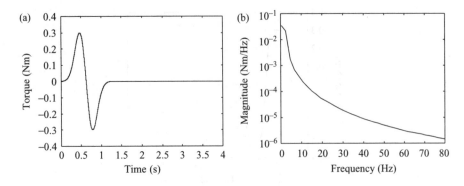

Figure 8.11   *The Gaussian shaped torque input: (a) time-domain, (b) spectral density*

bang-bang torque input for the first three modes was achieved as 38.06, 35.92 and 23.52 dB, respectively. The corresponding settling time and overshoot of the hub-angle response were obtained as 0.896 s and 0.6 per cent, respectively. As noted, a much faster hub-angle response with less overshoot, as compared to the unshaped input, was achieved.

The response of the manipulator to the shaped input with erroneous natural frequencies is used to examine the robustness of the technique. As noted, the level of reduction in the vibration of the manipulator is slightly less than the case without error. However, it is noted that significant vibration reduction was achieved. The level of vibration reduction for the end-point acceleration with erroneous natural frequencies is achieved as 26.02, 25.19 and 24.44 dB for the first three vibration modes. In this case, the settling time and overshoot of the hub-angle response were obtained as 0.882 s and 0.56 per cent, respectively. As noted, despite an increase in the time response parameters, as compared to the case without error, significant reduction in the settling time and overshoot were achieved as compared to the response of the system to the unshaped input.

8.6.3   *Gaussian shaped input*

A Gaussian shaped input torque was developed with a cut-off frequency at 10.0 Hz, $\sigma = 0.15$ and $\mu = 10\sigma$. The performance of the manipulator was studied experimentally with this Gaussian shaped torque input in comparison to a bang-bang input torque for a similar angular displacement, keeping the peak torque at a similar level in each case. The experimentation includes a comparative study, in the two cases, of the applied torque, angular displacement, and end-point acceleration and their respective spectral densities. The results of this study are shown in Figures 8.11 and 8.12.

Figure 8.11 shows the Gaussian shaped torque input. It is noted that with a cut-off frequency of 10.0 Hz, the spectral attenuations in the input in comparison to the bang-bang torque input, are by 70.53, 77.17 and 82.45 dB in the first, second and third modes, respectively. The corresponding system response is shown in Figure 8.12.

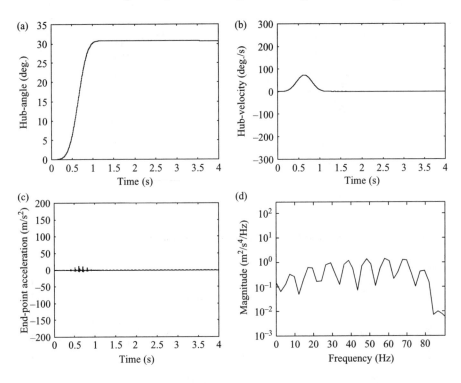

Figure 8.12   *System response with the Gaussian shaped torque input: (a) hub-angle, (b) hub-velocity, (c) end-point acceleration, (d) PSD of end-point accleration*

It is noted that the speed of response is slightly lower than that using shaped torque input developed earlier. The spectral attenuations achieved at the first three resonance modes with the end-point acceleration were 29.54, 29.54 and 23.88 dB.

8.6.4   Filtered input torque

Using the low-pass filter, the input energy at all frequencies above the cut-off frequency can be attenuated. In this study, sixth-order Butterworth low-pass filters with cut-off frequency at 75 per cent of the first vibration mode were designed. Thus, for the flexible manipulator, the cut-off frequencies of the filters were selected at 9 and 11.7 Hz for the two cases of exact and erroneous natural frequencies respectively. On the other hand, using the band-stop filter, the input energy at dominant modes of the system can be attenuated. In this study, band-stop filters with stop-bands of 5 Hz were designed for the first three modes. Similarly, the filters were designed with consideration of exact and 30 per cent error in natural frequencies. The filtered torque inputs with the low-pass and band-stop filters are shown in Figure 8.13.

  Figure 8.14 shows the response of the flexible manipulator with the PSD of the end-point acceleration to the low-pass filtered torque with exact and erroneous natural

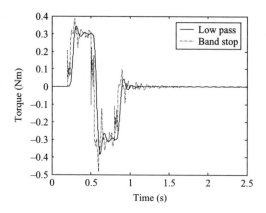

Figure 8.13   Lowpass and bandstop filtered torque inputs

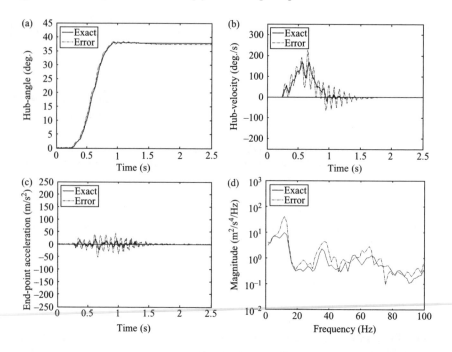

Figure 8.14   Response with lowpass filtered inputs with exact and erroneous natural frequencies: (a) hub-angle, (b) hub-velocity, (c) end-point acceleration, (d) PSD of end-point acceleration

frequencies. With exact frequencies, the system vibration at the natural frequencies has been considerably reduced in comparison to the unshaped input. The levels of reduction at the first three modes of vibration of the system with the end-point acceleration achieved were 20, 27.96 and 29.54 dB, respectively. The settling time and

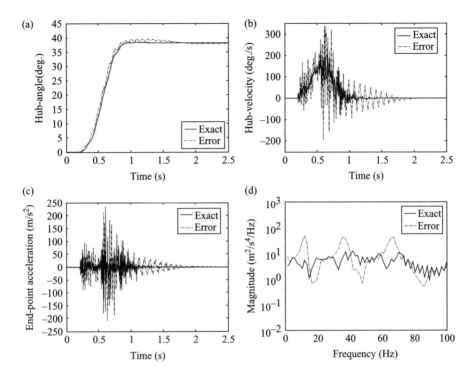

Figure 8.15   *Response with bandstop filtered inputs with exact and erroneous natural frequencies: (a) hub-angle, (b) hub-velocity, (c) end-point acceleration, (d) PSD of end-point acceleration*

overshoot of the hub-angle response to the low-pass filtered input were obtained as 0.894 s and 0.98 per cent, respectively.

The robustness of the technique is demonstrated with the system response to the filtered torque with erroneous natural frequencies. As evidenced in the magnitude of the time responses, relatively small reduction in the system vibration was achieved. The levels of reduction at the first three modes of vibration of the system with the end-point acceleration achieved were 8.52, 21.94 and 23.52 dB. The settling time and overshoot of the hub-angle response were obtained as 1.268 s and 2.83 per cent, respectively. In this case, the settling time increased to 80 per cent of the unshaped input. Moreover, higher overshoot than the response to unshaped input was noted.

The response of the flexible manipulator with sixth-order Butterworth band-stop filtered inputs with exact and erroneous natural frequencies is shown in Figure 8.15. With exact frequencies, it is noted that only small amount of vibration reduction at the first three vibration modes was achieved in comparison with the response to unshaped input. The levels of reduction at the first three modes of vibration with the end-point acceleration of the system obtained were 26.02, 18.42 and 17.5 dB. The corresponding settling time and overshoot of the hub-angle response to the filtered input were 0.836 s and 2.15 per cent respectively. It is noted that significant reduction

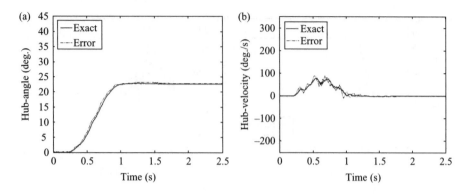

Figure 8.16   Response at the hub with payload using input shaping (with exact and erroneous natural frequencies): (a) hub-angle, (b) hub-velocity

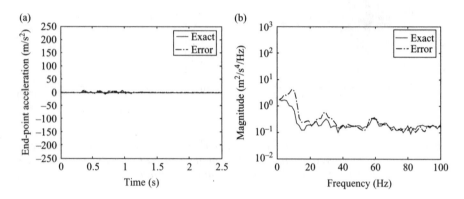

Figure 8.17   Response the end-point with 40 g payload using input shaping (with exact and erroneous natural frequencies): (a) end-point acceleration, (b) PSD of end-point acceleration

in settling time was achieved but relatively small reduction in the overshoot of the response. With erroneous natural frequencies, the level of vibration of the system at the natural frequencies was not significantly affected as compared with the unshaped input case. The levels of reduction with the end-point acceleration response for the first three modes were 6.02, 1.94 and 0 dB. The settling time and overshoot of the hub-angle response to the band-stop filtered input with erroneous natural frequencies are achieved as 1.474 s and 4.60 s, respectively. Thus, no improvement in the time response was achieved as compared to the unshaped input.

8.6.5   System with payload

Figures 8.16 and 8.17 show responses of the flexible manipulator with a payload of 40 g when driven with shaped inputs using a four-impulse sequence with exact and erroneous natural frequencies. It is revealed that the input shaping can handle

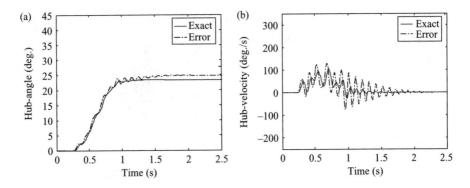

Figure 8.18   Response at the hub with 40 g payload using lowpass filtered inputs (with exact and erroneous natural frequencies): (a) hub-angle, (b) hub-velocity

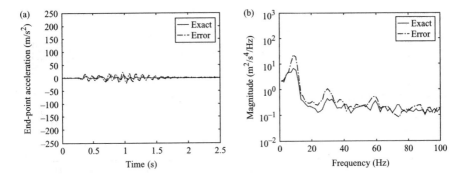

Figure 8.19   Response at the end-point with 40 g payload using lowpass filtered inputs with exact and erroneous natural frequencies: (a) end-point acceleration, (b) PSD of end-point acceleration

vibrations of the system with a payload. In both cases, the magnitudes of vibration at the resonance modes of the system have significantly been reduced as compared to the unshaped input. Significant vibration reduction has also been achieved using the shaped input with erroneous natural frequencies.

Similarly, responses of the system with a 40 g payload to the sixth-order low-pass and band-stop filtered inputs with exact and erroneous natural frequencies are shown in Figures 8.18–8.21. It is noted that considerable reduction in the system vibration was achieved using the filtered inputs with exact natural frequencies. However, the levels of vibration at the hub-angle, hub-velocity and end-point acceleration were higher as compared to the system without payload. Moreover, there was no major improvement with the low-pass and band-stop filtered inputs with erroneous natural frequencies. In this case, the system vibrations were higher.

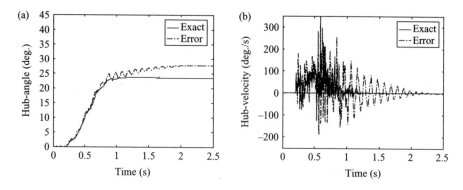

Figure 8.20   *Response at the hub with 40 g payload using bandstop filtered inputs (with exact and erroneous natural frequencies): (a) hub-angle, (b) hub-velocity*

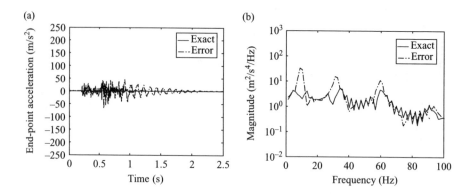

Figure 8.21   *Response at the end-point 40 g payload using bandstop filtered inputs (with exact and erroneous natural frequencies): (a) end-point acceleration, (b) PSD of end-point acceleration*

8.7   Comparative performance assessment

A comparison of the responses of the flexible manipulator using the control strategies reveals that the best performance in reduction of vibration of the flexible manipulator is achieved with the shaped torque input. This is observed as compared to the low-pass and band-stop filtered inputs at the first three modes of vibration. The performance of the technique is also evidenced in the magnitude of vibration with the hub-angle, hub-velocity and end-point acceleration responses in Figures 8.12, 8.14 and 8.15. For the level of vibration reduction with the end-point acceleration response, almost twofold improvement was achieved with input shaping as with the filtered inputs. It is noted that better performance in vibration reduction of the system is achieved with the low-pass filtered input than with the band-stop filtered input. This is mainly

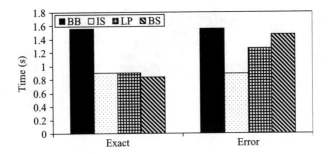

Figure 8.22   *Settling times of the hub-angle response with exact and erroneous natural frequencies using bang-bang torque (BB), Input shaping (IS), lowpass filter (LP) and bandstop filter (BS)*

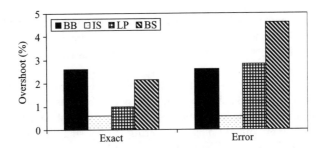

Figure 8.23   *Overshoot of the hub-angle response with exact and erroneous natural frequencies using BB, IS, LP and BS*

due to the higher level of input energy reduction achieved with the low-pass filter, especially at the second and third vibration modes. The settling time and overshoot of the hub-angle responses using the control strategies are summarised in Figures 8.22 and 8.23, respectively.

It is noted that significant reduction in the settling time was achieved with the control strategies. However, the differences in the settling time achieved with the shaped and filtered inputs are negligibly small. It is also revealed that the lowest overshoot was achieved with the input shaping technique.

A comparison of the system responses and the level of vibration reduction achieved with controllers with errors in natural frequencies reveals that the highest robustness to parameter uncertainty is achieved with the input shaping technique. It is noted that the shaped input can successfully handle errors in the natural frequencies as significant reduction in system vibration was achieved as compared with the other control techniques. This is revealed by comparing the magnitude of vibration of the system in Figures 8.12, 8.14 and 8.15. The settling times achieved using the techniques with erroneous natural frequencies is shown in Figure 8.22. As noted, the settling time with input shaping increased to 50 per cent of the settling time of the

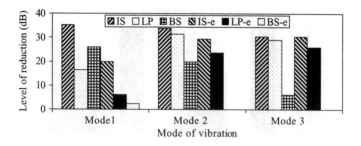

Figure 8.24   The level of vibration reduction with the end-point acceleration response of the system with 40 g payload using controllers with exact frequency (IS, LP and BS) and error in natural frequencies (IS-e, LP-e and BS-e)

unshaped input whereas settling times of 80–95 per cent of the unshaped input were achieved with other control techniques. Moreover, using input shaping technique, as demonstrated in Figure 8.23, the overshoot in the hub-angle response was not affected with the error.

On the other hand, higher overshoots were obtained using lowpass and bandstop filtered inputs. The input shaping technique is more robust, as significant reduction was achieved at the first mode of vibration, which is the most dominant mode. The bandstop filtered input did not handle the error as only small amount of reduction of the system vibration was achieved. On the other hand, using the lowpass filter, a significant amount of attenuation of the system vibration was achieved at the second and third modes. Moreover, the vibration reduction achieved with the lowpass filtered input was similar to the shaped input at these modes.

A comparison of the results using shaping techniques reveals that input shaping can successfully handle vibrations in the presence of payload in the system. With exact and erroneous natural frequencies, vibrations in the hub-angle, hub-velocity and end-point acceleration were significantly suppressed. On the other hand, considerable reduction in vibration of the system was achieved using filtered inputs. However, the filtered inputs were unable to handle payload with error in natural frequencies. Figure 8.24 shows the level of vibration reduction of the system at the end-point acceleration response with a 40 g payload. Similar to the case without payload, the input shaping technique has provided the highest level of vibration reduction and robustness.

The settling times achieved using the shaping techniques for the manipulator with a 40g payload are shown in Figure 8.25. Using the shaped inputs with exact and erroneous natural frequencies, significant reduction in settling times was achieved. However, using the filtered inputs with erroneous natural frequencies, the settling time increased to 90–100 per cent of the unshaped input case. However, with the shaped input, although the requirement to achieve the same steady-state hub-angle response level as the unshaped input was fulfilled, experimental results showed a steady-state error of 5° in the hub-angle response. Further experimental investigations showed

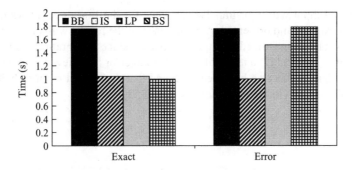

Figure 8.25   *Settling time of the hub-angle response of the system with 40 g payload using BB, IS, LP and BS*

that with higher levels of vibration, the experimental rig moves to a higher angle. This might be due to the effects of payload rotary inertia. In this case, to ensure the same steady-state level as the unshaped input, a feedback controller is required at the expense of settling time.

8.8   Summary

Investigations into the development of open-loop command-generation techniques for control of flexible manipulators using Gaussian shaped input, low-pass and band-stop filtered input, shaped torque input has been presented. The system response to unshaped bang-bang torque input has been identified and used to determine the parameters of the system for design and evaluation of the control strategies. The design approach of such an input torque for a given cut-off frequency and duty cycle has been presented. With a Gaussian shaped input, although the level of vibration reduces, the speed of response is slightly lower than that using a filtered torque input.

The filtered torque input functions have been developed on the basis of the identified resonance modes of the system through analytical, experimental, GA modelling and NN modelling. The developed filtered torques has been investigated in an open-loop control configuration for control of a single-link flexible manipulator. Comparing the results achieved with the low-pass and band-stop filtered torque inputs reveal that better performance at reduction of level of vibrations of the system is achieved with low-pass filtered torque inputs. This is due to the indiscriminate spectral attenuation in the low-pass filtered torque input at all the resonance modes of the system. Utilization of band-stop filters, however, is advantageous in that spectral attenuation in the input at selected resonance modes of the system can be achieved. Thus, an open-loop control strategy based on band-stop filters is optimum in this sense. Performances of the techniques have been evaluated in terms of level of vibration reduction, time response specifications and robustness to variation in natural frequencies. Significant improvement in the reduction of system vibrations has been achieved with all the filtered torque inputs as compared to a bang-bang torque input.

A shaped input with four-impulse sequence for the first three modes of vibration were designed and presented. A comparison of the results with other open-loop approaches has demonstrated that the best performance in vibration reduction and time response, especially in terms of robustness to error is achieved with the shaped input technique. It has also been demonstrated that the shaped input technique can successfully handle vibrations in the presence of payload.

Chapter 9

Control of flexible manipulators with input shaping techniques

W.E. Singhose and W.P. Seering

This chapter provides an overview of the use of real-time command shaping to limit vibration in flexible robotic systems. Smoothing of commands to reduce system vibration is an old idea. However, command shaping did not come into widespread use until inexpensive digital controllers were available to implement the techniques in real time. Shaped commands can address multiple resonant frequencies and can be designed to be very robust to system modelling errors. An array of examples illustrate the effect that these command shaping methods have on the performance of flexible systems.

9.1 Introduction

Robotic systems with flexible dynamics present a challenging class of real-time control problems. Typically, such systems can be represented by the block diagram shown in Figure 9.1. Four distinct system elements contribute to overall system capability: design of the physical hardware, feedback control, feedforward control and command generation. Each of the four blocks present unique opportunities for improving the system performance. For example, the feedback block can be designed to reject random disturbances, while the feedforward block cannot. However, the feedforward block can compensate for unwanted effects before they show up in the output, while the feedback controller cannot.

As Figure 9.1 shows, the command generation block receives the desired performance specifications. It then generates system reference commands for both the feedforward and feedback control blocks. These two elements send control effort directly into the plant. The plant then responds to this combination of commands. Unfortunately, this general structure is not universally accepted. Many references

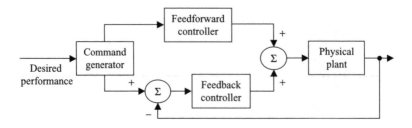

Figure 9.1　Generic system block diagram

confuse command generation with feedforward control. For clarity, the following nomenclature will be used here:

- The *command generation block* transforms the desired system performance objective into sets of reference command signals designed to direct the system to achieve the specified objective.
- The *feedback control block* compares the reference command signals from the command generation block with measurements of the system states and determines a force (torque, voltage, etc.) vector to act on the plant.
- The *feedforward control block* employs an internal model of the system to transform the reference commands from the command generation block into a force vector to act on the plant in addition to the force vector from the feedback control block. In the absence of a feedback control loop, the command generation and feedforward control blocks can be combined into a single block.

One reason for the inconsistent use of the term *feedforward* is that some techniques can be employed either as feedforward or as command generation. However, it should be noted that the strengths and weaknesses of the techniques change when their roles change.

Perhaps the least studied and utilized element in Figure 9.1 is the command generator. This component is of fundamental importance and can greatly influence the system performance. This chapter will document the importance of command generation for the control of flexible manipulators and describe command generation techniques that can greatly reduce vibration.

When performance is pushed to the limit in terms of motion, velocity and throughput, the problem of flexibility usually arises. Flexibility comes from physical deformation of the structure and/or compliance introduced by the feedback control system. For example, implementing a proportional and derivative (PD) controller is analogous to adding a spring and damper to the system. The controller 'spring' can lead to problematic flexibility in the system. Deformation of the structure can occur in the links, cables and joints. This can lead to problems with positioning accuracy, trajectory following, settling time, component wear, and stability, and may also introduce non-linear dynamics if the deflections are large.

An example of a robot whose flexibility is detrimental to its positioning accuracy is the long-reach manipulator, RALF, described in Figure 9.2(a). This robotic arm

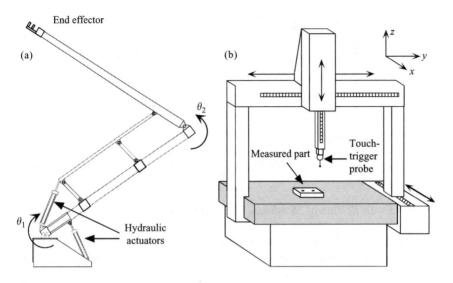

Figure 9.2   Examples of robotic manipulators: (a) long reach manipulator RALF, (b) coordinate measuring machine

was built to test methods for cleaning nuclear waste storage tanks (Magee and Book, 1993). The arm needs to enter a tank through an access hole and then reach long distances to clean the tank walls. These types of robots have mechanical flexibility in the links, the joints and, possibly, the base to which they are attached. Both the gross motion of the arm and the cleaning motion of the end-effector can induce vibration.

The long links that compose RALF make its flexibility obvious. However, all robotic systems will deflect if they are moved rapidly enough. Consider the moving-bridge coordinate measuring machine (CMM) described in Figure 9.2(b). The machine is composed of stiff components including a granite base and large cross-sectional structural members. The goal of the CMM is to move a probe throughout its workspace so that it can contact the surfaces of manufactured parts that are fixed to the granite base. In this way it can accurately determine the dimensions of the part. The position of the probe is measured by optical encoders that are attached to the granite base and the moving bridge. However, if a laser interferometer is used to measure the probe, it can be seen that its location will differ from that indicated by the encoders. This difference arises because the physical structure deflects between the encoders and the probe end-point. Figure 9.3 shows the micron-level deflection for a typical move where the machine rapidly approaches surface of a part and then slows down just before making contact with the part (Singhose *et al.*, 1996c). If the machine is attempting to measure with micron resolution, then the 20–25 micron vibration during the approach phase is a significant drawback.

Flexibility introduced by the control system is also commonplace. Feedback control works by detecting a difference between the actual response and the desired response. This difference is then used to generate a restoring force. In many cases,

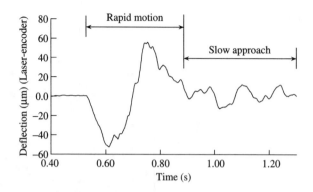

Figure 9.3   Deflection of coordinate measuring machine

this force acts like a spring. Increasing the gains may have the effect of stiffening the system, but it also increases actuator demand and noise problems, and can cause instability.

There are a wide range of command generators and feedforward controllers, so globally characterising their strengths and weaknesses is difficult. However, in general, command generators are less aggressive, rely less on an accurate system model and are consequently more robust to uncertainty and plant variation. Feedforward control action can produce better trajectory tracking than command generation, but it is usually less robust. There are also control techniques that can be implemented as command generation or as feedforward control, so it is not always obvious at first how a control action should be characterised.

9.2   Command generation

Creating specially shaped reference commands that move flexible systems in a desired fashion is an old idea (Smith, 1957, 1958). Commands can be created such that the motion of the system will cancel its own vibration. Some of these techniques require that the commands be pre-computed using boundary conditions before the move is initiated. Others can be implemented with command strings being created in real time. A significant difference between the various control methods is robustness to modelling errors. Some techniques require a very good system model to work effectively, while others only need rough estimates of the system parameters. This chapter concentrates on techniques that can be implemented in real time to accommodate time-varying performance requirements.

9.2.1   Gantry crane example

Automated overhead bridge cranes, like the one shown schematically in Figure 9.4, are difficult to position accurately. The payload is hoisted up by an overhead suspension cable. The upper end of the cable is attached to a trolley that travels along a bridge

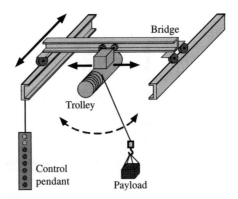

Figure 9.4   Overhead bridge crane

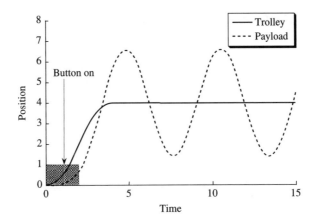

Figure 9.5   Response of bridge crane to a single button press

to position the payload. Furthermore, the bridge on which the trolley travels can also move perpendicular to the trolley motion, thereby providing three-dimensional positioning. These cranes are usually controlled by a human operator who presses buttons to cause the trolley to move, but there are also automated versions where a control computer drives the motors. If the control button is depressed for a finite time period, then the trolley will move a finite distance and come to rest. The payload on the other hand, will usually oscillate about the new trolley position. The planar motion of the payload for a typical trolley movement is shown in Figure 9.5.

Payload oscillations can cause problems when the payload is being transported through a cluttered work environment containing obstacles and human workers. Furthermore, it complicates final payload positioning. Figure 9.6 shows an overhead view of the position of a crane payload while it is being driven through an obstacle field by a novice operator. There is considerable payload sway both during the transport and at the final position. This data was obtained via an overhead camera that tracked the

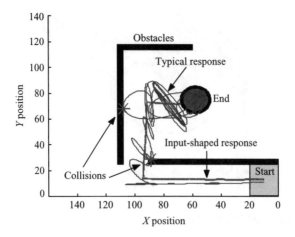

Figure 9.6   Payload response moving through an obstacle field

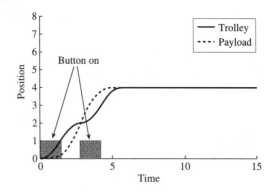

Figure 9.7   Response of bridge crane to a double button press

payload motion of a bridge crane at Georgia Tech (Singhose *et al.*, 2003). An experienced crane operator can often produce the desired payload motion with much less vibration by pressing the control buttons multiple times at the proper time instances. This two-stage operator command and payload response for planar motion is shown in Figure 9.7. Comparing Figures 9.5 and 9.7, the benefits of properly choosing the reference command are obvious. This is the type of effect that the command generator block in Figure 9.1 strives to achieve. Figure 9.6 shows a well-behaved payload response of the real crane when an input-shaping command generator is utilized.

The first suggested use of this technology on cranes was made in the late 1950s by Smith (1957, 1958) when he developed posicast control. However, given the lack of digital controllers at that time, it does not appear to have been implemented on a real crane-like structure until Starr (1985) reported some excellent results using robotic manipulation of suspended payloads. Since the inception of this two-stage command shaping procedure, its lack of robustness to modelling errors and plant variations has

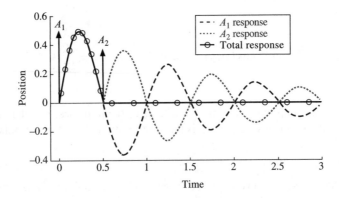

Figure 9.8   Vibration from two impulses can cancel out

been well known (Tallman and Smith, 1958). These effects occur in cranes when the payload is hoisted, the payload geometry changes, the motors exhibit non-linearity, or the suspension length is not known with certainty. Therefore, several researchers have sought to improve command shaping technology for cranes by developing more robust shaping methods (Feddema, 1993; Kenison and Singhose, 1999; Kress *et al.*, 1994; Lewis *et al.*, 1998; Singer *et al.*, 1997; Singhose *et al.*, 2000).

9.2.2   Generating zero vibration commands

As a first step to understanding how to generate commands that move flexible robots without vibration, it is helpful to start with the *simplest* such command. A fundamental building block for all commands is an impulse. This theoretical command is often a good approximation of a short burst of force such as that from a hammer blow, or from momentarily turning the actuator full on. Applying an impulse to a flexible robot will cause it to vibrate. However, if a second impulse is applied to the robot, some of the vibration induced by the first impulse can be cancelled. This concept is demonstrated in Figure 9.8. Impulse A_1 induces the vibration indicated by the dashed line, while A_2 induces the dotted response. Combining the two responses, using superposition, results in zero residual vibration. The second impulse must be applied at the correct time and must have the appropriate magnitude. Note that this two-impulse sequence is analogous to the two-pulse crane command shown in Figure 9.7.

In order to derive the amplitudes and time locations of the two-impulse command shown in Figure 9.8, a mathematical description of the residual vibration that results from a series of impulses must be utilized. If the system's natural frequency is ω, and the damping ratio is ζ, then the residual vibration that results from a sequence of impulses applied to a second-order system can be described by (Bolz and Tuve, 1973; Singer and Seering, 1990; Smith, 1958):

$$V(\omega, \zeta) = e^{-\zeta \omega t_n} \sqrt{[C(\omega, \zeta)]^2 + [S(\omega, \zeta)]^2} \tag{9.1}$$

where,

$$C(\omega,\zeta) = \sum_{i=1}^{n} A_i e^{\zeta\omega t_i} \cos(\omega_d t_i) \quad \text{and} \quad S(\omega,\zeta) = \sum_{i=1}^{n} A_i e^{\zeta\omega t_i} \sin(\omega_d t_i)$$

$$(9.2)$$

A_i and t_i are the amplitudes and time locations of the impulses, n is the number of impulses in the impulse sequence, t_n is the time location of the final impulse, and

$$\omega_d = \omega\sqrt{1-\zeta^2} \tag{9.3}$$

Note that equation (9.1) is expressed in a non-dimensional form. It is generated by taking the absolute amplitude of residual vibration from an impulse series and then dividing by the vibration amplitude from a single, unity-magnitude impulse. This expression predicts the percentage of residual vibration that will remain after input shaping has been implemented. For example, if $V = 0.05$ when the impulses and system parameters are entered into equation (9.1), then the input shaper will reduce the residual vibration to 5 per cent of the amplitude that occurs without input shaping. Of course, this result only applies to under-damped systems, because over-damped systems do not have residual vibration.

To generate an impulse sequence that causes no residual vibration, equation (9.1) is set equal to zero and solved for the impulse amplitudes and time locations. However, a few more restrictions must be placed on the impulses, because the solution can converge to zero-valued or infinitely valued impulses. To avoid the trivial solution of all zero-valued impulses, and to obtain a normalised result, the impulse amplitudes are required to sum to unity:

$$\sum_{i=1}^{n} A_i = 1 \tag{9.4}$$

At this point, the impulses could satisfy equation (9.4) by taking on very large positive and negative values. These large impulses would saturate the actuators. One way to obtain a bounded solution is to limit the impulse amplitudes to positive values:

$$A_i > 0, \quad i = 1, \dots, n \tag{9.5}$$

Limiting the impulses to positive values provides a good solution. However, performance can be pushed even further by allowing a limited number of negative impulses (Singhose et al., 1997e).

The problem to be solved can now be stated explicitly: find a sequence of impulses that makes equation (9.1) equal to zero, while also satisfying equation (9.4) and equation (9.5).[1] For the two-impulse sequence shown in Figure 9.8 satisfying the above specifications, there are four unknowns – the two-impulse amplitudes (A_1, A_2) and the two-impulse time locations (t_1, t_2).

[1] This problem statement and solution are similar to one first published by O.J.M. Smith in 1957.

Without loss of generality, the time location of the first impulse can be set equal to zero:

$$t_1 = 0 \qquad (9.6)$$

The problem is now reduced to finding three unknowns (A_1, A_2, t_2). In order for equation (9.1) to equal zero, the expressions in equation (9.2) must both equal zero independently because they are squared in equation (9.1). Therefore, the impulses must satisfy the following:

$$0 = \sum_{i=1}^{2} A_i e^{\zeta \omega t_i} \cos(\omega_d t_i) = A_1 e^{\zeta \omega t_1} \cos(\omega_d t_1) + A_2 e^{\zeta \omega t_2} \cos(\omega_d t_2) \qquad (9.7)$$

$$0 = \sum_{i=1}^{2} A_i e^{\zeta \omega t_i} \sin(\omega_d t_i) = A_1 e^{\zeta \omega t_1} \sin(\omega_d t_1) + A_2 e^{\zeta \omega t_2} \sin(\omega_d t_2) \qquad (9.8)$$

Substituting for t_1 from equation (9.6) into equations (9.7) and (9.8) yields

$$0 = A_1 + A_2 e^{\zeta \omega t_2} \cos(\omega_d t_2) \qquad (9.9)$$

$$0 = A_2 e^{\zeta \omega t_2} \sin(\omega_d t_2) \qquad (9.10)$$

In order for equation (9.10) to be satisfied in a non-trivial manner, the sine term must equal zero. This occurs when its argument contains a positive time value and equals a multiple of π,

$$\omega_d t_2 = p\pi, \qquad p = 1, 2, \ldots \qquad (9.11)$$

In other words,

$$t_2 = \frac{p\pi}{\omega_d} = \frac{pT_d}{2}, \qquad p = 1, 2, \ldots \qquad (9.12)$$

where T_d is the damped period of vibration. This result indicates that there are an infinite number of possible values for the location of the second impulse – they occur at multiples of the half period of vibration. Time considerations require using the smallest value for t_2, that is

$$t_2 = \frac{T_d}{2} \qquad (9.13)$$

The impulse time locations can be described very simply; the first impulse is at time zero and the second impulse is located at half the period of vibration. For this simple two-impulse case, the amplitude constraint given in equation (9.4) reduces to

$$A_1 + A_2 = 1 \qquad (9.14)$$

Using the expression for the damped natural frequency given in equation (9.3) and substituting for t_2 and $A_1 + A_2$ from equations (9.13) and (9.14) into equation (9.9) gives

$$0 = A_1 - (1 - A_1) e^{\left(\zeta \pi / \sqrt{1 - \zeta^2}\right)} \qquad (9.15)$$

Rearranging equation (9.15) and solving for A_1 gives

$$A_1 = \frac{e^{\left(\zeta\pi/\sqrt{1-\zeta^2}\right)}}{1 + e^{\left(\zeta\pi/\sqrt{1-\zeta^2}\right)}} \tag{9.16}$$

To simplify the expression, multiply top and bottom of the right-hand side by the inverse of the exponential term to get

$$A_1 = \frac{1}{1 + K} \tag{9.17}$$

where the inverse of the exponential term is

$$K = e^{\left(-\zeta\pi/\sqrt{1-\zeta^2}\right)} \tag{9.18}$$

Substituting for A_1 from equation (9.17) into equation (9.14) yields

$$A_2 = \frac{K}{1 + K} \tag{9.19}$$

The sequence of two impulses that leads to zero vibration (ZV) can now be summarised in matrix form as

$$\begin{bmatrix} A_i \\ t_i \end{bmatrix} = \begin{bmatrix} 1/(1+K) & K/(1+K) \\ 0 & 0.5T_d \end{bmatrix} \tag{9.20}$$

9.2.3   Using ZV impulse sequences to generate ZV commands

Real systems cannot be moved around with impulses, thus it is required to convert the properties of the impulse sequence given in equation (9.20) into a usable command. This can be done by simply convolving the impulse sequence with any desired command. The convolution product is then used as the command to the system. If the impulse sequence, also known as an input shaper, causes no vibration, then the convolution product will also cause no vibration (Bhat and Miu, 1990; Singer and Seering, 1990). This command generation process, called input shaping, is demonstrated in Figure 9.9 for an initial pulse function. Note that the convolution product in this case is the two-pulse command shown in Figure 9.7, which moved the crane with no residual vibration. In this case, the shaper duration is longer than the initial command, but in most cases the impulse sequence will be much shorter than the command profile. This is especially true when the baseline command is generated to move a robot through a complex trajectory and the periods of the system vibration are small compared to the duration of the move. When this is the case, the components of the shaped command that arise from the individual impulses run together to form a smooth continuous function as shown in Figure 9.10.

Since the inception of command shaping for vibration suppression, it has been noted that the process is effectively pole-zero cancellation. In their paper showing experimental implementation of posicast control, Tallman and Smith write, 'Posicast control can then be described as a process whereby complex zeros are generated so as to fall upon the complex poles of the system...' (Tallman and Smith, 1958). In his PhD

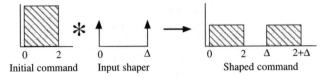

Figure 9.9   Input shaping a short pulse

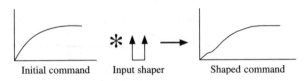

Figure 9.10   Input shaping a generic trajectory

thesis, Singer noted that all of his input shapers had an infinite number of zeros and that the Fourier transform of the input shapers for undamped systems were exactly zero at the natural frequency of the system (Singer, 1989). Bhat and Miu clearly stated the more general case for damped systems when they wrote, 'the necessary and sufficient condition for zero residual vibration is that the Laplace transform of the time bounded control input have zero component at the system poles' (Bhat and Miu, 1990).

9.2.4   Robustness to modelling errors

The amplitudes and time locations of the impulses depend on the system parameters (ω and ζ). If there are errors in these values (and there always are), then the input shaper will not result in ZV. In fact, when using the two-impulse sequence discussed above, there can be a noticeable amount of vibration for a relatively small modelling error. This lack of robustness was a major stumbling block for the original formulation of this idea that was developed in the 1950s (Tallman and Smith, 1958). This problem can be visualised by plotting the sensitivity curve of an input shaper. These curves show the amplitude of residual vibration caused by a shaper as a function of the system parameters. One such sensitivity curve for the ZV shaper given in equation (9.20) is shown in Figure 9.11. The horizontal axis is a normalised frequency and the percentage vibration is on the vertical axis. Note that as the actual frequency, ω, deviates from the modelling frequency, ω_m, the amount of vibration increases rapidly.

The first input shaper designed to have robustness to modelling errors was developed by Singer and Seering in the late 1980s (Singer and Seering, 1990; Singer *et al.*, 1990). This shaper was designed by requiring the derivative of the residual vibration, with respect to the frequency, to be equal to zero at the modelling frequency. Mathematically, this can be stated as:

$$\frac{\partial V\left(\omega,\zeta\right)}{\partial\omega}=0 \tag{9.21}$$

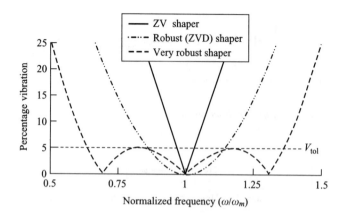

Figure 9.11　Sensitivity curves for several input shapers

Including this constraint has the effect of keeping the vibration near zero as the actual frequency starts to deviate from the modelling frequency. The sensitivity curve for this zero vibration and derivative (ZVD) shaper is also shown in Figure 9.11. Note that this shaper keeps the vibration at a low level over a much wider range of frequencies than the ZV shaper.

Since the development of the ZVD shaper, several other robust shapers have been developed. In fact, shapers can now be designed to have any amount of robustness to modelling errors (Singhose *et al.*, 1996*b*). Any real robotic system will have some amount of tolerable vibration; real machines are always vibrating at some level. Using this tolerance, a shaper can be designed to suppress any frequency range. The sensitivity curve for a very robust shaper is included in Figure 9.11 (Singhose *et al.*, 1996*b*, 1997*d*). This shaper is created by establishing a tolerable vibration limit, V_{tol}, and then restricting the vibration to below this value over a desired range of frequency errors. This design approach is related to traditional low-pass and band-stop filter design; however, the key difference is that no pass band range is required. Several studies have been conducted to document the advantages of input shapers over traditional filters (Mohamed and Tokhi, 2004; Singer *et al.*, 1999).

Input shaping robustness is not restricted to errors in the frequency value. Figure 9.12 shows a three-dimensional sensitivity curve for a shaper that was designed to suppress vibration between 0.7 and 1.3 Hz and also over the range of damping ratios between 0 and 0.2. Notice that the shaper is quite insensitive to changes in the damping ratio.

To achieve greater robustness, input shapers generally must contain more than two impulses and their durations must increase. For example, the ZVD shaper (Singer and Seering, 1990) obtained by satisfying equation (9.21) contains three impulses given by:

$$\begin{bmatrix} A_i \\ t_i \end{bmatrix} = \begin{bmatrix} 1/(1+K)^2 & 2K/(1+K)^2 & K^2/(1+K)^2 \\ 0 & 0.5T_d & T_d \end{bmatrix} \tag{9.22}$$

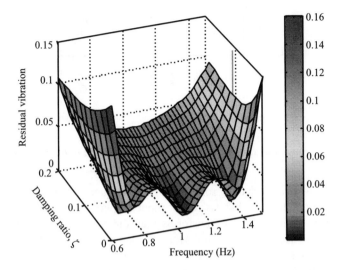

Figure 9.12   Three-dimensional sensitivity curve

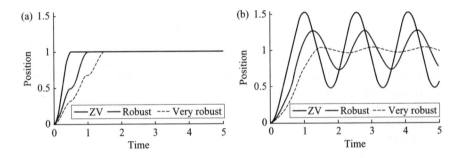

*Figure 9.13   Oscillator response to shaped step inputs: (a) perfect model, (b) 35%
            frequency error*

The increase in shaper duration means that the shaped command will also increase
in duration. Fortunately, even very robust shapers have fairly short durations. For
example, the ZVD shaper has a time duration of only one period of the natural
frequency. This time penalty is often a small cost in exchange for the improved
robustness to modelling errors. To demonstrate this tradeoff, Figure 9.13 shows the
response of a second-order harmonic oscillator to step commands shaped with the
three shapers used to generate Figure 9.11. Figure 9.13(a) shows the response when
the model is perfect and Figure 9.13(b) shows the case when there is a 35 per cent
error in the estimated system frequency. The increase in rise time caused by the
robust shapers is apparent in Figure 9.13(a), while Figure 9.13(b) shows the vast
improvement in vibration reduction that the robust shapers provide in the presence
of modelling errors. In this case the very robust shaper yields nearly zero vibration,
even with the 35 per cent frequency error.

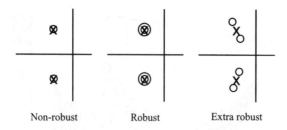

Non-robust   Robust   Extra robust

Figure 9.14   Command shaping as pole-zero cancellation in the s-plane

The analysis of input shaping robustness in the frequency domain was introduced by Singer. He pointed out that his robust input shapers had a wider trough in the amplitude plot of their Fourier transforms than did the less robust input shapers (Singer, 1989). Once again, Bhat and Miu provided a more general statement when they stated the sensitivity to modelling errors could be reduced by decreasing the magnitude of the Laplace transform in the neighbourhood of the system poles (Bhat *et al.*, 1991). They further suggested that Singer and Seering's technique of forcing the derivative of the Laplace transform to zero at the modelling parameters would achieve such a desired result.

A clearer explanation of this effect was provided by Singh and Vadali (1993) when they noted that the zero derivative constraint was adding a second zero on top of the resonant poles in the *s*-plane. This case is shown in the centre of Figure 9.14. Additional zeros placed over the poles corresponds to setting higher-order derivatives to zero and leads to additional improvement in robustness. However, a more efficient means of obtaining increased robustness is to place the zeros from the input shaping process in the neighbourhood of the flexible poles, rather than directly at the modelled locations (Pao and Singhose, 1998; Singhose *et al.*, 1994, 1997*c*). This approach is shown on the right side of Figure 9.14.

9.2.5   *Multi-mode input shaping*

Many robotic systems will have more than one flexible mode that can degrade system performance. Several methods have been developed for generating input shapers to suppress multiple modes of vibration (Hyde and Seering, 1991; Pao, 1999; Singh and Heppler, 1993; Singhose *et al.*, 1997*b*). These techniques can be used to solve the multiple vibration constraint equations sequentially, or concurrently. Suppressing multiple modes can lead to a large increase in the number of impulses in the input shaper, so methods have been developed to limit the impulses to a small number (Singh and Heppler, 1993). Furthermore, the nature of multi-input systems can be exploited to reduce the complexity and duration of a shaper for a multi-mode system (Pao, 1999).

Though there are more efficient ways (Hyde and Seering, 1991), a simple method for constructing a multi-mode input shaper is to design an input shaper independently for each problematic mode of vibration. Then, the individual input shapers can simply

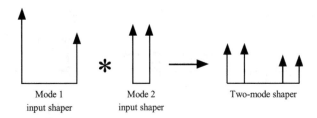

Figure 9.15   Forming a two-mode shaper through convolution

be convolved together to produce a multi-mode shaper. This straightforward process is shown in Figure 9.15. Note that information about mode shapes is not required for input shaping. This greatly simplifies vibration control solutions for multi-mode problems.

9.2.6   Real-time implementation

One of the strengths of command shaping is that it can often be implemented in a real-time control system. For example, the input shaping technique requires only a simple convolution that can usually be implemented with just a few multiplication and addition operations each time through the control loop.

Many motion control boards and digital signal processing (DSP) chips have built in algorithms for performing the real-time convolution that is necessary for input shaping. If these features are not available, then a very simple algorithm can be added to the control system. The algorithm starts by creating a command buffer, just a vector variable of a finite length. This buffer is used to store the command values for each time step. For example, the first value in the buffer would be the shaped command at the first time instance. A graphical representation of such a buffer is shown in the upper right-hand corner of Figure 9.16. The upper left-hand portion of the figure shows the unshaped baseline command in the digital domain. This baseline command can be created in real time, for example by reading a joystick position.

In order to fill the buffer with the input-shaped command, the algorithm acquires the baseline command each time through the control loop. Then, the algorithm multiplies the baseline command by the amplitude of the first impulse in the input shaper. This value is added to the current time location in the shaped-command buffer. The amplitude of the second impulse is then multiplied by the baseline command. However, this value is not sent directly out to the control loop. Rather, it is added to the future buffer slot that corresponds to location of the impulse in time. For example, assuming a 10 Hz sampling rate, if the time location of the second impulse was at 0.4 s, then this second value would be added to the buffer four slots ahead of the current position. This real-time process will build up the shaped command as demonstrated in Figure 9.16. The figure indicates the state of the algorithm at the first time step and at the eighth time step. To avoid having the index exceed the size of the buffer, a circular buffer is used where the index goes back to the beginning when it reaches the end of the buffer.

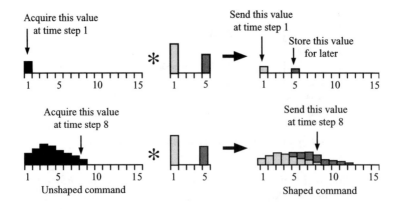

Figure 9.16   Real-time input shaping

9.2.7   Trajectory following

Although input shaping may not be specifically designed to optimise the ability of a system to follow a desired trajectory, the vibration reduction produced by input shaping allows the system to track a trajectory without continually oscillating around the trajectory path. In this respect, input shaping relies on a good baseline command for the desired trajectory. The trajectory needs to be nominally of the correct shape and needs to account for physical limitations of the hardware such as workspace boundaries and actuator limits. This baseline trajectory would normally be derived from these physical and kinematic requirements.

As a demonstration of input shaping trajectories, consider a painting robot that has flexible modes arising from a two-stage arm attached to an XY positioning stage (Singhose and Singer, 1996). The stages are positioned by PD controllers. The end-effector holds a compressed-air paint brush that paints on paper. No information about the paintbrush position is utilized in the control system. Experiments were conducted by turning on the flow of air to the paint brush, commencing the desired trajectory and then shutting off the flow of paint at the end of the move. Figure 9.17(a) shows the result when a fast circular trajectory of 7.6 m diameter was commanded. When input shaping was enabled, the system followed the desired trajectory much more closely, as shown in Figure 9.17(b).

9.2.8   Applications

The robustness and ease of use of input shaping has enabled its implementation on a large variety of systems and there have been a number of patents related to useful command shaping technology (Calvert and Gimpel, 1957; Feddema *et al.*, 1998; Jones *et al.*, 1991; Robinett *et al.*, 1999; Singer *et al.*, 1990; Singhose *et al.*, 1997*d*). When input shaping is implemented on a real machine it is often combined with feedback control. Several researchers have investigated this complementary relationship and have developed optimal methods for combining the two control

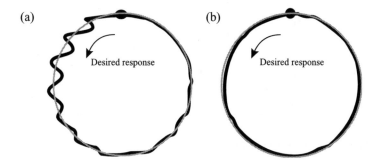

(a)   (b)

Desired response   Desired response

Figure 9.17   Response of painting robot: (a) unshaped circle trajectory, (b) shaped circle trajectory

techniques (Drapeau and Wang, 1993; Kenison and Singhose, 2002; Khorrami *et al.*, 1993; Muenchhof and Singh, 2002; Tokhi and Mohamed, 2003).

Input shaping has been found to be very useful in the high-tech manufacturing area. The repeatability of coordinate measuring machines can be significantly improved with input shaping (Jones and Ulsoy, 1999; Seth *et al.*, 1993; Singhose *et al.*, 1996c). The multiple modes of a silicon-handling robot were eliminated with input shaping (Rappole *et al.*, 1994). Shaping was an important part of a control system developed for a wafer stepper (deRoover *et al.*, 1996, 1998). The throughput of a disk-drive-head tester was significantly improved with shaping (Singhose *et al.*, 1997e). Many of the disk drive systems on the market today employ a form of input shaping. There are even reports of shaping being used on nanopositioners (Jordan, 2002).

Input shaping has also proven useful for moving large-scale machines. Shaping has been successfully implemented on a number of cranes and crane-like structures (Feddema, 1993; Kress *et al.*, 1994; Singer *et al.*, 1997; Singhose *et al.*, 2000). The performance of a long-reach manipulator was improved with command shaping (Jansen, 1992; Magee and Book, 1995). Researchers at Sandia Laboratory developed a command shaping scheme to reduce sloshing during the manipulation of liquid containers (Feddema *et al.*, 1997).

Command shaping has received considerable attention in the spacecraft field because these structures are always very flexible. A series of input-shaping experiments was performed on-board the space shuttle Endeavor using a flexible truss structure (Singhose *et al.*, 1997c; Tuttle and Seering, 1997). If on–off thruster-jets are in use, then command generation can be used to develop the time-optimal command profiles (Liu and Wie, 1992; Pao, 1996; Pao and Singhose, 1997; Singh and Vadali, 1994; Tuttle and Seering, 1999). When constructing reaction jet commands, the amount of fuel can be limited or set to a specific amount (Lau and Pao, 2002; Meyer and Silverberg, 1996; Singhose *et al.*, 1996a, 1999; Wie *et al.*, 1993). It is also possible to limit transient deflections (Kojima and Naka-jima, 2003; Singhose *et al.*, 1997a). Furthermore, vibration reduction and slewing can be completed in coordination with momentum dumping operations (Banerjee *et al.*, 2001).

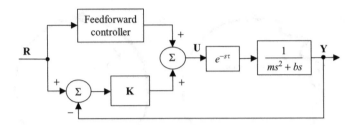

Figure 9.18   Feedforward compensation of a system with a time delay

9.3   Feedforward control action

Feedforward control is concerned with directly generating a control action (force, torque, voltage, etc.), rather than generating a reference command. By including an anticipatory corrective action before an error shows up in the response, a feedforward controller can provide better performance than a feedback controller alone. It can be used for a variety of cases such as systems with time delays, non-linear friction (Tung and Tomizuka, 1993), or systems performing repeated motions (Sadegh, 1995). Most feedforward control methods require an accurate system model, so robustness is an important issue to consider. Feedforward control can also be used to compensate for a disturbance, if the disturbance itself can be measured before the effect of the disturbance shows up in the system response.

9.3.1   Feedforward control of a simple system with time delay

To demonstrate a very simple feedforward control scheme, consider a system under proportional feedback control that can be modelled as a mass-damper system with a time delay. The block diagram for this case is shown in Figure 9.18. The desired output is represented by the reference signal, **R**. The actions of the feedforward control and the feedback control combine to produce the actuator effort, **U**, that then induces the actual output **Y**. To simplify this example the effect of the command generator in selecting **R** has been ignored.

Consider first the response of the system without feedforward compensation. Suppose that the feedback control system is running at 10 Hz, the time delay is 0.2 s, $m = 1$, and $b = 1$. The dynamic response of the system can be adjusted to some degree by varying the proportional gain, K. Figure 9.19 shows the oscillatory step response to a variety of K values. This is a case where the system flexibility results from the feedback control, rather than from the physical plant. Note that for low values of K, the system is sluggish. The system rise time can be improved by increasing K, but that strategy soon drives the system to instability. The corresponding control effort is shown in Figure 9.19(b).

Rather than performing a step motion, suppose the desired motion is the smooth function

$$r(t) = 1 - \cos(\omega t) \tag{9.23}$$

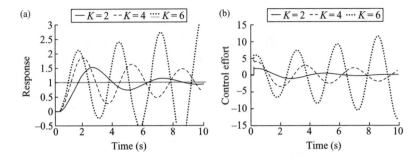

Figure 9.19 *Step response of time-delay system without feedforward compensation:*
(a) position response, (b) control effort

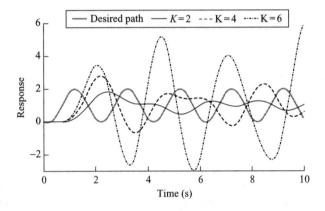

Figure 9.20 *Tracking a smooth function without feedforward compensation*

The simple proportional feedback controller might be able to provide adequate tracking if the frequency of the desired trajectory is very low. However, if the trajectory is demanding (relative to system frequency), then the feedback controller will provide poor tracking and possibly exhibit instability as shown in Figure 9.20. To improve performance, several options may be considered; redesigning the physical system to improve the dynamics and reduce the time delay, adding additional sensors, improving the feedback controller, using command shaping, or adding feedforward compensation. Only the role of feedforward compensation is examined here.

A simple feedforward control strategy would place the inverse of the plant in the feedforward block shown in Figure 9.18. If this controller can be implemented in real time, then the overall transfer function from desired response to system response would be unity. That is, the plant would respond exactly in the desired manner. There are, of course, limitations to what can be requested of the system in real time. But, let us proceed with this example and discuss the limitations after the basic concept is

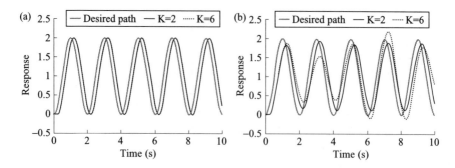

Figure 9.21   *Tracking a smooth function with feedforward compensation: (a) perfect model, (b) 10% modelling error*

demonstrated. In this case, the plant transfer function including the time delay is

$$G_P = \frac{e^{-s\tau}}{ms^2 + bs} \tag{9.24}$$

The feedforward controller would then be

$$G_{FF} = \frac{ms^2 + bs}{e^{-s\tau}} \tag{9.25}$$

When implemented in the digital domain, the time delay in the denominator becomes a time shift in the numerator. This time shift would be accomplished by essentially looking ahead at the desired trajectory. If the desired trajectory is acquired in real time, as would be the case for example with joystick driven trajectory control, then this process will delay the force output to the system. If this delay becomes too large, then this approach would not be appropriate for real-time implementation.

Figure 9.21(a) shows that under feedforward compensation, the system perfectly tracks the desired trajectory for various values of the feedback gain. This perfect result will not apply to real systems because there will always be modelling errors. Figure 9.21(b) shows the responses when there is a 10 per cent error in the system mass and damping parameters. With a low proportional gain, the tracking is still fairly good, but the system goes unstable for the higher gain.

One important issue to consider with feedforward control is the resulting control effort. Given that the feedback controller generates some effort and the feedforward adds to this effort, the result might be unrealistically high effort that saturates the actuators. Furthermore, the control effort can be highly dependent on the desired trajectory. Consider again the smooth trajectory given in equation (9.23). Sending the desired trajectory through the feedforward compensator, results in a feedforward control effort of

$$G_{FF}(s) \cdot R(s) = e^{s\tau} \frac{\omega^2 (ms + b)}{s^2 + \omega^2} \tag{9.26}$$

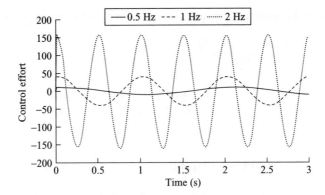

Figure 9.22   *Control effort for various frequencies with feedforward compensation*

Converting this into the time domain yields

$$m\omega^2 \cos\left[\omega\,(t + \tau)\right] + b\omega \sin\left[\omega\,(t + \tau)\right] \qquad (9.27)$$

or

$$C_1 \sin\left[\omega\,(t + \tau) + C_2\right] \qquad (9.28)$$

where,

$$C_1 = \omega\sqrt{m^2\omega^2 + b} \qquad \text{and} \qquad C_2 = \tan^{-1}\left(\tfrac{m\omega}{b}\right) \qquad (9.29)$$

Note that the control effort increases with the desired frequency of response. Therefore, requesting a very fast response will lead to a large control effort that will saturate the actuators. This effect is demonstrated in Figure 9.22. Finally, if the desired trajectory has discontinuous derivatives that cannot physically be realised – such as step and ramp commands for systems that have inertia – then the feedforward control effort would also saturate the actuators.

9.3.2   Zero phase error tracking control

The above feedforward control scheme is subject to some limitations. For example, only physically realisable desired trajectories can be utilized. An important limitation for model-inverting feedforward control schemes occurs when the model contains non-minimum phase zeros. These cannot be inverted because doing so would require implementing unstable poles in the feedforward control path. A way around this problem is to invert only the portion of the plant that yields a stable feedforward path.

To improve the performance of such partial-plant-inversion schemes, various extensions and scaling techniques have been developed. A good example of a feedforward controller that only inverts the acceptable parts of the plant is the zero phase error tracking controller (ZPETC) (Tomizuka, 1987). In the digital formulation of this controller, the plant transfer function is written in the following form:

$$G\left(z^{-1}\right) = \frac{z^{-d}B^a\left(z^{-1}\right)B^u\left(z^{-1}\right)}{A\left(z^{-1}\right)} \tag{9.30}$$

The numerator is broken down into three parts, a pure time delay, z^{-d}, a part that is acceptable for inversion, B^a, and a part that should not be inverted, B^u.

9.4   ZPETC as command shaping

If the ZPETC is designed to compensate for the entire closed-loop system, then the physical plant and the feedback controller can be reduced to a single transfer function:

$$G_c\left(z^{-1}\right) = \frac{z^{-d}B_c^a\left(z^{-1}\right)B_c^u\left(z^{-1}\right)}{A_c\left(z^{-1}\right)} \tag{9.31}$$

where $B_c^u\left(z^{-1}\right) = b_{c0}^u + b_{c1}^u z^{-1} + \cdots + b_{cs}^u z^{-p}$. The integer p represents the order of the unacceptable zeros. The c subscripts have been added to denote the transfer function that now represents the entire closed-loop system dynamics. The superscripts on the numerator again refer to parts that are acceptable and unacceptable for inversion. In this case, the output of the ZPETC is a reference signal for the closed-loop controller. It does not directly apply a force to the plant. It therefore does not try to overrule or add to the efforts of the feedback controller. Given this structure, the ZPETC should be considered a command generator, rather than a feedforward compensator.

To cancel the acceptable portion of the plant model, create the ZPETC so that the reference signal is given by

$$r\left(k\right) = \frac{A_c\left(z^{-1}\right)}{B_c^a\left(z^{-1}\right)B_c^u\left(1\right)}y_d^*\left(k+d\right) \tag{9.32}$$

Rather than invert the unacceptable zeros, the reference signal is formed using the unacceptable portion evaluated at $z = 1$. The $B_c^u(1)$ term is used to scale the steady-state gain of the controller, thereby compensating for the portion of the system that is not inverted. Note that the ZPETC operates on a term related to the desired trajectory denoted by, y_d^*. Choosing this function carefully can also help compensate for the incomplete model inversion. Note that for perfect trajectory tracking $y_d^*(k)$ would be chosen as

$$y_d^*\left(k\right) = \frac{B_c^u\left(1\right)}{B_c^u\left(z^{-1}\right)}y_d\left(k\right) \tag{9.33}$$

However, the term in the denominator would cause unacceptable oscillation or instability in the calculation of y_d^*. The natural choice would then be to simply choose:

$$y_d^*\left(k\right) = y_d\left(k\right) \tag{9.34}$$

However, this would lead to a phase lag between the desired response and the actual response. In order to reduce the phase lag, the ZPETC uses

$$y_d^* (k) = \frac{B_c^u (z)}{B_c^u (1)} y_d (k) \tag{9.35}$$

The total effect of the ZPETC can then be summarised as

$$r (k) = \frac{A_c \left(z^{-1}\right) B_c^{u^*} \left(z^{-1}\right)}{B_c^a \left(z^{-1}\right) \left[B_c^u (1)\right]^2} y_d (k + d + p) \tag{9.36}$$

Recall that d comes from the delay and p comes from the unacceptable zeros. The starred term in the numerator is $B_c^{u^*} \left(z^{-1}\right) = b_{cs}^u + b_{c(s-1)}^u z^{-1} + \cdots + b_{c0}^u z^{-p}$. Note that this technique requires knowing the desired trajectory for a number of time steps into the future. Therefore, its use as a real-time control element will be limited for some systems.

9.5 Summary

Command shaping operates by taking the baseline command and changing its shape slightly so that it will not excite the flexible modes in the system. Many command shaping methods have good robustness to modelling errors. Therefore, they can be used on a wide variety of systems, even if their dynamics are somewhat uncertain or change over time.

A highly successful robotic system would likely be composed of four well-designed components; hardware, feedback control, feedforward control and command shaping. Each of the components has their strengths and weaknesses. Luckily, they are all very compatible with each other and thus a good solution will make use of all of them when necessary.

This chapter has presented techniques to implement a command generator that can operate in real time and a number of applications have been described.

Chapter 10

Enhanced PID-type classical control of flexible manipulators

S.P. Goh and M.D. Brown

Industrial control applications have been dominated by the use of proportional, integral, derivative (PID)-type algorithms and are likely to remain so in the near future. This chapter adds a modern flavour to the traditional usage by describing recent PI–PD control strategies, and extending the methodology into a discrete-time multi-input multi-output (MIMO) version. The algorithm can be cast as a general second-order linear regulator and its parameters can thus be systematically chosen by pole-placement. More importantly, the design permits the derivation of an appropriate multivariable decoupling strategy. This is a key improvement and the effectiveness of the proposed multivariable controller has been validated experimentally in real-time motion tracking and vibration control of an asymmetrical flexible manipulator moving in the horizontal and vertical planes under gravity. The PI–PD control performance can be comparable to that of a decoupling MIMO pole-placement controller while being less sensitive to initial transient errors and having a simpler implementation.

10.1 Introduction

It is well claimed that proportional, integral, derivative (PID) or PID-like controllers have been the most widely used in process control. Their popularity can be attributed mainly to their relatively simple structure and the fact that conventional PID controllers are directly defined in terms of the three principal control effects (Ziegler and Nichols, 1942), enabling them to be applied with acceptable performance in most practical applications. Wang *et al.* (1997) therefore suggested that with the advantages of simplicity, PID controllers are always preferred to more advanced controllers unless there is evidence showing that PID control is not adequate and cannot meet the required specifications.

A review of the literature shows that relatively few PID controllers have been applied to motion tracking and vibration control of flexible manipulators although they have been successfully applied to rigid manipulators. This lack of confidence has been aggravated by both simulated and experimental results reported in the literature. For instance, He *et al.* (1993) and Friman and Waller (1994) have, respectively, given simulation results to show that the Ziegler–Nichols (ZN) or refined ZN tuned and internal model control theory tuned PID controllers give sluggish responses when applied to non-minimum-phase systems. Moreover, Åström and Hägglund (1995) have suggested using a more complex controller to improve control performance when a system to be controlled has high order, long dead-time and oscillatory modes. Similarly, Misir *et al.* (1996) have concluded that conventional PID controllers cannot effectively control high-order and time-delayed linear systems as well as nonlinear systems. Passino and Yurkovich (1998) have also implied that conventional controllers like PID are more suitable for linear time-invariant systems whereas nonlinear systems are more effectively controlled by more advanced control approaches. Kwak *et al.* (1997) and Atherton and Majhi (1999) have reported inherent limitations in the achievable performance when using conventional PID controllers to control integrating systems.

Hitherto, among the various forms of modified PID control structures, the PI–PD structure introduced by Atherton and co-workers (Atherton, 1999; Atherton and Majhi, 1999) offers several advantages, as described later in this chapter, which can alleviate some of the inherent limitations encountered by a conventional PID controller. This chapter therefore extends the continuous-time PI–PD controller design to the discrete multivariable domain.

Nevertheless, poor decoupling in a multivariable system has been ranked as the principal common control problem in industry (Wang *et al.*, 1997). Consequently, dynamic interactions among different channels have to be considered during the controller design if the performance of the controller is to be improved (Johansson *et al.*, 1998; Zhuang and Atherton, 1994). In fact, Johansson *et al.* (1998) have noted that most researchers have limited their control structures to a decentralised configuration of single-input single-output (SISO) PID controllers. Consequently, simulation results of decentralised controllers presented by Loh and Vasnani (1992), Palmor *et al.* (1995), Wang *et al.* (1997) and Zhuang and Atherton (1994) have all exhibited significant couplings in plant output responses.

Friman and Waller (1994) have therefore proposed applying a decoupling matrix with unity on the diagonal to decouple a multi-input multi-output (MIMO) system. However, the design of an appropriate decoupling matrix is neither always guaranteed to exist nor easy to find since it is related to the system characteristics. On the other hand, Wang *et al.* (1997) have proposed to decouple a controlled system by revising the design matrix equations, separating those involving diagonal elements from those involving off-diagonal ones. In particular, their objective is to ensure that the equivalent diagonal elements of the plant are independent from the off-diagonal elements of the controller. However, they have noted that the exact solutions for the controller parameters are very complex even if all plant data can be identified.

Owing to the aforementioned shortcomings in the existing limited number of decoupling approaches, this chapter therefore proposes a novel approach based on the discrete-time structure of a MIMO PI–PD controller.

Apart from a suitable control structure, it is also essential to select optimal parameters to render a PID-like controller effectively. However, Åström and Hägglund (1984) have noted that simple robust methods for tuning PID controllers have not been available. This is mainly because the parameter tuning is not a straightforward process since the effects of changing PID parameters are not independent. Moreover, the parameter tuning can be further complicated by non-linearity, dominant delays, time-varying and high-order dynamics (Bueno *et al.*, 1991; Rad *et al.*, 1997), temperature influence (Loron, 1997), ageing, multiple loops (Daley and Liu, 1999) and couplings in the plants. Furthermore, although a multivariable PID controller is readily implementable within the existing control software and hardware (Wang *et al.*, 1997) Zhuang and Atherton (1994) have noted that the procedures for PID tuning in the SISO case may not be suitable for a MIMO system. Consequently, the theory of MIMO PID tuning has been lacking although many industrial processes are inherently of multivariable nature (Palmor *et al.*, 1995; Wang *et al.*, 1997).

A comprehensive review on approaches that have been proposed to improve PID tuning has been conducted by Goh (2001). By casting the discrete PI–PD controller as a general second-order linear regulator, this chapter proposes a pole-placement tuning approach because it allows the controller parameters to be calculated based on the desired closed-loop poles. Such a tuning method is therefore very effective and relatively more straightforward besides simplifying analyses of the closed-loop system. Moreover, the limitation of having to restrict the plant model to an assumed standard form can also be eliminated in the proposed approach. This is a significant extension of the tuning method adopted by various researchers, such as Isermann (1989), Bueno *et al.* (1991) and Wellstead and Zarrop (1991), who used pole-placement to tune PI and PID controllers on restricted plant models. More importantly, the proposed approach enables algebraic derivation of a decoupling strategy that can be verified analytically.

The proposed MIMO PI–PD controller has been applied to the motion tracking and vibration control of an asymmetrical flexible manipulator in real time. On the basis of the observed open-loop responses, the flexible manipulator system is non-linear with cross-couplings in its horizontal and vertical motions. Hence, it is essential that the proposed multivariable controller is effectively decoupling.

A SISO PI–PD control algorithm is briefly described and derived in the next section as a prelude to the design methodologies of a discrete-time MIMO PI–PD controller presented in Section 10.3. In particular, strategies to achieve complete decoupling are suggested and analysed. Section 10.4 describes the experimental flexible manipulator used in this study. The effectiveness of the proposed decoupling MIMO PI–PD controller as it is applied to the asymmetrical flexible manipulator is validated in Section 10.5 where both simulation and experimental results are presented. Moreover, the relative performance of the controller against a multivariable pole-placement (MPP) controller is also compared.

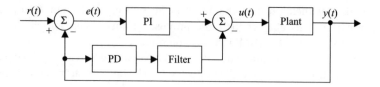

Figure 10.1   Block diagram of a PI–PD controller in closed-loop

10.2   Single-input single-output PI–PD

10.2.1   Basic algorithm

This section presents the features of a SISO PI–PD controller. Instead of filtering the derivative term only, as proposed by Atherton and colleagues (1999), the total output of the PD term is filtered for the purposes of anti-aliasing in sampled signals and removing high-frequency noise due to the derivative action. The proposed PI–PD controller is represented by the block diagram in Figure 10.1 and described as

$$u(t) = G_{PI}e(t) - G_F G_{PD} y(t) \tag{10.1}$$

where G_{PI}, G_{PD} and G_F are transfer functions for the PI, PD and filter blocks, respectively, $e(t) = r(t) - y(t)$ is the control error signal with the usual definitions for all terms.

Compared with a standard PID algorithm, the PI–PD algorithm has the following advantages:

- The introduction of the PD internal feedback loop can be used to change the poles of the plant transfer function to more desirable locations for control by a PI controller in the outer loop (Atherton, 1999; Atherton and Boz, 1998; Atherton and Majhi, 1998).
- Moving the derivative term from the forward loop to the feedback loop results in the same root locus but with different closed-loop zeros (Atherton and Boz, 1998; Atherton and Majhi, 1998).
- The derivative 'kick' effect can be avoided by having the derivative term acting on the plant output only (Atherton and Majhi, 1998).

Moreover, the integrator in the PI block can be implemented as a limited integrator to avoid integral 'windup'.

10.2.2   Discrete-time algorithm

It is desirable to derive a discrete-time version of the PI–PD algorithm so that it can be used to predict and analyse real-time responses directly. Using the backward-difference method of discretisation, the discrete-time PI–PD algorithm can be derived as follows:

$$G_{PI}(s) = K_p \left(1 + \frac{1}{sT_i} \right)$$
$$= K_p + \frac{K_i}{s}$$

where $K_i = K_p/T_i$ is the integral gain of the PID block in Simulink®. The discrete equivalent of the above is given as

$$
\begin{aligned}
G_{\text{PI}}\left(z^{-1}\right) &= K_p + \frac{K_i \tau_s}{1 - z^{-1}} \\
&= k_p + \frac{k_i}{1 - z^{-1}}
\end{aligned}
\tag{10.2}
$$

where τ_s represents the sample period and

$$
\begin{aligned}
k_p &= K_p \\
k_i &= K_i \tau_s
\end{aligned}
\tag{10.3}
$$

Similarly, the continuous PD formulation is given as

$$
\begin{aligned}
G_{\text{PD}}\left(s\right) &= K_f\left(1 + sT_d\right) \\
&= K_f + sK_d
\end{aligned}
$$

where $K_d = K_f T_d$ is the derivative gain of the PID block in Simulink®. This is expressed in discrete time as

$$
G_{\text{PD}}\left(z^{-1}\right) = K_f + \frac{K_d\left(1 - z^{-1}\right)}{\tau_s}
\tag{10.4}
$$

Using a first-order lowpass filter with pole at q in the z-plane, the filter transfer function can be expressed as

$$
G_{\text{F}}\left(z^{-1}\right) = \frac{1 - q}{1 - qz^{-1}}
\tag{10.5}
$$

Thus, using equations (10.3) and (10.5) yields

$$
\begin{aligned}
G_{\text{F}}\left(z^{-1}\right) G_{\text{PD}}\left(z^{-1}\right) &= \frac{1 - q}{1 - qz^{-1}}\left[K_f + \frac{K_d\left(1 - z^{-1}\right)}{\tau_s}\right] \\
&= \frac{K_f\left(1 - q\right) + \frac{K_d}{\tau_s}\left(1 - q\right)\left(1 - z^{-1}\right)}{1 - qz^{-1}}
\end{aligned}
$$

or

$$
G_{\text{F}}\left(z^{-1}\right) G_{\text{PD}}\left(z^{-1}\right) = \frac{k_f + k_d\left(1 - z^{-1}\right)}{1 - qz^{-1}}
\tag{10.6}
$$

where

$$
\begin{aligned}
k_f &= K_f\left(1 - q\right) \\
k_d &= \frac{K_d}{\tau_s}\left(1 - q\right)
\end{aligned}
$$

Equation (10.1) can be re-written as follows:

$$u(t) = G_{\text{PI}}\left(z^{-1}\right) r(t) - \left[G_{\text{PI}}\left(z^{-1}\right) + G_{\text{F}}\left(z^{-1}\right) G_{\text{PD}}\left(z^{-1}\right)\right] y(t) \qquad (10.7)$$

Substituting for $G_{\text{PI}}\left(z^{-1}\right)$ and $G_{\text{F}}\left(z^{-1}\right) G_{\text{PD}}\left(z^{-1}\right)$ from equations (10.2) and (10.6) into equation (10.7) and simplifying yields

$$u(t) = \frac{\left[(k_p + k_i) + (-k_p - k_p q - k_i q) z^{-1} + (k_p q) z^{-2}\right] r(t)}{(1 - z^{-1})(1 - q z^{-1})} -$$

$$\frac{\left[(k_p + k_i + k_f + k_d) + (-k_p - k_p q - k_i q - k_f - 2k_d) z^{-1} + (k_p q + k_d) z^{-2}\right] y(t)}{(1 - z^{-1})(1 - q z^{-1})}$$

$$(10.8)$$

Equation (10.8) can be generalized as

$$u(t) = \frac{\left(h_0 + h_1 z^{-1} + h_2 z^{-2}\right) r(t) - \left(g_0 + g_1 z^{-1} + g_2 z^{-2}\right) y(t)}{(1 - z^{-1})(1 - q z^{-1})} \qquad (10.9)$$

where,

$$g_0 = k_p + k_i + k_f + k_d, \quad g_1 = -k_p - k_p q - k_i q - k_f - 2k_d, \quad g_2 = k_p q + k_d$$
$$h_0 = k_p + k_i, \quad h_1 = -k_p - k_p q - k_i q, \quad h_2 = k_p q$$

Assuming the plant model in Figure 10.1 can be expressed as $B\left(z^{-1}\right)/A\left(z^{-1}\right)$, the closed-loop transfer function, using equation (10.9) can thus be obtained as

$$\frac{y(t)}{r(t)} = \frac{B\left(z^{-1}\right)\left(h_0 + h_1 z^{-1} + h_2 z^{-2}\right)}{(1 - z^{-1})(1 - q z^{-1}) A\left(z^{-1}\right) + B\left(z^{-1}\right)\left(g_0 + g_1 z^{-1} + g_2 z^{-2}\right)}$$

$$(10.10)$$

The transfer function in equation (10.10) can be written more succinctly as

$$\frac{y(t)}{r(t)} = \frac{B\left(z^{-1}\right) H\left(z^{-1}\right)}{F\left(z^{-1}\right) A\left(z^{-1}\right) + B\left(z^{-1}\right) G\left(z^{-1}\right)} \qquad (10.11)$$

where

$$F\left(z^{-1}\right) = (1 - z^{-1})(1 - q z^{-1})$$
$$G\left(z^{-1}\right) = g_0 + g_1 z^{-1} + g_2 z^{-2}$$
$$H\left(z^{-1}\right) = h_0 + h_1 z^{-1} + h_2 z^{-2}$$

The closed-loop transfer function in equation (10.11) shows that the PI–PD control algorithm can be made equivalent to a general 2 degrees of freedom (DOF) linear regulator described by Åström and Wittenmark (1997) and Landau (1998). Hence, design and analysis tools for a linear regulator can also be applied to a PI–PD controller. In particular, the PI–PD controller can be more systematically analysed and tuned using a pole-placement cum sensitivity-function shaping approach as described by Landau *et al.* (1996) and Goh *et al.* (2000*b*). More features of the discrete-time SISO PI–PD controller, including its practical design and application example, can be found in Goh, 2001.

10.3 Multi-input multi-output PI–PD

It is desirable to design a completely decoupling MIMO controller since many practical processes experience dynamic interactions among their different channels. However, the main difficulties have been how to model the processes including the dynamic interactions and to decouple the interacting channels so that they can be controlled independently. With the basis of the discrete-time SISO PI–PD algorithm developed in the previous section, this section will concentrate on decoupling issues and solving the diophantine equation for a multivariable controller.

10.3.1 Basic notations

Consider the same PI–PD control structure illustrated in Figure 10.1. Assuming an accurate plant model for the flexible manipulator system has been identified, then the transfer functions of the open-loop plant and the first-order lowpass filter can be defined, respectively, as follows:

$$\mathbf{y}(t) = \mathbf{A}^{-1}\left(z^{-1}\right)\mathbf{B}\left(z^{-1}\right)\mathbf{u}(t)$$

$$\mathbf{G}_{\text{Filter}} = \frac{\mathbf{I} - \mathbf{q}}{\mathbf{I} - \mathbf{q}z^{-1}}$$

Since a MIMO system is considered here, all the polynomial functions are matrices and the signal variables are vectors, while the filter pole location is a diagonal matrix. These are typically defined as

$$\mathbf{A}(z^{-1}) = \begin{bmatrix} a_{11,0} + a_{11,1}z^{-1} + \cdots + a_{11,n}z^{-n} & \cdots & a_{1p,0} + a_{1p,1}z^{-1} + \cdots + a_{1p,n}z^{-n} \\ \vdots & & \vdots \\ a_{p1,0} + a_{p1,1}z^{-1} + \cdots + a_{p1,n}z^{-n} & \cdots & a_{pp,0} + a_{pp,1}z^{-1} + \cdots + a_{pp,n}z^{-n} \end{bmatrix}$$

or

$$\mathbf{A}\left(z^{-1}\right) = \mathbf{a}_0 + \mathbf{a}_1 z^{-1} + \cdots + \mathbf{a}_n z^{-n}$$

$$\mathbf{y}(t) = \begin{bmatrix} y_1(t) \\ y_2(t) \\ \vdots \\ y_p(t) \end{bmatrix}$$

$$\mathbf{q} = \begin{bmatrix} q_1 & 0 & \cdots & 0 \\ 0 & q_2 & \ddots & \vdots \\ \vdots & \ddots & \ddots & 0 \\ 0 & \cdots & 0 & q_p \end{bmatrix}$$

where p represents the number of channels, n the polynomial order, and $\mathbf{I}$ the $p \times p$ identity matrix. $\mathbf{K}_p$, $\mathbf{K}_i$, $\mathbf{K}_f$ and $\mathbf{K}_d$ are the respective $p \times p$ matrix gains of the MIMO PI–PD controller.

Similar to the SISO case, $\mathbf{u}(t)$ (a $p \times 1$ matrix) can be generalized into the form:

$$\mathbf{u}(t) = \left[\left(\mathbf{I} - z^{-1} \right) \left(\mathbf{I} - \mathbf{q}z^{-1} \right) \right]^{-1} \left[\left(\mathbf{h}_0 + \mathbf{h}_1 z^{-1} + \mathbf{h}_2 z^{-2} \right) \mathbf{r}(t) \right.$$
$$\left. - \left(\mathbf{g}_0 + \mathbf{g}_1 z^{-1} + \mathbf{g}_2 z^{-2} \right) \mathbf{y}(t) \right] \tag{10.12}$$

As such, the closed-loop transfer function can be obtained as

$$\mathbf{y}(t) = \left[\left(\mathbf{I} - \mathbf{I}z^{-1} \right) \left(\mathbf{I} - \mathbf{q}z^{-1} \right) \mathbf{A} \left(z^{-1} \right) + \mathbf{B} \left(z^{-1} \right) \left(\mathbf{g}_0 + \mathbf{g}_1 z^{-1} + \mathbf{g}_2 z^{-2} \right) \right]^{-1}$$
$$\times \mathbf{B} \left(z^{-1} \right) \left(\mathbf{h}_0 + \mathbf{h}_1 z^{-1} + \mathbf{h}_2 z^{-2} \right) \mathbf{r}(t) \tag{10.13}$$

For subsequent discussions on developing an appropriate decoupling algorithm, let $\mathbf{A}_m \left(z^{-1} \right)$, be a polynomial matrix containing all the desired closed-loop poles:

$$\mathbf{A}_m \left(z^{-1} \right) = \mathbf{k} \left(\mathbf{I} - \mathbf{p}_1 z^{-1} \right) \left(\mathbf{I} - \mathbf{p}_2 z^{-1} \right) \cdots \left(\mathbf{I} - \mathbf{p}_j z^{-1} \right)$$

where $\mathbf{k}$ is a matrix multiplier to give unity steady-state gain, $\mathbf{p}_i$, $i = 1, 2, \ldots, j$ represent matrices containing the locations of the j desired closed-loop poles in the z-plane.

10.3.2   Decoupling algorithm

To completely decouple the closed-loop system of equation (10.13), it is essential that the polynomials of the transfer function are specified as diagonal matrices $\mathbf{N}(z^{-1})$ and $\mathbf{D}(z^{-1})$, so that the MIMO system can be expressed as multiple independent SISO systems (Dutton *et al.*, 1997);

$$\mathbf{y}(t) = \mathbf{D}^{-1} \left(z^{-1} \right) \mathbf{N} \left(z^{-1} \right) \mathbf{r}(t) = \begin{bmatrix} d_1 & 0 & \cdots & 0 \\ 0 & d_2 & \ddots & \vdots \\ \vdots & \ddots & \ddots & 0 \\ 0 & \cdots & 0 & d_p \end{bmatrix}^{-1} \begin{bmatrix} n_1 & 0 & \cdots & 0 \\ 0 & n_2 & \ddots & \vdots \\ \vdots & \ddots & \ddots & 0 \\ 0 & \cdots & 0 & n_p \end{bmatrix} \mathbf{r}(t)$$

$$\left. \begin{aligned} d_1 y_1(t) &= n_1 r_1(t) \\ d_2 y_2(t) &= n_2 r_2(t) \\ &\vdots \\ d_p y_p(t) &= n_p r_p(t) \end{aligned} \right\} \tag{10.14}$$

where n_i and d_i, $i = 1, 2, \ldots, p$, are appropriate scalar polynomials.

There are various strategies to transform equation (10.13) to achieve the condition in equation (10.14).

10.3.2.1 Strategy A

In this strategy, the desired closed-loop poles are not selected to cancel the extra zeros introduced by $\mathbf{H}\left(z^{-1}\right)$. Equation (10.13) thus becomes

$$\mathbf{y}\left(t\right) = \mathbf{A}_m^{-1}\left(z^{-1}\right)\mathbf{B}\left(z^{-1}\right)\left(\mathbf{h}_0 + \mathbf{h}_1 z^{-1} + \mathbf{h}_2 z^{-2}\right)\mathbf{r}\left(t\right)$$

Hence, comparing with equation (10.14), the following conditions have to be met:

$$\mathbf{B}\left(z^{-1}\right)\left(\mathbf{h}_0 + \mathbf{h}_1 z^{-1} + \mathbf{h}_2 z^{-2}\right) = \mathbf{N}\left(z^{-1}\right) \qquad (10.15)$$

$$\left(\mathbf{I} - \mathbf{I}z^{-1}\right)\left(\mathbf{I} - \mathbf{q}z^{-1}\right)\mathbf{A}\left(z^{-1}\right) + \mathbf{B}\left(z^{-1}\right)\left(\mathbf{g}_0 + \mathbf{g}_1 z^{-1} + \mathbf{g}_2 z^{-2}\right)$$

$$= \mathbf{A}_m\left(z^{-1}\right) = \mathbf{D}\left(z^{-1}\right) \qquad (10.16)$$

- Since $\mathbf{A}_m\left(z^{-1}\right)$ contains the desired closed-loop poles, it can always be specified as diagonal. Then, the diophantine equation (10.16) can be solved for the unknown parameters, although each term is a matrix instead of a scalar. A method of solving this can be found in (Goh, 2001).
- However, to obtain a unique solution, there must be an equal number of unknowns and simultaneous equations. Since there are only four unknowns ($\mathbf{g}_0, \mathbf{g}_1, \mathbf{g}_2, \mathbf{q}$) in equation (10.16), this strategy is only applicable to second-order plants.
- For the condition in equation (10.15), $\mathbf{H}\left(z^{-1}\right) = \left(\mathbf{h}_0 + \mathbf{h}_1 z^{-1} + \mathbf{h}_2 z^{-2}\right)$ can be specified as adj $\left[\mathbf{B}\left(z^{-1}\right)\right]$ since, for a non-singular square matrix $\mathbf{X}$, $\mathbf{X} \cdot$ adj $(\mathbf{X}) = |\mathbf{X}|\,\mathbf{I} = x\mathbf{I}$, where x is a scalar and therefore $x\mathbf{I}$ is diagonal. Hence, $\mathbf{B}\left(z^{-1}\right)\mathbf{H}\left(z^{-1}\right) = \left|\mathbf{B}\left(z^{-1}\right)\right|\mathbf{I} = b\left(z^{-1}\right)\mathbf{I}$, where $b\left(z^{-1}\right)$ is a scalar polynomial.
- Once all unknowns ($\mathbf{g}_0, \mathbf{g}_1, \mathbf{g}_2, \mathbf{q}, \mathbf{h}_0, \mathbf{h}_1, \mathbf{h}_2$) have been found, they can be used to calculate the matrix gains ($\mathbf{K}_p, \mathbf{K}_i, \mathbf{K}_f$ and $\mathbf{K}_d$) of the MIMO PI–PD controller based on the relationships given in Section 10.2.2.

Some advantages of this strategy include the following:

1. Different sets of closed-loop poles can be specified for each channel. This is a very desirable feature because the poles can be independently selected to give the most optimal performance for each channel.
2. The decoupling algorithm is very straightforward without requiring additional data.
3. The controller design and implementation is simple and without intensive mathematical calculation.
4. The controller structure of Figure 10.1 can be maintained.

However, this strategy has some limitations, which include the following:

1. $\mathbf{B}\left(z^{-1}\right)$ must be non-singular so that $b\left(z^{-1}\right)$ is non-zero.
2. $\mathbf{B}\left(z^{-1}\right)$ must be of second order so that its adjoint matrix is also of second order to agree with $\mathbf{H}\left(z^{-1}\right)$, which is limited to second order only. Hence, compared to the SISO case, this excludes strategy A from being applied to a second-order plant but with a first-order numerator.
3. It is not guaranteed that $\mathbf{H}\left(z^{-1}\right)$ can always be specified as $adj\left[\mathbf{B}\left(z^{-1}\right)\right]$ since its coefficients still have to meet other conditions such as ensuring zero steady-state gain and validity of inherent relationships when casting derived $\mathbf{u}\left(t\right)$ into the generalized format of equation (10.12).
4. Since it is assumed that the plant is causal, $b\left(z^{-1}\right)$ inherently introduces extra time delays. Note that if this causes problems in the closed-loop system, then an arbitrary forward shift z^k, which does not make $b\left(z^{-1}\right)$ non-causal, can be introduced as described in strategy B.

On the other hand, if the desired closed-loop poles have been selected to cancel the extra zeros introduced by $\mathbf{H}\left(z^{-1}\right)$, then it is required that $\mathbf{B}\left(z^{-1}\right) = \mathbf{N}\left(z^{-1}\right)$, which is impossible unless the numerator of the plant is already decoupled. In this case, an extra term, $\mathbf{E}\left(z^{-1}\right)$, can be introduced such that $\mathbf{B}\left(z^{-1}\right)\mathbf{E}\left(z^{-1}\right) = \mathbf{N}\left(z^{-1}\right)$. A suggestion for $\mathbf{E}\left(z^{-1}\right)$ together with its effects on the closed-loop system is described in strategy B.

10.3.2.2   Strategy B

If it is desirable to cancel the extra zeros introduced by $\mathbf{H}\left(z^{-1}\right)$, one way to avoid the problematic requirement of $\mathbf{B}\left(z^{-1}\right) = \mathbf{N}\left(z^{-1}\right)$ is by adding an extra term to the system forward path in Figure 10.1 such that the PI–PD control structure is modified to that of Figure 10.2.

The revised closed-loop transfer function can be derived as follows:

$$\mathbf{y}\left(t\right) = \mathbf{A}^{-1}\left(z^{-1}\right)\mathbf{B}\left(z^{-1}\right)\mathbf{E}\left(z^{-1}\right)\left[\left(\mathbf{I} - \mathbf{I}z^{-1}\right)\left(\mathbf{I} - \mathbf{q}z^{-1}\right)\right]^{-1}$$
$$\times\left[\left(\mathbf{h}_0 + \mathbf{h}_1 z^{-1} + \mathbf{h}_2 z^{-2}\right)\mathbf{r}\left(t\right) - \left(\mathbf{g}_0 + \mathbf{g}_1 z^{-1} + \mathbf{g}_2 z^{-2}\right)\mathbf{y}\left(t\right)\right]$$

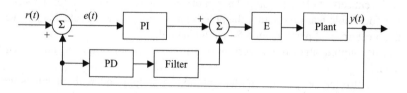

Figure 10.2   Block diagram of a modified MIMO PI–PD controller in closed-loop

Assuming $\mathbf{B}\left(z^{-1}\right)\mathbf{E}\left(z^{-1}\right)$ is diagonal, then

$$\mathbf{y}\left(t\right) = \left[\left(\mathbf{I} - \mathbf{I}z^{-1}\right)\left(\mathbf{I} - \mathbf{q}z^{-1}\right)\mathbf{A}\left(z^{-1}\right) + \mathbf{B}\left(z^{-1}\right)\mathbf{E}\left(z^{-1}\right)\left(\mathbf{g}_0 + \mathbf{g}_1 z^{-1} + \mathbf{g}_2 z^{-2}\right)\right]^{-1}$$
$$\times \mathbf{B}\left(z^{-1}\right)\mathbf{E}\left(z^{-1}\right)\left(\mathbf{h}_0 + \mathbf{h}_1 z^{-1} + \mathbf{h}_2 z^{-2}\right)\mathbf{r}\left(t\right) \tag{10.17}$$

Hence, the conditions to be met to decouple the closed-loop system of equation (10.17) to the form of equation (10.14) are

$$\mathbf{B}\left(z^{-1}\right)\mathbf{E}\left(z^{-1}\right) = \mathbf{N}\left(z^{-1}\right)$$

$$\left(\mathbf{I} - \mathbf{I}z^{-1}\right)\left(\mathbf{I} - \mathbf{q}z^{-1}\right)\mathbf{A}\left(z^{-1}\right) + \mathbf{B}\left(z^{-1}\right)\mathbf{E}\left(z^{-1}\right)\left(\mathbf{g}_0 + \mathbf{g}_1 z^{-1} + \mathbf{g}_2 z^{-2}\right)$$
$$= \left(\mathbf{h}_0 + \mathbf{h}_1 z^{-1} + \mathbf{h}_2 z^{-2}\right)\mathbf{A}_m\left(z^{-1}\right) = \mathbf{D}\left(z^{-1}\right) \tag{10.18}$$

- Note that $\left(\mathbf{h}_0 + \mathbf{h}_1 z^{-1} + \mathbf{h}_2 z^{-2}\right)$ is included in equation (10.18) since $\mathbf{B}\left(z^{-1}\right)\mathbf{E}\left(z^{-1}\right)$ will be made diagonal.
- To make $\mathbf{B}\left(z^{-1}\right)\mathbf{E}\left(z^{-1}\right)$ diagonal, one effective way is to define $\mathbf{E}\left(z^{-1}\right)$ as follows (Plummer and Vaughan, 1997):

$$\mathbf{E}\left(z^{-1}\right) = \mathrm{adj}\left[\mathbf{B}\left(z^{-1}\right)\right]z^k$$

where z^k is the maximum forward shift, which does not make $\mathbf{E}\left(z^{-1}\right)$ non-causal. Hence,

$$\mathbf{B}\left(z^{-1}\right)\mathbf{E}\left(z^{-1}\right) = \left|\mathbf{B}\left(z^{-1}\right)\right|z^k\mathbf{I} = b_k\left(z^{-1}\right)\mathbf{I} \tag{10.19}$$

where $b_k\left(z^{-1}\right)$ is a scalar polynomial.

- z^k is introduced to ensure that $b_k\left(z^{-1}\right)$ has the same time delay as $\mathbf{B}\left(z^{-1}\right)$.
- Substituting equation (10.19) into equation (10.18) gives

$$\left(\mathbf{I} - \mathbf{I}z^{-1}\right)\left(\mathbf{I} - \mathbf{q}z^{-1}\right)\mathbf{A}\left(z^{-1}\right) + b_k\left(z^{-1}\right)\mathbf{I}\left(\mathbf{g}_0 + \mathbf{g}_1 z^{-1} + \mathbf{g}_2 z^{-2}\right)$$
$$= \left(\mathbf{h}_0 + \mathbf{h}_1 z^{-1} + \mathbf{h}_2 z^{-2}\right)\mathbf{A}_m\left(z^{-1}\right) = \mathbf{D}\left(z^{-1}\right) \tag{10.20}$$

- Since the order of $b_k(z^{-1})\mathbf{I}$ is higher than that of $\mathbf{B}\left(z^{-1}\right)$, the diophantine equation (10.20) can no longer be solved for a unique solution if the order of $b_k(z^{-1})\mathbf{I}$ is higher than four because there will be more equations than unknowns. This problem arises because the controller polynomials are fixed as second order (due to the PI–PD nature) whereas, in a general linear regulator design by pole-placement, the order of the controller polynomials can vary according to the order of the plant polynomials. One way of overcoming this problem is to introduce extra unknowns to the system in the form of diagonal closed-loop poles as

$$\left(\mathbf{I} - \mathbf{I}z^{-1}\right)\left(\mathbf{I} - \mathbf{q}z^{-1}\right)\mathbf{A}\left(z^{-1}\right) + b_k(z^{-1})\mathbf{I}\left(\mathbf{g}_0 + \mathbf{g}_1 z^{-1} + \mathbf{g}_2 z^{-2}\right)$$
$$= \left(\mathbf{h}_0 + \mathbf{h}_1 z^{-1} + \mathbf{h}_2 z^{-2}\right)\mathbf{A}_m\left(z^{-1}\right)\mathbf{A}'_m\left(z^{-1}\right) = \mathbf{D}\left(z^{-1}\right) \tag{10.21}$$

where $\mathbf{A}'_m\left(z^{-1}\right)$ is a diagonal polynomial matrix whose order depends on the order of $b_k\left(z^{-1}\right)\mathbf{I}$.

- To ensure unity steady-state gain, the following condition has to be met:

$$\mathbf{A}_m\left(1\right)\mathbf{A}'_m\left(1\right) = b_k\left(1\right)\mathbf{I} \tag{10.22}$$

- As a result, $\mathbf{A}_m\left(z^{-1}\right)$ can still contain all desired closed-loop poles and be specified as diagonal but not all poles of the resulting closed-loop system can be selected arbitrarily.

Consequently, this strategy has the following drawbacks:

1. The resulting system may contain inappropriate poles.
2. Solving of the diophantine equation (10.21) has become more complicated since extra conditions need to be specified to ensure that $\mathbf{A}'_m\left(z^{-1}\right)$ is diagonal.

On the other hand, this strategy gives the following advantages:

1. It can be applicable to any order of plant model. This flexibility is a significant advantage especially that the plant models for flexible manipulator applications usually have orders higher than two.
2. Compared to strategy A, relatively more closed-loop poles can be specified if needed. Goh (2001) has shown that the additional poles can be exploited to reduce control sensitivities of the closed-loop system particularly at high frequencies. This is a very desirable advantage because it can reduce the magnitudes of the control signals and hence increase the robustness of the system against measurement noise.
3. Different sets of closed-loop poles can be specified for each channel. This is a very desirable feature because the poles can be independently selected to give the most optimal performance for each channel.
4. The decoupling algorithm is still straightforward without requiring additional data.

10.3.2.3 Strategy C

A third strategy, which is a hybrid between strategies A and B, may also be proposed. The following conditions are to be met to achieve complete decoupling of the closed-loop system:

$$\mathbf{B}\left(z^{-1}\right)\mathbf{E}\left(z^{-1}\right)\mathbf{H}\left(z^{-1}\right) = \mathbf{N}\left(z^{-1}\right) \tag{10.23}$$

$$\left(\mathbf{I} - \mathbf{I}z^{-1}\right)\left(\mathbf{I} - \mathbf{q}z^{-1}\right)\mathbf{A}\left(z^{-1}\right) + \mathbf{B}\left(z^{-1}\right)\mathbf{E}\left(z^{-1}\right)\left(\mathbf{g}_0 + \mathbf{g}_1 z^{-1} + \mathbf{g}_2 z^{-2}\right)$$
$$= \mathbf{A}_m\left(z^{-1}\right) = \mathbf{D}\left(z^{-1}\right) \tag{10.24}$$

- As in strategy A, the extra zeros due to $\mathbf{H}\left(z^{-1}\right)$ are not cancelled. On the other hand, the extra term recommended in strategy B, $\mathbf{E}\left(z^{-1}\right)$, is still included to diagonalise the closed-loop numerator as in equation (10.24).
- This strategy effectively extends strategy A to plant models with orders higher than two, where $\mathbf{E}\left(z^{-1}\right)\mathbf{H}\left(z^{-1}\right)$ can take on any order as long as it is causal, using the definition,

$$\mathbf{E}\left(z^{-1}\right)\mathbf{H}\left(z^{-1}\right) = \mathrm{adj}\left[\mathbf{B}\left(z^{-1}\right)\right]z^k \tag{10.25}$$

where z^k is the maximum forward shift that does not make $\mathbf{E}\left(z^{-1}\right)\mathbf{H}\left(z^{-1}\right)$ non-causal. Hence,

$$\mathbf{B}\left(z^{-1}\right)\mathbf{E}\left(z^{-1}\right)\mathbf{H}\left(z^{-1}\right) = \left|\mathbf{B}\left(z^{-1}\right)\right|z^k\mathbf{I} = b_k\left(z^{-1}\right)\mathbf{I} \tag{10.26}$$

where $b_k\left(z^{-1}\right)$ is a scalar polynomial.

- Moreover, unlike strategy A, equation (10.25) can usually be achieved while $\mathbf{H}\left(z^{-1}\right)$ meets other conditions. This is because $\mathbf{E}\left(z^{-1}\right)$ can be selected arbitrarily.
- Since $b_k\left(z^{-1}\right)\mathbf{I}$ is diagonal, it follows from equation (10.26) that

$$\mathbf{B}\left(z^{-1}\right)\mathbf{E}\left(z^{-1}\right) = b_k\left(z^{-1}\right)\mathbf{I}\mathbf{H}^{-1}\left(z^{-1}\right) = \mathbf{H}^{-1}\left(z^{-1}\right)b_k\left(z^{-1}\right)\mathbf{I} \tag{10.27}$$

- Now that there are more simultaneous equations than unknowns, using the idea from strategy B, so that a unique solution can still be obtained, the diophantine equation (10.24) can be modified as

$$\left(\mathbf{I} - \mathbf{I}z^{-1}\right)\left(\mathbf{I} - \mathbf{q}z^{-1}\right)\mathbf{A}\left(z^{-1}\right) + \mathbf{B}\left(z^{-1}\right)\mathbf{E}\left(z^{-1}\right)\left(\mathbf{g}_0 + \mathbf{g}_1 z^{-1} + \mathbf{g}_2 z^{-2}\right)$$
$$= \mathbf{A}_m\left(z^{-1}\right)\mathbf{A}'_m\left(z^{-1}\right) = \mathbf{D}\left(z^{-1}\right) \tag{10.28}$$

where $\mathbf{A}'_m\left(z^{-1}\right)$ is as defined in strategy B and meets the condition in equation (10.22).

- Substituting equation (10.27) into equation (10.28) and simplifying yields

$$\mathbf{H}\left(z^{-1}\right)\left(\mathbf{I} - \mathbf{I}z^{-1}\right)\left(\mathbf{I} - \mathbf{q}z^{-1}\right)\mathbf{A}\left(z^{-1}\right) + b_k\left(z^{-1}\right)\mathbf{I}\left(\mathbf{g}_0 + \mathbf{g}_1 z^{-1} + \mathbf{g}_2 z^{-2}\right)$$
$$= \mathbf{H}\left(z^{-1}\right)\mathbf{A}_m\left(z^{-1}\right)\mathbf{A}'_m\left(z^{-1}\right) \tag{10.29}$$

Comparing equations (10.21) and (10.29), it can be seen that

1. The latter is more complicated to solve.
2. For a given plant, there are always two more simultaneous equations that can be formed from equation (10.29) due to the presence of $\mathbf{H}\left(z^{-1}\right)$ on the left-hand-side of the equation. As a result, the order of $\mathbf{A}'_m\left(z^{-1}\right)$ also has to be increased by two to give two more unknowns in the equation. Consequently, there are two more poles that cannot be selected arbitrarily in the closed-loop system. Hence, this is not a good design approach because there are more closed-loop poles that cannot be guaranteed to be appropriate.

In summary, the following remarks can be made about the three different decoupling strategies described above:

- Strategy A is the simplest but is only applicable to second-order plants. Although the resulting system can contain the fewest closed-loop poles, and all of them can be selected arbitrarily.
- Strategy B is applicable to any order of plants but its diophantine equation is more difficult to solve. Compared to strategy A, the resulting system can contain more closed-loop poles but some of them cannot be selected arbitrarily. The allowance to specify more closed-loop poles can be exploited to reduce the control sensitivities of the system and thus increase its performance robustness.
- Strategy C is also applicable to any order of plants but its diophantine equation is the most difficult to solve among the three strategies and it contains more closed-loop poles, which cannot be selected arbitrarily.
- Consequently, strategy B is recommended as a compromise strategy.
- In all the strategies, different sets of closed-loop poles can be independently selected to optimise the performance of each channel in a MIMO system.

10.4   Experimental set-up

This section describes the experimental set-up used in this study. A single-link 2-DOF flexible manipulator, as shown in Figure 10.3, has been constructed according to the specifications shown in Table 10.1. The 2-DOF allow the manipulator to move in both the horizontal and vertical planes under gravity, allowing any coupling between the motions to be studied.

The flexible link is a homogeneous, cylindrical, aluminium rod with constant properties along its length. Unlike many other experimental set-ups, the motion of the link is not artificially constrained in any plane within the safe working envelope. The link is symmetrical about its longitudinal axis, and this axis intersects both horizontal and vertical drive axes at the hub.

Horizontal and vertical tip (end-point) displacements of the link are calculated based on two measurements. First, the position of the end-point in relation to the hub axis (θ_{flex}) is measured using a precision potentiometer. This is combined with the angle of the hub relative to horizontal or vertical axis (θ) measured by an incremental

Figure 10.3   The experimental flexible manipulator

Table 10.1   The experimental flexible manipulator specifications

Flexible link length (L)	1.0 m
Flexible link mass (M_l)	0.35 kg
Flexible link offset from motor axes	0.06 m
Flexible link flexural rigidity (β)	72.2 Nm^2
Payload mass (M_p)	0.79 kg
Payload C. G. offset	0.25 m
Horizontal inertia of the hub (I_{hy})	0.468 kgm^2
Vertical inertia of the hub (I_{hz})	0.249 kgm^2
Maximum torque of torque motors	33 Nm
Rotor inertia of torque motors	0.013 kgm^2
Torque constant of torque motors	1.2 Nm/A
Maximum control signal of torque motors	± 8.5 V
Line count of encoders	3600
(Incremental) resolution ($\times 4$ counting) of encoders	4.4×10^{-4} rad
Linearity of potentiometers	$\pm 0.2\%$
Linearity of potentiometers with respect to link end-point position	$\pm 3 \times 10^{-3}$ rad
A/D (0–10 V differential input)	14 bit
D/A (± 10 V outputs)	16 bit
Encoder interface card	24 bit

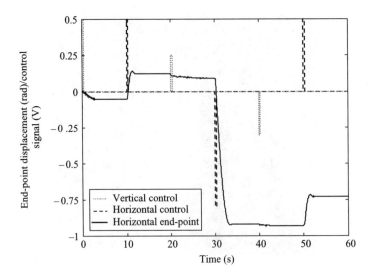

Figure 10.4   The experimental flexible manipulator

optical encoder. These signals are interfaced to a PC that is used to implement the control algorithm.

It is important to note that the payload weighs more than twice the weight of the flexible link itself. This is a very high payload-to-link-mass ratio compared to those used by other researchers.

Due to asymmetry of the payload, the motions are coupled and different polarities of control signals give different dynamics. Moreover, contrary to specification, it has been found that the resistance of the motor armature is not consistent throughout. Since torque is proportional to current, the same input voltage may give a different torque, resulting in different magnitude of displacement per volt of input voltage or, in other words, different output gain. Consequently, the manipulator system is non-linear as noted in Figure 10.4, which shows the experimental open-loop horizontal end-point displacement under the actuation of a series of horizontal and vertical control signals.

10.5   Simulation and experimental results

A fourth-order autoregressive with exogenous inputs (ARX) MIMO plant model for the asymmetrical flexible manipulator described in Section 10.4 has been identified by Goh (2001). The multivariable PI–PD (MPIPD) gains and filter parameters required to achieve the desired closed-loop characteristics can then be calculated according to the algorithm presented in Section 10.3. In particular, strategy B has been applied.

In order to exhibit clearly any residual coupling between the motions, the reference demands for horizontal and vertical motions were purposely offset in time. Based on a sampling time of 0.03 s, experimental results were collected from the flexible manipulator. The end-point displacements and control signals for the horizontal and

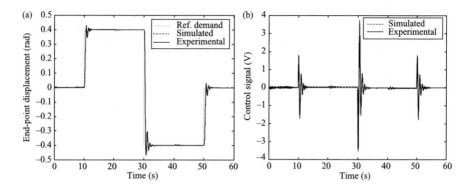

Figure 10.5   *Experimental and simulated end-point displacements and control signals for horizontal motion using MPIPD controller: (a) end-point displacement, (b) control signal*

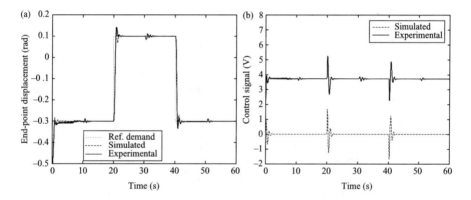

Figure 10.6   *Experimental and simulated end-point displacements and control signals for vertical motion using MPIPD controller: (a) end-point displacement, (b) control signal*

vertical motions along with their corresponding simulation results, for comparison purposes, are shown in Figures 10.5 and 10.6 respectively.

Compared to the open-loop end-point displacements in Figure 10.4, it can be seen that the MPIPD controller has significantly reduced the amount of cross-couplings, thus justifying the decoupling multivariable design. Moreover, the experimental and simulation results match closely, validating the design analyses and the identified plant model. The large discrepancy in Figure 10.6(b) is due to the extra torque provided by an explicit control loop for supporting the weight of the mechanical sensing arrangement (Goh *et al.*, 2000*b*).

The slight overshoots and oscillations at step changes can be minimised by optimising the choice of closed-loop poles and thus the controller parameters. Alternatively, the set-point weighting method proposed by Åström and Hägglund (1995)

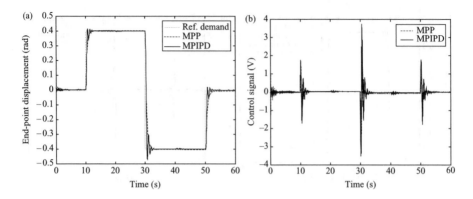

Figure 10.7 *Experimental end-point displacements and control signals for horizontal motion using MPIPD and MPP controllers: (a) end-point displacement, (b) control signal*

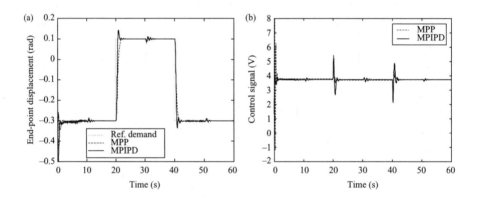

Figure 10.8 *Experimental end-point displacements and control signals for vertical motion using MPIPD and MPP controllers: (a) end-point displacement, (b) control signal*

can be used. Nevertheless, none of these have been thoroughly performed during the controller design so that the expected characteristic features can be exhibited for validating the design analyses, particularly in casting the PI–PD control algorithm as a general 2-DOF linear regulator and thence deriving an appropriate decoupling algorithm.

As a further illustration, the experimental performance of the proposed MPIPD was compared against other multivariable techniques. Figure 10.7 compares the experimental end-point displacements and control signals for the horizontal motion with those obtained from the same flexible manipulator using a MPP controller (Goh, 2001; Goh *et al.*, 2000a). Figure 10.8 shows a similar comparison for the vertical motion.

Table 10.2   Experimental ISTE values

Controller type	Horizontal motion	Vertical motion	Overall
MPP	0.1893	0.2882	0.4775
MPIPD	0.1908	0.2920	0.4828

Based on Figures 10.7 and 10.8, as expected, both the output and control performances of the MPP controller are generally better than those of the MPIPD controller. This is because the controller polynomials of MPP are not limited to second order only and thus all the closed-loop poles can be selected arbitrarily based on a minimal-order solution. On the other hand, MPIPD performs better during start-up due to having less initial transient. Moreover, with a relatively simpler implementation, the performance of MPIPD is acceptable especially in terms of decoupling efficiency.

Besides qualitative visual comparisons, a numerical performance index based on the integral squared timed error (ISTE) criterion was also calculated for both the controllers. The indices, based on the entire 60 s of operation, are shown in Table 10.2 for reference.

10.6   Summary

This chapter has reviewed some limitations in applying standard PID control to motion tracking and vibration control of flexible manipulators, difficulties in tuning PID control parameters and also shortcomings in some of the existing multivariable controllers. In addition, the derivations of discrete-time SISO and MIMO PI–PD control algorithms have been presented.

In particular, it has been illustrated how the discrete-time PI–PD control algorithm can be cast into the form of a general second-order linear regulator. Besides allowing more systematic parametric tuning via a pole-placement approach, the form provides a significant advantage in enabling an appropriate decoupling strategy to be derived for a multivariable system. Apart from the fundamental need of eliminating dynamic interactions among channels, the decoupling feature allows different sets of closed-loop poles to be selected independently to better optimise the performance of each channel of the multivariable system. Moreover, the linear-regulator form relates the controller parameters to the overall closed-loop dynamics and enables the use of analysis tools for linear regulators. As an example, closed-loop sensitivity-function analyses for a multivariable system, as described by Goh (2001), can be applied to gauge the performance and robustness of the PI–PD controller during the design stage. Furthermore, the proposed algorithm does not restrict the orders of applicable plant models nor require the plant models to be in certain standard forms.

In comparison to existing multivariable decoupling controllers, the structure and implementation of the proposed MPIPD controller is relatively simpler. Besides that, with thoroughly optimised parameters, the MPIPD controller can perform on a par with more complex multivariable controllers in motion and vibration control of a flexible manipulator. Furthermore, having low sensitivity to initial transient errors can be an important feature during system start-up. As such, the proposed MPIPD control algorithm may be more acceptable in an industrial context than more complex control techniques.

It has also been verified experimentally that the MPIPD controller can effectively decouple dynamic interactions between motions of the asymmetrical flexible manipulator, even at a very high payload-to-link-mass ratio and under the influence of gravity. The effectiveness of the proposed decoupling MPIPD controller will be of direct relevance to industrial applications, and means that future robotic manipulators can be designed to have asymmetrical links and also can handle asymmetrical payloads.

Chapter 11

Force and position control of flexible manipulators

B. Siciliano and L. Villani

While several control schemes have been proposed for force and position control of rigid robot manipulators, only a few related to flexible manipulators have been published so far. In this chapter the main force and position control strategies for flexible manipulators are surveyed and two different approaches are illustrated in depth. One achieves force and position regulation in an indirect way, by computing the joint and deflection variables in the presence of an external contact via a suitable closed-loop inverse kinematics scheme. The other exploits singular perturbation techniques to design force and position control schemes similar to those adopted for rigid robot manipulators, with an additional control action used to stabilise the fast dynamics related to link flexibility. A planar two-link flexible manipulator in contact with a compliant surface is considered, and simulation studies demonstrating the performance of the control techniques are presented and discussed.

11.1 Introduction

In a wide number of applications, such as polishing, debarring, machining or assembling, it is necessary to control the interaction between the robot manipulator and the environment. During the interaction, the environment sets constraints on the geometric paths that can be followed by the end-effector. In such a case, purely motion control strategies for controlling the interaction will fail. In fact, any motion planning or position tracking error may give rise to a contact force that, if not controlled, may produce an unstable behaviour and will inflect damage either to the robot or to the environment. The higher the environment stiffness and position control accuracy are, the easier such a situation can occur.

On the other hand, the intrinsic compliance of a flexible-link manipulator may contribute to reduction in the value of the forces that can be generated when the interaction task is executed by a rigid robot. This means that by using flexible robots to perform interaction tasks, some benefits may be gained (Sur and Murray, 1997), even though the distributed flexibility of the links makes the interaction control problem more complex than for rigid robots.

The most common solution to interaction control is to use a force/torque sensor, mounted between the last link and the end-effector, to provide force measurements that can be used to achieve force control.

Robot force control has attracted a wide number of researchers in the past two decades. Several control schemes have been proposed, and a state of the art on force control of rigid robot manipulators can be found in Gorinevski *et al.* (1997) and Siciliano and Villani (1999). For the specific case of flexible manipulators, however, only few papers have been published.

Interaction control strategies can be grouped into two categories: those performing indirect force controls and those performing direct force controls. The main difference between the two categories is that the former achieves force control via motion control, without explicit closure of a force feedback loop, and the latter offers the possibility of controlling the contact force to a desired value with the closure of a force feedback loop.

The first category includes compliance control (Salisbury, 1980) and impedance control (Hogan, 1985), where the end-effector is made compliant by relating the position error to the contact force through a static or dynamic relationship of adjustable parameters. If a detailed model of the environment is available then a widely adopted strategy, which belongs to the second category, known as the hybrid position/force control may be used. This strategy aims to control position along the unconstrained task directions and force along the constrained task directions. A selection matrix acting on both desired and feedback quantities serves this purpose for typically planar contact surfaces (Raibert and Craig, 1981), whereas the explicit constraint equations have to be taken into account for general curved contact surfaces (McClamroch and Wang, 1988; Yoshikawa, 1987).

In most practical situations, a detailed model of the environment is not available. In such a case, an effective strategy, still in the second category, that may be adopted is the inner/outer motion/force control where an outer force control loop is closed around the inner motion control loop, which is typically available in a robot manipulator (De Schutter and Van Brussel, 1988). In order to embed the possibility of controlling motion along the unconstrained task directions, the desired motion of the end-effector can be input to the inner loop of an inner/outer motion/force control scheme. The resulting parallel control is composed of a force control action and a motion control action, where the former is designed so as to dominate the latter in order to ensure force control along the constrained task directions (Chiaverini and Sciavicco, 1993).

For the case of flexible manipulators, early works addressing the stability problems of force control are reported in Chiou and Shahinpoor (1990) and Mills (1992). Hybrid force/position control is adopted in Matsuno and Yamamoto (1994), Matsuno *et al.* (1994) and Rocco and Book (1996), and is used in Hu and Ulsoy (1994) and

Yang *et al.* (1995) to design robust and adaptive control strategies, respectively. The problem of controlling the interaction of flexible macro-manipulators carrying a rigid micro-link is considered in Lew and Book (1993) and Yoshikawa *et al.* (1996) in the framework of hybrid control as well. Force and position control strategies conceived to manage the interaction with more or less compliant environments, without requiring a detailed model, are proposed in Siciliano and Villani, (1999, 2001).

The inherent difficulty of force control of flexible manipulators is due to problems similar to those arising in motion control (Book, 1993; Canudas de Wit *et al.,* 1996). Moreover, the kinematics and the dynamics of the robot cannot be stated independent of the forces acting on the robot tip (end-point).

In fact, when a robot interacts with the environment, the additional deflections caused by contact forces must be suitably taken into account for the computation of inverse kinematic solution. This can be done by adding a corrective term to the Jacobian, as in the solution based on the closed-loop inverse kinematics (CLIK) algorithm developed in Siciliano (1999) for the case of contact with an infinitely stiff environment and in Siciliano (1998) for the case of a compliant environment.

As for the dynamics of a flexible manipulator in contact with the environment, Matsuno and Yamamoto (1994) proposed a model derived using the Hamilton's principle, where the boundary condition of the link in contact with the environment depends on the contact force and input torque, as well as on the contact position. This makes the flexible manipulator equation very difficult to solve and simplification must be made. For example, in Matsuno *et al.* (1994), the boundary conditions are simplified by considering a quasi-static model derived on the basis of the static relationship between the elastic deformations and the contact force.

On the other hand, if the assumed mode technique is adopted to model the flexible manipulator, the mode functions must satisfy the geometric boundary conditions, which are not altered by the contact with the environment, while the natural boundary conditions (i.e. those involving the balance of forces and moments at the ends of the links) are automatically taken into account by the Lagrange formulation of the mathematical model (Book, 1984; Meirovitch, 1967; Rocco and Book, 1996). This modelling approach is also pursued in Hu and Ulsoy (1994), Kim *et al.* (1996), Lin (2003), Mills (1992), Siciliano and Villani (1999) and Yang *et al.* (1995).

Another difficulty in controlling flexible robots is the problem of damping the vibrations that are naturally excited during the task execution. An effective approach is based on singular perturbation theory (Kokotovic *et al.*, 1986). When the link stiffness is large, a two-timescale model of the flexible manipulator can be derived (Siciliano and Book, 1988) consisting of a slow subsystem corresponding to the rigid-body motion and a fast subsystem describing the flexible motion. A composite control strategy can then be applied, based on a slow control designed for the equivalent rigid manipulator and a fast control, which stabilises the fast subsystem. Further developments of perturbation techniques can be found in Fraser and Daniel (1991), Moallem *et al.* (1997), Siciliano *et al.* (1992) and Vandegrift *et al.* (1994) for the case of flexible manipulators moving in free space, and in Matsuno and Yamamoto (1994), Rocco and Book (1996), Siciliano and Villani (2000) and Yang *et al.* (1995) for the case of contact with the environment.

The focus of this chapter is on force control strategies for flexible manipulators, which are conceived to manage the interaction with a more or less compliant environment without requiring an accurate model thereof. Two different approaches are presented.

The first approach, based on Siciliano and Villani (2001) achieves force and position regulation in an indirect way as long as the arm kinematic model, the mass distribution and stiffness of the links as well as the environment stiffness and position are known. In detail, assuming a simple elastic model for the contact surface, a position set-point is assigned, corresponding to the desired force applied to the desired point on the surface. Then a closed-loop inverse kinematics algorithm based on a Jacobian transpose scheme described in (Siciliano, 1998) is adopted to compute the joint and deflection variables. These are input to a simple proportional, derivative (PD) joint regulator (De Luca and Siciliano, 1993).

The second approach, based on Siciliano and Villani (2000), exploits singular perturbation techniques to design interaction control schemes in the framework of parallel force and position control with an additional control action used to stabilise the fast dynamics related to link flexibility.

Simulation results are presented for a two-link planar manipulator under gravity in contact with an elastically compliant surface.

The chapter is organized as follows. In Section 11.2 the model of a planar n-link flexible manipulator in contact with an elastic environment is presented. The two-stage algorithm achieving indirect force and position control is presented in Section 11.3. In Section 11.4 a singular perturbed model for the flexible manipulator is developed and two different parallel control schemes are considered for the slow dynamics: the first ensures force and position regulation and the second guarantees force regulation and position tracking. Section 11.5 provides concluding remarks.

11.2   Modelling

For the purpose of this chapter, planar n-link flexible manipulators with revolute joints are considered. The links are subject to bending deformation in the plane of motion only, that is, torsional effects are neglected. A sketch of a two-link arm, with coordinate frame assignment, is shown in Figure 11.1. The rigid motion is described by the joint-angle θ_i, while $w_i(x_i)$ denotes the transversal deflection of link i at x_i, $0 \le x_i \le L_i$, with L_i as the length of the link.

Let ${}^i\mathbf{p}_i(x_i) = \begin{bmatrix} x_i & w_i(x_i) \end{bmatrix}^T$ be the position of a point along the deflected link i with respect to frame (X_i, Y_i) and $\mathbf{p}_i$ the position of the same point in the base frame. Also let ${}^i\mathbf{r}_{i+1} = {}^i\mathbf{p}_i(L_i)$ be the position of the origin of frame (X_{i+1}, Y_{i+1}) with respect to frame (X_i, Y_i), and $\mathbf{r}_{i+1}$ its position in the base frame.

The joint (rigid) rotation matrix $\mathbf{R}_i$ and the rotation matrix $\mathbf{E}_i$ of the (flexible) link at the end-point are, respectively,

$$\mathbf{R}_i = \begin{bmatrix} \cos\theta_i & -\sin\theta_i \\ \sin\theta_i & \cos\theta_i \end{bmatrix}$$

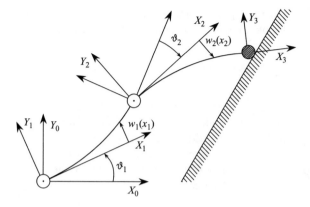

Figure 11.1   Planar two-link flexible manipulator

and

$$\mathbf{R}_i = \begin{bmatrix} 1 & -w'_{ie} \\ w'_{ie} & 1 \end{bmatrix}_i$$

where $w'_{ie} = (\partial w_i/\partial x_i)\,|_{x_i=L_i}$, and the small deflection approximation $\tan^{-1}\left(w'_{ie}\right) \cong w'_{ie}$ has been made. Hence the above absolute position vectors can be expressed as

$$\mathbf{p}_i = \mathbf{r}_i + \mathbf{W}_i{}^i\mathbf{p}_i$$

and

$$\mathbf{r}_{i+1} = \mathbf{r}_i + \mathbf{W}_i{}^i\mathbf{r}_{i+1}$$

where $\mathbf{W}_i$ is the global transformation matrix from the base frame to the frame (X_i, Y_i) given by the recursive equation,

$$\mathbf{W}_i = \mathbf{W}_{i-1}\mathbf{E}_{i-1}\mathbf{R}_i = \hat{\mathbf{W}}_{i-1}\mathbf{R}_i$$

with

$$\hat{\mathbf{W}}_0 = \mathbf{I}$$

On the basis of the above relations, the kinematics of any point along the manipulator is completely specified as a function of joint angles and link deflections.

A finite-dimensional model (of order m_i) of link flexibility can be obtained by the assumed modes technique. By exploiting the separability in time and space of solutions to the Euler–Bernoulli equation for flexible beams;

$$(EI)_i \frac{\partial^4 w_i\,(x_i, t)}{\partial x_i^4} + \frac{\partial^2 w_i\,(x_i, t)}{\partial t^2} = 0$$

for $i = 1, \ldots, n$ where ρ_i is the uniform mass density and $(EI)_i$ is the constant flexural rigidity of link i, the link deflection can be expressed as

$$w_i (x_i, t) = \sum_{j=1}^{m_i} \phi_{ij} (x_i) \, \delta_{ij} (t) \tag{11.1}$$

where $\delta_{ij} (t)$ are the time-varying variables associated with the assumed spatial mode shapes $\phi_{ij} (x_i)$ of link i. The mode shapes have to satisfy proper boundary conditions at the base (clamped) and at the end of each link (mass).

In view of equation (11.1), a direct kinematics equation can be derived expressing the position $\mathbf{p}$ of the manipulator end-point as a function of the $(n \times 1)$ joint variable vector $\boldsymbol{\theta}$ and the $(m \times 1)$ deflection variable vector $\boldsymbol{\delta}$ that is,

$$\mathbf{p} = \mathbf{k} \, (\boldsymbol{\theta}, \boldsymbol{\delta}) \tag{11.2}$$

where $m = \sum_{i=1}^{n} m_i$. For later use in the inverse kinematics scheme, the differential kinematics is also needed. The absolute linear velocity of a point on the arm is

$$\dot{\mathbf{p}}_i = \mathbf{P}_i + \dot{\mathbf{W}}_i{}^i\mathbf{p}_i + \mathbf{W}_i{}^i\dot{\mathbf{p}}_i \tag{11.3}$$

with ${}^i\mathbf{r}_{i+1} = {}^i\mathbf{p}_i (L_i)$. Since the links are assumed to be inextensible ($\dot{x}_i = 0$), then ${}^i\mathbf{p}_i (x_i) = \begin{bmatrix} 0 & \dot{w}_i (x_i) \end{bmatrix}^{\mathrm{T}}$. The computation of equation (11.3) takes advantage of the recursion

$$\mathbf{W}_i = \dot{\mathbf{W}}_{i-1}\mathbf{R}_i + \hat{\mathbf{W}}_{i-1}{}^i\dot{\mathbf{R}}_i \tag{11.4}$$

with

$$\dot{\mathbf{W}}_i = \dot{\mathbf{W}}_i\mathbf{E}_i + \mathbf{W}_i{}^i\dot{\mathbf{E}}_i \tag{11.5}$$

Also, note that

$$\dot{\mathbf{R}}_i = \mathbf{S}\mathbf{R}_i\dot{\theta}_i, \qquad \dot{\mathbf{E}}_i = \mathbf{S}\dot{w}'_{ie} \tag{11.6}$$

with

$$\mathbf{S} = \begin{bmatrix} 0 & -1 \\ 1 & 0 \end{bmatrix} \tag{11.7}$$

In view of equations (11.2)–(11.7), it is not difficult to show that the differential kinematics equation expressing the end-point velocity $\dot{\mathbf{p}}$ as a function of $\dot{\boldsymbol{\theta}}$ and $\dot{\boldsymbol{\delta}}$ can be written in the form:

$$\dot{\mathbf{p}} = \mathbf{J}_\theta \, (\boldsymbol{\theta}, \boldsymbol{\delta}) \, \dot{\boldsymbol{\theta}} + \mathbf{J}_\delta \, (\boldsymbol{\theta}, \boldsymbol{\delta}) \, \dot{\boldsymbol{\delta}} \tag{11.8}$$

where $\mathbf{J}_\theta = \partial \mathbf{k}/\partial \boldsymbol{\theta}$ and $\mathbf{J}_\delta = \partial \mathbf{k}/\partial \boldsymbol{\delta}$.

Assume that the manipulator is in contact with the environment. By virtue of the virtual work principle, the vector $\mathbf{f}$ of the forces exerted by the manipulator on the environment performing work on $\mathbf{p}$ has to be related to the $(n \times 1)$ vector $\mathbf{J}_\theta^{\mathrm{T}}\mathbf{f}$ of joint torques performing work on $\boldsymbol{\theta}$ and the $(m \times 1)$ vector $\mathbf{J}_\delta^{\mathrm{T}}\mathbf{f}$ of the elastic reaction forces performing work on $\boldsymbol{\delta}$.

A finite-dimensional Lagrangian dynamic model of the planar manipulator in contact with the environment can be obtained in terms of θ and δ in the form (De Luca and Siciliano, 1991):

$$\mathbf{M}_{\theta\theta}\,(\theta,\delta)\,\ddot{\theta} + \mathbf{M}_{\theta\delta}\,(\theta,\delta)\,\ddot{\delta} + \mathbf{c}_\theta\,(\theta,\delta,\dot{\theta},\dot{\delta}) + \mathbf{g}_\theta\,(\theta,\delta) = -\mathbf{J}_\theta^T\,(\theta,\delta)\,\mathbf{f} \qquad (11.9)$$

$$\mathbf{M}_{\theta\delta}^T\,(\theta,\delta)\,\ddot{\theta} + \mathbf{M}_{\delta\delta}\,(\theta,\delta)\,\ddot{\delta} + \mathbf{c}_\delta\,(\theta,\delta,\dot{\theta},\dot{\delta}) + \mathbf{g}_\delta\,(\theta,\delta) + \mathbf{D}\dot{\delta} + \mathbf{K}\delta = -\mathbf{J}_\delta^T\,(\theta,\delta)\,\mathbf{f}$$
$$(11.10)$$

where $\mathbf{M}_{\theta\theta}$, $\mathbf{M}_{\theta\delta}$ and $\mathbf{M}_{\delta\delta}$ are the matrix blocks of the positive definite symmetric inertia matrix, $\mathbf{c}_\theta$, $\mathbf{c}_\delta$ are the vectors of Coriolis and centrifugal forces, $\mathbf{g}_\theta$, $\mathbf{g}_\delta$ are the vector of gravitational forces, $\mathbf{K}$ is the diagonal and positive definite link stiffness matrix, $\mathbf{D}$ is the diagonal and positive semi definite link damping matrix, and τ is the vector of the input joint torques.

To analyse the performance of the position and force control algorithms, a model of the contact force is required. A real contact is a naturally distributed phenomenon in which the local characteristics of both the end-effector and the environment are involved. Moreover, friction effects between parts typically exist, which greatly complicate the nature of the contact itself. A simplified analysis can be pursued by considering a frictionless and planar surface, which is locally a good approximation to surfaces of regular curvature, and considering an elastic force given by

$$\mathbf{f} = k_e \mathbf{n}\mathbf{n}^T\,(\mathbf{p} - \mathbf{p}_e) = k_e \mathbf{n}\mathbf{n}^T\,(\mathbf{k}\,(\theta,\delta) - \mathbf{p}_e) \qquad (11.11)$$

where k_e is the surface stiffness, $\mathbf{p}_e$ is the un-deformed (constant) position of the surface, $\mathbf{n}$ is the (constant) unit vector of the direction normal to the surface, and the direct kinematics equation (11.2) has been used to express the position of the contact point in terms of joint and deflection variables. Also, it is assumed that contact is not lost.

Notice that for the derivation of the dynamic model in equations (11.9) and (11.10), the presence of the contact with the environment does not affect the choice of the mode shapes $\phi_{ij}\,(x_i)$ in equation (11.1) and the force enters into the equations of motion through the Jacobian (Rocco and Book, 1996).

Also, in this chapter only the interaction with a more or less compliant environment described by relations of the form in equation (11.11) is considered. On the other hand, in case of interaction with an infinitely rigid environment, kinematic constraints are imposed on the coordinates of robot end-point and a constraint force must be considered in the dynamic model of the flexible manipulator, expressed in terms of Lagrange multipliers (McClamroch and Wang, 1988). As in the case of rigid robots, the presence of the constraints reduces the number of degrees of freedom of the system. Moreover, it is possible to reduce the number of differential equations by resorting to a coordinate partitioning procedure (Hu and Ulsoy, 1994; Matsuno and Yamamoto, 1994; Matsuno *et al.*, 1994; Mills, 1992; Rocco and Book, 1996; Yang *et al.*, 1995).

11.3   Indirect force and position regulation

The interaction of a flexible-link manipulator with a compliant environment can be managed by controlling both the contact force and the end-point position. In view of the model of the contact force in equation (11.11), the control objective can be specified in terms of a desired force $f_d \mathbf{n}$ aligned with $\mathbf{n}$ and a desired position $\mathbf{p}_d$ on the contact plane. Nevertheless, the quantities f_d and $\mathbf{p}_d$ cannot be assigned independently, since they have to be consistent with the model in equation (11.11). In other words, the desired value of the force f_d can be achieved only if the component normal to the plane of the desired position $\mathbf{p}_d$ is chosen as

$$p_{dn} = \mathbf{n}^T \mathbf{p}_d = k_e^{-1} f_d + p_{en} \tag{11.12}$$

Hence, force control can be realised indirectly via position control, provided that the surface stiffness k_e and the component p_{en} of the un-deformed position of the surface are known.

In this section, a force and position regulator is presented, which achieves a desired position on the contact plane as well as a desired force, provided that equation (11.12) is satisfied, without requiring direct measurement of the contact force.

The controller is based on a two-stage algorithm. The first stage is in charge of solving the inverse kinematics problem to compute the desired vectors of joint variables $\boldsymbol{\theta}_d$ and deflection variables $\boldsymbol{\delta}_d$ that place the end-point of the flexible arm at a desired position $\mathbf{p}_d$; the component p_{dn} of $\mathbf{p}_d$ is chosen according to equation (11.12) to achieve a desired force f_d, while the components of the desired position tangential to the plane can be freely chosen. In the second stage, which constitutes a joint regulator, the variables $\boldsymbol{\theta}_d$ and $\boldsymbol{\delta}_d$ are used as set-points.

11.3.1   First stage

The first stage of the algorithm computes the inverse kinematics solution. To derive a Jacobian-based inverse kinematics scheme, the differential kinematic equation accounting for link deflections caused by gravity and contact with the environment must be considered.

For the regulation problem, a static situation can be considered. By virtue of equation (11.10), in a static situation the deflections satisfy the equation

$$\mathbf{g}_\delta\,(\boldsymbol{\theta}, \boldsymbol{\delta}) + \mathbf{K}\boldsymbol{\delta} = -\mathbf{J}_\delta^T\,(\boldsymbol{\theta}, \boldsymbol{\delta})\,\mathbf{f}$$

According to the small deflection approximation, it can be assumed that $\mathbf{g}_\delta$ is only a function of $\boldsymbol{\theta}$ (De Luca and Siciliano, 1993) and so is the case for $\mathbf{J}_\delta$ in equation (11.8) and $\mathbf{p}$ in equation (11.11). Hence, the deflection variables can be computed from equation (11.12) as

$$\boldsymbol{\delta} = -\mathbf{K}^{-1}\,(k_e \mathbf{j}_{\delta n}\,(\boldsymbol{\theta})\,(p_n\,(\boldsymbol{\theta}) - p_{en}) + \mathbf{g}_\delta\,(\boldsymbol{\theta})) \tag{11.13}$$

where

$$\mathbf{j}_{\delta n}\,(\boldsymbol{\theta}) = \mathbf{J}^T \mathbf{n}, \qquad p_n = \mathbf{n}^T \mathbf{p}, \qquad p_{en} = \mathbf{n}^T \mathbf{p}_e$$

For later use in the inverse kinematics scheme, differentiating equation (11.13) with respect to time gives

$$\dot{\delta} = \mathbf{J}_{fg}(\boldsymbol{\theta})\,\dot{\boldsymbol{\theta}} \tag{11.14}$$

where

$$\mathbf{J}_{fg} = -\mathbf{K}^{-1}\left(k_e\mathbf{J}_f(\boldsymbol{\theta}) + \mathbf{J}_g(\boldsymbol{\theta})\right)$$

with

$$\mathbf{J}_f = \frac{\partial \mathbf{j}_{\delta n}}{\partial \boldsymbol{\theta}}(p_n - p_{en}) + \mathbf{j}_{\delta n}\frac{\partial p_n}{\partial \boldsymbol{\theta}}$$

and $\mathbf{J}_g = \partial \mathbf{g}_\delta/\partial\boldsymbol{\theta}$. Substituting for $\dot{\delta}$ from equation (11.14) into equation (11.8) yields

$$\dot{\mathbf{p}} = \mathbf{J}_p(\boldsymbol{\theta}, \delta)\,\dot{\boldsymbol{\theta}} \tag{11.15}$$

where

$$\mathbf{J}_p = \mathbf{J}_\theta + \mathbf{J}_\delta\mathbf{J}_{fg}$$

is the overall Jacobian matrix relating joint velocity to end-point velocity. Notice that the Jacobian in equation (11.15) is obtained by modifying the rigid-body Jacobian $\mathbf{J}_\theta$ with two terms that account for the deflections induced by the contact force and gravity, respectively.

The attractive feature of the differential kinematics in equation (11.15) is its formal analogy with the differential kinematics equation for a rigid arm. Therefore, any Jacobian-based inverse kinematics scheme can be adopted in principle. In this respect, one of the most effective schemes is the CLIK scheme (Siciliano, 1990) that reformulates the inverse kinematics problem in terms of the convergence of a suitable closed-loop dynamic system.

According to the Jacobian transpose scheme, the joint variables vector is computed by integrating the joint velocity vector chosen as

$$\dot{\boldsymbol{\theta}} = \mathbf{J}_p^T(\boldsymbol{\theta}, \delta)\,\mathbf{K}_p(\mathbf{p}_d - \mathbf{p}) \tag{11.16}$$

Using a Lyapunov argument (Sicilano, 1990) it can be shown that, as long as the vector $\mathbf{K}_p(\mathbf{p}_d - \mathbf{p})$ in equation (11.16) is outside the null space of $\mathbf{J}_p^T$, the end-point position error $\mathbf{p}_d - \mathbf{p}$ asymptotically tends to zero. In fact, a suitable choice of the matrix $\mathbf{K}_p$ can be made to avoid that the scheme gets stuck with $\mathbf{p}_d - \mathbf{p} \neq \mathbf{0}$ and $\dot{\boldsymbol{\theta}} = \mathbf{0}$. In summary, $\boldsymbol{\theta}$ and δ tend asymptotically to the constant values $\boldsymbol{\theta}_d$ and δ_d such that $\mathbf{p}_d = \mathbf{k}(\boldsymbol{\theta}_d, \delta_d)$.

Notice that one of the attractive features of this approach is that, similar to the rigid arm case, any Jacobian-based inverse kinematics scheme can be adopted in principle, as well as any joint-space control law. The solution chosen in this work for kinematic inversion does not require the inverse of the Jacobian and thus it works well in the neighbourhood of singularities.

11.3.2   Second stage

The second stage of the algorithm is in charge of regulating the joint and deflection variables to the values θ_d and δ_d computed in the first stage. To this aim, the simple PD regulator presented in (De Luca and Siciliano, 1993) can be adopted:

$$\tau = \mathbf{K}_1 \left(\theta_d - \theta\right) - \mathbf{K}_2 \dot{\theta} + \mathbf{g}_\theta \left(\theta_d, \delta_d\right) + \mathbf{J}_\theta^T \left(\theta_d, \delta_d\right) f_d \mathbf{n} \tag{11.17}$$

where $\mathbf{K}_1$ and $\mathbf{K}_2$ are suitable positive definite matrix gains. The feedforward terms $\mathbf{g}_\theta \left(\theta_d, \delta_d\right)$ and $\mathbf{J}_\theta^T \left(\theta_d, \delta_d\right) f_d \mathbf{n}$ are required to compensate for the gravity torque and contact force respectively, at steady state.

The control law in equation (11.17) ensures asymptotic convergence of θ and δ to the corresponding set-points. Hence, the two-stage control scheme in equations (11.16) and (11.17) guarantees that $\mathbf{p} \rightarrow \mathbf{p}_d$ and $\mathbf{f} \rightarrow f_d \mathbf{n}$ as $t \rightarrow \infty$.

Notice that the PD regulator ensures asymptotic stability only in the presence of significant damping. When passive damping is too low, active vibration damping can be achieved by using full state-feedback (Canudas de Wit *et al.*, 1996).

It is also worth noting that the scheme only makes use of joint position and velocity measurements. Obviously, any joint position control law for flexible arms may be used in the second stage of the scheme in lieu of the simple PD regulator in equation (11.17). In any case, the overall performance in terms of end-point position and force errors strongly depends on the accuracy of the static model of the flexible arm, as well as on the accuracy of the available estimates of the stiffness and position of the contact surface.

11.3.3   Simulation

To illustrate the performance of the two-stage algorithm, a planar two-link flexible manipulator is considered (see Figure 11.1):

$$\theta = \begin{bmatrix} \theta_1 & \theta_2 \end{bmatrix}^T$$

and a payload of 0.1 kg is assumed to be placed at the end-point of the manipulator. An expansion with two clamped-mass assumed modes is taken for each link:

$$\delta = \begin{bmatrix} \delta_{11} & \delta_{12} & \delta_{21} & \delta_{22} \end{bmatrix}^T$$

The parameters of the manipulator are given in Table 11.1.

The resulting natural frequencies of vibration are

$$f_{11} = 1.40 \,\text{Hz}, \qquad f_{12} = 5.10 \,\text{Hz}$$
$$f_{21} = 536.09 \,\text{Hz}, \qquad f_{22} = 20792.09 \,\text{Hz}.$$

The stiffness matrix $\mathbf{K}$ is

$$\mathbf{K} = \begin{bmatrix} 38.79 & 513.37 \\ 536.09 & 20792.09 \end{bmatrix}$$

Table 11.1   Link parameters

Parameter (unit)	Link 1	Link 2
Density (kg/m)	1	1
Length (m)	0.5	0.5
Centre of mass (m)	0.25	0.25
Mass (kg)	0.5	0.5
Hub mass (kg)	1	1

The dynamic model of the manipulator and the missing numerical data can be found in (De Luca and Siciliano, 1991) while the direct and differential kinematics equations are reported in (Siciliano and Villani, 2001).

The contact surface is a vertical plane, thus the normal vector in equation (11.11) is $\mathbf{n} = \begin{bmatrix} 1 & 0 \end{bmatrix}^T$; a point of the un-deformed plane is

$$\mathbf{p}_e = \begin{bmatrix} 0.55 & 0 \end{bmatrix}^T \text{m}$$

and the contact stiffness is $k_e = 50\text{N/m}$.

The feedback gain matrix $\mathbf{K}_p$ of the CLIK algorithm in equation (11.16) is chosen as

$$\mathbf{K}_p = \text{diag}\left\{ 500 \quad 500 \right\}$$

and the inverse kinematics scheme is discretised at a sampling time $T_c = 1\text{ms}$, using the Euler integration rule. In particular, according to equation (11.16), the joint variables vector $\boldsymbol{\theta}_d$ is computed as

$$\boldsymbol{\theta}_d\,(t_{k+1}) = \boldsymbol{\theta}_d\,(t_k) + T_c \mathbf{J}_p^T\,(\boldsymbol{\theta}_d\,(t_k)\,,\boldsymbol{\delta}_d\,(t_k))\, \mathbf{K}_p\,(\mathbf{p}_d\,(t_k) - \mathbf{p}\,(t_k))$$

and, according to equation (11.13), the deflection variables vector $\boldsymbol{\delta}_d$ is computed as

$$\boldsymbol{\delta}_d\,(t_{k+1}) = -\mathbf{K}^{-1}\,(k_e \mathbf{j}_{\delta n}\,(\boldsymbol{\theta}_d\,(t_k))\,(p_n\,(\boldsymbol{\theta}\,(t_k)) - p_{en}) + \mathbf{g}_\delta\,(\boldsymbol{\theta}_d\,(t_k)))$$

The feedback matrix gains in (11.17) are chosen as:

$$\mathbf{K}_1 = \text{diag}\left\{ 25 \quad 25 \right\}, \qquad \mathbf{K}_2 = \text{diag}\left\{ 3 \quad 3 \right\}$$

Numerical simulations were performed using Matlab with Simulink. The arm was initially placed with the end-point in contact with the un-deformed plane at the position

$$\mathbf{p}_e = \begin{bmatrix} 0.55 & -0.55 \end{bmatrix}^T \text{m}$$

with null contact force; the corresponding generalized coordinates of the manipulator, computed using the CLIK algorithm in equation (11.16), are

$$\boldsymbol{\theta} = \begin{bmatrix} -1.396 & 1.462 \end{bmatrix}^T \text{rad}$$

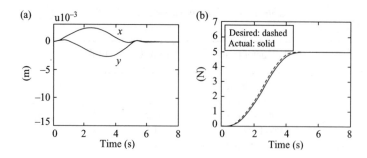

Figure 11.2   Time histories of the position error and of the contact force for the first example: (a) position error, (b) control force

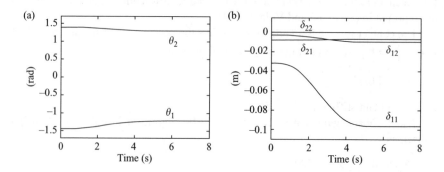

Figure 11.3   Time histories of the joint angles and of the link deflections for the first example: (a) joint angle, (b) link deflection

$$\delta = \begin{bmatrix} -0.106 & 0.001 & -0.009 & -0.0001 \end{bmatrix}^T m$$

It is desired to reach the end-point position

$$\mathbf{p}_d = \begin{bmatrix} 0.55 & -0.50 \end{bmatrix}^T m$$

and a fifth-order polynomial trajectory with null initial and final velocity and acceleration is imposed from the initial to the final position with a duration of 5 s.

In the first example, it is assumed that the stiffness of the environment is known, hence the desired force corresponding to the desired position is

$$\mathbf{f}_d = \begin{bmatrix} 5 & 0 \end{bmatrix}^T N$$

The time histories of the position errors and of the actual and desired contact forces are shown in Figure 11.2, and the time histories of the joint angles and link deflections are shown in Figure 11.3. It can be seen that the tracking error along the trajectory is small, although the scheme was conceived as a regulator. Moreover, both the desired force and position are achieved at steady state. Note also that, because of gravity and contact force, the arm has to bend to reach the desired end-point position properly.

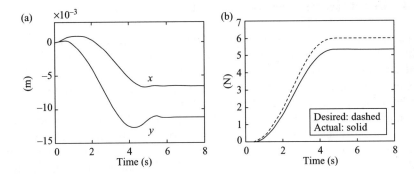

Figure 11.4 Time histories of the position error and of the contact force for the second example: (a) position error, (b) contact force

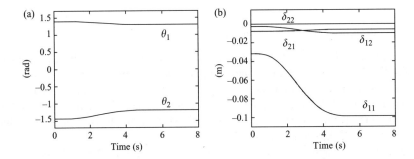

Figure 11.5 Time histories of the joint angles and of the link deflections for the second example: (a) joint angle, (b) link deflection

Actually the bending is much larger on the first link as expected (the links have the same parameters).

In the second numerical example, all the data are the same except for the estimated contact stiffness, which is assumed to be 60 Nm in lieu of the true value of 50 N/m. Hence, the desired force

$$\mathbf{f}_d = \begin{bmatrix} 6 & 0 \end{bmatrix}^T \mathrm{N}$$

is expected, with the same desired position.

The resulting time histories of the position errors and of the actual and desired contact forces are shown in Figure 11.4, and the time histories of the joint angles and link deflections are shown in Figure 11.5. It can be seen that the tracking error along the trajectory is limited, but a constant offset remains at steady state. Accordingly, the contact force reaches a constant value that is lower than the desired one, due to the fact that the contact stiffness was overestimated.

11.4 Direct force and position control

If a force sensor is available at the end-point of the manipulator, it is possible to achieve direct force control without requiring an exact estimate of the stiffness and of the position of the environment at rest. Moreover, if the dynamics related to link flexibility are suitably taken into account, tracking of a time-varying desired position can be achieved as well as regulation to a constant force.

In this section two different force and position control algorithms are presented, based on the parallel force and position approach of Chiaverini and Sciavicco (1993). The first scheme only requires partial knowledge of the model of the manipulator and guarantees force and position regulation. The second scheme achieves force regulation and position tracking by using more modelling information. Both schemes are part of a composite control strategy based on a two-time scale model of the flexible manipulator.

11.4.1 Composite control strategy

When stiffness of the link is large, it is reasonable to expect that the dynamics related to link flexibility are much faster than the dynamics associated with rigid motion of the robot so that the system naturally exhibits a two-timescale dynamic behaviour in terms of rigid and flexible variables. This feature can be conveniently exploited for control design (Matsuno and Yamamoto, 1994; Mills, 1992; Rocco and Book, 1996; Siciliano and Book, 1988; Yang et al., 1995).

Following the approach proposed in (Siciliano and Book, 1988), the system can be decomposed into slow and fast subsystems by using singular perturbation theory; this leads to a composite control strategy for the full system based on separate control designs for the two reduced-order subsystems.

Assuming that full-state measurement is available and that a force sensor is mounted at the end-point of the manipulator, the joint torques can be conveniently chosen as

$$\boldsymbol{\tau} = \mathbf{g}_\theta \left(\boldsymbol{\theta}, \boldsymbol{\delta}\right) + \mathbf{J}_\theta^T \left(\boldsymbol{\theta}_d, \boldsymbol{\delta}_d\right) \mathbf{f} + \mathbf{u} \qquad (11.18)$$

to cancel out the effects of the static torques acting on the rigid part of the manipulator dynamics; the vector $\mathbf{u}$ is the new control input to be designed on the basis of the singular perturbation approach.

The timescale separation between the slow and fast dynamics can be determined by defining the singular perturbation parameter $\varepsilon = 1/\sqrt{k_m}$, where k_m is the smallest coefficient of the diagonal stiffness matrix $\mathbf{K}$, and the new variable

$$\mathbf{z} = \mathbf{K}\boldsymbol{\delta} = \frac{1}{\varepsilon^2}\hat{\mathbf{K}}\boldsymbol{\delta}$$

corresponds to the elastic force, where $\mathbf{K} = k_m\hat{\mathbf{K}}$. Considering the inverse $\mathbf{H}$ of the inertia matrix $\mathbf{M}$, the dynamic model in equations (11.9) and (11.10) with control law in equation (11.18), can be rewritten in terms of the new variable $\mathbf{z}$ as

$$\ddot{\boldsymbol{\theta}} = \mathbf{H}_{\theta\theta}^T\left(\boldsymbol{\theta},\varepsilon^2\mathbf{z}\right)\left(\mathbf{u}-\mathbf{c}_\theta\left(\boldsymbol{\theta},\varepsilon^2\mathbf{z},\dot{\boldsymbol{\theta}},\varepsilon^2\dot{\mathbf{z}}\right)\right)-\mathbf{H}_{\theta\delta}\left(\boldsymbol{\theta},\varepsilon^2\mathbf{z}\right)\left[\mathbf{c}_\delta\left(\boldsymbol{\theta},\varepsilon^2\mathbf{z},\dot{\boldsymbol{\theta}},\varepsilon^2\dot{\mathbf{z}}\right)\right.$$

$$\left.+\mathbf{g}_\delta\left(\boldsymbol{\theta},\varepsilon^2\mathbf{z}\right)+\varepsilon^2\mathbf{D}\hat{\mathbf{K}}^{-1}\dot{\mathbf{z}}+\mathbf{z}+\mathbf{J}_\delta^T\left(\boldsymbol{\theta},\varepsilon^2\mathbf{z}\right)\mathbf{f}\right] \tag{11.19}$$

$$\varepsilon^2\mathbf{z} = \hat{\mathbf{K}}\mathbf{H}_{\theta\delta}^T\left(\boldsymbol{\theta},\varepsilon^2\mathbf{z}\right)\left(\mathbf{u}-\mathbf{c}_\theta\left(\boldsymbol{\theta},\varepsilon^2\mathbf{z},\dot{\boldsymbol{\theta}},\varepsilon^2\dot{\mathbf{z}}\right)\right)-\hat{\mathbf{K}}\mathbf{H}_{\delta\delta}\left(\boldsymbol{\theta},\varepsilon^2\mathbf{z}\right)\left[\mathbf{c}_\delta\left(\boldsymbol{\theta},\varepsilon^2\mathbf{z},\dot{\boldsymbol{\theta}},\varepsilon^2\dot{\mathbf{z}}\right)\right.$$

$$\left.+\mathbf{g}_\delta\left(\boldsymbol{\theta},\varepsilon^2\mathbf{z}\right)+\varepsilon^2\mathbf{D}\hat{\mathbf{K}}^{-1}\dot{\mathbf{z}}+\mathbf{z}+\mathbf{J}_\delta^T\left(\boldsymbol{\theta},\varepsilon^2\mathbf{z}\right)\mathbf{f}\right] \tag{11.20}$$

where a suitable partition of $\mathbf{H}$ has been considered:

$$\mathbf{H} = \mathbf{M}^{-1} = \begin{bmatrix} \mathbf{H}_{\theta\theta} & \mathbf{H}_{\theta\delta} \\ \mathbf{H}_{\theta\delta}^T & \mathbf{H}_{\delta\delta} \end{bmatrix}$$

Equations (11.19) and (11.20) represent a singularly perturbed form of the flexible manipulator model; when $\varepsilon \to \infty$, the model of an equivalent rigid manipulator is recovered. In fact, setting $\varepsilon = 0$ and solving for $\mathbf{z}$ in equation (11.20) gives

$$\mathbf{z}_s = \overline{\mathbf{H}}_{\delta\delta}^{-1}\left(\boldsymbol{\theta}_s\right)\overline{\mathbf{H}}_{\theta\delta}^T\left(\boldsymbol{\theta}_s\right)\left(\mathbf{u}_s-\overline{\mathbf{c}}_\theta\left(\boldsymbol{\theta}_s,\dot{\boldsymbol{\theta}}_s\right)\right)-\overline{\mathbf{c}}_\delta\left(\boldsymbol{\theta}_s,\dot{\boldsymbol{\theta}}_s\right)-\overline{\mathbf{g}}_\delta\left(\boldsymbol{\theta}_s\right)-\overline{\mathbf{J}}_\delta^T\left(\boldsymbol{\theta}_s\right)\mathbf{f}_s \tag{11.21}$$

where the subscript s indicates that the system is considered in the *slow* timescale and the bar denotes that a quantity is computed with $\varepsilon = 0$. Substituting equation (11.21) into equation (11.19) with $\varepsilon = 0$ yields

$$\ddot{\boldsymbol{\theta}}_s = \overline{\mathbf{M}}_{\theta\theta}^{-1}\left(\boldsymbol{\theta}_s\right)\left(\mathbf{u}_s-\overline{\mathbf{c}}_\theta\left(\boldsymbol{\theta}_s,\dot{\boldsymbol{\theta}}_s\right)\right) \tag{11.22}$$

where the equality:

$$\overline{\mathbf{M}}_{\theta\theta}^{-1}\left(\boldsymbol{\theta}_s\right) = \left(\overline{\mathbf{M}}_{\theta\theta}\left(\boldsymbol{\theta}_s\right)-\overline{\mathbf{M}}_{\theta\delta}\left(\boldsymbol{\theta}_s\right)\overline{\mathbf{M}}_{\delta\delta}^{-1}\left(\boldsymbol{\theta}_s\right)\overline{\mathbf{M}}_{\theta\delta}^T\left(\boldsymbol{\theta}_s\right)\right)$$

has been exploited, where $\overline{\mathbf{M}}_{\theta\theta}\left(\boldsymbol{\theta}_s\right)$ represents the inertia matrix of the equivalent rigid manipulator and $\overline{\mathbf{c}}_\theta\left(\boldsymbol{\theta}_s,\dot{\boldsymbol{\theta}}_s\right)$ the vector of the corresponding Coriolis and centrifugal torques.

The dynamics of the system in the *fast* timescale can be obtained by setting $t_f = t/\varepsilon$, treating the slow variables as constants in the fast timescale, and introducing the fast variables $\mathbf{z}_f = \mathbf{z} - \mathbf{z}_s$. Thus, the fast system in equation (11.20) is

$$\frac{\mathrm{d}^2\mathbf{z}_f}{\mathrm{d}t_f^2} = -\hat{\mathbf{K}}\overline{\mathbf{H}}_{\delta\delta}\left(\boldsymbol{\theta}_s\right)\mathbf{z}_f+\hat{\mathbf{K}}\overline{\mathbf{H}}_{\theta\delta}^T\left(\boldsymbol{\theta}_s\right)\mathbf{u}_f \tag{11.23}$$

where the fast control $\mathbf{u}_f = \mathbf{u} - \mathbf{u}_s$ has been introduced accordingly.

On the basis of the above two-timescale model, the design of a feedback controller for the system in equations (11.19) and (11.20) can be performed according to a composite control strategy, that is,

$$\mathbf{u} = \mathbf{u}_s\left(\boldsymbol{\theta}_s,\dot{\boldsymbol{\theta}}_s\right)+\mathbf{u}_f\left(\mathbf{z}_f,\mathrm{d}\mathbf{z}_f/\mathrm{d}t_f\right) \tag{11.24}$$

with the constraint that $\mathbf{u}_f (0,0) = 0$ so that $\mathbf{u}_f$ is inactive along the equilibrium manifold specified by equation (11.21).

To design the slow control for the rigid nonlinear system in equation (11.22), it is useful to derive the slow dynamics corresponding to the end-point position. Differentiating equation (11.8) gives the end-point acceleration;

$$\dot{\mathbf{p}} = \mathbf{J}_\theta \left(\theta, \delta\right) \ddot{\theta} + \mathbf{J}_\delta \left(\theta, \delta\right) \ddot{\delta} + \mathbf{h} \left(\theta, \delta, \dot{\theta}, \dot{\delta}\right)$$

where $\mathbf{h} = \dot{\mathbf{J}}_\theta \dot{\theta} + \dot{\mathbf{J}}_\delta \dot{\delta}$. Hence, the corresponding slow system is

$$\dot{\mathbf{p}}_s = \overline{\mathbf{J}}_\theta \left(\theta_s\right) \overline{\mathbf{M}}_{\theta\theta}^{-1} \left(\theta_s\right) \left(\mathbf{u}_s - \overline{\mathbf{c}}_\theta \left(\theta_s, \dot{\theta}_s\right)\right) + \overline{\mathbf{h}} \left(\theta_s, \dot{\theta}_s\right) \tag{11.25}$$

where equation (11.22) has been used. The slow dynamic models in equations (11.22) and (11.25) enjoy the same notable properties of the rigid robot dynamic models (Canudas de Wit *et al.*, 1996), hence the control strategies used for rigid manipulators can be adopted.

The fast system in equation (11.23) is a marginally stable linear slowly time-varying system that can be stabilised to the equilibrium manifold $\dot{z}_f = 0$ $(\dot{z} = 0)$ and $z_f = 0$ $(z = z_s)$ by a proper choice of the control input $\mathbf{u}_f$. A reasonable way to achieve this goal is to design a state-space control law of the form:

$$\mathbf{u}_f = \mathbf{K}_1 \dot{z}_f + \mathbf{K}_2 z_f \tag{11.26}$$

where, in principle, the matrices $\mathbf{K}_1$ and $\mathbf{K}_2$ should be tuned for every configuration θ_s. However, the computational burden necessary to perform this strategy can be avoided by using constant matrix gains tuned with reference to a given robot configuration (Siciliano and Book, 1988); any state-space technique can be used, for example, based on classical pole placement algorithms.

11.4.2   Force and position regulation

The control objective consists of simultaneous regulation of the contact force $\mathbf{f}$ to a constant set point $\mathbf{f}_d$ and of the position $\mathbf{p}$ to a constant set-point $\mathbf{p}_d$.

In case of contact with an elastically compliant surface, a viable strategy is the parallel control approach (Chiaverini and Sciavicco, 1993), which is especially effective in the case of inaccurate contact modelling. The key feature is to have a force control loop working in parallel to a position control loop along each task space direction. The logical conflict between the two loops is managed by imposing dominance of the force control action over position control, that is, force regulation is always guaranteed at the expense of a position error along the constrained directions.

A force/position parallel regulator controller for rigid robots was proposed in Chiaverini *et al.* (1994), based on position PD position control, gravity compensation, desired force feedforward and PI force control.

For the case of the flexible-link manipulator in equations (11.9) and (11.10), with reference to the slow system in equation (11.25), the following parallel regulator can be adopted:

$$\mathbf{u}_s = \overline{\mathbf{J}}_\theta^T \left(\theta_s\right) k_P \left(\mathbf{p}_r - \mathbf{p}_s\right) - k_D \dot{\theta}_s \tag{11.27}$$

where $\mathbf{p}_r$ is defined as:

$$\mathbf{p}_r = \mathbf{p}_d + k_P^{-1} \left(k_F \left(\mathbf{f}_d - \mathbf{f}_s \right) + k_I \int_0^t \left(\mathbf{f}_d - \mathbf{f}_s \right) d\tau \right) \tag{11.28}$$

and $k_P, k_D, k_F, k_F > 0$ are suitable feedback gains.

A better insight into the behaviour of the system during the interaction can be achieved by considering a model of the compliant environment. For the purpose of this work, it is assumed that the same equation can be established in terms of the slow variables, that is,

$$\mathbf{f}_s = k_e \mathbf{n} \mathbf{n}^T \left(\mathbf{p}_s - \mathbf{p}_o \right)$$

The above elastic model shows that the contact force is normal to the plane, and thus a null force error can be obtained only if the desired force $\mathbf{f}_d$ is aligned with $\mathbf{n}$. Also, it can be recognised that null position errors can be obtained only on the contact plane while the component of the position along $\mathbf{n}$ has to accommodate the force requirement specified by $\mathbf{f}_d$.

The stability analysis for the slow system in equation (11.25) with the control law in equations (11.27) and (11.28) can be carried out with the same arguments used in Chiaverini *et al.* (1994) for the case of rigid robots. In particular, it can be shown that if the Jacobian $\bar{\mathbf{J}}_\theta \left(\theta_s \right)$ of the equivalent rigid manipulator is full-rank, then the closed-loop system has an exponentially stable equilibrium at

$$\mathbf{p}_{s\infty} = \left(\mathbf{I} - \mathbf{n} \mathbf{n}^T \right) \mathbf{p}_d + \mathbf{n} \mathbf{n}^T \left(k_e^{-1} \mathbf{f}_d + \mathbf{p}_o \right) \tag{11.29}$$

$$\mathbf{f}_{s\infty} = k_e \mathbf{n} \mathbf{n}^T \left(\mathbf{p}_{s\infty} - \mathbf{p}_o \right) = \mathbf{f}_d \tag{11.30}$$

where the matrix $\left(\mathbf{I} - \mathbf{n} \mathbf{n}^T \right)$ projects the vectors on the contact plane. The equilibrium position is depicted in Figure 11.6. It can be seen that $\mathbf{p}_{s\infty}$ differs from $\mathbf{p}_d$ by a vector aligned along the normal to the contact plane whose magnitude is that necessary to

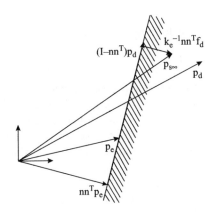

Figure 11.6   Equilibrium position with parallel force and position control

guarantee $\mathbf{f}_{s\infty} = \mathbf{f}_d$ in view of equation (11.30). Therefore, (for the slow system) force regulation is ensured while a null position error is achieved only for the component parallel to the contact plane.

If $\mathbf{f}_d$ is not aligned with $\mathbf{n}$, then it can be found that a drift motion of the end-point of the manipulator is generated along the plane; for this reason, if the contact geometry is unknown, it is advisable to set $\mathbf{f}_d = \mathbf{0}$.

As a final step, the full-order system in equations (11.9) and (11.10) and the composite control law in equation (11.27) with $\mathbf{u}_s$ in equation (11.27) and $\mathbf{u}_f$ in equation (11.26) have to be analysed. By virtue of Tikhonov's theorem it can be shown that regulation of the force $\mathbf{f}$ and of the position $\mathbf{p}$ is achieved with an order ε approximation.

11.4.3   Force regulation and position tracking

If tracking of a time-varying position $\mathbf{p}_d(t)$ on the contact plane is desired (with an order ε approximation), an inverse dynamics parallel control scheme can be adopted for the slow system, that is,

$$\mathbf{u}_s = \overline{\mathbf{B}}_{\theta\theta}\left(\boldsymbol{\theta}_s\right)\overline{\mathbf{J}}_\theta^{-1}\left(\boldsymbol{\theta}_s\right)\left(\mathbf{a}_s - \overline{\mathbf{h}}\left(\boldsymbol{\theta}_s,\dot{\boldsymbol{\theta}}_s\right)\right) + \overline{\mathbf{c}}_\theta\left(\boldsymbol{\theta}_s,\dot{\boldsymbol{\theta}}_s\right) \tag{11.31}$$

where $\mathbf{a}_s$ is a new control input and a non-redundant manipulator has been considered.

Substituting equation (11.31) into equation (11.25) gives

$$\dot{\mathbf{p}}_s = \mathbf{a}_s$$

Hence, the control input $\mathbf{a}_s$ can be chosen as

$$\mathbf{a}_s = \dot{\mathbf{p}}_r + k_D\left(\dot{\mathbf{p}}_r - \dot{\mathbf{p}}_s\right) + k_P\left(\mathbf{p}_r - \mathbf{p}_s\right) \tag{11.32}$$

where

$$\mathbf{p}_r = \mathbf{p}_d + \mathbf{p}_C \tag{11.33}$$

and $\mathbf{p}_C$ is the solution of the differential equation

$$k_A\dot{\mathbf{p}}_C + k_V\dot{\mathbf{p}}_C = \mathbf{f}_d - \mathbf{f}_s \tag{11.34}$$

where $k_P, k_D, k_A, k_V > 0$ are suitable feedback gains.

Using the same arguments developed in Siciliano and Villani (1999) for rigid robots, it can be easily shown for the slow system that the control law in equations (11.31)–(11.34) ensures regulation of the contact force to the desired set-point $\mathbf{f}_d$ and tracking of the time-varying component of the desired position on the contact plane $\left(\mathbf{I} - \mathbf{nn}^T\right)\mathbf{p}_d(t)$.

As before, Tikhonov's theorem has to be applied to the full-order system in equations (11.9) and (11.10) with the composite control law in equations (11.24), (11.26) and (11.31)–(11.34), and it can be shown that force regulation and position tracking are achieved with an order ε approximation.

11.4.4   Simulation

The above control laws are tested in simulation on the planar two-link flexible manipulator considered in Section 11.3, placed in the same initial position with the end-point in contact with the plane and null contact force. It is desired to reach the end-point position

$$\mathbf{p}_d = \begin{bmatrix} 0.55 & -0.35 \end{bmatrix}^T \text{m}$$

and a fifth-order polynomial trajectory with null initial and final velocity and acceleration is imposed from the initial to the final position with a duration of 5 s.

The desired force is taken from zero to the desired value

$$\mathbf{f}_d = \begin{bmatrix} 5 & 0 \end{bmatrix}^T \text{N}$$

according to a fifth-order polynomial trajectory with null initial and final velocity and acceleration and duration of 1s.

The fast control law $\mathbf{u}_f$ has been implemented with $\varepsilon = 0.1606$ and the matrix gains in equation (11.26) were tuned by solving a linear quadratic (LQ) problem for the system in equation (11.23) with the configuration dependent terms computed in the initial manipulator configuration. The matrix weights of the index performance have been chosen so that to preserve the timescale separation between slow and fast dynamics for both the control schemes. The resulting matrix gains are

$$\mathbf{K}_1 = \begin{bmatrix} -0.0372 & -0.0204 & -0.0375 & 0.1495 \\ 0.0573 & 0.0903 & 0.0080 & -0.7856 \end{bmatrix}$$

$$\mathbf{K}_2 = \begin{bmatrix} -0.1033 & -0.0132 & -0.0059 & -0.0053 \\ -0.0882 & 0.0327 & -0.0537 & -0.0217 \end{bmatrix}$$

In order to reproduce in simulation a real situation of a continuous-time system with a digital controller, the control laws are discretised with 5 ms sampling time, while the equations of motion are integrated using a variable step Runge–Kutta method with a minimum step size of 1 ms.

In the first case study, the slow controller in equations (11.27) and (11.28) is considered with the composite control law in equation (11.24). The actual force $\mathbf{f}$ and position $\mathbf{p}$ are used in the slow control law instead of the corresponding slow values, assuming that direct force measurement is available and that the end-point position is computed from joint angles and link deflection measurements via the direct kinematics equation (11.2). The control gains were set to $k_P = 100$ and $k_D = 4$.

Figure 11.7 shows the position error together with the time histories of the desired and actual contact force. It is easy to see that the contact force remains close to the desired value during the end-point motion (notice that the commanded position trajectory has a duration of 5 s) and reaches the desired set-point at steady state. The y-component of the desired position, which corresponds to a direction parallel to the contact plane, is regulated to the desired value. On the other hand, significant error occurs for the x-component; which corresponds to the direction normal to the contact plane, as expected. Notice that the steady-state value of the position error is exactly

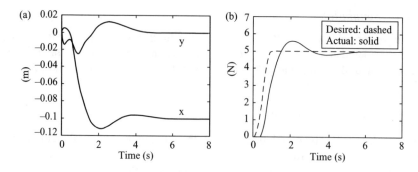

Figure 11.7   Time histories of the position error and of the contact force for the first case study: (a) position error, (b) contact force

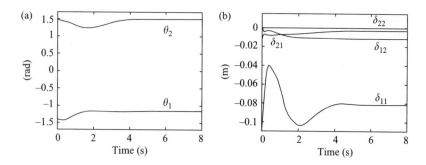

Figure 11.8   Time histories of the joint angles and of the link deflections for the first case study: (a) joint, angle, (b) link deflection

that required to achieve null force error, according to the equilibrium equations (11.29) and (11.30).

It can be seen from the time histories of the joint angles and link deflections shown in Figure 11.8 that the link response oscillations are well damped; moreover, because of gravity and contact force, the manipulator has to bend to reach the desired force and position.

Figure 11.9 shows the time history of the joint torque **u** and the first 0.5 s of the time history of the fast torque $\mathbf{u}_f$. It can be observed that the control effort keeps limited values during task execution; remarkably, the control torque $\mathbf{u}_f$ converges to zero with a transient much faster than the transient of **u** as expected.

In the second case study, the slow system in equations (11.31)–(11.34) is considered with the composite control law in equation (11.24). As before, the actual force **f** and position **p** are used in the controller in lieu of the corresponding slow variables. The control gains were set to $k_P = 100$, $k_D = 22$, $k_A = 0.7813$ and $k_V = 13.75$.

The time histories of the contact force and position errors are shown in Figure 11.10. This time the desired force set-point is reached after about 3 s, before

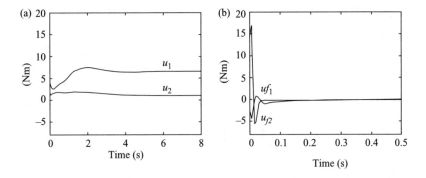

Figure 11.9   *Time histories of the joint torques and fast control for the first case study: (a) joint torque, (b) fast control*

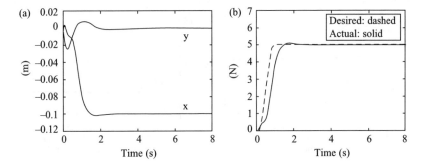

Figure 11.10   *Time histories of the position error and of the contact force for the second case study: (a) position error, (b) contact force*

the completion of the end-point motion. Moreover, the tracking performance for the y-component of the desired position is better than that in the previous case study.

The time histories of the joint angles and of the link deflections are shown in Figure 11.11, while the time histories of the components of the joint torque vector $\mathbf{u}$ and of the fast torque vector $\mathbf{u}_f$ are shown in Figure 11.12. It can be seen that although the performance is better than that in the previous case study, a similar control effort is required.

It is worth pointing out that the simulation of both the slow control laws without the fast control action in equation (11.26) has revealed an unstable behaviour; the results have not been reported for brevity.

11.5   Summary

The force control problem for a flexible manipulator in contact with a compliant environment has been considered. A dynamic model for a planar manipulator has been presented, which takes into account the forces acting on the end-point of the

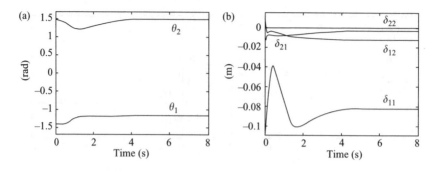

Figure 11.11   *Time histories of the joint angles and of the link deflections for the second case study: (a) joint angle, (b) link deflection*

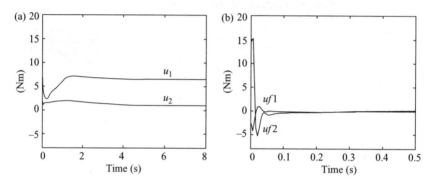

Figure 11.12   *Time histories of the joint torques and fast control for the second case study: (a) joint torque, (b) fast control*

robot. Two different approaches have been presented: a two-stage algorithm and a composite control law.

The attractive feature of the two-stage scheme is that it does not require force and deflection measurements. The price to pay is that an exact knowledge of the arm kinematics as well as the stiffness and position of the environment at rest are required to guarantee regulation of the force and of end-point position to a constant value.

In the composite control law, on the other hand, the additional objective of damping the vibrations that are naturally excited during the task execution is explicitly considered. By using the singular perturbation theory, the system has been split into a slow subsystem describing the rigid motion dynamics and a fast subsystem describing the flexible dynamics. This allows designing a fast control action for vibration damping as well as adopting algorithms designed for the rigid motion to achieve force and position control. Two different controllers have been considered, in the framework of parallel force and position control.

The simulation case studies developed on the model of a two-link flexible manipulator under gravity in contact with a compliant environment have confirmed the theoretical results.

Chapter 12

Collocated and non-collocated control of flexible manipulators

M.O. Tokhi, A.K.M. Azad, M.H. Shaheed and H. Poerwanto

This chapter presents the development of closed-loop control strategies for flexible manipulator systems based on collocated and non-collocated control. A closed-loop control strategy using hub-angle and hub-velocity feedback for rigid-body motion control and end-point acceleration feedback for flexural motion control is considered. This is then extended to an adaptive collocated and non-collocated control mechanism using online modelling and controller design. Proportional, integral, derivative (PID) type as well as inverse-model control techniques are considered for flexural motion control. The non-minimum phase behaviour of the plant in the latter case is addressed through conventional techniques. This is further addressed through development of an adaptive neuro-inverse model strategy. The control strategies thus developed are verified and their performances assessed through simulated and experimental exercises.

12.1 Introduction

The advantages of flexible manipulators can be realised in practice only at the expense of rendering accurate controllers for its vibration control. Vibration control techniques, as discussed in Chapter 1, are generally classified into two categories: passive and active control. Active control utilizes the principle of wave interference. This is realised by artificially generating anti-source(s) (actuator(s)) to destructively interfere with the unwanted disturbances and thus result in reduction in the level of vibration. Active control of flexible manipulator systems can in general be divided into two categories: open-loop and closed-loop control. Open-loop control involves altering the shape of actuator commands by considering the physical and vibrational properties of the flexible manipulator system. The approach does not account for changes in the system once the control input is developed. Closed-loop control differs from the open-loop control in that it uses measurements of the system's state and changes the actuator input accordingly to reduce the system oscillation. Effective control of a system always depends on accurate real-time monitoring and the corresponding control effort in a close-loop manner.

This chapter focuses on the development of closed-loop controllers for flexible manipulator systems. The difficulties in controlling a flexible manipulator stem from several factors, such as flexibility of the structure, coupling effects, non-linearities, parameter variations and unmodelled dynamics. Ignoring such factors could not attain satisfactory closed-loop control performance. A common strategy in the control of flexible manipulator systems involves the utilization of proportional and derivative (PD) feedback of collocated sensor signals, such as hub-angle and hub-velocity. Such a strategy is initially adopted in this chapter through joint-based collocated (JBC) control. The JBC controllers are capable of reducing the vibration at the end-point of the manipulator as compared to a response with open-loop bang-bang input torque. However, for effective control of end-point vibration, it is necessary to use a further control loop accounting for flexural motion control of the system. A hybrid collocated and non-collocated control structure for control of a single-link flexible manipulator has previously been reported (Tokhi and Azad, 1996a), where a PD configuration has been applied for control of the rigid-body motion and a proportional, integral, derivative (PID) control scheme with end-point displacement feedback has been used for vibration suppression of the manipulator. These two loops are then summed to give a command torque input. The control scheme has been tested and investigated within a simulation environment and also with an experimental rig.

The proposed collocated JBC and hybrid collocated and non-collocated control structures are capable of reducing the level of vibration at the end-point for fixed operating conditions. In the case of a flexible manipulator system any change in pay-load mass will affect the system dynamics for which such a fixed controller will not be adequate. This problem can be addressed by making the developed controllers adaptive so that they can be adjusted according to changes in the system dynamics.

A self-tuning scheme is initially implemented using the pole assignment technique with JBC control. The hybrid collocated and non-collocated control scheme is then realised with an adaptive JBC position controller and an inverse end-point-model vibration controller. A recursive least-squares (RLS) algorithm is utilized to obtain an inverse model of the plant in parametric form. The problem of controller instability arising from the non-minimum phase characteristics exhibited in the plant model is resolved by reflecting the non-invertible zeros into the stability region. The performances of both schemes are investigated within a flexible manipulator simulation environment.

It is evident from the literature that neural networks (NNs) have enjoyed a great deal of success in solving many complex control problems that are extremely difficult to solve with other approaches (Blum and Li, 1991; Jain and Mohiuddin, 1996; Leshno, 1993; Yesildierk et al., 1994). Thus, to address the end-point vibration control of the manipulator as well as account for stability of the non-collocated control loop, a neuro-inverse model approach is adopted. A non-linear auto-regressive process with exogeneous input model (NARX) along with a multi-layered perceptron (MLP) NN based on Marquardt–Levenberg modified version of the backpropagation learning algorithm (Hagan and Menhaj, 1994; Marquardt, 1963) is used. The neuro-inverse controller is realised within a MATLAB/Simulink based finite difference simulation (Poerwanto, 1998) environment of the system (see Chapter 3 for details) and its performance is accordingly assessed and discussed. The flexible manipulator system considered in this chapter is the Sheffield manipulator described in Chapter 1.

12.2   JBC control

The task of the JBC control loop is to move or position the flexible arm to the specified angle of demand. The control objective here is for the output $y(t)$ (hub-angle) to follow a reference signal (angle) $r(t)$ in some predetermined way. This can be achieved through the utilization of PD feedback of collocated sensor signals. The basic structure of the scheme is shown in Figure 12.1 where A_c represents gain of the motor amplifier, K_p and K_v the proportional and derivative gains of the controller respectively, R_f the reference hub-angle, $\theta(t)$ and $\dot{\theta}(t)$ the hub-angle and hub-velocity and $u(t)$ the control signal.

To design the controller so as to result in the closed-loop roots with maximum negative real parts, it has been suggested to choose the ratio $K_p/K_v \approx 2.0$ and set the derivative gain K_v to vary within 0–1.2 (Azad, 1994). This is based on information derived from a simulation environment of the flexible manipulator.

In a practical environment, however, effects of uncontrolled dynamics of the flexible manipulator, actuators, filters and sensor as well as delays caused by measuring and sampling of feedback signal have to be taken into account. This type of controller gives optimum performance for a certain loading condition of the flexible manipulator. It performs adequately well in controlling the position of the flexible manipulator for different angle demands. However, in a practical environment, where the loading conditions of the flexible manipulator change, causing changes in the characteristics of the system, the system performance with a fixed JBC control scheme will not be satisfactory. This makes fixed JBC control impractical, as any load change will require the controller to be modified for the desired performance to be achieved.

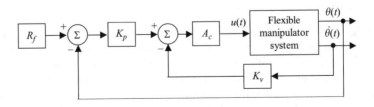

Figure 12.1   Fixed JBC controller for a flexible manipulator system

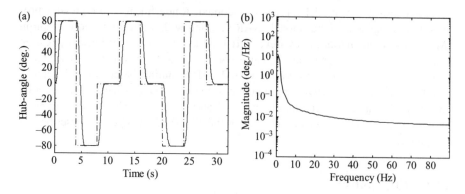

Figure 12.2   *Hub-angle with fixed JBC controller (dashed = demanded): (a) time-domain, (b) spectral density*

Table 12.1   *Fixed JBC controller parameters for various loads*

Payload (g)	0.0	20	40	60	80	100
K_p	1.0	1.0	1.0	1.0	1.0	1.0
K_v	0.3	0.34	0.37	0.42	0.45	0.47

12.2.1   Simulation results

For the purposes of this section the proportional gain (K_p), derivative gain (K_v), and overall controller gain (A_c) are set to 0.4, 0.35 and 0.01, respectively. The structural damping is set to $D_s = 0.148$. The arm is moved from one position to another within the range of ±80°. The system with this set-up was simulated without a payload at the end-point of the manipulator and the system response was observed at the hub and end-point. Figure 12.2 shows the angular displacement as measured at the hub. It is noted that the manipulator reached the demanded position from +80° to −80° within less than 2.0 s with no significant overshoot.

12.2.2   Experimental results

The performance of the fixed JBC controller for positioning the flexible manipulator with various payloads was investigated, and the controller parameters were adjusted offline. The controller parameters thus utilized with various loads are shown in Table 12.1 and the corresponding hub-angle is shown in Figure 12.3, where the demanded position in each case is shown with the dashed-line curve. It is noted that, with suitable control parameters, the manipulator achieves the demanded angular position smoothly and within reasonable timescales.

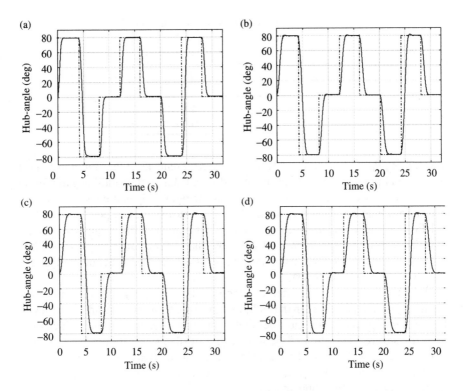

Figure 12.3 *Performance of the fixed JBC-controlled system with various payloads:*
(a) 0 g, (b) 40 g, (c) 80 g, (d) 100 g payload

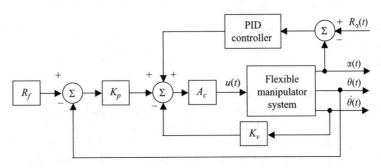

Figure 12.4 *Hybrid collocated non-collocated control*

12.3 Collocated and non-collocated feedback control involving PD and PID

A block diagram of the hybrid collocated non-collocated control structure is shown
in Figure 12.4, where $\alpha(t)$ and $R_\alpha(t)$ represent the end-point acceleration and
corresponding reference signal, respectively. Note that for minimum end-point

Table 12.2   Parameters of non-collocated controller

Payload (g)	K_p	K_i	K_d
0.0	10	400	0.2
50.0	5	1000	0.1

acceleration, $R_\alpha(t)$ is set to zero. To satisfy the requirements of the vibration control loop, the end-point vibration signal must be decoupled from the rigid-body motion so that the low-frequency movement representing the rigid-body motion is not included in the measurement of the end-point acceleration. Note that the end-point acceleration is defined as the acceleration due to the residual motion of the manipulator's end-point. The end-point oscillation is related to the elastic deflection $u(x, t)$ as described in Chapter 2 and hence minimising the end-point acceleration is the main objective of the vibration control process.

A high-pass filter can be used to decouple the rigid body dynamics from the flexural dynamics contained in the end-point measurement. The cut-off frequency of the required high-pass filter must be set to a sufficient frequency value, which does not cause severe attenuation of the lowest first flexible mode. In this case, the use of a high-pass filter with 8.0 Hz cut-off frequency with 20 dB/decade attenuation in the stopband would be adequate. A low-pass filter is applied to limit the system's bandwidth up to the first three modes.

12.3.1   Simulation results

To investigate the performance of the hybrid collocated–non-collocated control scheme in Figure 12.4, the flexible manipulator simulation environment is utilized. The flexible manipulator is set with a structural damping $D_s = 0.024$. The Ziegler–Nichols method (Ogata, 2001) is used in obtaining the PID controller parameters. The set of non-collocated PID controller parameters thus obtained is shown in Table 12.2.

The performance of the system with the hybrid collocated–non-collocated controller with no load at the hub and the end-point is shown in Figures 12.5 and 12.6. The corresponding input torque is shown in Figure 12.7. For reasons of comparison, the uncompensated and compensated end-point accelerations are superimposed, where the dotted-line graph shows the uncompensated end-point acceleration and the solid-line graph shows the compensated end-point acceleration. Figures 12.8–12.10 show the system performance with a 50 g load at the end-point. It can be seen from Figure 12.6 that the end-point accelerations has been reduced significantly after 0.2 s, and it takes only 0.5 s for the oscillations to reach a minimum level. The reduction at the first, second and third resonance modes of oscillation are 23, 12 and 7.4 dB, respectively. A similar trend is observed with a 50 g load at the end-point, see Figure 12.9. The system has shown a good hub-angle tracking performance and the input torque in each case remains within practical limits.

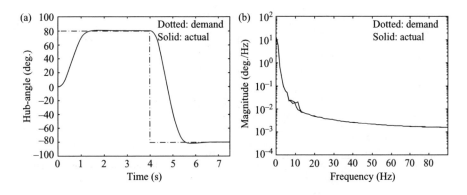

Figure 12.5   Hub-angle with hybrid collocated non-collocated controller without payload: (a) time-domain, (b) spectral density

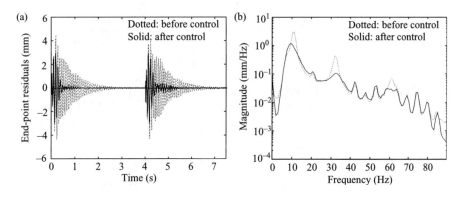

Figure 12.6   End-point acceleration with hybrid collocated non-collocated controller without payload: (a) time-domain, (b) spectral density

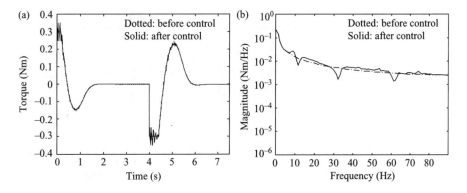

Figure 12.7   Torque input with hybrid collocated non-collocated controller without payload: (a) time-domain, (b) spectral density

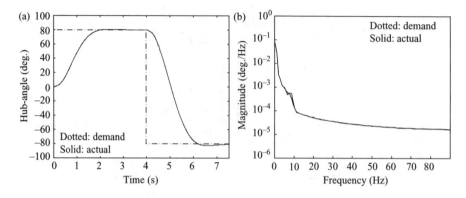

Figure 12.8   Hub-angle with hybrid collocated non-collocated controller with 50 g payload: (a) time-domain, (b) spectral density

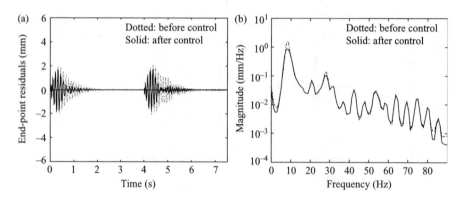

Figure 12.9   End-point accelerations with hybrid collocated non-collocated controller with 50 g payload: (a) time-domain, (b) spectral density

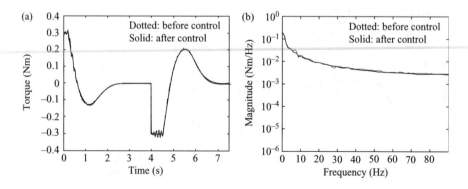

Figure 12.10   Torque input with hybrid collocated non-collocated controller with 50 g payload: (a) time-domain, (b) spectral density

The choice of a suitable control mechanism for a flexible manipulator system under movement from one point to another is affected by the extent of flexibility of the manipulator. An oscillatory behaviour (vibration) results due to the flexible nature of a manipulator. For rigid manipulators, the control scheme employs sensors that are collocated with actuators. The collocation of sensors and actuators provide a stable servo control (Gevarter, 1970). Non-collocated systems, on the other hand, lack these inherent stability characteristics. To achieve end-point position control, the desired end-point location is converted through real-time kinematics computation to the equivalent angles and then the actuator is activated to drive the end-point to these angles using a servo loop. Note that this controller will be working as long as the arm is stiff enough so that the end-point will remain in the intended location. The inherent flexibility of flexible manipulators may lead to such a control method not achieving accurate positioning. The use of a non-collocated control scheme, where the end-point of the flexible manipulator system is controlled by measuring its position, can be applied to improve the overall performance. The non-collocation of the sensor and the actuator can provide more reliable output measurement and increase the bandwidth of the overall closed-loop system.

It has been shown that PID non-collocated scheme is effective with a fixed set of controller parameters to actively reduce the vibration at the end-point for a certain loading condition only. Thus, if the characteristics of the flexible manipulator change, for example, due to loading conditions, the PID parameters must be adjusted accordingly to achieve and maintain the desired performance. In this manner, an adaptive inverse-dynamic controller is considered later in this chapter and its performance investigated in comparison to the fixed PID non-collocated control scheme.

12.4 Adaptive JBC control

Essentially, the task of the JBC control loop is to move or position the flexible arm to some predefined angle of demand. It is assumed that the flexible manipulator can possibly carry a load at the end-point. Based on this assumption, it was demonstrated earlier that fixed JBC control cannot sufficiently compensate for the changes in the characteristics of the system that may be caused by variation in the loading conditions, unless some adjustments on the control parameters are made. An adaptive JBC control scheme is expected to overcome this problem. The expectation is that an adaptive JBC controller could identify changes in the characteristics of the flexible manipulator and, thus, provide a better response in terms of speed and settling time by feeding suitable amount of torque to the flexible manipulator. A block diagram of an adaptive JBC control scheme is shown in Figure 12.11, where $\hat{K}_p = A_c K_p$ and $\hat{K}_v = A_c K_v$ with $\hat{K}_p$ and $\hat{K}_v$ representing the proportional and derivative gains, respectively. The hub-angle is represented by $\theta(t)$, R_f is the desired hub-angle and A_c is the amplifier gain. The input–output signals are sampled at a sampling period $\tau_s = 5.0$ ms, and a pole-placement approach is adopted with a second-order desired closed-loop characteristic, in designing the controller.

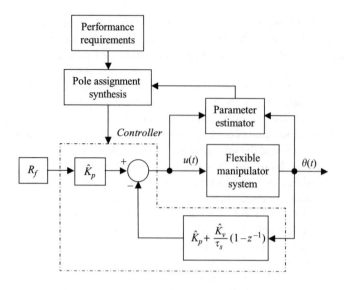

Figure 12.11   Adaptive joint based collocated control

The control signal, $u(z)$ in Figure 12.11 can be written as

$$u(z) = \widehat{K}_p R_f - \left[\widehat{K}_p + \frac{\widehat{K}_v}{\tau_s}(1 - z^{-1}) \right] \theta(z) \tag{12.1}$$

where z^{-1} represents the backward shift operator.

In a general design setting where the underlying system may have complex dynamics, a number of rules of thumb exist which assist in the selection of the controller parameters A_c, K_p and K_v. In a synthestic situation, however, the requirements on the underlying system are quite strict. In particular, with a second-order closed-loop response, in order to synthesise exactly the controller parameters, the system to be controlled must be assumed to have a special structure of the form

$$\theta(z) = \frac{b_0 z^{-1}}{1 + a_1 z^{-1} + a_2 z^{-2}} u(z) \tag{12.2}$$

This restriction on the system model is to ensure that only one set of controller parameters arise from the pole-placement design.

Combining the system model in equation (12.2) with the controller equation (12.1) yields the closed-loop equation relating R_f and θ as

$$\frac{\theta}{R_f} = \frac{A_c K_p b_0 z^{-1}}{(1 + a_1 z^{-1} + a_2 z^{-2}) + A_c \left[K_v \tau_s^{-1}(1 - z^{-1}) + K_p \right] b_0 z^{-1}} \tag{12.3}$$

The characteristic equation (Ch − Eqn) of the controlled system is thus

$$\text{Ch} - \text{Eqn} = 1 + \left(a_1 + A_c K_p b_0 + A_c K_v \tau_s^{-1} b_0 \right) z^{-1} + \left(a_2 - A_c K_v \tau_s^{-1} b_0 \right) z^{-2}$$

(12.4)

The controller coefficients can be determined by equating the characteristic equation with a desired closed-loop characteristic equation given by a T polynomial;

$$T = 1 + t_1 z^{-1} + t_2 z^{-2}$$

(12.5)

where

$$t_1 = -2 \exp\left(-\zeta \omega_n \tau_s \right) \cos \left\{ \tau_s \omega_n \left(1 - \zeta^2 \right)^{1/2} \right\}$$
$$t_2 = \exp\left(-2\zeta \omega_n \tau_s \right)$$

with ζ, τ_s and ω_n representing the damping factor, sampling period and natural frequency of the desired closed-loop second-order transient response, respectively. This yields the controller settings as

$$\hat{K}_p = \frac{t_1 + t_2 - a_1 - a_2}{b_0}$$

(12.6)

$$\hat{K}_v = \frac{(a_2 - t_2) \tau_s}{b_0}$$

(12.7)

Equations (12.6) and (12.7) are the controller design rules, realisation of which results in a pole assignment self-tuning control scheme, see Figure 12.11.

12.4.1   Simulation results

In the simulation studies presented in this section the flexible manipulator simulation algorithm is utilized as a test platform with a structural damping of $D_s = 0.024$. The torque input signal is limited to ± 0.3 Nm. The adaptive JBC control begins with identifying a parametric model of the hub-angle. A time space of 2.5 s has been specified with an expectation that, by the end of this period, all identified parameters have converged. For better presentation purposes, the outputs during the identification process are not plotted.

Figure 12.12 shows the hub-angle of the flexible manipulator. Figure 12.12(a) shows the hub-angle without a payload at the end-point. The first angle destination of 80° is achieved within 4.0 s, which is considered to be too slow compared to that achieved with the fixed JBC control scheme. However, better settling times are achieved for subsequent demanded angles after 10.0 s and it can be shown from the graph that the average settling time is found as 2.0 s, which is 0.5 s slower compared to the settling time achieved with the fixed JBC control. As noted, similar performances were achieved with a 50 g payload. The settling time for the two conditions is also found to be consistent and about 2.0 s. Thus, in terms of positioning control performance, the developed adaptive JBC control system performed very well with $D_s = 0.024$.

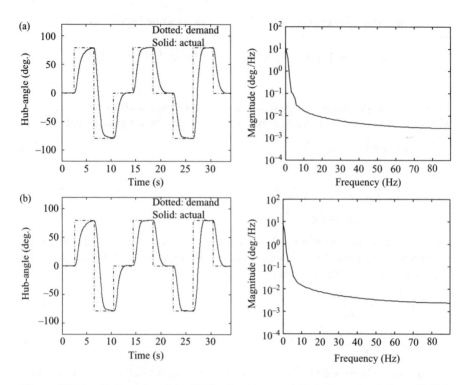

Figure 12.12   Hub-angle of the JBC-adaptive controlled system with $D_i = 0.024$: (a) without payload, (b) with 50 g payload

The corresponding end-point residuals are shown in Figure 12.13. The end-point residual represents the flexible motion of the end-point of the manipulator. It can be noted that the resonance modes are at 12.15, 32.71 and 61.68 Hz for the first, second and third modes respectively. The levels of the end-point vibration reach 3.5 mm for the first mode, 0.31 mm for the second mode and 0.052 mm for the third mode and the movement itself tends to oscillate for a longer time. It is noted that the end-point residuals reach their maximum level when the given torque to the flexible manipulator changes sharply. This is an important observation, which is useful for designing a suitable controller in minimising the vibrations of the flexible manipulator. Furthermore, it is noted that the level of vibration at the third mode is very small compared to that at the first and second modes.

12.4.2   Experimental results

The performance of the adaptive JBC-controlled system with various loads is shown in Figure 12.14. It can be seen that the flexible manipulator comfortably reaches the demanded angular position with the adaptive JBC controller. A closer observation of the system response with the fixed and adaptive JBC controllers reveals that both perform poorly with heavier loads. The dead-zone compensator algorithm could not

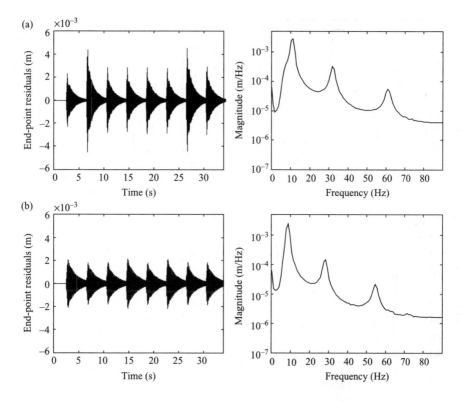

Figure 12.13 *End-point residuals of the JBC-adaptive controlled system with $D_i =$*
0.024: (a) without payload, (b) with 50 g payload

bring the hub-angle to the demanded angle. The amount of the required compensation
torque which was set by the value of c (refer to the dead-zone compensator developed
in Chapter 2), was not sufficient to ensure the flexible arm with a heavier load to keep
moving until the demanded angle is achieved. The amount of c can be increased
but this will have a significant effect on system vibration when the compensator is
activated. This requires further investigation to be carried out.

12.5 Adaptive collocated and non-collocated control

The combined collocated and non-collocated adaptive inverse-dynamic scheme is
described in Figure 12.15. The basic idea of adaptive inverse control is to drive
the flexible manipulator with an additional signal from a controller whose transfer
function is the inverse of that of the plant itself. Note that the adaptive inverse control
is active when the flexible manipulator is in motion, so that the computed torque is
used to reduce the end-point vibration. Since the plant is generally unknown, it is
necessary to adapt or adjust the parameters of the controller in order to create the true
plant inverse.

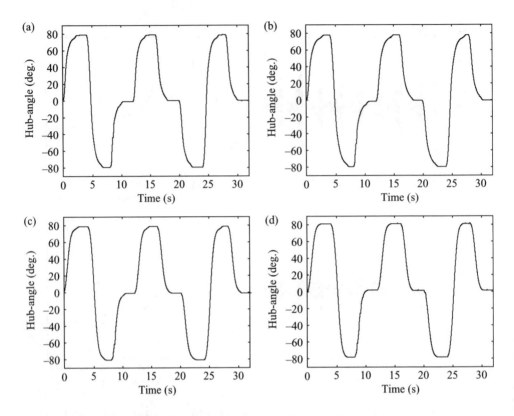

Figure 12.14   Performance of adaptive JBC-controlled system with various payloads: (a) 0 g, (b) 40 g, (c) 80 g, (d) 100 g payload

In implementing the adaptive inverse-dynamic control scheme, in addition to the practical issues related to properties of the disturbance signal, robustness of the estimation and control, system stability and processor-related issues such as word length, speed and computational power, a problem commonly encountered is that of instability of the system, specially, when the plant model is non-minimum phase. Thus, to avoid this problem of instability either the estimated model can be made minimum phase by reflecting its non-invertible zeros into the stability region and using the resulting minimum phase model to design the controller or once the controller has been designed the poles that are outside the stability region can be reflected into the stability region. In this manner, a factor $(1 - pz^{-1})$ corresponding to a controller-pole/model-zero at $z = p$, in the complex z-plane that is outside the stability region can be reflected into the stability region by replacing the factor with $(p - z^{-1})$. This strategy is used in obtaining the inverse-dynamic controller for realising the vibration control loop in the scheme shown in Figure 12.15, where the high-pass filter is used to decouple the vibration control loop from the rigid-body motion control loop.

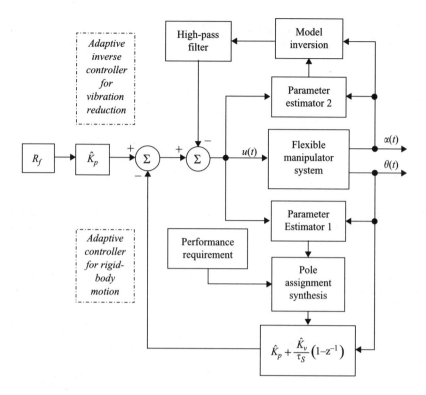

Figure 12.15   Adaptive collocated and non-collocated control scheme

12.5.1   Simulation results

Figures 12.16–12.18 show the performance of the adaptive inverse model control scheme without a payload. It is noted that the system tracked the demanded hub-angle well. The level of vibration at the end-point, on the other hand, was reduced by 13.6 and 6.1 dB at the first and second modes, respectively. Moreover, the input torque to the system remained within practically safe limits.

Figures 12.19–12.21 show the performance of the adaptive inverse-model controlled system with a payload of 50 g at the end-point. As noted in Figure 12.19, vibration reduction with a 50 g payload at the end-point of the manipulator was slightly less than that with no payload. It is noted that a 5.6 dB reduction at the first flexible mode and a 5.4 dB reduction at the second mode was achieved.

12.5.2   Experimental results

The adaptive inverse controller is implemented here and tested with the experimental flexible manipulator. Owing to limitations in the computing power of the platform used, the JBC control loop was fixed and a second-order system was chosen for the end-point acceleration model. This will allow only the first resonance mode of the

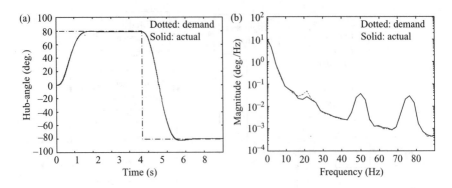

Figure 12.16   Hub-angle response of the adaptive collocated and non-collocated controlled system without a payload: (a) time-domain, (b) spectral density

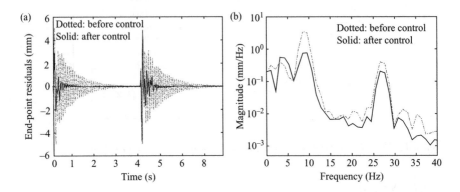

Figure 12.17   End-point acceleration response of the adaptive collocated and non-collocated controlled system without a payload: (a) time-domain, (b) spectral density

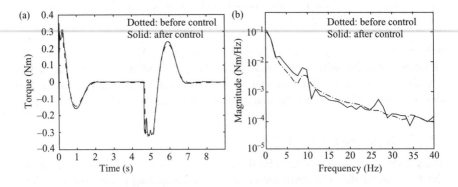

Figure 12.18   Input torque profile of the adaptive collocated and non-collocated controlled system without a payload: (a) time-domain, (b) spectral density

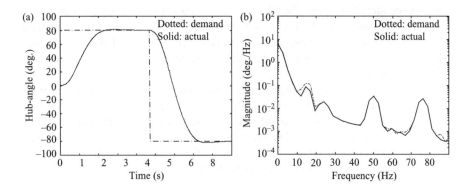

Figure 12.19 Hub-angle response of the adaptive collocated and non-collocated controlled system with 50 g payload: (a) time-domain, (b) spectral density

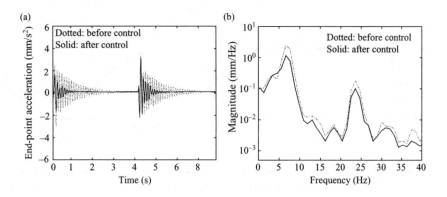

Figure 12.20 End-point acceleration response of the adaptive collocated and non-collocated controlled system with 50 g payload: (a) time-domain, (b) spectral density

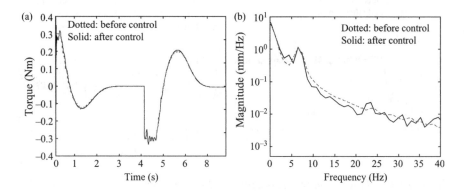

Figure 12.21 Input torque to the adaptive collocated and non-collocated controlled system with 50 g payload: (a) time-domain, (b) spectral density

system to be identified. Thus, the adaptive inverse model is considered to control the end-point vibration up to the first mode. The end-point acceleration measurements should not contain modes above the first resonance frequency, so that to avoid observation spillover. A low-pass filter with 20 Hz cut-off frequency was thus used to suppress the higher modes.

The performance of the adaptive inverse-model controlled system at the end-point vibration reduction is shown in Figure 12.22. The dotted line shows the uncompensated end-point acceleration and the solid line shows the compensated end-point acceleration. It is noted that using the adaptive inverse-model controller, the end-point vibration was reduced by 11.48 dB for the first resonance mode. Figure 12.23 shows the corresponding torque input to the system.

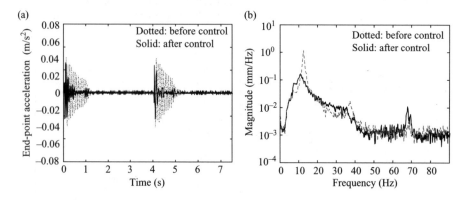

Figure 12.22   End-point acceleration response of the adaptive collocated and non-collocated controlled system (experimental result): (a) time-domain, (b) spectral density

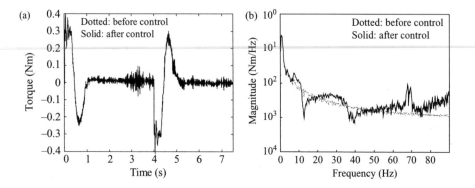

Figure 12.23   Input torque of the adaptive collocated and non-collocated controlled system (experimental result): (a) time-domain, (b) spectral density

12.6 Collocated and non-collocated feedback control with PD and neuro-inverse model

An alternative to the parametric approach described in the previous section, a neuro-inverse modelling approach can be adopted to realise the inverse plant model. This results in a neuro-inverse model control scheme. The neuro-controller thus obtained is used along with the adaptive JBC control to achieve both trajectory tracking and vibration suppression as illustrated in Figure 12.24.

It has been shown that an MLP NN employing the backpropagation algorithm can approximate a wide range of non-linear functions to any desired accuracy (Blum and Li, 1991). Owing to these capabilities, the classic backpropagation based MLP NN combined with the learning abilities based on the gradient descent method is used to realise the neuro-controller.

The backpropagation learning algorithm is commonly used with MLP NNs. Derivation of the algorithm can be found in a number of books (Luo and Unbehauen, 1997, Omatu *et al.*, 1996). The standard backpropagation algorithm may get stuck in a shallow local minimum as the algorithm is based on the steepest descent (gradient) approach. The problem can be solved by using Marquardt–Levenberg (Marquardt, 1963) modified version of backpropagation. While backpropagation is a steepest

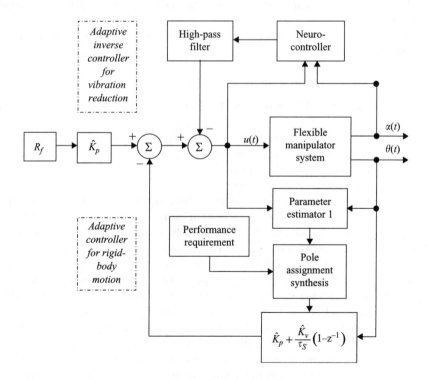

Figure 12.24   Adaptive neuro-inverse model control scheme

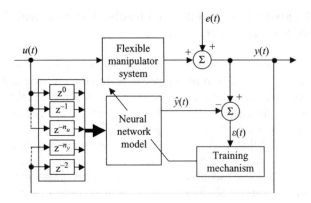

Figure 12.25   MLP neural network with NARX model scheme

descent algorithm, the Marquardt–Levenberg algorithm is an approximation to New-ton's method. The Marquardt-Levenberg algorithm can be considered a trust region modification to Gauss–Newton. The key step in this algorithm is the computation of the Jacobian matrix. For the NN mapping problem the terms in the Jacobian matrix can be computed by a simple modification to the backpropagation algorithm (Hagan and Menhaj, 1994).

Data is presented to the network using the NARX model structure. The model is described as

$$\hat{y}(t) = f[(y(t-1), y(t-2), \ldots, y(t-n_y),$$
$$u(t-1), u(t-2), \ldots, u(t-n_u)] + e(t) \tag{12.8}$$

where $\hat{y}(t)$ is the output vector determined by the past values of the system input vector and output vector with maximum lags n_u and n_y, respectively, $f(\cdot)$ is the system mapping constructed through MLP NN Marquardt modified backpropagation algorithm. The model is shown in Figure 12.25.

It is noted that the output of the model is a function of the plant input and output. This implies that the network output consists of the one-time-step-ahead estimates of the system output. The training data is recorded from the simulation of the adaptive JBC collocated control system. A section of the data is used for training the network and the remaining portion of the data for validating and testing the network. The model is further cross-validated using a separate set of data generated by adding a load to the manipulator. These steps ensure that the model is generalizing adequately the input–output behaviour of the real plant. In this investigation the lagged inputs of the neuro model are the end-point acceleration output of the plant along with the torque applied to the system. The output of the model is, thus, the torque. A high-pass filter of cut-off frequency 5 Hz is used to eliminate the rigid-body response and the resulting torque containing only the flexural motion response is subtracted from the input torque applied to the system.

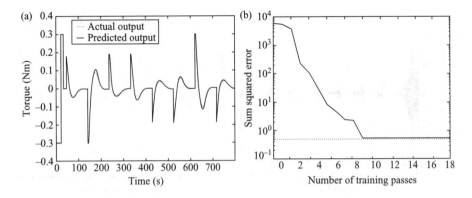

Figure 12.26   *Performance of the MLP neural network in inverse modelling of the system: (a) actual and predicted outputs, (b) sum squared error*

12.6.1   Simulation results

The neuro-inverse controller was realised using the MATLAB-Simulink environment. The flexible manipulator is moved from one point to another as specified by the angle of demand R_f using an adaptive JBC controller. The adaptive neuro-inverse controller could not be realised in the simulation environment due to the discrepancy between the time constraints required for training the network and the controller response. Thus, the network was trained off-line and the trained network was utilized to realise the neuro-inverse controller. The network was trained using 8 000 data points. After training for 20 epochs, the neuro-inverse model was able to predict the one-time-step-ahead values of the outputs of interest. The predictions and sum-squared error are shown in Figure 12.26. As noted, the NN successfully learned the inverse dynamics of the system.

The performance of the neuro-inverse controller in suppressing the vibration of the system without payload is shown in Figures 12.27 and 12.28. The controller is designed here for vibration suppression up to the second vibration mode of the system, as these contribute dominantly to the overall vibration of the system as mentioned before. It is noted that the level of vibration in the end-point acceleration was reduced by 3.65 and 5.11 dB for the first and second modes, respectively.

Figures 12.29 and 12.30 show the performance of the neuro-inverse controller in suppressing the vibration of the system with a payload of 50 g. As noted, vibration reduction with a payload of 50 g was slightly less than that with no payload. It is noted that the reductions in the level of vibration were by 1.95 and 3.08 dB for the first and second modes, respectively.

12.7   Summary

The developments of various classical and advanced controllers along with a neuro-inverse model controller for vibration reduction during the motion of flexible

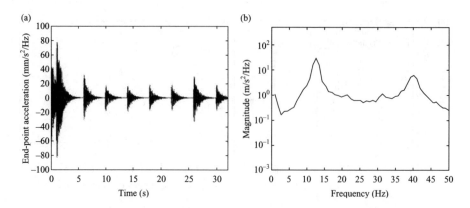

Figure 12.27   End-point acceleration without inverse neuro-control (without payload): (a) time-domain, (b) spectral density

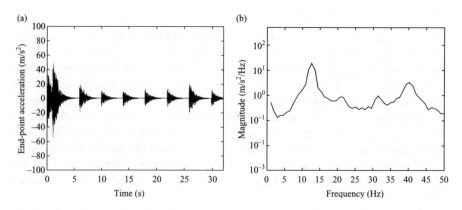

Figure 12.28   End-point acceleration with inverse neuro-control (without payload): (a) time-domain, (b) spectral density

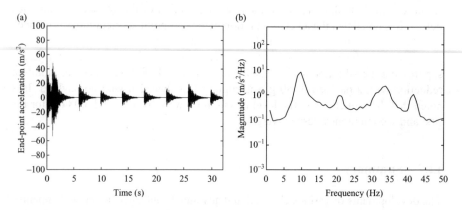

Figure 12.29   End-point acceleration without inverse neuro-control (with 50 g payload): (a) time-domain, (b) spectral density

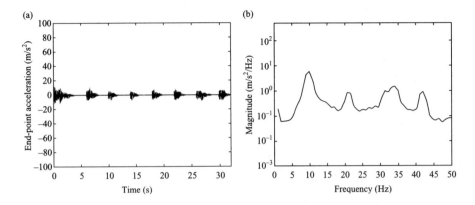

Figure 12.30   End-point acceleration with inverse neuro-control (with 50 g payload): (a) time-domain, (b) spectral density

manipulators have been presented. While fixed controllers, such as in JBC control and collocated and non-collocated hybrid control are capable of reducing vibration to some extent, they are not capable of handling system uncertainties with a payload change. An adaptive collocated control scheme with pole placement has been shown, however, to perform well in positioning the manipulator to the angle of demand. An adaptive hybrid collocated and non-collocated control scheme with PID and RLS based inverse-control scheme has been shown to perform well in reducing the vibration at the end-point with and without payload. The level of vibration reduction achieved with the neuro-controller is comparable to that achieved with PID based collocated and RLS-based inverse-modelling method. However, a significant achievement of the neuro-inverse model approach is that, the problem of controller instability due to the non-minimum phase behaviour of the plant model has been accounted for and resolved. This is important and provides further opportunities for control systems of this nature.

Chapter 13

Decoupling control of flexible manipulators

G. Fernández, J.C. Grieco and M. Armada

It is well known that conventional robot manipulators made of rigid links incorporate strongly coupled dynamics, and this situation is the source of many control problems. This difficulty is more pronounced when considering the control of flexible link manipulators. Taking into account the physical properties of flexible manipulators and their inherent vibratory behaviour it is proposed in this chapter to consider the control problem in the multivariable frequency domain. Such an approach is first introduced for analysing the coupling of a mixed rigid–flexible two-dimensional (2D) manipulator. The approach is then extended to the case of a 2D flexible manipulator. Following these analyses, a figure of merit to quantify the interaction between links is introduced, and the resultant designs for obtaining row and column dominance of both 2D manipulators are presented. Finally, an approach involving the manipulator Jacobian is proposed to define the control law. Finite element analysis is used in all cases to simulate the manipulators, and the models are experimentally validated.

13.1 Introduction

Over the last few years the authors have been working on a new approach for robot control using multivariable classical techniques, which have widely been tested and have proved to be useful in different applications (Rosenbrock 1974). Previous work by Armada (1987) and Ower and van de Vegte (1987) using classical approaches for controlling rigid and flexible robots have motivated the authors to continue with research in this direction (Fernandez *et al.*, 1993, 1994*a, b*, 2002*a, b*). The ultimate reason for this attempt is very simple and of practical relevance: positive results would lead to the application of best-known industrial control technology to complex plants such as flexible robots. Moreover, success in this area would make it easier to tune conventional proportional, integral, derivative (PID) controllers for robot link manoeuvering.

Classical multivariable approach for chemical plants and other industrial applications is based on the decoupling of different loops, so that the control design problem is transformed to the design of a decoupling filter. The advantage of such an approach is that many theorems exist that prove the existence of the decoupling filter, thus assuring the stability of the control loops (Maciejowsky, 1991).

The authors have applied these control techniques to rigid robots, to rigid–flexible robots and to flexible one-dimensional (1D) and two-dimensional (2D) beams. In this chapter, the rigid–flexible case is considered and the decoupling filter design for the 2D flexible link is presented. Gravity effects are assumed negligible as a particular beam that minimises the torsion effects and bending caused by the gravity is used. The links are modelled using the FE method and the models are properly validated with experimental set-ups.

The chapter provides a brief outline of the basics of multivariable control along with the essentials of modelling of flexible links and the associated problems in the decoupling control of flexible manipulators. Previous results have clearly established that non-collocated control issues and strong non-causality effects cause severe difficulties for tip (end-point) control of the link. Accordingly, a new control strategy is presented based on pre-filter design for decoupling effects. Finally, a further new control approach, based on calculation of the Jacobian of a pseudo-link is introduced, avoiding the problems associated with non-collocation. Following the pre-filter design, a single loop control strategy can be adopted assuring stability, in accordance to Rosenbrock criteria.

13.2   Multivariable control basics

Coupling effects of one input over other outputs make the multivariable control of a plant a complicated task. This problem has been studied in the past in the case of large plants, with many inputs and outputs. Total decoupling is possible using complex filters; in order to avoid such complication, some degree of diagonal dominance in the matrix transfer function (MTF) is required and decoupling to some extent is achieved. In this case, stability can be assured and tested using the Nyquist criterion. Each loop of the plant is thus analysed independently and a graphical analysis of the stability and interaction is used. In this chapter a quantitative measure of the interaction is obtained using the coupling factor, analysed with Bode plots. With such a design the total system consists of a multivariable pre-compensator in the direct loop and a diagonal cascade control designed for each loop independently.

Following the work of Rosenbrock (1974), diagonal dominance has become a very important concept for multivariable systems. Consider the multivariable feedback system shown in Figure 13.1.

Defining $\mathbf{Q} = \mathbf{G}(s)\mathbf{K_p}(s)$, then $\hat{\mathbf{Q}} = \mathbf{Q}^{-1} = \mathbf{K_p}^{-1}\hat{\mathbf{G}}(s)$ will be diagonal dominant if the following condition holds:

$$\sum_{\substack{j=1 \\ j \neq i}}^{m} |q_{ij}(s)| < |q_{ii}(s)| \quad i = 1, 2, \ldots, m \tag{13.1}$$

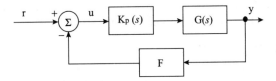

Figure 13.1   Multivariable feedback system

for any $s \in C$, where C is the Nyquist contour, refer to Maciejowsky (1991) for details. The matrix $\hat{\mathbf{Q}}$ can thus be described as

$$\hat{\mathbf{Q}} = \begin{bmatrix} \hat{q}_{11}(s) & \cdots & \hat{q}_{1m}(s) \\ \vdots & \vdots & \vdots \\ \hat{q}_{m1}(s) & \cdots & \hat{q}_{mm}(s) \end{bmatrix} \tag{13.2}$$

Defining d_i as

$$d_i = \sum_{\substack{j=1 \\ j \neq i}}^{m} |q_{ij}(s)|, \quad i = 1, 2, \ldots, m \tag{13.3}$$

diagonal dominance can be verified graphically by drawing circles of radii d_i on the Nyquist diagram for direct transfer functions or on the Nyquist diagram of inverse transfer functions resulting in the direct Nyquist array (DNA) or the inverse Nyquist array (INA) method, respectively. The authors have studied the robot coupling problem for rigid structures (Armada, 1987; Fernandez *et al.*, 1993) and have designed constant pre-filters for such robots (Fernandez *et al.*, 1994a,b).

In this chapter a pre-compensator is designed using the Perron–Fröbenius eigenvalue method, to achieve row dominance and it is shown that decoupling is realised using a constant pre-filter. However, as it is well known that row dominance is not helpful for decoupling control when using direct MTF, and column dominance must be achieved. For achieving such diagonal dominance, numerous methods have been attempted; in this case it is not possible to achieve dominance with a constant filter and the ALIGN algorithm, according to the approach proposed by Ford and Daly, see Maciejowsky (1991), must be used. With this design method, each element of the pre-compensator is calculated using an nth-order frequency polynomial; a second-order polynomial was used in this case. Frequency variations must be taken into account to deal with inter-link vibration coupling.

One of the main problems associated with the application of these methodologies to robot control is the variation of the MTF with the operating point. The rigid–flexible robot and a 2D flexible arm, presented here, were each linearised around a trajectory and a varying coefficient polynomial was designed for each robot. Variations in the polynomial coefficients were studied depending on the trajectory. The robots were simulated and coupling analyses are presented with and without pre-compensator.

To achieve row dominance, the Perron–Fröbenius approach can be used; details can be found in Maciejowsky (1991). It can be demonstrated that a constant diagonal pre-compensator $\mathbf{K_f}$ exists that allows row dominance if the normalised comparison matrix $\mathbf{C}(j\omega_i)$, its Perron–Fröbenius eigenvalue, at frequency ω_i is less than two:

$$\mathbf{C}(s) = \text{abs}\left(\mathbf{G}_{\text{diag}}^{-1}(s)\,\mathbf{G}(s)\right) \tag{13.4}$$

then $\mathbf{K_f}(s) = \text{diag}\left\{k_1(s) \quad \cdots \quad k_m(s)\right\}$ exists if $\lambda_P(j\omega) < 2$, where $\lambda_P(j\omega)$ is the Perron–Fröbenius eigenvalue of $\mathbf{C}$.

The constant, diagonal, pre-compensator can be designed using a matrix $\mathbf{T}(j\omega)$, defined as

$$t_{ij} = \max_{\omega}\left\{c_{ij}(j\omega)\right\} \tag{13.5}$$

where $c_{ij}(j\omega)$ is the element in row i and column j of matrix $\mathbf{C}$.

Finally, $\mathbf{K_f}$ can be obtained by locating the eigenvector for the maximum eigenvalue of $\mathbf{T}$ in its diagonal; if λ_i represents the eingenvalues of $\mathbf{T}$, for $i = 1, 2, \ldots, m$, then $\mathbf{K_f}$ can be obtained as

$$\upsilon_{\mathbf{max}} = \upsilon\left(k, \max(\lambda_i)\right) \tag{13.6}$$

where υ is the eigenvector matrix of $\mathbf{T}$ and $k = 1, 2, \ldots,$ numberofrows.

$$\mathbf{K_f} = \text{diag}(\upsilon_{\mathbf{max}}) \tag{13.7}$$

This pre-compensator makes the system row dominant at frequencies where the Perron–Fröbenius eigenvalue of matrix $\mathbf{C}$ is less than two, as indicated earlier. If column dominance is required, as it is the case for direct transfer matrix function, then the ALIGN algorithm can be used (Maciejowski, 1991). Here the minimisation of a functional J_j is pursued. This is given as

$$J_j = \frac{\sum\limits_{k} p_k\left\{\sum_{i\neq j}\left|q_{ij}(j\omega_k)\right|^2\right\}}{\sum_{k=1}^{m} p_k\left|q_{jj}(j\omega_k)\right|^2} \tag{13.8}$$

Ford and Daly (1979) proposed to use the ALIGN algorithm in order to design a dynamic pre-compensator as

$$\mathbf{K_P}(s) = \left[\begin{array}{ccc} \mathbf{k_1}(s) & \cdots & \mathbf{k_m}(s) \end{array}\right] \tag{13.9}$$

where each column has the form

$$\mathbf{k_j}(s) = \mathbf{k_{oj}} + \mathbf{k_{1j}}s + \mathbf{k_{2j}}s^2 \cdots + \mathbf{k_{\beta_j}}s^\beta \tag{13.10}$$

13.3   Modelling a flexible link

13.3.1   *Rigid–flexible robot case*

The coupling problem for a 2D rigid–flexible robot is analysed in this section. Mixing rigid and flexible modes have helped to reduce structural vibration (Fernández *et al.*, 2002*b*; Hara and Yoshida, 1994; Suzuki *et al.*, 1993). The robot considered here

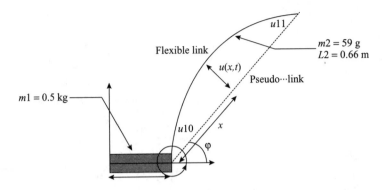

Figure 13.2   Schematic diagram of the 2D mixed rigid-flexible robot

moves in the horizontal plane; the rigid link is prismatic and the flexible link is on a rotational joint. For modelling purposes gravity effects and torsion phenomena are not considered. Modelling is performed using the finite element (FE) approach; see Chapra and Canale (1988) for an exposition of the FE method. In this case it is shown that one element per link is enough to obtain good results; link deformations are modelled using third-order polynomials. A schematic view of the robot is shown in Figure 13.2.

A pseudo-link based on previous work (Bayo, 1987; Bayo *et al.*, 1989; López-Linares, 1993) is defined here. In this approach the link deformation is modelled using a third-order shape function vector $\mathbf{N}(x)$. The movement of a point on the beam is modelled using variations of angle φs and time variations of deformation u, see López-Linares (1993) for details. A vector of generalized variables is defined as

$$\mathbf{v} = \begin{bmatrix} \rho \\ \varphi \\ u_{10} \\ u_{11} \end{bmatrix} \tag{13.11}$$

13.3.2   Modelling the 2D flexible robot

As before, the Euler–Bernoulli model is used here for the link. The deformation $u(x,t)$ is characterised as

$$u(x,t) = \sum_{i=1}^{n} N_i(x)q_i(t) \tag{13.12}$$

The full model derivation, including friction and actuator models, is given in Fernandez (1997). The dynamic non-linear equation for the 2D manipulator is obtained as

$$\left(\mathbf{M} + \mathbf{L}J_{eq}\mathbf{L}^T\right)\ddot{\mathbf{q}} + \left(\mathbf{C} + \mathbf{L}B_{eq}\mathbf{L}^T\right)\dot{\mathbf{q}} + \mathbf{L}C_v\mathbf{L}^T\dot{\mathbf{q}} + \mathbf{L}C_F\,\mathrm{sgn}(\omega_m) + \mathbf{K}\mathbf{q}$$

$$= \mathbf{L}K_{eq}\mathbf{V_a} \tag{13.13}$$

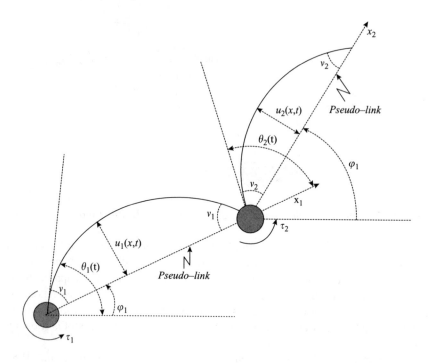

Figure 13.3   Variables used in the dynamic equation for 2D flexible beam

where **V** is the armature voltage of the d.c. motors acting on the robot joints. Equation (13.13), with $\mathbf{q} = v$, constitutes the general governing dynamic equation of the system, including actuators (d.c. motors), see Fernández (1997) for details and full determination of matrices **M**, **C** and **K**.

The vector **q** is defined as

$$\mathbf{q} = \begin{bmatrix} \varphi_1(t) \\ \varphi_2(t) \\ v_1(t) \\ v_2(t) \end{bmatrix} = \begin{bmatrix} \varphi_1(t) \\ \varphi_2(t) \\ v_{10}(t) \\ v_{11}(t) \\ v_{20}(t) \\ v_{21}(t) \end{bmatrix} \tag{13.14}$$

where, the variables involved are shown in Figure 13.3.

Equation (13.13) can be linearised and the MTF obtained at the operating point. As an example, for the case of the 2D manipulator around $\mathbf{q} = 0$, the MTF is obtained as described below.

$$\mathbf{G}(s) = \begin{bmatrix} g_{11}(s) & g_{12}(s) \\ g_{21}(s) & g_{22}(s) \end{bmatrix} \tag{13.15}$$

with

$$g_{ij}(s) = \frac{n_{ij}(s)}{d_j(s)} \tag{13.16}$$

and

$$n_{11}(s) = \begin{pmatrix} -0.002326s^8 + 44.71s^7 + \\ +9.5 \times 10^4 s^6 + 1.47 \times 10^7 s^5 + 1.08 \times 10^9 s^4 + \\ +3.36 \times 10^{10} s^3 + 4.4 \times 10^{10} s^2 + 2.8 \times 10^{12} s + 1.7 \times 10^{13} \end{pmatrix}$$

$$d_1(s) = s \begin{pmatrix} s^{10} + 4615.0 s^9 + 1.01 \times 10^6 s^8 + \\ +9.66 \times 10^7 s^7 + 4.51 \times 10^9 s^6 + 1.29 \times 10^{11} s^5 + \\ +2.17 \times 10^{12} s^4 + 1.99 \times 10^{13} s^3 + 1.21 \times 10^{14} s^2 + \\ +4.73 \times 10^{14} s + 7.72 \times 10^{14} \end{pmatrix} \tag{13.17}$$

$$n_{12}(s) = \begin{pmatrix} -0.02769 s^8 - 554.9 s^7 - 3.09 \times 10^5 s^6 - \\ -3.31 \times 10^7 s^5 - 1.65 \times 10^9 s^4 - \\ -5.56 \times 10^{10} s^3 - 8.63 \times 10^{11} s^2 - 3.81 \times 10^{12} s - 1.94 \times 10^{12} \end{pmatrix}$$

$$d_2(s) = \begin{pmatrix} s^{10} + 4615.0 s^9 + 1.01 \times 10^6 s^8 + \\ +9.66 \times 10^7 s^7 + 4.51 \times 10^9 s^6 + 1.29 \times 10^{11} s^5 + \\ +2.17 \times 10^{12} s^4 + 1.99 \times 10^{13} s^3 + \\ +1.21 \times 10^{14} s^2 + 4.73 \times 10^{14} s + 7.72 \times 10^{14} \end{pmatrix} \tag{13.18}$$

$$n_{21}(s) = \begin{pmatrix} 9.6 \times 10^{-4} s^9 - 21.99 s^8 - 1.87 \times 10^4 s^7 - \\ -4.2 \times 10^6 s^6 - 3.61 \times 10^8 s^5 - \\ -1.13 \times 10^{10} s^4 - 1.0 \times 10^{11} s^3 - \\ -8.69 \times 10^{11} s^2 + 1.60 \times 10^{13} s + 6.1 \times 10^{13} \end{pmatrix} \tag{13.19}$$

$$n_{22}(s) = \begin{pmatrix} 0.01064 s^8 + 290.2 s^7 + 2.388 \times 10^4 s^6 + \\ +2.84 \times 10^6 s^5 + 3.82 \times 10^8 s^4 + \\ +2.48 \times 10^{10} s^3 + 6.88 \times 10^{11} s^2 + \\ +1.04 \times 10^{13} s + 7.89 \times 10^{13} \end{pmatrix} \tag{13.20}$$

Once linearised, the variable state description can be obtained and matrices **A** and **B** derived;

$$\mathbf{A} = \begin{bmatrix} \mathbf{A_{11}} & \mathbf{A_{12}} \\ \mathbf{I_{6x6}} & \mathbf{0_{6x6}} \end{bmatrix} \tag{13.21}$$

where

$$
A_{11} = \begin{bmatrix}
-0.0139 & 0.0431 & 102.21 & 131.6 & 2.8 & 0.93 \\
0.0045 & -0.0165 & -47.0 & -75.3 & -6.3 & -9.57 \\
-6.3369 & -0.0410 & -108.8 & -131.7 & -2.79 & -0.93 \\
-5.6459 & 5.0964 & -1986.6 & -3816.1 & -320.75 & -300.54 \\
0.0940 & -0.4995 & -1831.5 & -3591.9 & -317.39 & -290.14 \\
0.0630 & -0.3645 & -1390.0 & -2792.8 & -318.48 & -372.84
\end{bmatrix}
$$

(13.22)

and

$$
A_{12} = \begin{bmatrix}
0 & 0 & 0.2116 & 0.2725 & 0.0160 & 0.0054 \\
0 & 0 & -0.0973 & -0.1560 & -0.0366 & -0.0556 \\
0 & 0 & -0.2122 & -0.2728 & -0.0160 & -0.0054 \\
0 & 0 & -4.1129 & -7.8923 & -1.8922 & -1.7453 \\
0 & 0 & -3.7919 & -7.4394 & -1.8402 & -1.6848 \\
0 & 0 & -2.8781 & -5.7844 & -1.8473 & -2.1651
\end{bmatrix} \times 10^4
$$

(13.23)

The input matrix **B** is similarly determined as

$$
B = \begin{bmatrix} B_1 \\ 0_{6\times 2} \end{bmatrix}
$$

(13.24)

where

$$
B_1 = \begin{bmatrix}
-0.0023 & -0.0277 \\
0.001 & 0.0106 \\
0.4955 & 0.0277 \\
0.0356 & -3.2767 \\
0.0322 & 0.3213 \\
0.0239 & 0.2344
\end{bmatrix}
$$

(13.25)

13.4   Pre-compensator design

13.4.1   *Rigid–flexible robot case*

The system model of the rigid–flexible manipulator was obtained and properly linearised around the 'worst case' point, from the coupling point of view. The state variable expression for the system was found, with φ and ρ as outputs (see Figure 13.2), and voltages at both links as inputs. As such, column and row dominance was studied, and the results shown in Figure 13.4 were obtained for the first and second column of the direct MTF. All computations were done in Matlab.

As it can be seen in Figure 13.4, the Gershgorin plots show that the zero is surrounded by circles, but it is very difficult to quantify the degree of non-dominance. For this reason another measure of dominance is required. Let the dominance ratio

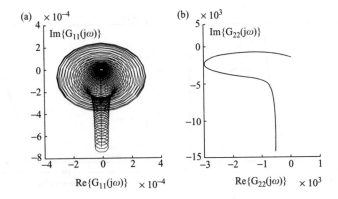

Figure 13.4   Gershgorin's circles for the (a) first and (b) second columns of the robot

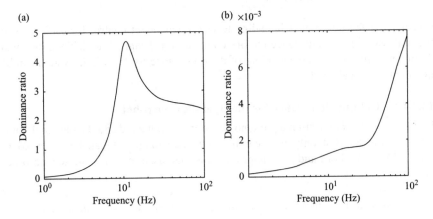

Figure 13.5   Bode diagrams for the dominance ratio: (a) first column, (b) second column

be defined as

$$r_i = \frac{\displaystyle\sum_{\substack{j=1 \\ j \neq i}}^{m} |q_{ij}(s)|}{|q_{ii}(s)|} \qquad i = 1, 2, \ldots, m \qquad (13.26)$$

The ratio in equation (13.26) is calculated for any $s \in C$, and the terms q_{ij} are elements in column j of matrix $\mathbf{Q}$. In this manner, it is possible to draw the Bode diagram for such ratio, as shown in Figure 13.5. For all those frequencies where the ratio is greater than one, the system is not column diagonal dominant.

This provides the designer with a tool that shows clearly the dependence of dominance with the frequency and the designer can determine the magnitude of the dominance, and the potential coupling problems that can arise for control purposes;

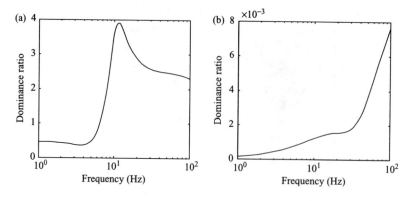

Figure 13.6 Dominance ratio of the system after filtering: (a) first column, (b) second column

especially if the designer is intending to use independent control for each loop of the multivariable system. Once the pre-compensator designed, to achieve diagonal dominance, the frequency behaviour of the dominance for the whole system can be quantitatively evaluated.

13.4.1.1   Column dominance for the rigid–flexible robot

Using the ALIGN algorithm a pre-compensator was obtained and the diagonal dominance achieved, but only in a certain frequency range, as can be observed in Figure 13.6. The Ford and Daly approach was then used to obtain the second-order pre-filter given as

$$
K_p = \begin{bmatrix} 1.00 + (0.9159 \times 10^{-3})s - (0.2034 \times 10^{-4})s^2 & \vdots & 0.0023 + 0.0002s + (0.0059 \times 10^{-3})s^2 \\ \cdots & \vdots & \cdots \\ 0.0035 - (0.7584 \times 10^{-3})s - (02345 \times 10^{-4})s^2 & \vdots & -1 - 0.0078 - (0.4355 \times 10^{-3})s^2 \end{bmatrix}
$$

(13.27)

The dominance ratios thus obtained are shown in Figure 13.7.

The enhancement achieved is significant when the first column is observed, where complete dominance is achieved in the frequency range considered, and a high degree of decoupling is established. Problems remain for the second column to achieve the desired dominance, but at this stage the designer knows that if some modes beyond the indicated frequencies are excited during control, then system stability cannot be assured in accordance with Rosenbrock's criterion. The authors have attempted the approach with pre-filters of higher order, and good results, in terms of dominance for both columns, have been achieved. However, in this sense the problem is transferred from the control law to the pre-filter implementation. The advantage of the proposed method, however, is that the designer is provided with a complete set of well-tested tools for the pre-filter design, and simultaneously control stability and accessibility for single-loop control is granted.

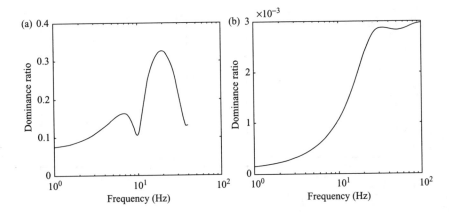

*Figure 13.7   Dominance ratios after filtering using the Ford-Daly design approach:
(a) first column, (b) second column*

13.4.1.2   Column dominance for rigid–flexible robot workspace

A filter that depends on the operating point of the flexible link was designed. The link angle was varied between 0 and π radians; and the Ford–Daly approach was used to design a filter for each point of the workspace. The variations in the filter coefficients as a function of the operating point thus obtained are shown in Figure 13.8. The horizontal axis in each case is linearly graduated in accordance with different evaluating points for the flexible-link angle, between 0 and π. The graphs in Figure 13.8 show the filter coefficients C_{ij}, where j represents the column number of the filter and i is the respective power of s.

Each element of the filter is defined as

$$\mathbf{k}_j = \mathbf{k}_{oj} + \mathbf{k}_{1j}s + \mathbf{k}_{2j}s^2 \tag{13.28}$$

and each $\mathbf{k}_{ij}$ is a column vector given as $\mathbf{k}_{ij} = \begin{bmatrix} C_{ij} & C'_{ij} \end{bmatrix}^{\mathrm{T}}$. Application of the filter to the TFM resulted in a dominance ratio for all the robot workspace. This is shown in Figure 13.9.

13.4.2   2D flexible robot case

Results obtained for the case of a 2D flexible robot are presented in this section. As indicated earlier, the application of multivariable frequency-domain techniques seems very appropriate for the analysis and design of controllers applied to flexible robots because the flexibility is characterised mainly by the frequency response of the link. The high degree of coupling among the inputs and the vibration modes makes it possible to design filter(s) that help to diminish such coupling and eventually remove it.

13.4.2.1   Design of the decoupling filter for the 2D flexible robot

In this case, the input–output pairing selected was the input voltage to the motors and the pseudo-link angle of each manipulator link (see Figure 13.3). It can be noticed in

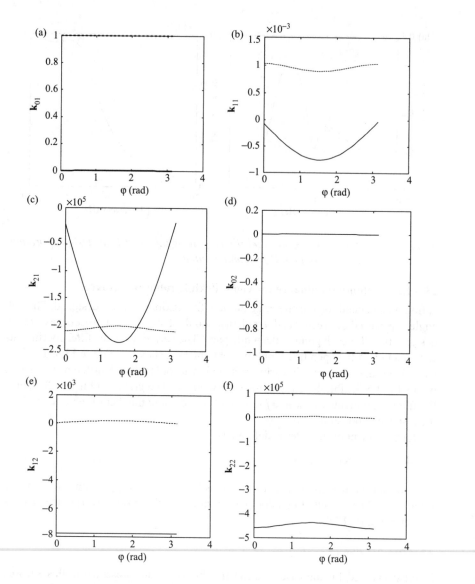

Figure 13.8   *Pre-compensator coefficient variations for the whole workspace:*
(a) k_{01}, (b) k_{11}, (c) k_{21}, (d) k_{02}, (e) k_{12}, (f) k_{22}

Figures 13 10 an 13 11 that excitation o the irst motor has a signi icant in luence
on the irst vi ration mo e o the secon link It can e o serve that the coupling
values are even greater or the secon link

In or er to control the manipulator, it is necessary to esign a ilter that imin-
ishes the egree o coupling among the links Using the For –Daly esign strategy,
mentione earlier, a irst-or er ilter was o taine The pre- ilter thus o taine is

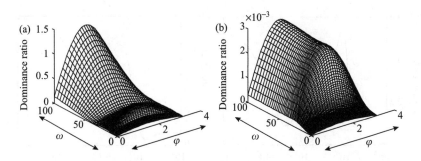

Figure 13.9   Dominance ratios for the (a) first and (b) second columns of the MTF

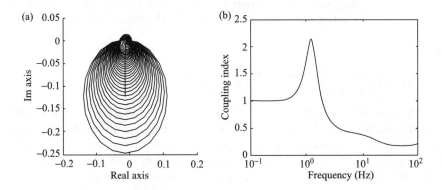

Figure 13.10   Coupling analyses for the first link of the 2D flexible robot: (a) Gershgorin circles, (b) coupling index without pre-filter

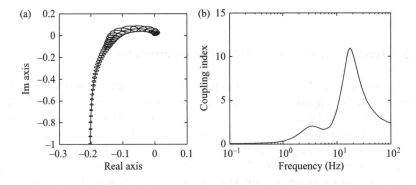

Figure 13.11   Coupling analyses for the second link of the 2D flexible robot: (a) Gershgorin circles for the second column of the MTF, (b) coupling index without pre-filter for the second column of the MTF

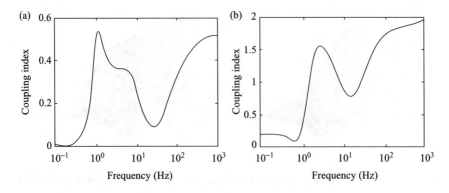

Figure 13.12 Measure of the decoupling achieved for the 2D flexible robot: (a) coupling index after compensation using Ford–Daly design technique, (b) coupling index with compensation for the second column of the MTF

given as

$$\mathbf{K_P} = \begin{bmatrix} 0.992 - 0.124s & -0.803 + 0.593s \\ -0.029 + 0.007s & 0.057 + 0.019s \end{bmatrix}$$

(13.29)

The theoretical degree of decoupling reached is shown in Figure 13.12. The result was significantly good for the first link and the coupling diminished for the second joint, but at some frequencies, near the vibration mode, it was not possible to go under one, at least using the pre-filter previously shown. Some successful tests were carried out with higher-order filters but the inclusion of high-order derivatives in the filter design caused noise in the control loop. In any case, the high degree of improvement in the dominance obtained for the first column and the high reduction obtained for the second column, using a simple first-order filter are very interesting and promising results. Some experimental results have been presented in Fernández *et al.* (2002*b*), using the designed filters, combined with the control strategies presented in the next section.

13.5 Jacobian control of a 1D flexible manipulator

A new approach for the control of flexible links is presented in this section. The intention here is to apply control techniques developed for rigid robots to flexible manipulators. In the Jacobian strategy, the 'virtual' collocated nature of the end-point control is a key feature of the global strategy. The control results are compared with the non-collocated strategy where the motor position is controlled using information on the end-point position. The vibration reduction is clearly achieved even when the flexibility of the beam is strongly overstated. The non-minimum phase nature of flexible beams and the problems associated with collocated and non-collocated systems have been treated in previous chapters of this book; further details are also provided in Cannon and Schmitz (1984), and Eppinger and Seering (1988). The control problem of non-collocated systems is one of the main problems researched in the context of control of flexible manipulators. There is a natural limitation in the maximum bandwidth reachable by the control owing to the time that the vibratory signals travel along the beam. This limitation can be solved only if a collocated control scheme is employed and a vibration control strategy to cancel the vibrations is implemented. The proposed Jacobian control is shown in this respect to cancel the vibrations using a non-collocated scheme.

There are significant problems associated with the control of flexible manipulators. The non-minimum phase behaviour creates a problem when the inverse dynamic for the robot is pursued. Feliú *et al.* (1995) have reported a control scheme similar to the Jacobian control studied here, but much more difficult to implement. The scheme uses the inverse Jacobian to transform the end-point deflections into motor axis differences. The work presented in Trautman and Wang (1996) is similar to the technique studied here.

13.5.1 *Jacobian control*

This Jacobian control strategy is based on the end-point position feedback. Two control loops are used: a simple PD internal loop to control the motor axis position, where the motor axis reference angle is obtained under the premise of a rigid beam, and an outer loop for control of the end-point position, where the end-point position is measured using infrared three-dimensional (3D) cameras. Position errors measured on the x–y Cartesian space are transformed to errors on the joint angle space using the inverse Jacobian for the virtual rigid beam formed between the motor axis and end-point of the beam. Using the measured error, a control signal that modifies the operating point for the internal loop is generated in order to correct those errors produced by end-point vibrations or due to gravity action. This control strategy was validated experimentally, and the results are presented below. Schematically, the control variables are shown in Figure 13.13.

In the proposed scheme an external loop is added in order to reduce the end-point vibration. In that loop the x–y end-point position is compared with a reference trajectory, generating an error in the Cartesian space. This error is converted to the joint space (angle differences) using the Jacobian transformation as

$$\mathbf{J}^+ \left(\boldsymbol{\theta}_{\text{ref}} \right) \mathbf{e}_{xy} = \mathbf{e}_{\theta} \tag{13.30}$$

where $\mathbf{J}^+$ is the Jacobian pseudo-inverse,

$$\mathbf{J}^+ \left(\theta_{\text{ref}} \right) = \left[-\sin \left(\theta_{\text{ref}} \right)/l \quad \cos \left(\theta_{\text{ref}} \right)/l \right] \tag{13.31}$$

and l is the beam length.

The control law is given by

$$u(t) = K_{\text{P}} \left(\theta_{\text{ref}} - \theta \right) + K_{\text{V}} \left(\dot{\theta}_{\text{ref}} - \dot{\theta} \right) + K_{\text{PJ}} \left(\mathbf{J}^+ \mathbf{e}_{xy} \right) + K_{\text{VJ}} \left(\mathbf{J}^+ \dot{\mathbf{e}}_{xy} \right) \tag{13.32}$$

The Jacobian is computed for each point on the reference trajectory; any non-reached position will result in error in the x–y space. These errors are produced by vibration or by gravity effects on the beam. The Jacobian term on the control law

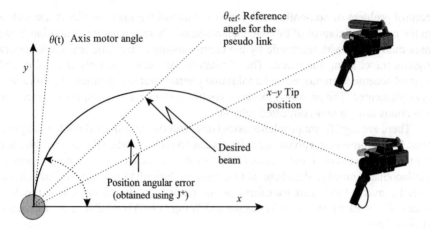

Figure 13.13   *Controlled variables in the Jacobian control*

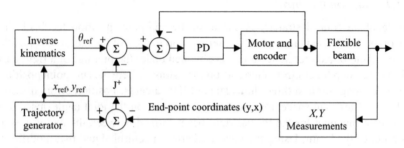

Figure 13.14   *Pseudo-link Jacobian control scheme*

compensates for such errors. The inverse kinematics are calculated for the virtual rigid beam so as to avoid the non-minimum phase behaviour. The proposed control scheme avoids the non-collocated problems because the control problem is handled with a collocated approach by adding a correction signal associated with the Cartesian error. The control scheme is shown in Figure 13.14.

Note that in Figure 13.14, all phenomena related to the beam flexibility are included in the box titled 'Flexible beam'. End-point position measurements are absolute measurements with respect to an inertial reference system located on the motor axis. The measurement of the position of the motor axis is obtained using an incremental optical encoder. End-point position measurements were performed using infrared LEDs and cameras from SELCOM (Selcom Selective Electronics, 1994). The SELSPOT system was employed for measurement of the position of terminal elements (Oussama *et al.*, 1989). The system resolution varies with the focal distance of the employed lenses; a resolution of up to 0.125 mm has been reported. In this work, lenses with focal distances of 50 mm were used. It is possible to obtain resolutions to 0.16 mm. The main limitations to reach such resolutions are the reflections on floor, ceiling, walls, and so on.

13.5.2 Control results

The results obtained with the Jacobian control method were compared with collocated control of the joint angle and also with non-collocated control of end-point position. Figures 13.15–13.17 show the performance of the proposed control method. The trajectory errors in the x–y space and the effort control of the Jacobian term are shown in Figure 13.17.

It is observed in these results that the error during transient response is less than 40 mm, while the input excitation reaches $\pi/2$ in 2 s. This is a strong excitation to test the control strategy with. Such good results are due to the double control loop strategy. The inner loop tries to compensate for deviations sensed over the motor hub, as measured by the motor encoder; the second control loop measures the end-point position in the x–y plane, using infrared cameras as sensors, and compensates for

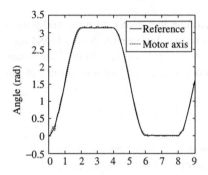

Figure 13.15   Response of the joint angle with the reference under Jacobian control

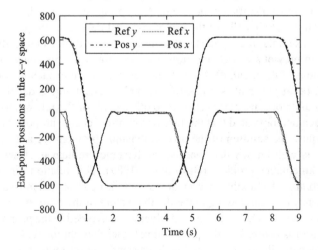

Figure 13.16   End-point position in the x–y space under Jacobian control

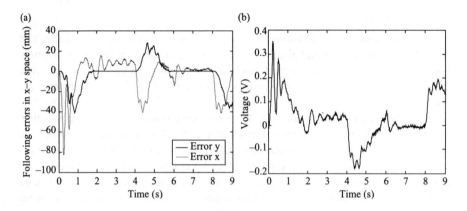

Figure 13.17   (a) Error and (b) effort control with the proposed control scheme

the errors using the Jacobian to convert from the end-point space to the hub-space. Only a PD controller is needed because no gravity effect was considered; the gravity effect was minimised by physical construction of the beam, otherwise the gravity force would produce a constant perturbation over the end-point.

13.6   Summary

Several design techniques have been tested to achieve diagonal dominance and, furthermore, decoupling control. A constant pre-compensator has been designed using Perron–Fröbenius eigenvalue with which only row dominance was achieved. Using the approach proposed by Ford and Daly (1979) a position-varying polynomial has been designed to achieve column dominance throughout the robot workspace, and it has been shown that a sufficient degree of decoupling is achieved between the rigid and flexible modes. This allows the use of classical design techniques and stability analyses, for example, the Nyquist criterion for each loop, independently. As it has been stated in previous works, and further demonstrated in this study through coupling behaviour of the plant, dNA is still the best tool to study such interaction. This is important because it makes it possible to apply the techniques developed for 1D flexible manipulators and SISO schemes to other kinds of flexible robots.

It has been demonstrated that the Jacobian control scheme proposed in this study does not require the solution of the inverse dynamic problem for a flexible manipulator; as it is well known that the inverse dynamic problem is closely related to the inverse kinematics problem (Bayo et al., 1989). This relationship makes non-collocated control of a flexible manipulator a complicated task that must be addressed in the frequency domain. Owing to the flexible nature of the beam the plant exhibits non-minimum phase behaviour. These problems are avoided in the proposed scheme because control is centred on the pseudo-link rather than on the real flexible beam, and vibration cancellation is pursued with a second control loop. A simple strategy was implemented for a 1D flexible beam. This strategy is based on the conversion of the end-point position error from the x–y space to the joint angle space, using the Jacobian of the rigid pseudo-link. An internal control loop, based on measurements of the error signals supplied by the position encoder, has been complemented with the control signals provided by the inverse Jacobian. The control strategy has been proved to be effective in the x–y end-point control. It is necessary to extend the strategy to the 2D case, for which it will be required to apply the multivariable control schemes discussed in this chapter to flexible manipulators.

Chapter 14

Modelling and control of space manipulators with flexible links

K. Senda

This chapter describes modelling and control of space manipulators with flexible links. The chapter begins with the mathematical model of space manipulators, highlighting differences between control problems of flexible manipulators in space and on the ground. A methodology of stable manipulation-variable feedback control of space manipulators with flexible links for positioning control to a static target and continuous path tracking control is discussed. To avoid instability of direct manipulation-variable feedback, a virtual rigid manipulator (VRM) concept is introduced and a pseudo-resolved-motion-rate control (pseudo-RMRC) for flexible manipulators is derived from the RMRC for rigid manipulators. Using the VRM, other controls for rigid manipulators are extended to those for flexible manipulators, which can be transformed into joint-variable feedback controls, and are robust stable. For the path tracking control, their orbital stability is discussed in terms of the singular perturbation method. Numerical simulations and hardware experiments of flexible space manipulators successfully demonstrate the effectiveness and feasibility of the proposed method.

14.1 Introduction

In case of the shuttle remote manipulator system (RMS), approximately one-third of the time in operating the RMS is spent in waiting for vibrations to decay (Longman and Lindberg, 1990). The lightweight requirement is also strict for smaller manipulators on the ground, and their structural flexibility becomes problematic. Some control methods for flexible manipulators are realised as control of joint variables, but control of manipulation variables is really desirable. For space manipulators, manipulation-variable controls are especially essential. Hence, stable manipulation-variable feedback controls are required for flexible manipulators, whereas the direct manipulation-variable feedback easily becomes unstable (Murotsu *et al.*, 1990) because it does not satisfy the so-called collocation conditions.

This chapter proposes a methodology for designing stable manipulation-variable feedback control for flexible manipulators, which enables positioning control to a static target and tracking control in a continuous path. The proposed methods are also effective for flexible manipulators with fixed base because the equation of motion of the space manipulators has the same form as that of flexible manipulators with fixed base. This will be explained later in this chapter.

A controlled plant is considered in this chapter as a space manipulator with flexible links mounted on a free-floating satellite vehicle as illustrated in Figure 14.1. This is similar to the Shuttle RMS. The situation illustrated in Figure 14.2 may be observed because the position and attitude of the free-floating satellite base changes when the manipulator moves. Figure 14.2(a) and (b) shows the motion in joint-variable space with (θ_1, θ_2) and the path of the hand in the manipulation-variable space, respectively. Assume that the hand has tracked path (i) and reached the desired manipulation variables in Figure 14.2(b) when the manipulator has tracked trajectory (i) in Figure 14.2(a) without elastic deflection (vibration) in the links.

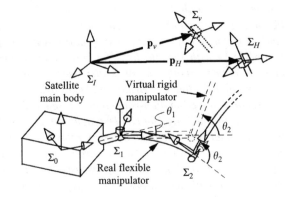

Figure 14.1   Space manipulator with flexible links and virtual rigid manipulator

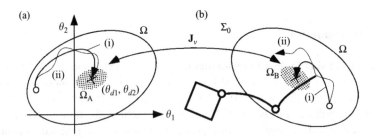

Figure 14.2   Difference between joint variable control and manipulation variable control: (a) joint variable trajectory, (b) manipulation variable path

The desired joint variables (θ_{d1}, θ_{d2}) are finally to be achieved when an appropriate joint-variable control is applied even if the joint variables have tracked trajectory (ii) in Figure 14.2(a) caused by vibration in the links, modelling error, disturbance, and so on. But the desired manipulation variable must not realise, for example, path (ii) in Figure 14.2(b) even though the final joint variables are the same as those of trajectory (i) since the manipulation variables are dependent on the history of the joint and vibration variables as well as the instantaneous joint variables. This is a similar phenomenon as the one pointed out by Longman (1990), which is that the structural vibration in the satellite-mounted manipulator would be attitude disturbances trying to tumble the spacecraft. Therefore, it is not the joint-variable error but manipulation-variable error that must be fed back in the manipulator control. However, it is hard to realise robustly stable controllers because the manipulation-variable error feedback does not satisfy the so-called collocation condition of inputs and outputs.

A few studies on flexible space manipulators have been reported (Komatsu *et al.*, 1990; Longman and Lindberg, 1990; Murotsu *et al.*, 1990, 1992), but two similar problems exist, for example, the control of rigid space robots (Dubowsky and Papadopoulos, 1993; Longman and Lindberg, 1990) and slewing control of satellites with flexible appendages (Azam *et al.*, 1992; Junkins and Bang, 1993). In the slewing control, the spacecraft attitude can be controlled directly by the attitude control system mounted on the spacecraft. Hence, the latter is different from the problem addressed in this chapter, where the manipulation variables must be controlled indirectly by the joint input torques.

The basic strategy for designing controllers and the structure of this chapter are as follows. The equation of motion of a space manipulator with flexible links is formulated in Section 14.2. A virtual rigid manipulator (VRM) concept is then introduced in Section 14.3. The kinematic relations of a real flexible manipulator and its VRM are discussed in detail. Kinematic relations are essential for the stability of following manipulation-variable feedback controls. Section 14.4 shows asymptotic stability of a proportional, derivative (PD) control applied to the original non-linear system, where joint variables are directly fed back to joint input torques using collocated sensors and actuators. Asymptotic stability of the PD-control yields a basis for stability of subsequent control methods in this chapter. In Section 14.5 a pseudo-resolved-motion-rate control (pseudo-RMRC) for flexible manipulators is derived from the RMRC for rigid manipulators. The pseudo-RMRC with manipulation-variable feedback of the VRM can be transformed into PD-control of joint variables satisfying the collocation condition. Using the VRM, other controls for rigid manipulators are extended to those for flexible manipulators, for example, an extended local PD-control and a pseudo-resolved-acceleration control. Those manipulation-variable feedback controls using VRM are robustly stable. Combining these control approaches with a reduced-order modal control yields a composite control. The asymptotic stability of the proposed control schemes is verified for position control to the static target. Orbital stability for continuous path tracking via a singular perturbation method is also discussed in this section. In Section 14.6, numerical simulations and hardware experiments of space flexible manipulators successfully demonstrate the effectiveness and feasibility of the proposed methods. Section 14.7 provides an overall summary of the chapter. Details of some omitted theorems and proofs can be found in Senda (1993).

14.2 Model of flexible manipulators

A typical formulation of equations of motion for flexible manipulators on the ground has been presented by Book (1984). The formulation for flexible manipulators in space is given below.

The system illustrated in Figure 14.1 consists of a main body satellite and manipulators composed of n flexible or rigid links connected in the form of an open-loop chain. The system is floating and does not contact the environment. External forces and torques are not applied to the system, and thus linear and angular momentum conservation strictly holds. Links are connected by revolute joints with 1 degree of freedom (DOF) and form a tree-like open-loop system.

The satellite base is named link 0. Links of the manipulator are numbered in sequential order from link 1 to link n. Joints are similarly numbered in sequence. For example, the joint between link $i-1$ and link i is referred to as joint i. As illustrated in Figure 14.1, the inertial reference coordinate frame is Σ_I, and its origin O_I is located at the centre of mass of the whole system. The origin O_i of link i in the coordinate frame Σ_i is fixed to each link at joint i. The third axis $\mathbf{k}_i$ corresponds to that of joint i. The origin O_0 of link 0 in the fixed coordinate frame Σ_0 is located at its centre of mass.

A coordinate frame Σ_i has orthogonal basis $\Sigma_i^T = [\mathbf{i}_i \quad \mathbf{j}_i \quad \mathbf{k}_i]$ whose third axis $\mathbf{k}_i$ corresponds to the rotation axis of joint i. A direction cosine matrix ${}^I\mathbf{R}_i$ denotes attitude of Σ_i relative to Σ_I as

$$\Sigma_i^T = \Sigma_I^{T}{}^I\mathbf{R}_i$$

For an arbitrary vector $\mathbf{a}$, its coordinate vector ${}^i\mathbf{a}$ expressed in Σ_i satisfies

$$\mathbf{a} = \Sigma_i^{T}{}^i\mathbf{a}$$

where superscript I on the left shoulder for Σ_I can be omitted.

Let $\mathbf{p}_0$ be the position vector of O_0 from O_I, $\mathbf{v}_0$ its velocity, $\boldsymbol{\theta}_0$ the attitude vector, for example, Eulerian angles, and $\boldsymbol{\omega}_0$ the angular velocity.

The elastic deflection of a link is described by a finite number of vibration modes, for example, component modes (Guyan, 1965; Hurty, 1965), based on the assumption of small elastic motion. For example, the translational flexible motion of point X_i in link i is described by $\boldsymbol{\delta}_i(\mathbf{x}_i, t)$, where the point was at $\mathbf{x}_i$ from O_i before link i was elastically deformed, as

$$
{}^i\boldsymbol{\delta}_i(\mathbf{x}_i, t) = \sum_{j=1}^{m_i} {}^i\boldsymbol{\phi}_{ij}(\mathbf{x}_i)\xi_{ij}(t) = {}^i\boldsymbol{\Phi}_i(\mathbf{x}_i)\boldsymbol{\xi}_i(t) \tag{14.1}
$$

where t represents time, $\boldsymbol{\phi}_{ij}(\mathbf{x}_i)$ translational displacement function of mode j of deformation of link i, $\xi_{ij}(t)$ modal coordinate of mode j of link i and m_i the number of mode shape functions expressing deformation of link i, $m = \Sigma_{i=0}^n m_i$ total number of mode functions, and

$$
{}^i\boldsymbol{\Phi}_i(\mathbf{x}_i) = \begin{bmatrix} {}^i\boldsymbol{\phi}_{i1}(\mathbf{x}_i) & \cdots & {}^i\boldsymbol{\phi}_{im_i}(\mathbf{x}_i) \end{bmatrix}, \qquad \boldsymbol{\xi}_i(t) = \begin{bmatrix} \xi_{i1}(t) & \cdots & \xi_{im_i}(t) \end{bmatrix}^T
$$

Hence the position $\mathbf{p}_i(\mathbf{x}_i)$ of point X_i from O_I expressed in Σ_I is

$$
{}^I\mathbf{p}_i(\mathbf{x}_i) = {}^I\mathbf{p}_0 + {}^I\mathbf{R}_0 \left({}^0\mathbf{l}_0 + {}^0\boldsymbol{\delta}_0 \left(\mathbf{l}_0 \right) + {}^0\hat{\mathbf{R}}_0{}^0\mathbf{R}_1 \left({}^1\mathbf{l}_1 + \cdots + {}^{i-1}\hat{\mathbf{R}}_{i-1}{}^{i-1}\mathbf{R}_i \left({}^i\mathbf{x}_i + {}^i\boldsymbol{\delta}_i \left(\mathbf{x}_i \right) \right) \cdots \right) \right)
$$

(14.2)

where ${}^{i-1}\mathbf{R}_i \left(\theta_i \right)$ is the rotational transformation matrix or the direction cosine matrix denoting rotation of joint i, θ_i is angle of joint i, $\mathbf{l}_i$ the position vector of O_{i+1} from O_i, ${}^i\hat{\mathbf{R}}_i \left(\boldsymbol{\psi}_i \left(\mathbf{l}_i \right) \right)$ the rotational transformation matrix denoting rotational deformation of link i at O_{i+1} and $\boldsymbol{\psi}_i \left(\mathbf{l}_i \right)$ the rotational deflection vector at O_{i+1} caused by deformation of link i. To simplify the description, time t is omitted whereas $\mathbf{p}_i \left(\mathbf{x}_i \right)$ and others are explicit functions of t. The velocity $\mathbf{v}_i \left(\mathbf{x}_i \right)$ of point X_i is obtained as the time derivative of equation (14.2). Let dm_{xi} and dv_{xi} represent the differential mass and the differential volume at $\mathbf{x}_i$. Thus, kinetic energy T and elastic potential V of the whole system are

$$
T(\boldsymbol{\theta}, \boldsymbol{\xi}, \boldsymbol{\omega}_0, \mathbf{v}_0, \dot{\boldsymbol{\theta}}, \dot{\boldsymbol{\xi}}) = \frac{1}{2} \sum_{i=0}^{n} \int_{\text{link } i} \mathbf{v}_i \cdot \mathbf{v}_i \, dm_{xi}
$$

(14.3)

$$
V(\boldsymbol{\xi}) = \frac{1}{2} \sum_{i=0}^{n} \int_{\text{link } i} \boldsymbol{\varepsilon}_i \cdot \boldsymbol{\sigma}_i dv_{xi}
$$

(14.4)

where $\boldsymbol{\theta} = \begin{bmatrix} \theta_1 & \cdots & \theta_n \end{bmatrix}^{\mathrm{T}} \in \mathbf{R}^{n \times 1}$ is the joint-variable vector, $\boldsymbol{\xi} = \begin{bmatrix} \boldsymbol{\xi}_0^{\mathrm{T}} & \cdots & \boldsymbol{\xi}_n^{\mathrm{T}} \end{bmatrix}^{\mathrm{T}} \in \mathbf{R}^{m \times 1}$ the modal coordinate vector, $\boldsymbol{\varepsilon}_i \left(\boldsymbol{\xi}_i \right)$ the strain vector and $\boldsymbol{\sigma}_i \left(\boldsymbol{\xi}_i \right)$ the stress vector. The equation of motion for the system is obtained by the Lagrangian method as

$$
\mathbf{M}_0 \ddot{\mathbf{q}}_0 + \mathbf{h}_0 + \mathbf{K}_0 \mathbf{q}_0 = \mathbf{f}_0
$$

(14.5)

where $\mathbf{M}_0 \in \mathbf{R}^{(N+6) \times (N+6)}$ is the inertia matrix, $\mathbf{K}_0 \in \mathbf{R}^{(N+6) \times (N+6)}$ the stiffness matrix, $\mathbf{q}_0 = \begin{bmatrix} \boldsymbol{\theta}_0^{\mathrm{T}} & \mathbf{p}_0^{\mathrm{T}} & \boldsymbol{\theta}^{\mathrm{T}} & \boldsymbol{\xi}^{\mathrm{T}} \end{bmatrix}^{\mathrm{T}} \in \mathbf{R}^{(N+6) \times 1}$, $\mathbf{h}_0 \in \mathbf{R}^{(N+6) \times 1}$ the centrifugal force and Coriolis force, $\mathbf{f}_0 \in \mathbf{R}^{(N+6) \times 1}$ the generalized force corresponding to $\mathbf{q}_0$, $m = \sum_{i=1}^{n} m_i$ and $N = n + m$. Equation (14.5) is different for manipulators with a fixed base because the generalized coordinates contain the position $\mathbf{p}_0$ and attitude $\boldsymbol{\theta}_0$ of the satellite base, which are cyclic.

Assume that the translational momentum $\mathbf{P}$ and the angular momentum $\mathbf{L}$ about O_I are conserved as

$$
\left. \begin{array}{l} \mathbf{P}(\boldsymbol{\theta}, \boldsymbol{\xi}, \boldsymbol{\omega}_0, \mathbf{v}_0, \dot{\boldsymbol{\theta}}, \dot{\boldsymbol{\xi}}) = \mathbf{0} \\ \mathbf{L}(\boldsymbol{\theta}, \boldsymbol{\xi}, \boldsymbol{\omega}_0, \mathbf{v}_0, \dot{\boldsymbol{\theta}}, \dot{\boldsymbol{\xi}}) = \mathbf{0} \end{array} \right\}
$$

(14.6)

Solving these for angular velocity $\boldsymbol{\omega}_0$ and translational velocity $\mathbf{v}_0$, substituting into equation (14.3) and eliminating $\boldsymbol{\omega}_0$ and $\mathbf{v}_0$ reduce to $T = T(\boldsymbol{\theta}, \boldsymbol{\xi}, \dot{\boldsymbol{\theta}}, \dot{\boldsymbol{\xi}})$. Thus, the equation of motion for the system is obtained by the Lagrangian method as

$$
\mathbf{M}\ddot{\mathbf{q}} + \mathbf{h} + \mathbf{K}\mathbf{q} = \mathbf{B}_0 \boldsymbol{\tau}
$$

(14.7)

where

$$q = \begin{bmatrix} \theta \\ \xi \end{bmatrix}, \qquad h = \dot{M}\dot{q} - \frac{\partial}{\partial q}\left[\frac{1}{2}\dot{q}^T M\dot{q} \right] = \begin{bmatrix} h_1 \\ h_2 \end{bmatrix}$$

$$M = \begin{bmatrix} M_{11} & M_{12} \\ M_{12}^T & M_{22} \end{bmatrix}, \qquad K = \begin{bmatrix} 0 & 0 \\ 0 & K_\xi \end{bmatrix}, \qquad B_0 = \begin{bmatrix} I^{n \times n} \\ 0 \end{bmatrix}$$

$M \in R^{N \times N}$, $K_\xi \in R^{m \times m}$, $h \in R^{N \times 1}$ and $\tau \in R^{N \times 1}$ is the joint input torque vector. The equation of motion for the space manipulators with flexible links is obtained in the same form as that for the base-fixed manipulators by containing the linear and angular momentum conservation. Since ω_0 and v_0 can be computed by the non-holonomic constraint, equation (14.6), the manipulation variables depend on the trajectories of the joint variables and the elastic deformations. Equation (14.7) is considered as the controlled plant in this chapter. Hence all results are also effective for base-fixed flexible manipulators.

The inertial matrix M and the stiffness matrix K_ξ have the properties

$$M^T = M > 0, \qquad K_\xi^T = K_\xi > 0 \tag{14.8}$$

Note that the following discussion does not need concrete terms of the equations of motion, equation (14.7), while the forms and the natural properties in equation (14.8) are assumed. In order to clarify that all the vibration modes are stabilised irrespective of damping, the system is considered in this chapter without natural damping.

14.3   VRM concept

14.3.1   *Definition of VRM*

In this section, the VRM concept is introduced to derive stable manipulation variable feedback controllers for flexible manipulators (Senda, 1993; Senda and Murotsu, 1994b). As illustrated in Figure 14.1, VRM is defined as a hypothetical rigid manipulator that constitutes the un-deformed links and the same joint angles of the real flexible manipulator (RFM).

Kinematic relations of RFM and VRM are essential for the stability of manipulation feedback controls introduced later. The kinematic relations are summarised below, where these will be discussed in detail in Section 14.3.2.

The position vector p_H of the RFM hand, expressed in Σ_I, is derived using equation (14.2) as

$$^I p_H = {}^I p_v + {}^I p_\xi + o(\xi) \tag{14.9}$$

where p_v is the position vector of the VRM hand, p_ξ the relative position of the RFM hand from the VRM hand that is caused by an elastic deformation ξ, and $o(\xi)$ the higher-order infinitesimal of ξ. In the same manner, the rotational transformation matrix $^I R_H$ denoting the orientation of the RFM hand is given as

$$^I R_H = {}^I R_0 {}^0 \hat{R}_0 {}^0 R_1 {}^1 \hat{R}_1 \cdots {}^{n-1} R_n {}^n \hat{R}_n = {}^I \hat{R}_\xi {}^I R_v + o(\xi) \tag{14.10}$$

where $^I\mathbf{R}_v$ is a rotational transformation matrix denoting the orientation of VRM hand and $^I\hat{\mathbf{R}}_\xi$ is a rotational transformation matrix expressing the relative orientative deflection of the RFM hand to the VRM hand that is caused by the elastic deformation $\boldsymbol{\xi}$.

Assume that the linear and angular momenta of the VRM about the centre of mass of the system are conserved in the state of zero because the elastic deformation is small. The following differential relation is thus obtained from equations (14.9) and (14.10):

$$d^I\mathbf{y}_H = \mathbf{J}_v(\boldsymbol{\theta}_0, \boldsymbol{\theta})d\boldsymbol{\theta}_e + \mathbf{J}_\xi(\boldsymbol{\theta}_0, \boldsymbol{\theta})d\boldsymbol{\xi} \tag{14.11}$$

where $d^I\mathbf{y}_H = [\,d^I\mathbf{p}_H^T \quad d^I\boldsymbol{\vartheta}_H^T\,]^T$, $d\boldsymbol{\vartheta}_H^T$ differential vector satisfying $\boldsymbol{\omega}_H = d\boldsymbol{\vartheta}_H/dt$ for angular velocity $\boldsymbol{\omega}_H$ of RFM hand, $\mathbf{J}_v$ the generalised Jacobian matrix (Umetani and Yoshida, 1989) for VRM, and $\mathbf{J}_\xi$ the Jacobian matrix relating $\boldsymbol{\xi}$ to $\mathbf{y}_H$. For a base-fixed RFM on the ground, an equation equivalent to equation (14.11) holds, whereas it is independent of $\boldsymbol{\theta}_0$.

14.3.2 Kinematic relations of RFM and VRM

The coordinate vector $^i\boldsymbol{\psi}_i(\mathbf{x}_i, t) = [\psi_{ix} \quad \psi_{iy} \quad \psi_{iz}]^T$ expressed in Σ_i is the rotational deflection of point X_i at $\mathbf{x}_i$ from O_i caused by deformation of link i, where ψ_{iy} is a rotation angle about $\mathbf{j}_i$ of Σ_i, for example. Because small elastic deformation is considered, ψ_{ix}, ψ_{iy} and ψ_{iz} are equivalent to the 1–2–3 Eulerian angles or the roll–pitch–yaw angles. As in equation (14.1), these can be described by

$$^i\boldsymbol{\psi}_i(\mathbf{x}_i, t) = \sum_{j=1}^{m_i} {}^i\boldsymbol{\psi}_{ij}(\mathbf{x}_i)\xi_{ij}(t) = {}^i\boldsymbol{\Psi}_i(\mathbf{x}_i)\boldsymbol{\xi}_i(t) \tag{14.12}$$

where $^i\boldsymbol{\psi}_{ij}(\mathbf{x}_i)$ is rotational displacement function of mode j of link i, and

$$^i\boldsymbol{\Psi}_i(\mathbf{x}_i) = [\,{}^i\boldsymbol{\psi}_{i1}(\mathbf{x}_i) \quad \cdots \quad {}^i\boldsymbol{\psi}_{imi}(\mathbf{x}_i)\,]$$

The following simple descriptions are used for translational and rotational deflections at $\mathbf{x}_i = \mathbf{l}_i$:

$$^i\boldsymbol{\delta}_i = {}^i\boldsymbol{\delta}_i(\mathbf{l}_i, t) = {}^i\boldsymbol{\Phi}_i(\mathbf{l}_i)\boldsymbol{\xi}_i(t), \qquad {}^i\boldsymbol{\psi}_i = {}^i\boldsymbol{\psi}_i(\mathbf{l}_i, t) = {}^i\boldsymbol{\Psi}_i(\mathbf{l}_i)\boldsymbol{\xi}_i(t) \tag{14.13}$$

The position vector $\mathbf{p}_H$ of the RFM hand expressed in Σ_I is obtained from equation (14.2) as

$$^I\mathbf{p}_H = {}^I\mathbf{p}_0 + {}^I\mathbf{R}_0 \left({}^0\mathbf{l}_0 + {}^0\boldsymbol{\delta}_0 + {}^0\hat{\mathbf{R}}_0{}^0\mathbf{R}_1 \left({}^1\mathbf{l}_1 + \cdots + {}^{n-1}\hat{\mathbf{R}}_{n-1}{}^{n-1}\mathbf{R}_n \left({}^n\mathbf{l}_n + {}^n\boldsymbol{\delta}_n \right) \cdots \right) \right)$$

$$= {}^I\mathbf{p}_0 + {}^I\mathbf{R}_0 \left({}^0\mathbf{l}_0 + {}^0\hat{\mathbf{R}}_0{}^0\mathbf{R}_1 \left({}^1\mathbf{l}_1 + \cdots \left({}^{n-1}\mathbf{l}_{n-1} + {}^{n-1}\hat{\mathbf{R}}_{n-1}{}^{n-1}\mathbf{R}_n{}^n\mathbf{l}_n \right) \cdots \right) \right)$$

$$+ {}^I\mathbf{R}_0 \left({}^0\boldsymbol{\delta}_0 + {}^0\hat{\mathbf{R}}_0{}^0\mathbf{R}_1 \left({}^1\boldsymbol{\delta}_1 + \cdots \left({}^{n-1}\boldsymbol{\delta}_{n-1} + {}^{n-1}\hat{\mathbf{R}}_{n-1}{}^{n-1}\mathbf{R}_n{}^n\boldsymbol{\delta}_n \right) \cdots \right) \right)$$

$$\tag{14.14}$$

Since the rotational deflection $\boldsymbol{\psi}_i$ (l_i) is small, ${}^i\hat{\mathbf{R}}_i(\boldsymbol{\psi}_i)$ satisfies

$$ {}^i\hat{\mathbf{R}}_i(\boldsymbol{\psi}_i) = \mathbf{I} + [{}^i\boldsymbol{\psi}_i \times] \tag{14.15} $$

where $\mathbf{I} \in \mathbf{R}^{3\times3}$ is the identity matrix, $[{}^i\boldsymbol{\psi}_i \times]$ is an outer product matrix;

$$ [{}^i\boldsymbol{\psi}_i \times] = \begin{bmatrix} 0 & -\psi_{iz} & \psi_{iy} \\ \psi_{iz} & 0 & -\psi_{ix} \\ -\psi_{iy} & \psi_{ix} & 0 \end{bmatrix} $$

satisfying the following relation for any vector ${}^i\mathbf{a} \in \mathbf{R}^{3\times1}$:

$$ {}^i\boldsymbol{\psi}_i \times {}^i\mathbf{a} = [{}^i\boldsymbol{\psi}_i \times]{}^i\mathbf{a} $$

Hence, the second term on the right-hand side of equation (14.14) can be rearranged as

$$ {}^I\mathbf{R}_0 \left({}^0l_0 + [\mathbf{I} + {}^0\boldsymbol{\psi}_0 \times]{}^0\mathbf{R}_1 \left({}^1l_1 + \cdots \left({}^{n-1}l_{n-1} + [\mathbf{I} + {}^{n-1}\boldsymbol{\psi}_{n-1} \times]{}^{n-1}\mathbf{R}_n{}^nl_n \right) \cdots \right) \right) $$

$$ = {}^I\mathbf{p}_{v,0} + {}^I\boldsymbol{\psi}_0 \times {}^I\mathbf{p}_{v,1} + \cdots + {}^I\boldsymbol{\psi}_{n-1} \times {}^I\mathbf{p}_{v,n} + \mathbf{o}(\boldsymbol{\psi}) \tag{14.16} $$

where

$$ {}^I\mathbf{p}_{v,i} = {}^I\mathbf{p}_0 + {}^I\mathbf{R}_0 \left({}^0l_0 + {}^0\hat{\mathbf{R}}_0{}^0\mathbf{R}_1 \left({}^1l_1 + \cdots \left({}^{n-1}l_{n-1} + {}^{n-1}\hat{\mathbf{R}}_{n-1}{}^{n-1}\mathbf{R}_n{}^nl_n \right) \cdots \right) \right) $$

position vector of VRM hand from O_i expressed in Σ_I

$$ {}^I\mathbf{R}_i = {}^I\mathbf{R}_0(\theta_0){}^0\mathbf{R}_1(\theta_1) \cdots {}^{i-1}\mathbf{R}_i(\theta_i) $$

In the same manner, the third term on the right-hand side of equation (14.14) is rearranged as

$$ {}^I\mathbf{R}_0{}^0\boldsymbol{\delta}_0 + {}^I\mathbf{R}_1{}^1\boldsymbol{\delta}_1 + \cdots + {}^I\mathbf{R}_n{}^n\boldsymbol{\delta}_n + \mathbf{o}(\boldsymbol{\delta}) = {}^I\boldsymbol{\delta}_0 + {}^I\boldsymbol{\delta}_1 + \cdots + {}^I\boldsymbol{\delta}_n + \mathbf{o}(\boldsymbol{\delta}) \tag{14.17} $$

Substituting equations (14.16) and (14.17) into equation (14.14) yields

$$ {}^I\mathbf{p}_H = {}^I\mathbf{p}_v + {}^I\mathbf{p}_\xi + \mathbf{o}(\boldsymbol{\xi}) = {}^I\mathbf{p}_0 + {}^I\mathbf{p}_{v,0} + {}^I\mathbf{p}_\xi + \mathbf{o}(\boldsymbol{\xi}) \tag{14.18} $$

that is, equation (14.9), where $\mathbf{p}_\xi$ is obtained by using equations (14.14)–(14.17) as

$$ {}^I\mathbf{p}_\xi = {}^I\boldsymbol{\delta}_0 + {}^I\boldsymbol{\delta}_1 + \cdots + {}^I\boldsymbol{\delta}_n - {}^I\mathbf{p}_{v,1} \times {}^I\boldsymbol{\psi}_0 - \cdots - {}^I\mathbf{p}_{v,n} \times {}^I\boldsymbol{\psi}_{n-1} $$

$$ = {}^I\boldsymbol{\Phi}_0\boldsymbol{\xi}_0 + {}^I\boldsymbol{\Phi}_1\boldsymbol{\xi}_1 + \cdots + {}^I\boldsymbol{\Phi}_n\boldsymbol{\xi}_n - \left[{}^I\mathbf{p}_{v,1} \times\right]{}^I\boldsymbol{\Psi}_0\boldsymbol{\xi}_0 - \cdots - \left[{}^I\mathbf{p}_{v,n} \times\right]{}^I\boldsymbol{\Psi}_{n-1}\boldsymbol{\xi}_{n-1} $$

$$ = \left[\left({}^I\boldsymbol{\Phi}_0 - \left[{}^I\mathbf{p}_{v,1} \times\right]{}^I\boldsymbol{\Psi}_0\right) \cdots \left({}^I\boldsymbol{\Phi}_{n-1} - \left[{}^I\mathbf{p}_{v,n} \times\right]{}^I\boldsymbol{\Psi}_{n-1}\right) {}^I\boldsymbol{\Phi}_n \right]\boldsymbol{\xi} $$

$$ \equiv \mathbf{J}_{p\xi}(\boldsymbol{\theta}_0, \boldsymbol{\theta})\boldsymbol{\xi} \tag{14.19} $$

where ${}^I\boldsymbol{\Phi}_j = {}^I\mathbf{R}_j{}^j\boldsymbol{\Phi}_j$ and ${}^I\boldsymbol{\Psi}_j = {}^I\mathbf{R}_j{}^j\boldsymbol{\Psi}_j$. Equation (14.9) or (14.18) shows that the position of the RFM hand is given as the position of the VRM hand added with the displacements caused by the elastic deformation.

In the same manner, the direction cosine matrix $^I\mathbf{R}_H$ of equation (14.10) representing the orientation of the RFM hand is obtained as

$$
\begin{aligned}
^I\mathbf{R}_H &= {}^I\mathbf{R}_0{}^0\hat{\mathbf{R}}_0{}^1\mathbf{R}_1{}^1\hat{\mathbf{R}}_1 \cdots {}^{n-1}\mathbf{R}_n{}^n\hat{\mathbf{R}}_n \\
&= {}^I\mathbf{R}_0[\mathbf{I} + {}^0\boldsymbol{\psi}_0\times]{}^0\mathbf{R}_1[\mathbf{I} + {}^1\boldsymbol{\psi}_1\times] \cdots {}^{n-1}\mathbf{R}_n[\mathbf{I} + {}^n\boldsymbol{\psi}_n\times] \\
&= {}^I\mathbf{R}_0{}^0\mathbf{R}_1{}^1\mathbf{R}_2 \cdots {}^{n-1}\mathbf{R}_n \\
&\quad + {}^I\mathbf{R}_0[{}^0\boldsymbol{\psi}_0\times]{}^0\mathbf{R}_1{}^1\mathbf{R}_2 \cdots {}^{n-1}\mathbf{R}_n \\
&\quad + {}^I\mathbf{R}_0{}^0\mathbf{R}_1[{}^1\boldsymbol{\psi}_1\times]{}^1\mathbf{R}_2 \cdots {}^{n-1}\mathbf{R}_n \\
&\quad + \cdots + {}^I\mathbf{R}_0{}^0\mathbf{R}_1{}^1\mathbf{R}_2 \cdots {}^{n-1}\mathbf{R}_n[{}^n\boldsymbol{\psi}_n\times] + \mathbf{o}(\boldsymbol{\psi}) \\
&= [\mathbf{I} + {}^I\boldsymbol{\vartheta}_\xi\times]{}^I\mathbf{R}_v + \mathbf{o}(\boldsymbol{\xi}) \\
&\equiv {}^I\hat{\mathbf{R}}_\xi{}^I\mathbf{R}_v + \mathbf{o}(\boldsymbol{\xi}) = {}^I\hat{\mathbf{R}}_\xi{}^I\mathbf{R}_0{}^0\mathbf{R}_v + \mathbf{o}(\boldsymbol{\xi})
\end{aligned}
\tag{14.20}
$$

where $^I\mathbf{R}_v = {}^I\mathbf{R}_n$ and

$$
^I\boldsymbol{\vartheta}_\xi = {}^I\boldsymbol{\psi}_0 + {}^I\boldsymbol{\psi}_1 + \cdots + {}^I\boldsymbol{\psi}_n = [{}^I\boldsymbol{\psi}_0 \quad {}^I\boldsymbol{\psi}_1 \quad \cdots \quad {}^I\boldsymbol{\psi}_n]\boldsymbol{\xi} = \mathbf{J}_{\vartheta\xi}(\boldsymbol{\theta}_0, \boldsymbol{\theta})\boldsymbol{\xi}
\tag{14.21}
$$

The above equation shows that the orientation of the RFM hand is the orientation of the VRM hand rotated by the sum of the rotational deflections caused by the elastic deformation.

The following remark is reasonable to make from the discussions above.

Remark 14.1: The manipulation variables, that is, the position and the orientation, of the RFM are manipulation variables of VRM added with the manipulation variable components caused by the elastic deformation $\boldsymbol{\xi}$. The manipulation variables of VRM can be computed from attitude angles $\boldsymbol{\theta}_0$ of link 0 and joint-variables $\boldsymbol{\theta}$ of RFM. Equations (14.19) and (14.21) show that the manipulation variable components caused by the elastic deformation are given by the Jacobian matrices $\mathbf{J}_{p\xi}$ and $\mathbf{J}_{\theta\xi}$ multiplied by the deformation variables $\boldsymbol{\xi}$, where the Jacobian matrices are functions of only $\boldsymbol{\theta}_0$ and $\boldsymbol{\theta}$.

Differential relations between the RFM and the VRM are derived from the above results. Some differential relations for equations (14.9) and (14.18) are derived below. The following equations for a differential of VRM hand position vector $\mathbf{p}_v$ are derived as well as those of a rigid manipulator:

$$
\begin{aligned}
d^I\mathbf{p}_v &= d^I\mathbf{p}_0 + d^I\boldsymbol{\vartheta}_0 \times {}^I\mathbf{p}_{v,0} + d\theta_1{}^I\mathbf{k}_1 \times {}^I\mathbf{p}_{v,1} + \cdots + d\theta_n{}^I\mathbf{k}_n \times {}^I\mathbf{p}_{v,n} \\
&= \begin{bmatrix} \mathbf{I} & [{}^I\mathbf{p}_{v,0}\times]^\mathrm{T} \end{bmatrix} \begin{bmatrix} d^I\mathbf{p}_0 \\ d^I\boldsymbol{\vartheta}_0 \end{bmatrix} + \begin{bmatrix} {}^I\mathbf{k}_1 \times {}^I\mathbf{p}_{v,1} & \cdots & {}^I\mathbf{k}_n \times {}^I\mathbf{p}_{v,n} \end{bmatrix} d\boldsymbol{\theta} \quad (14.22) \\
&\equiv \mathbf{J}_{p0}(\boldsymbol{\theta}_0, \boldsymbol{\theta})d\mathbf{q}_0 + \mathbf{J}_{p\theta}(\boldsymbol{\theta}_0, \boldsymbol{\theta})d\boldsymbol{\theta}
\end{aligned}
$$

where $d\mathbf{q}_0 \equiv [\, d^l\mathbf{p}_0^\mathrm{T} \quad d^l\boldsymbol{\vartheta}_0^\mathrm{T}\,]^\mathrm{T}$, $\boldsymbol{\vartheta}_i$ is the pseudo-coordinate vector defining the angular velocity of Σ_i with respect to Σ_I as

$$^I\boldsymbol{\omega}_i = \frac{d^l\boldsymbol{\vartheta}_i}{dt} \quad (i = 1, 2, \ldots, n) \tag{14.23}$$

Equation (14.19) gives the differential of the displacement $\mathbf{p}_\xi$ caused by the elastic deformation.

$$d^l\mathbf{p}_\xi = \mathbf{J}_{p\xi}(\boldsymbol{\theta}_0, \boldsymbol{\theta})d\boldsymbol{\xi} \tag{14.24}$$

Substituting equations (14.22) and (14.24) into equation (14.9) or (14.18) yields the differential of RFM hand position $\mathbf{p}_H$ as

$$d^l\mathbf{p}_H = d^l\mathbf{p}_v + d^l\mathbf{p}_\xi = \mathbf{J}_{p0}(\boldsymbol{\theta}_0, \boldsymbol{\theta})d\mathbf{q}_0 + \mathbf{J}_{p\theta}(\boldsymbol{\theta}_0, \boldsymbol{\theta})d\boldsymbol{\theta} + \mathbf{J}_{p\xi}(\boldsymbol{\theta}_0, \boldsymbol{\theta})d\boldsymbol{\xi} \tag{14.25}$$

Some differential relations for equation (14.10) or (14.20) are derived below. The derivative of the direction cosine matrix $^I\mathbf{R}_H$ under the definition of equation (14.23) is

$$d^l\mathbf{R}_H = [d^l\boldsymbol{\vartheta}_H \times]^I\mathbf{R}_H \tag{14.26}$$

where $d^l\boldsymbol{\vartheta}_H$ is the differential rotation of the RFM hand that satisfies the following relation with $d^l\boldsymbol{\vartheta}_v$ of VRM hand and $d^l\boldsymbol{\vartheta}_\xi$ caused by the elastic deformation:

$$d^l\boldsymbol{\vartheta}_H = d^l\boldsymbol{\vartheta}_v + d^l\boldsymbol{\vartheta}_\xi = d^l\boldsymbol{\vartheta}_0 + d^l\boldsymbol{\vartheta}_{v,0} + d^l\boldsymbol{\vartheta}_\xi \tag{14.27}$$

The differential rotation $d^l\boldsymbol{\vartheta}_v$ of VRM hand is obtained as well as that of a rigid manipulator as

$$\begin{aligned}
d^l\boldsymbol{\vartheta}_v &= d^l\boldsymbol{\vartheta}_0 + d\theta_1\,^I\mathbf{k}_1 + \cdots + d\theta_n\,^I\mathbf{k}_n \\
&= \begin{bmatrix} 0 & \mathbf{I} \end{bmatrix} d\mathbf{q}_0 + \begin{bmatrix} ^I\mathbf{k}_1 & \cdots & ^I\mathbf{k}_n \end{bmatrix} d\boldsymbol{\theta} \\
&\equiv \mathbf{J}_{\vartheta 0}d\mathbf{q}_0 + \mathbf{J}_{\vartheta\theta}(\boldsymbol{\theta}_0, \boldsymbol{\theta})d\boldsymbol{\theta}
\end{aligned} \tag{14.28}$$

The differential rotation $d^l\boldsymbol{\vartheta}_v$ of VRM hand is derived from equation (14.21) as

$$d^l\boldsymbol{\vartheta}_\xi = d^l\boldsymbol{\psi}_0 + d^l\boldsymbol{\psi}_1 + \cdots + d^l\boldsymbol{\psi}_n = [\,^I\boldsymbol{\psi}_0 \quad ^I\boldsymbol{\psi}_1 \quad \cdots \quad ^I\boldsymbol{\psi}_n\,]d\boldsymbol{\xi} = \mathbf{J}_{\vartheta\xi}(\boldsymbol{\theta}_0, \boldsymbol{\theta})d\boldsymbol{\xi} \tag{14.29}$$

Equations (14.27)–(14.29) yield

$$d^l\boldsymbol{\vartheta}_H = d^l\boldsymbol{\vartheta}_v + d^l\boldsymbol{\vartheta}_\xi = \mathbf{J}_{\vartheta 0}d\mathbf{q}_0 + \mathbf{J}_{\vartheta\theta}(\boldsymbol{\theta}_0, \boldsymbol{\theta})d\boldsymbol{\theta} + \mathbf{J}_{\vartheta\xi}(\boldsymbol{\theta}_0, \boldsymbol{\theta})d\boldsymbol{\xi} \tag{14.30}$$

Combining equations (14.25) and (14.30) results in the following differential relation between the RFM and the VRM:

$$d^l\mathbf{y}_H = d^l\mathbf{y}_v + d^l\mathbf{y}_\xi = \mathbf{J}_0(\boldsymbol{\theta}_0, \boldsymbol{\theta})d\mathbf{q}_0 + \mathbf{J}_\theta(\boldsymbol{\theta}_0, \boldsymbol{\theta})d\boldsymbol{\theta} + \mathbf{J}_\xi(\boldsymbol{\theta}_0, \boldsymbol{\theta})d\boldsymbol{\xi} \tag{14.31}$$

where

$$d^l\mathbf{y}_H = \begin{bmatrix} d^l\mathbf{p}_H \\ d^l\boldsymbol{\vartheta}_H \end{bmatrix}, \qquad d^l\mathbf{y}_v = \begin{bmatrix} d^l\mathbf{p}_v \\ d^l\boldsymbol{\vartheta}_v \end{bmatrix}, \qquad d^l\mathbf{y}_\xi = \begin{bmatrix} d^l\mathbf{p}_\xi \\ d^l\boldsymbol{\vartheta}_\xi \end{bmatrix}$$

$$\mathbf{J}_0 = \begin{bmatrix} \mathbf{J}_{p0} \\ \mathbf{J}_{\vartheta 0} \end{bmatrix}, \qquad \mathbf{J}_\theta = \begin{bmatrix} \mathbf{J}_{p\theta} \\ \mathbf{J}_{\vartheta\theta} \end{bmatrix}, \qquad \mathbf{J}_\xi = \begin{bmatrix} \mathbf{J}_{p\xi} \\ \mathbf{J}_{\vartheta\xi} \end{bmatrix}$$

Assume that the linear and angular momenta of the VRM about the centre of mass of the system are conserved in the state of zero because the elastic deformation is small. Equation (14.31) can be rearranged as equation (14.11) under the conservation of momentum:

$$d^l\mathbf{y}_H = d^l\mathbf{y}_v + d^l\mathbf{y}_\xi = \mathbf{J}_v(\boldsymbol{\theta}_0, \boldsymbol{\theta})d\boldsymbol{\theta}_e + \mathbf{J}_\xi(\boldsymbol{\theta}_0, \boldsymbol{\theta})d\boldsymbol{\xi} \tag{14.32}$$

where $\mathbf{J}_v$ is a generalized Jacobian matrix for VRM. For a base-fixed RFM on the ground, the following equation, equivalent to equation (14.11) or (14.32), holds without considering the conservation of momentum:

$$d^l\mathbf{y}_H = d^l\mathbf{y}_v + d^l\mathbf{y}_\xi = \mathbf{J}_v(\boldsymbol{\theta})d\boldsymbol{\theta}_e + \mathbf{J}_\xi(\boldsymbol{\theta})d\boldsymbol{\xi} \tag{14.33}$$

The following remark can be made from equation (14.11) or (14.32) and equation (14.33).

Remark 14.2: Differential manipulation variables $d^l\mathbf{y}_H$ of the RFM are the sum of VRM's $d^l\mathbf{y}_v$ and the components $d^l\mathbf{y}_\xi$ caused by the elastic deformation. Differential manipulation variables $d^l\mathbf{y}_v$ of VRM can be computed from attitude angles $\boldsymbol{\theta}_0$ of link 0 and joint-variables $\boldsymbol{\theta}$ of RFM. The differential manipulation-variable components caused by the elastic deformation are given by the Jacobian matrix $\mathbf{J}_\xi$ multiplied by the differential deformation variables $d\boldsymbol{\xi}$, where the Jacobian matrix is a function of only $\boldsymbol{\theta}_0$ and $\boldsymbol{\theta}$.

14.4 PD-control

14.4.1 PD-control for joint variables

This section discusses the asymptotic stability of a PD-control scheme as a basis for control methods in subsequent sections, where PD-control cannot directly be applied to the manipulation-variable control of space manipulators.

The PD-control for joint variables of the flexible manipulator is given as

$$\boldsymbol{\tau} = -\mathbf{G}_D\dot{\boldsymbol{\theta}}_e - \mathbf{G}_P\boldsymbol{\theta}_e \tag{14.34}$$

where $\boldsymbol{\theta}_e = \boldsymbol{\theta} - \boldsymbol{\theta}_d$ is the error of joint-variable vector, $\boldsymbol{\theta}_d$ the time-invariant desired joint-variable, $\mathbf{G}_P$ and $\mathbf{G}_D$ symmetric positive and definite matrices. The stability conditions of the PD-control are summarised below (Senda, 1993; Senda and Murotsu, 1993).

14.4.2   Stability of linearised system

The asymptotic stability of $\mathbf{q} = \dot{\mathbf{q}} = \mathbf{0}$ controlled by the PD-control has been proven under the assumption that the system in equation (14.7) has natural damping $\mathbf{D}\dot{\mathbf{q}}$, and $\mathbf{D} + \mathbf{B}_0\mathbf{G}_D\mathbf{B}_0^T > 0$ and $\mathbf{K} + \mathbf{B}_0\mathbf{G}_P\mathbf{B}_0^T > 0$ are true (Balas, 1979). While this proof gives a sufficient condition for asymptotic stability, $n \geq N$ must be satisfied when $\mathbf{D} = \mathbf{0}$ holds because $\mathbf{G}_D \in \mathbf{R}^{n \times n}$ and $\mathbf{B}_0 \in \mathbf{R}^{N \times n}$. Since $n < N$ is true, flexible manipulators cannot satisfy the above sufficient condition if natural damping does not exist.

To obtain applicable conditions, the plant in equation (14.7) with the control input in equation (14.34) is linearised about $\boldsymbol{\theta} = \boldsymbol{\theta}_d$ and $\boldsymbol{\xi} = \mathbf{0}$. Decomposing the control input as $\boldsymbol{\tau} = \boldsymbol{\tau}_P + \boldsymbol{\tau}_D$, and transposition of the proportional control term $\boldsymbol{\tau}_P = -\mathbf{G}_P\boldsymbol{\theta}_e$ yields

$$\mathbf{M}\ddot{\mathbf{q}} + \mathbf{K}_P\mathbf{q} = \mathbf{B}_0\boldsymbol{\tau}_D \tag{14.35}$$

where $\mathbf{K}_P = \text{diag}\left[\mathbf{B}_0\mathbf{G}_P\mathbf{B}_0^T, \mathbf{K}_\xi\right] > 0$ and $\mathbf{q}$ is redefined as $\mathbf{q} \equiv [\boldsymbol{\theta}_e^T \quad \boldsymbol{\xi}^T]^T$. The output is $\boldsymbol{\theta} = \mathbf{B}_0^T\mathbf{q}$ or $\dot{\boldsymbol{\theta}} = \mathbf{B}_0^T\dot{\mathbf{q}}$. A non-singular transformation matrix $\boldsymbol{\Phi}_P$ exists satisfying

$$\boldsymbol{\Phi}_P^T\mathbf{M}\boldsymbol{\Phi}_P = \mathbf{I}^{N \times N} \quad \text{and} \quad \boldsymbol{\Phi}_P^T\mathbf{K}\boldsymbol{\Phi}_P = \text{diag}\,[\lambda_1 \cdots \lambda_N] = \boldsymbol{\Lambda}_P$$

where $0 < \lambda_1 \leq \cdots \leq \lambda_N$. Equation (14.35) is diagonally transformed using $\mathbf{q} \equiv \boldsymbol{\Phi}_P\boldsymbol{\chi}$ as

$$\ddot{\boldsymbol{\chi}} + \boldsymbol{\Lambda}_P\boldsymbol{\chi} = \boldsymbol{\Gamma}_P\boldsymbol{\tau}_D$$

where $\boldsymbol{\Gamma}_P \equiv \boldsymbol{\Phi}_P^T\mathbf{B}_0$. The obtained equation can be written in the following state and output equations can be written by defining $\mathbf{x} \equiv [\boldsymbol{\chi}^T \quad \dot{\boldsymbol{\chi}}^T]^T$ and $\mathbf{y} \equiv \boldsymbol{\theta}$ or $\mathbf{y} \equiv \dot{\boldsymbol{\theta}}$.

$$\dot{\mathbf{x}} = \mathbf{A}_P\mathbf{x} + \mathbf{B}_P\boldsymbol{\tau}_D, \qquad \mathbf{y} = \mathbf{C}_P\mathbf{x} \tag{14.36}$$

Substitute the derivative control

$$\boldsymbol{\tau}_D = -\mathbf{G}_D\dot{\boldsymbol{\theta}}_e = -\mathbf{G}_D\boldsymbol{\Gamma}_P^T\dot{\boldsymbol{\chi}} = -\mathbf{G}_D\mathbf{B}_P^T\mathbf{x} \tag{14.37}$$

into equation (14.36) and consider the system energy as

$$E = \frac{1}{2}(\dot{\boldsymbol{\chi}}^T\dot{\boldsymbol{\chi}} + \boldsymbol{\chi}^T\boldsymbol{\Lambda}_P\boldsymbol{\chi}) \tag{14.38}$$

$$\dot{E} = -\dot{\boldsymbol{\chi}}^T\boldsymbol{\Gamma}_P\mathbf{G}_D\boldsymbol{\Gamma}_P^T\dot{\boldsymbol{\chi}} \tag{14.39}$$

The energy function $E(\mathbf{x})$ can be a Lyapunov function, that is, E is positive definite, $\dot{E} \leq 0$, E and its partial derivatives are continuous with respect to $\mathbf{x}$ on an open domain Ω that includes the target state $\mathbf{x} = \mathbf{0}$. On the other hand, consider Assumption 14.1.

Assumption 14.1: The system in equation (14.37) is completely controllable or completely observable, where system satisfies the controllability and the observability conditions simultaneously.

In the system of equations (14.36) and (14.37), the maximal invariant set satisfying $\dot{E} = 0$ on the open domain is the origin $\mathbf{x} = \mathbf{0}$ if Assumption 14.1 is satisfied. Therefore, the following results are true due to La Salle's theorem (La Salle and Lefschetz, 1961).

Theorem 14.1: In the system of equations (14.36) and (14.37), the target state $\mathbf{x} = \mathbf{0}$ is asymptotically stable if the system satisfies Assumption 14.1.

Remark 14.3: In the system of equations (14.36) and (14.37), if the controllability (or the observability) condition is satisfied and $\mathbf{x} \neq \mathbf{0}$ holds essentially along the trajectory of the system, then $\dot{E} < 0$, that is, $\mathbf{\Gamma}_P^T \dot{\mathbf{\chi}} = \dot{\mathbf{\theta}}_e \neq \mathbf{0}$, is essentially true. Therefore, $\mathbf{x} \neq \mathbf{0}$ and $\dot{\mathbf{\theta}}_e = \mathbf{0}$ cannot essentially hold. The reverse is obvious, that is, $\mathbf{x} \neq \mathbf{0}$ holds when $\dot{\mathbf{\theta}}_e \neq \mathbf{0}$. The word 'essentially' means 'excluding enumerable points on the open domain'.

14.4.3   Stability of original non-linear system

The non-linear system of equations (14.7) and (14.34) can be transformed into the following system of first-order equation on a neighbourhood open domain Ω_A around $\mathbf{q} = \dot{\mathbf{q}} = \mathbf{0}$:

$$\dot{\mathbf{x}} = \mathbf{A}_0 \mathbf{x} + \mathbf{r} \tag{14.40}$$

where $\mathbf{A}_0$ is a time-invariant matrix, $\mathbf{r}$ is a convergent Taylor series of $\mathbf{x}$ with second or higher-order terms, that is, $\|\mathbf{r}(\mathbf{x})\| = o\,(\|\mathbf{x}\|)$, and the coefficients of $\mathbf{x}$ are bounded functions of time t. The origin $\mathbf{x} = \mathbf{0}$ of the non-linear equation is asymptotically stable if all eigenvalues of $\mathbf{A}_0$ have negative real parts, and all eigenvalues of $\mathbf{A}_0$ have negative real parts when Theorem 14.1 is satisfied. The following theorem is obtained for the original non-linear system.

Theorem 14.2: The target state $\mathbf{q} = \dot{\mathbf{q}} = \mathbf{0}$ of the system in equations (14.7) and (14.34) is asymptotically stable if the linearised system in equation (14.36) satisfies Assumption 14.1.

The stability condition is effective on the open domain Ω_A near the origin (see Figure 14.2(a)) because the proof uses equation (14.40). One can show asymptotic stability of the non-linear system on the bigger open domain Ω under the assumption that is equivalent to the controllability condition, that is, $\mathbf{x} \neq \mathbf{0}$ and $\dot{\mathbf{\theta}}_e = \mathbf{0}$ cannot hold essentially during the control action.

14.5   Control using VRM concept

14.5.1   Control methods using the VRM concept

The manipulation-variable error $\mathbf{e}_H$ of RFM is obtained, based on the assumption of small error, from equation (14.11) or (14.32) as

$$\mathbf{e}_H = \mathbf{e}_v + \mathbf{e}_\xi \equiv \mathbf{J}_v\,(\mathbf{\theta}_0, \mathbf{\theta})\,\mathbf{\theta}_e + \mathbf{J}_\xi\,(\mathbf{\theta}_0, \mathbf{\theta})\,\mathbf{\xi} \tag{14.41}$$

where $\mathbf{e}_v$ is the manipulation-variable error of VRM and e_ξ is that of RFM from VRM caused by the elastic deformation $\boldsymbol{\xi}$. If $\mathbf{J}_v \in \mathbf{R}^{n \times n}$ is non-singular, the joint-variable error is given by equation (14.41) as

$$\boldsymbol{\theta}_e = \mathbf{J}_v^{-1}(\boldsymbol{\theta}_0, \boldsymbol{\theta})\, \mathbf{e}_v \equiv \boldsymbol{\theta} - \boldsymbol{\theta}_d \tag{14.42}$$

Equation (14.42) gives definition of the desired joint-variable $\boldsymbol{\theta}_d$. Without considering VRM, manipulation-variable error $\mathbf{e}_H$ of RFM is often transformed by pre-multiplying $\mathbf{J}_v^{-1}$, and the following joint-variable error $\boldsymbol{\theta}_{eH}$ is obtained:

$$\boldsymbol{\theta}_{eH} = \mathbf{J}_v^{-1}\mathbf{e}_H = \mathbf{J}_v^{-1}(\mathbf{e}_v + \mathbf{J}_\xi \boldsymbol{\xi}) = \boldsymbol{\theta}_e + \mathbf{J}_v^{-1}\mathbf{e}_\xi \tag{14.43}$$

The errors $\mathbf{e}_v$ and $\mathbf{e}_H$ yield two RMRCs (Whitney, 1969):

$$\boldsymbol{\tau}_v = -\mathbf{G}_D \mathbf{J}_v^{-1}\dot{\mathbf{e}}_v - \mathbf{G}_P \mathbf{J}_v^{-1}\mathbf{e}_v = \mathbf{G}_D \dot{\boldsymbol{\theta}}_e - \mathbf{G}_P \boldsymbol{\theta}_e \tag{14.44}$$

$$\boldsymbol{\tau}_H = -\mathbf{G}_D \mathbf{J}_v^{-1}\dot{\mathbf{e}}_H - \mathbf{G}_P \mathbf{J}_v^{-1}\mathbf{e}_H = \boldsymbol{\tau}_v - \mathbf{G}_D \mathbf{J}_v^{-1}\mathbf{J}_\xi \dot{\boldsymbol{\xi}} - \mathbf{G}_P \mathbf{J}_v^{-1}\mathbf{J}_\xi \boldsymbol{\xi} \tag{14.45}$$

Even though $\boldsymbol{\tau}_v$ is a manipulation-variable feedback named the pseudo-RMRC, the closed-loop system is robustly asymptotically stable because $\boldsymbol{\tau}_v$ is equivalent to the PD-control of equation (14.14) satisfying the collocation condition of sensors and actuators in the joints. But due to the second and third terms in equation (14.45), the stability of $\boldsymbol{\tau}_H$ cannot be guaranteed.

The VRM concept allows constructing stable control schemes for flexible manipulators as extension to controls for rigid space manipulators, for example, the sensory feedback control (Masutani *et al.*, 1989) and the resolved acceleration control (RAC) (Yamada and Tsuchiya, 1990). These are named as extended local PD-control (extended LPDC) and the pseudo-RAC, and are expressed as

$$\boldsymbol{\tau} = -\mathbf{J}_v^{\mathrm{T}}\mathbf{G}_P \mathbf{e}_v - \mathbf{G}_D \dot{\boldsymbol{\theta}} \tag{14.46}$$

$$\boldsymbol{\tau} = \mathbf{M}_v \left(\ddot{\boldsymbol{\theta}}_d - \mathbf{G}_D \dot{\boldsymbol{\theta}}_e - \mathbf{G}_P \boldsymbol{\theta}_e \right) + \mathbf{h}_v \tag{14.47}$$

where $\mathbf{M}_v$ and $\mathbf{h}_v$ are the inertial matrix and the centrifugal and Coriolis forces of the VRM, respectively. The desired acceleration of joint variable is

$$\ddot{\boldsymbol{\theta}}_d = \mathbf{J}_v^{-1}\left[\ddot{\mathbf{y}}_d - \dot{\mathbf{J}}_v \left(\dot{\boldsymbol{\theta}} - \mathbf{J}_v^{-1}\dot{\mathbf{e}}_v \right) \right]$$

where it is assumed that $\mathbf{J}_v$ is always non-singular during control action.

To achieve higher performance, the composite control introduced later combines a reduced-order state feedback control with one of the aforementioned controllers in a hierarchical manner.

14.5.2   *Asymptotic stability of positioning control*

Although asymptotic stability to a static target of all the proposed controls has been proved (Senda, 1993; Senda and Murotsu, 1994*a*), only results for the pseudo-RMRC are presented here. Equivalent to Remark 14.3, assume that $\dot{\boldsymbol{\theta}}_e = \mathbf{0}$ is essentially true, only if $\begin{bmatrix} \mathbf{q}^{\mathrm{T}} & \dot{\mathbf{q}}^{\mathrm{T}} \end{bmatrix}^{\mathrm{T}} = \mathbf{0}$ essentially holds along the trajectory of equation (14.7) with control. Define the open domain Θ of $\left[\begin{array}{cc} \theta_0^T & \theta^T \end{array}\right]^T$ in which any point satisfying $L_{ev}(\theta_0, \theta) = 0$ locates, where the function is

$$L_{ev} = \frac{1}{2} e_v^T J_v^{-T} G_P J_v^{-1} e_v$$

Furthermore, Ξ is an open domain of $\left[\begin{array}{cc} \xi^T & \dot{q}^T \end{array}\right]^T$ around $\xi = 0$ and $\dot{q} = 0$. The control target set in an open domain $\Omega = \Theta \times \Xi$ of $x = \left[\begin{array}{ccc} \theta_0^T & q^T & \dot{q}^T \end{array}\right]^T$ is

$$\Pi = \{x \mid e_H = 0, \xi = 0, \dot{q} = 0\}$$

Theorem 14.2 below is thus obtained using La Salle's theorem (La Salle and Lefschetz, 1961).

Theorem 14.2: The solution $x(t)$ in the open domain Ω converges to the target set Π as $t \to \infty$ if the system of equations (14.7) and (14.44) satisfies the assumptions above.

14.5.3 Stability of continuous path control

The validity and stability of the proposed control schemes is explained through the following singularly perturbed model (Siciliano and Book, 1988). Since the inertia matrix M in equation (14.7) is positive definite, it can be inverted and denoted by H, which can be partitioned as

$$M^{-1} = H = \begin{bmatrix} H_{11}^{n \times n} & H_{12}^{n \times m} \\ H_{21}^{m \times n} & H_{22}^{m \times m} \end{bmatrix}$$

Extract a common scale factor k, for example, the smallest diagonal element of K_ξ, and define

$$\mu \equiv \frac{1}{k}, \quad \varepsilon \equiv \sqrt{\mu}$$

The following slow subsystem is obtained from equation (14.7) as $\mu \to 0$:

$$M_{11}(\bar{\theta}, 0)\ddot{\bar{\theta}} + h_1(\bar{\theta}, \dot{\bar{\theta}}, 0, 0) = \bar{\tau} \tag{14.48}$$

which is equivalent to the equation of motion of the VRM. On the other hand, a fast time scale $t_f = t/\varepsilon$ can be introduced and the following fast subsystem obtained as $\varepsilon \to 0$:

$$H_{22}^{-1}\ddot{\xi}_f + K_\xi \xi_f = H_{22}^{-1} H_{21} \tau_f \tag{14.49}$$

where $\xi_f \equiv \xi - \bar{\xi}$ and $\tau_f \equiv \tau - \bar{\tau}$. This is a linear system parameterised in the slow variables $\bar{\theta}$, and is equivalent to a vibration equation used for designing a reduced-order compensator in the composite control. Equation (14.49) is rearranged in the state-space form as

$$\varepsilon \dot{z}_f = A_f z_f + B_f \tau_f \tag{14.50}$$

The composite control (Chow and Kokotovic, 1978; Suzuki, 1981) is a hierarchical control combining an ideal full-state feedback control for the fast subsystem with a path tracking control for the slow subsystem. The following lemma (Kokotovic, 1984) with respect to the composite control is known.

Lemma 14.1: Consider $\bar{\theta}(t)$ as the trajectory of the slow subsystem controlled by a continuous path control. Further, suppose that equation (14.50) is completely controllable at any point on a slow trajectory $\bar{\theta}(t)$. If the pair $\mathbf{A}_f$ and $\mathbf{B}_f$ in equation (14.50) is uniformly asymptotically stabilised for any slow trajectory $\bar{\theta}(t)$ by a full-order state feedback, there exists $\varepsilon > 0$ that approximates the state of the original non-linear system of equations (14.7) as

$$\theta = \bar{\theta} + \mathbf{O}(\varepsilon), \quad \dot{\theta} = \dot{\bar{\theta}} + \mathbf{O}(\varepsilon),$$
$$\xi = \bar{\xi} + \xi_f + \mathbf{O}(\varepsilon), \quad \dot{\xi} = \dot{\xi}_f + \mathbf{O}(\varepsilon) \tag{14.51}$$

where $\mathbf{O}(\varepsilon)$ is the same infinitesimal order of ε. In this case, the closed-loop system is called approximately orbitally stable.

The above lemma cannot guarantee the approximate orbital stability of the closed-loop system because a full-order state feedback is impossible in a real composite controller.

The duality of control and observation is true because sensors, for example, encoders, tacho-generators and actuators are collocated on the same axes of joints. Hence, the following relation for the fast subsystem is obtained from equation (14.49)

$$\theta_f \equiv \theta - \bar{\theta} = \mathbf{B}_\xi^T \xi_f = \mathbf{H}_{22}^{-1} \mathbf{H}_{21} \xi_f \tag{14.52}$$

and the pseudo-RMRC of equation (14.44) becomes a full-order state feedback for the fast subsystem given by

$$\tau_{vf} = -\mathbf{G}_D \mathbf{B}_\xi^T \dot{\xi}_f - \mathbf{G}_P \mathbf{B}_\xi^T \xi_f \tag{14.53}$$

Consider the fast subsystem in equation (14.49) with the proportional control of the second term on the right-hand side of equation (14.53) and define $\tau_{vfD} \equiv -\mathbf{G}_D \mathbf{B}_\xi^T \dot{\xi}_f$. The obtained fast subsystem is rearranged in the state-space form as

$$\dot{\mathbf{x}}_f = \mathbf{A}_{fP} \mathbf{x}_f + \mathbf{B}_{fP} \tau_{vfD} \tag{14.54}$$

The following results are obtained using Lemma 14.1.

Lemma 14.2: If the pair $\mathbf{A}_{fP}$ and $\mathbf{B}_{fP}$ in equation (14.49) is completely controllable for any slow trajectory $\bar{\theta}(t)$, then the state $\xi_f = \dot{\xi}_f = 0$ is uniformly asymptotically stable.

Theorem 14.3: The RFM of equation (14.7) controlled by the pseudo-RMRC of equation (14.44) is approximately orbitally stable.

In the same way, approximate orbital stability of the extended LPDC and the pseudo-RAC is proved. The composite control combining the reduced-order compensator with a path tracking control, for example, the extended LPDC, the pseudo-RMRC and the pseudo-RAC, can be a full-order state feedback of the augmented fast subsystem because the path tracking control is a full-order state feedback of the fast subsystem. Hence, Lemma 14.1 guarantees that the composite control with a reduced-order compensator approximately orbitally stabilises the original non-linear system when the pair $\mathbf{A}_f$ and $\mathbf{B}_f$ is uniformly asymptotically stabilised by the resultant full-order state feedback.

14.6   Control examples

14.6.1   Positioning control

A numerical example is given here when the manipulation variables are controlled to a time-invariant target. The extended LPDC is applied to a planar space manipulator moving in a two-dimensional xy-plane. The manipulator consists of a two-link flexible manipulator and a rigid satellite base. The satellite base has a 100 kg mass, and is square with side lengths of 0.5 m. The flexible manipulator consists of two flexible links, where each link is a uniform beam of length 1.0 m, mass 1.0 kg and flexural rigidity 1.0 Nm^2.

Figures 14.3 and 14.4 illustrate (a) the motion of the whole system every second, (b) time history of the hand of the manipulator and (c) time history of translational deflection at the hand of the manipulator. All figures show responses for 20 s from time $t = 0$. Figure 14.3 is a simulation result when the LPDC for rigid manipulators is applied to the flexible manipulator. The closed-loop system is unstable because RFM's manipulation-variable error is fed back. Figure 14.4 shows the results when the extended LPDC for flexible manipulators is applied. The manipulator hand asymptotically converges to the time-invariant target. In the same way, a closed-loop system easily becomes unstable when a manipulation-variable feedback control for rigid manipulators is applied to flexible manipulators, whereas the VRM concept can ensure stability.

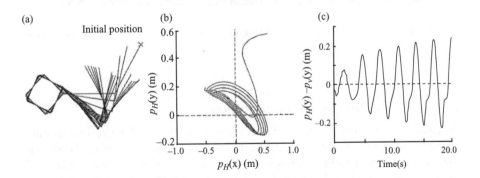

Figure 14.3   *Instability of local PD-control applied to a flexible manipulator* ($\mathbf{G}_P =$ diag[2.0, 2.0] (Nm), $\mathbf{G}_D =$ diag[1.1, 1.1] (Nm^2/rad)): *(a) motion, (b) hand position, (c) flexure deflection at hand*

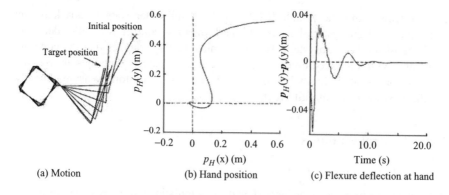

(a) Motion   (b) Hand position   (c) Flexure deflection at hand

Figure 14.4   *Extended local PD-control applied to a flexible manipulator ($\mathbf{G}_P =$ diag[2.0, 2.0] (Nm), $\mathbf{G}_D =$ diag[1.1, 1.1] (Nm^2/rad))*

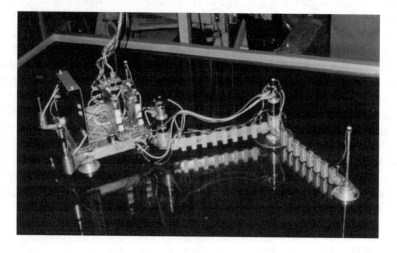

Figure 14.5   *Photograph of two-link flexible manipulator mounted on free-floating vehicle*

14.6.2   Path control: hardware experiment

A hardware space robot model (Figure 14.5) is composed of a two-link flexible manipulator and a free-floating vehicle. The model is supported on the horizontal table without friction using air-pads, and its free motion in rotational and translational directions is realised on the plane. Each link of the manipulator is 0.5 m long. The positions of LED (light-emitting-diode) markers on the robot model are measured using an optical position sensor system with a CCD camera. See Murotsu *et al.* (1993) and Senda (1993) for further details of the experimental system.

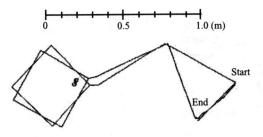

Figure 14.6   Motion of virtual rigid manipulator controlled by pseudo-RAC

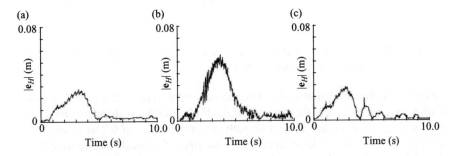

Figure 14.7   Experimental result of path control using extended LPDC, pseudo-RMRC, and pseudo-RAC: (a) extended LPDC, (b) pseudo-RMRC, (c) pseudo-RAC

In the initial state, the robot model is stationary and the joint variables are $\theta_1 = 30°$ and $\theta_2 = -60°$. In the following experiments, a desired path of manipulation variables, that is, xy-position of the manipulator hand, is a straight line of length 0.33 m for 5 s.

Figure 14.6 shows the motion of the VRM when the pseudo-RAC is used. Figures 14.7(a) to 14.7(c) show an Euclidian norm of $|e_H|$ when the robot model is controlled using the extended LPDC, the pseudo-RMRC and the pseudo-RAC. Each control scheme brings the hand along the desired path and stable closed-loop system is realised in each case. The control performances are almost equivalent. The results show that the hand can track the desired path with an error below 2 per cent of the link length (0.5 m). The steady-state errors of the final states are caused by the measurement errors of the optical position sensor.

14.6.3   Composite control

Control performances of the pseudo-RAC and a composite control using pseudo-RAC are compared in this section. Figure 14.8(a) illustrates a free-floating space manipulator composed of a rigid satellite and a four-link flexible manipulator with three rotational degrees at the shoulder and one rotational degree at the elbow. The longest link between the elbow and the hand is 2 m. Seven vibration modes are

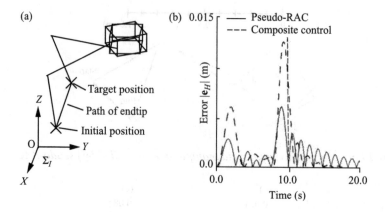

*Figure 14.8   Pseudo-RAC and composite control: (a) desired path, (b) end-point
error relative to target path*

considered in the simulation model. The reduced-order compensator of the composite
control scheme considers the lower four modes, while the higher three modes are
treated as residual modes. The target path is a straight line of 0.7 m in length and
lasting 10 s. It has an acceleration period for the first 2.5 s, deceleration period for the
final 2.5 s and the rest being standstill. Figure 14.8(b) shows $|e_H|$ of positional error
of the end-point of the manipulator from the target path. The maximum error of the
pseudo-RAC was smaller than that of the composite control. But the composite control
dampened the vibration more quickly after 10 s. Thereby, the reduced-order state
feedback resulted in superior vibration suppression performance of the composite
control. An inverse dynamics problem must be considered for finding a suitable $\bar{\xi}$.

14.7   Summary

This chapter has described modelling and control methodology of flexible manipulators for stable manipulation-variable feedback control. Differences between control
problems of flexible manipulators in space and on the ground have been highlighted.
When the momentum of the system is conserved, the equations of motion for space
flexible manipulators can be obtained in the same form as for base-fixed manipulators. To avoid instability of manipulation-variable feedback, the VRM concept
has been introduced. Several control schemes for flexible manipulators have been
developed using VRM, for example, the pseudo-resolved-motion-rate control, the
extended local PD-control, the pseudo-resolved-acceleration control and composite
control. The control laws using VRM have been stable because they are equivalent
to the joint-variable feedback control. Their asymptotic stability of the static target has been verified and their approximate orbital stability has been shown with
the singular perturbation method. Numerical simulations and hardware experiments
of space flexible manipulators have successfully demonstrated the effectiveness
and feasibility of the proposed methods. The proposed methods are effective for

both base-fixed manipulators on the ground and free-flying space manipulators with flexible links.

14.8 Acknowledgement

Part of this chapter was written as the author's PhD dissertation. The author expresses his gratitude to his former advisor Professor Y. Murotsu for his guidance. The author also gives his sincere thanks to Mr M. Hayashi for his help in the numerical simulations and Mr A. Mitsuya, Mr T. Nunohara and Mr K. Yamane for their help in the experimental work. This work was partially financially supported by a Grant-in-Aid for Scientific Research from the Ministry of Education, Culture, Sports, Science and Technology of Japan.

Chapter 15

Soft computing approaches for control of a flexible manipulator

S.K. Sharma, M.N.H. Siddique, M.O. Tokhi and G.W. Irwin

The aim of this chapter is to show the effectiveness of soft computing in the design of adaptive controllers for a single-link, flexible manipulator. The first part utilizes fuzzy logic (FL) and genetic algorithms (GAs) in the design of an offline modular neural network controller while the second part combines fuzzy logic and GAs in the construction of online proportional derivative (PD), proportional, integral (PI) and proportional, integral, derivative (PID)-type FL controllers. Experimental results, demonstrating the effectiveness of the methods for an experimental single-link, flexible manipulator are presented and discussed.

15.1   Introduction

The term soft computing (SC) has been introduced, in conjunction with the concept of machine intelligence quotient (MIQ), to describe a research trend towards producing computers with an increasingly higher MIQ (Zadeh, 1994). The latter is being achieved in two totally distinct ways. Conventional computers, based on classical logic and thus on precision, are evolving into faster-and-faster machines with increasing degrees of parallelism. On the other hand, new concepts such as fuzzy logic (FL), neural networks (NNs) and probabilistic reasoning (PR), comprising genetic algorithms (GAs), are being developed whose strengths lie in their capacity to deal with uncertainty and imprecision. In conventional computers very precise solutions are produced at a high computational cost. This is referred to as hard computing (HC). By contrast, SC leads to imprecise or uncertain solutions with a much lower computational effort.

The excessive precision of conventional computers is not useful in some cases and a non-traditional approach is preferable. With some problems, this is in fact the

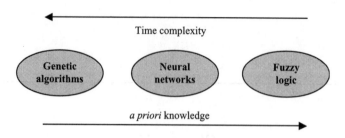

Figure 15.1   Soft computing components

only solution because the computational complexity of a classical approach would be prohibitive. Figure 15.1 compares three components of SC in terms of learning time and required *a priori* knowledge. FL is not capable of learning, while, NNs and GAs do have this capacity. Pure GAs in general require longer learning times. From the viewpoint of *a priori* knowledge, the order is reversed. GAs do not need *a priori* knowledge, NNs need very little knowledge while FL can require quite detailed knowledge of the problem to be solved.

In effect each of the three areas of SC has its advantages and disadvantages. FL does not share the inherent automatic learning capacity of NNs and so it is not possible to use FL when experts are not available. FL does, however, have a great advantage over the other two techniques in that the knowledge base is computationally much less complex and the linguistic representation is very close to human reasoning.

Neural networks are quite different from FL in terms of learning. They learn from example solutions to a problem. There are two evident disadvantages in using the NN learning process. What is learnt is not easy for humans to understand and hence the knowledge base extracted from NNs does not have such an intuitive representation as that provided, for example, by FL. Further, the activation functions that can be used in NNs have to possess precise regularity features and their derivatives must be known *a priori*.

While the learning speed of GAs (Goldberg, 1989; Holland, 1992) is usually slow, they have two important advantages over NNs. The functions used in GAs can be much more general in nature and knowledge of their gradients is usually not required. Finally, as these algorithms explore the solution space in several directions at the same time, they are much less affected by the problem of local extrema. Thus, a GA has far less likelihood than a NN of finding a local extreme than a global one. Even if the extreme found is not indeed a global one, it is likely to correspond to a less significant learning error.

On the basis of these considerations, it is a generally held view that a combination of these SC approaches would be an interesting and useful prospect. Such hybrid technique would inherit all the advantages, but not the less desirable features of the individual SC components. This chapter therefore utilizes some of the advantages of these SC components in the design of an adaptive controller for a single-link flexible manipulator.

The rest of the chapter is organised as follows. Section 15.2 explains the experimental flexible manipulator system used as the target application. Section 15.3 describes off-line construction of a modular neural network (MNN) and a genetic learning process for the MNN. A set of neural control results from the flexible manipulator are also presented in this section. Section 15.4 deals with the design of online FL controllers where previous knowledge of the plant model or dynamics is not required. Proportional, derivative (PD)-type, and proportional, integral (PI)-type FL controllers are described. An on-line switching mechanism from PD-type to PI-type fuzzy controllers is outlined, and unification of the common rule-base and optimisation of membership function parameters using GAs is described. Section 15.15 provides a summary of the chapter.

15.2   The flexible manipulator system

Figure 15.2 shows the experimental rig consisting of a flexible arm and associated measuring devices, used in this chapter as the target application. This is the Sheffield manipulator described in Chapter 1 with physical properties and parameters further given in Chapter 2. As noted, the arm constitutes a flexible link driven by a printed armature motor at the hub. The measuring devices are a shaft encoder, a tachometer, an accelerometer and strain gauges along the length of the arm. Measurements from the strain gauges were not employed in this study. The shaft encoder measures the hub-angle and the tachometer measures the hub velocity of the manipulator. The accelerometer at the end-point of the flexible arm measures the end-point acceleration.

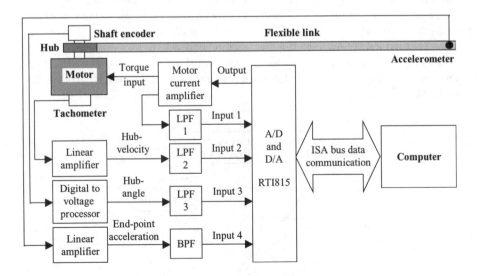

Figure 15.2   Experimental set-up of a flexible manipulator

It has been found that the vibration of the flexible arm is dominated by the first few (2–3) resonance modes (Sharma, 2000). The hub-angle defines the angular position, hub-velocity determines its angular velocity while end-point acceleration provides a measure of vibration associated with the flexible manipulator movement. For the work presented in this chapter, the flexible manipulator was modelled using three separate single-input single-output (SISO) models with input torque/voltage as a common input and hub-angle, hub-velocity and end-point acceleration as the outputs. The next section explains offline construction of an MNN controller for the flexible manipulator system.

15.3   Modular NN controller

This section is based on previous work reported by Sharma *et al.* (2003) where it was shown that a NN controller in modular form is simple to design and produces superior performance compared to a conventional one. Embedding modularity into a NN to perform local and encapsulated computation produces many advantages over a single network. It is also easier to encode *a priori* knowledge in a modular framework.

In general, a MNN is constructed from two types of network, as shown in Figure 15.3, namely expert networks and a gating network (Hodge *et al.*, 1999; Jacobs and Jordan, 1993). Expert networks compete to learn the training patterns, while the gating network mediates this competition. During training, the weights of the expert and gating networks are adjusted simultaneously using the backpropagation algorithm (Rumelhart *et al.*, 1986). The MNN learns to partition an input task into subtasks, allocating a different NN to learn each one, as shown in Figure 15.4. However, accuracy of the MNN depends greatly on accurate fusion of the individual networks as decided by a gating network (Hodge *et al.*, 1999). A new SC method, using GAs, which removes the need for a gating network, is presented here. Fusion of the individual networks is decided by optimum slope selection of the activation function. The GA also optimises the structure and weights of the individual networks in the MNN.

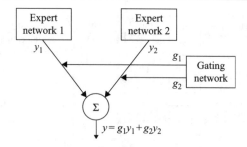

Figure 15.3   General modular neural network architecture

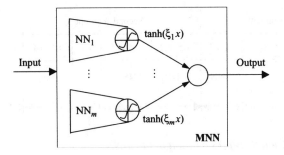

Figure 15.4   A modular neural network architecture

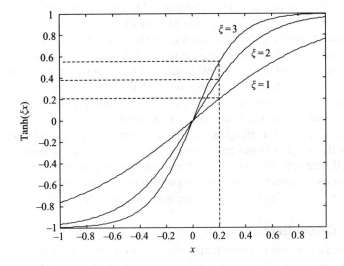

Figure 15.5   Activation function, tanh(ξx), at the NN output

15.3.1   Genetic representation of MNN architecture

In the MNN architecture of Figure 15.4, tanh (x) is used as the activation function in the hidden neurons, while different activation functions tanh $(\xi_i x)$, $i = 1, 2, \ldots, m$ are used in the output neurons. The overall MNN output is the sum of the outputs from all these subnetworks.

The hyperbolic activation function tanh $(\xi_i x)$ illustrated in Figure 15.5 for $\xi = 3.0$, 2.0 and 1.0, produces a larger neural response for the same input value by increasing ξ. Thus, changing the ξ_i values in the MNN regulates the individual contributions from each NN to the output.

15.3.1.1   Genetic encoding of NNs

Structured genetic coding is used to encode both the structure and the parameters of the MNN in a single chromosome. The genetic representation of the MNN is shown in

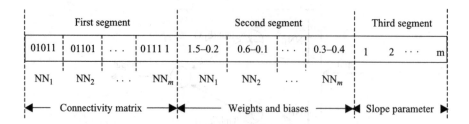

Figure 15.6   Genetic representation of the MNN

Figure 15.6. Each individual in the population is composed of three segments of strings to represent the modular network architecture. The first defines the connectivity, the second encodes the connection weights and biases, while the last segment denotes the activation function slope parameters ξ_i. In the first segment, each bit represents an individual connection, taking the value 1 or 0 to indicate whether a connection exists or not. The total number of bits in this first segment equals the total number of possible connections between the input and the hidden layer and between the hidden and the output layer. Since the weights, biases and slope parameters are numerical values, the second and third segments are encoded as a fixed-length real-number string rather than as a bit string. Initially the connections are encoded randomly as 1 or 0 while the weights, biases and slope parameters are chosen randomly within the range ±1.0. The optimal network architecture and corresponding weight, biases and slope parameters can now be determined simultaneously using GAs. The next section describes the genetic learning scheme applied to the encoded chromosomes.

15.3.1.2   Genetic learning for NN

The encoded chromosomes representing the NN parameters are randomly initialised for genetic evaluation. The GA is used here to find an optimal architecture as well as the weights, biases and slopes of the activation functions of the NN architecture. Hence, the fitness function in GA must include not only a mean-square error, a measure of accuracy, but also a feasibility measure of the network structure and complexity in terms of the number of nodes and their connectivity. The resulting fitness function should therefore be able to select a feasible network, which satisfies these criteria.

Here four constraints, namely, number of correctly classified data points in training (λ), change in error ($\dot{e}$), error (e) and number of connections (ψ), are chosen to define the fitness function for design of an optimum NN architecture. Here the number of correctly classified data points are defined when the actual output is within a certain desired limit from the required output. The first three of these constraints are responsible for tuning the NN weights, whereas the last one is for optimising the neural structure. Sharma *et al.* (2003) have shown that, in this situation, a hierarchical fuzzy method can work better than alternative conventional methods and is simple to apply. A hierarchical fuzzy method dynamically allocates the priority required as well as optimising the direction of movement of a chromosome by dividing the multiple

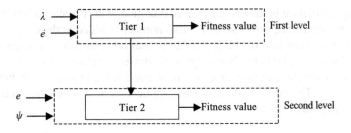

Figure 15.7   Hierarchical fuzzy logic approach

constraints into different tiers on the basis of priority. Figure 15.7 shows a hierarchical fuzzy structure containing two such tiers. In the first FL level the number of correctly classified data points (λ) and the change in error $(\dot{e})$ are used as antecedents, while the fitness value forms the consequent.

The error (e) and the number of MNN connections (ψ) constitute antecedents at the second level. Here the change in error is defined as

$$\dot{e}_n = e_n - e_{n-1} \qquad (15.1)$$

where e_n and e_{n-1} are the mean-square errors at the nth and $(n-1)$th generations, respectively.

The first level of FL selects all chromosomes with acceptable fitness in terms of the number of correctly classified data points and decreasing error. From these chromosomes those with fewer connections and less error are selected at the next FL level. The second level in the fuzzy hierarchy thus handles chromosomes, which have already been refined previously. Since the first priority is to select the chromosome with the maximum number of correctly classified data points and with decreasing error, the best chromosomes in every generation always come with a feasible solution. Here all four constraints, input to the hierarchical fuzzy methods, are fuzzified into different ranges and represented by triangular membership functions. The number of data points to be classified and the maximum number of connections are known at the beginning and ranges of their membership function can be pre-selected. The ranges for the membership function e and $\dot{e}$ are set dynamically in such a way that a considerable number of chromosomes is always near the solution. In this case, the dynamic range of e is three times the minimum error in the population at every generation, whereas it is four times the minimum value for $\dot{e}$. The minimum and maximum values for the membership function of ψ and λ are 30 per cent and 70 per cent of the total. Whenever any chromosome goes out of range a penalty is automatically provided by the shape of the membership function. The details for selection of a fitness function in hierarchical fuzzy methods can be found in Sharma and Tokhi (2000) and Sharma *et al.* (2003).

A steady-state GA (Michalewicz, 1994) is applied to selected chromosomes and the whole population is subjected to crossover and mutation. Uniform crossover (Wasserman, 1993) is applied to the binary coded part of a chromosome and arithmetic crossover is applied to the decimal coded part (Michalewicz, 1994). Half of the remaining population is subjected to biased mutation, with 1 replaced by 0 and vice-versa for binary coding and a small random number is added in the case of decimal encoding. The other half of the population is subjected to unbiased mutation with selected bits and weights randomly replaced. Biased mutation provides uniformity and unbiased mutation ensures diversity in the population, leading to a better overall representation. The next section describes application of this new SC technique to flexible manipulator control.

15.3.2   Implementation and simulation results

A flexible manipulator is a highly non-linear system where position, velocity and acceleration feedback can be grouped in modular form, thus offering a suitable case study for MNNs. In this chapter, dynamic feedback from the manipulator is used as the basis for MNN-based non-linear controller design. Some manipulator control schemes incorporating dynamic feedback for stabilisation and trajectory control have been reported in DeLuca *et al.* (1985*a*,*b*), Kotnik *et al.* (1988), Sira-Ramirez *et al.* (1992), Spong *et al.* (1987), and Stadenny and Belanger (1986). In these schemes, decoupling of dynamic feedback in modular form has not been considered, which is important for achieving optimum effects of individual feedback loops on the overall control trajectory. Here an MNN using decoupled feedback on the basis of position, velocity and end-point acceleration of the flexible manipulator system is considered. The flexible manipulator is considered with no payload, and it is to follow desired end-point position with low vibration.

15.3.2.1   End-point position tracking

The end-point acceleration model provides information on the vibration associated with the flexible manipulator. To reduce system vibration, absolute value of the end-point acceleration model output is required to be within prescribed minimum limit. Accordingly, the correctly classified data (λ) includes information from the output of the end-point acceleration model for reduction of vibration, along with the hub-angle model output required for tracking of a desired end-point position.

A NN controller will be deemed to have a larger λ value if it classifies the desired end-point position correctly and the absolute value of end-point acceleration output to less than say 1 per cent. Accordingly, a NN controller, which meets these criteria more often will have larger value of λ, and thus will be more acceptable at the first level of the hierarchical fuzzy selection process (Figure 15.7). Thus, the first level filters the chromosomes, which follow the desired end-point position closely and with less vibration. The second level then selects the chromosomes from this subset, which have fewer connections and less error.

Figure 15.8 shows the MNN as a non-linear controller for the flexible manipulator system. Here three small subnetworks were employed. The MNN controller was used in series with the flexible manipulator SISO model and genetic learning was used for optimisation of the MNN parameters. The grouping of dynamic feedback in modular form was achieved as follows: The desired position $\theta_d(k)$ and hub-angle

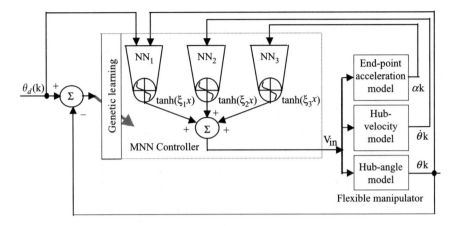

Figure 15.8   Modular neural network controller for the flexible manipulator

feedback signal $\theta(k)$ constituted the inputs to the first NN controller (NN$_1$). The hub-velocity $\dot{\theta}(k)$ formed the input to the second NN controller (NN$_2$), and the end-point acceleration $\alpha(k)$ constituted the input to the third NN controller (NN$_3$). Each NN was a multi-layer perceptron with one hidden layer. A suitable combination of ξ_1, ξ_2 and ξ_3 was required to achieve the desired closed-loop control performance.

The MNN had eight inputs. The first neural controller (NN$_1$) had four inputs $\bar{x}_1$ where $\bar{x}_1 = [\theta_k^d, \theta_{k-1}^d, \theta_{k-1}, \theta_{k-2}]$. The second neural controller (NN$_2$) had $\bar{x}_2 = [\dot{\theta}_{k-1}, \dot{\theta}_{k-2}]$ and the third controller (NN$_3$) had $\bar{x}_3 = [\alpha_{k-1}, \alpha_{k-2}]$ as inputs. The number of hidden neurons was set to three for each NN in the MNN controller. As shown in Figure 15.6, 20 chromosomes were evaluated for 5 runs of 200 generations each. The resulting best chromosome was then selected to decode the controller and the ξ_1, ξ_2 and ξ_3 parameters. A steady-state GA with crossover probability $p_c = 0.65$, biased mutation with $p_m = 0.05$ and unbiased mutation with $p_m = 0.3$ was used to optimise the controller parameters. The non-linear controller was then validated online for different trajectories.

15.3.2.2   Performance of MNN controller

The MNN was trained to track the desired position with low vibration. Thus, λ included the desired input–output pairs for end-point position and the absolute value of the end-point acceleration model for GA selection. A total elapsed time of 6 s with a sample time of 5 ms was used for testing. The trajectory followed and the vibration associated with the movement was recorded in each case. The required trajectory was set at 0 rad for the first 1.5 s, then to 0.8 rad for the next 1.5 s, to −0.8 rad from 3 to 4.5 s, and back to 0 rad beyond 4.5 s.

The reference trajectory and the actual trajectory followed by the flexible manipulator are shown in Figure 15.9. It is clear that the flexible manipulator was able to follow the desired trajectory very closely under MNN control. The end-point acceleration of the flexible manipulator with, and without, the MNN controller is given

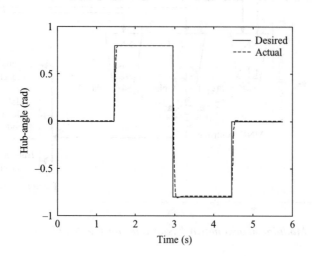

Figure 15.9   Tracking the reference hub-angle trajectory

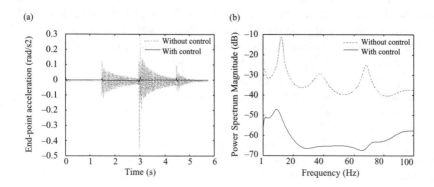

Figure 15.10   End point acceleration: (a) time-domain, (b) spectral density

in Figure 15.10. This shows that a significant reduction in the vibration was achieved with the MNN controller.

The results in Figures 15.9 and 15.10 confirm that design by SC methods while optimising ξ_1, ξ_2 and ξ_3 by genetic evaluation, combined with a priority-based approach through FL-based selection, will result in an MNN controller that produces better trajectory regulation with reduced system vibration.

This section concludes the offline construction and application of the MNN controller on a single-link flexible manipulator. In the next section, FL and GAs are used to produce an online controller for the same application where prior knowledge of the plant model or dynamics is not required.

15.4   FL control of a flexible-link manipulator

The current literature includes various efforts on modelling and control of flexible manipulators, from both theoretical and experimental points of view. Researchers have investigated a variety of techniques for representing flexible and rigid dynamic models of such mechanisms. These models are derived on the basis of different assumptions (small deflection, for example) and mode shape functions. Even if a relatively accurate model of the flexible-link manipulator can be developed, it is often too complex to use in controller development. There are then two choices available:

1. Make further effort to deal with the non-linear mathematical models.
2. Employ non-conventional techniques that do not require a mathematical model.

Different model-based conventional control techniques for flexible-link manipulators, their difficulties and limitations are discussed in other chapters of this book. Among the few non-conventional control approaches that do not require mathematical models, FL, NNs and hierarchical schemes are characterised by their ease of incorporation of experiential knowledge. Non-linearity can be constructed as a fuzzy limit (Lee *et al.*, 1994), so that it is natural to follow the second choice above.

The conventional approach to design of FL controller is to generate a fuzzy rule set based on the system states such as error, change of error or sum of error. This produces a two-input single-output PD-, or PI-type or a three-input, single-output PID-type control rule base. The performance of PI-type FL controllers is quite satisfactory for linear first-order systems (Lee, 1993; Mudi and Pal, 1999) but degrades for higher-order and non-linear systems due to large overshoots and excessive oscillation during transients (Lee, 1993). On the other hand, good performance is achieved with a PD-type fuzzy controller during the transient state but produces a large steady-state error or sustained oscillations (Chao and Teng, 1997; Chung *et al.*, 1998). In fact, a PID-type fuzzy controller would be a better choice, but unfortunately it requires a huge rule-base, which is time consuming to process. A trade-off is therefore required between PD-, PI-, and PID-type fuzzy controllers.

A comparative assessment of the performance of these controllers with different parameters and rule-bases is carried out in subsequent sections of this chapter. The effect of integral wind-up on the PI-type controller is also investigated and finally a switching-type PD-PI-type controller is proposed for a single-link flexible manipulator.

15.4.1   PD-type fuzzy logic control

A PD-type fuzzy logic control (FLC) scheme can be developed by using error and change of error as inputs. Thus,

$$u(k) = k_p e(k) + k_d \Delta e(k) \tag{15.2}$$

where k_p and k_d are the proportional and differential gain coefficients, $e(k)$ is the error and $\Delta e(k)$ is the change of error at sample number k.

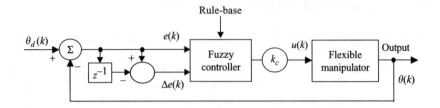

Figure 15.11   PD-type FLC with hub-angle error and change of hub-angle error

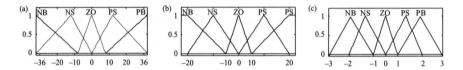

Figure 15.12   Membership functions for inputs and output: (a) hub-angle error (deg.), (b) change of hub-angle error (deg.), (c) torque (volts)

Here hub-angle error and change of hub-angle error are considered as two inputs for the PD-type FLC. Only hub-angle, $\theta\,(k)$, is measured from the system and the error and change-of-error are derived from $\theta\,(k)$. The hub-angle error and change of error are defined as

$$e\,(k) = \theta_d\,(k) - \theta\,(k) \tag{15.3}$$

$$\Delta e\,(k) = e\,(k) - e\,(k-1) \tag{15.4}$$

where $\theta_d(k)$ is the desired hub-angle. Figure 15.11 shows a block diagram of the PD-type FLC with error and change of error as inputs. Triangular membership functions are chosen for inputs and output. The membership functions for hub-angle error, change of hub-angle error and torque input are shown in Figure 15.12. The associated rule-base is shown in Table 15.1, where NB, NS, ZO, PS, PB are linguistic variables representing negative big, negative small, zero, positive small and positive big, respectively.

The membership functions defined in Figure 15.12 and the rule-base defined in Table 15.1 form the control surface of the controller, which is shown in Figure 15.13. The controller was applied to the single-link flexible manipulator. The performance of the system is shown in Figure 15.14.

For a demanded hub-angle of 36°, the controller response had a maximum overshoot of 50°. The PD-type FLC system shows rapid response at transient state, that is, a rise time of 2.38 s and a settling time of 6.16 s. The performance of the PD-type FLC system is very promising in respect of rise time, maximum overshoot and settling time but it shows a significant amount of steady-state error of 2.56°. The next section analyses the performance of PI-type FLC for the same demanded hub-angle.

Table 15.1   FLC rule-base with hub-angle error
and change of hub-angle error

	Change of error				
Error	**NB**	**NS**	**ZO**	**PS**	**PB**
NB	PB	PB	PB	PS	ZO
NS	PB	PS	PS	ZO	NS
ZO	PS	ZO	ZO	ZO	NS
PS	PS	ZO	NS	NS	NB
PB	ZO	NS	NB	NB	NB

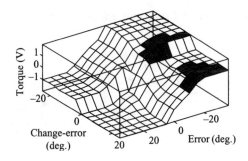

Figure 15.13   *Control surface with hub-angle error and change of hub-angle error*

15.4.2   PI-type fuzzy logic control

It is well known that a PI-type FLC system has good performance at the steady state, like the traditional PI-type controllers. That is, the PI-type FLC reduces steady-state error, but yields penalised rise time and settling time (Chao and Teng, 1997). It produces large overshoot while attempting to reduce the rise time for higher-order systems (Lee, 1993). These undesirable characteristics of fuzzy PI controllers are caused by integral wind-up action of the controller.

A conventional continuous-time PI-control law is described by

$$u(t) = k_p e(t) + k_I \int e(t)\, dt \tag{15.5}$$

where k_p and k_I are the proportional and the integral gain coefficients. Expressing equation (15.5) in discrete-time yields an absolute PI-type FL controller, represented mathematically as

$$u(k) = k_P e(k) + k_I \sum_{k=1}^{N} e(k) \tag{15.6}$$

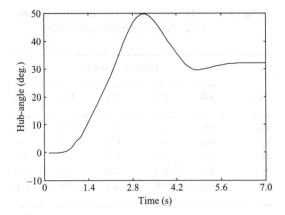

Figure 15.14   Hub-angle with FLC using hub-angle error and change of hub-angle error

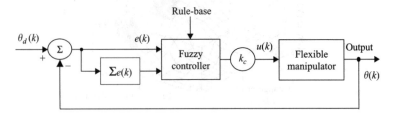

Figure 15.15   Block diagram of a PI-type FLC

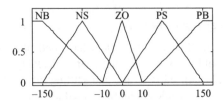

Figure 15.16   Membership function for sum of hub-angle error (deg/s)

A block diagram of the absolute PI-type FLC is shown in Figure 15.15. The membership functions for sum of hub-angle error are shown in Figure 15.16 and the rule-base for the PI-type FLC is given in Table 15.2.

It is difficult to decide on the number of samples, N, to use in calculating the sum in equation (15.6). Even the literature on conventional control theory tends to be somewhat vague on this point and the reason for this is that in conventional control, the integral term is traditionally approximated by analogue circuitry, and limits of integration cannot be stated precisely (Lewis, 1997). Experimental results suggested that a value of $N = 10$ works well for this application.

The control surface of the PI-type FLC with hub-angle error and sum of hub-angle error is shown in Figure 15.17.

Table 15.2   Rule-base for PI-type FLC with error and sum of error

Error	Sum of error				
	NB	**NS**	**ZO**	**PS**	**PB**
NB	PB	PB	PB	PS	ZO
NS	PB	PS	ZO	ZO	NS
ZO	PS	ZO	ZO	ZO	NS
PS	PS	ZO	ZO	NS	NB
PB	ZO	NS	NB	NB	NB

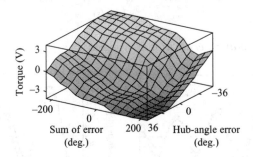

Figure 15.17   Control surface of the PI-type FLC with error and sum of error

The controller was implemented and applied to the single-link manipulator. The response of the flexible-link manipulator with the absolute PI-type FLC thus obtained is shown in Figure 15.18. It can be seen that the response has a very good performance for a demanded hub-angle of 36° with a small steady-state error of −0.34°. The rise time is 1.68 s, which is less than that of the PD-type FLC presented in the previous section, and a larger overshoot of 66.45° with excessive oscillation around the set point. The oscillations led to a prolonged settling time of 11.9 s. The next section considers the effect of integral wind-up action on PI-type FLC.

15.4.2.1   Integral wind-up action

Many techniques have been developed for dealing with the problem of integral wind-up, such as using fixed limits on the integral term, stopping summation on saturation, integral subtraction, the use of a velocity algorithm and analytical methods. Implementation of these algorithms can be found in Bennett (1994). In this study, the integral summation is calculated as follows:

$$\text{Calculate } \sum_{k=1}^{N} e\,(k) = \sum e\,(k-1) + e\,(k)$$
$$\text{If } k = N \text{ stop summation}$$
$$\text{Initialise } \sum e\,(k) \text{ to zero}$$

(15.7)

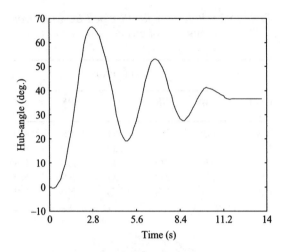

Figure 15.18   Hub-angle error with PI-type FLC

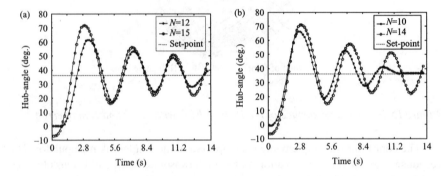

Figure 15.19   Integral wind- up action in PI-type FLC

The integral wind-up action, for different values of N, was investigated for the absolute PI-type FLC. Figure 15.19(a) shows the integral wind-up action for $N = 12$ and 16. Neither of them settled within 14 s. The overshoot with $N = 15$ was higher ($72.12°$) than that with $N = 12$ ($61.73°$). Figure 15.19(b) shows the integral wind-up action for $N = 10$ and 14. It was observed that the controller with $N = 10$ achieved the set point at 11.48 s with an overshoot of $66.34°$ whereas the controller with $N = 14$ did not settle within 14 s and had a larger overshoot of $71.23°$. The next section studies the application of PID-type FLC on the single-link flexible manipulator.

15.4.3   PID-type fuzzy logic controller

Generally, PD-type FLC cannot eliminate steady-state error whereas PI-type FLC can but at the expense of a slower response due to the integral control variable. These characteristics were studied and verified in earlier sections. In order to achieve short rise time, minimum overshoot, shorter settling time and zero steady-state error, a further option is to use a PID-type FLC, where short rise time, smaller overshoot and smaller settling time are obtained from the PD part and minimum steady-state error is from the PI part. The generic fuzzy PID controller is a four-dimensional (three input-one output) fuzzy system. The control law for a PID controller is of the form

$$u_{PID}(k) = k_P e(k) + k_D \Delta e(k) + k_I \sum_{k=1}^{N} e(k) \tag{15.8}$$

Theoretically, the number of rules R necessary to cover all possible input variations in a three-term fuzzy controller is $R = n_1 \times n_2 \times n_3$, where n_1, n_2 and n_3 represent the number of linguistic labels of the three input variables. In particular, if $n_1 = n_2 = n_3 = 7$, then $R = 343$. In practical applications, the design and implementation of such a large rule-base is a tedious task and it will take a substantial amount of memory space and reasoning time. Because of the latter, the response of such a generic PID-type FLC will be too slow and hence not suitable for the flexible-link manipulator where a fast response is desired. New switching-type PD and PI-type fuzzy controllers are therefore now described below, which have a smaller rule-base.

15.4.3.1 PD–PI-type fuzzy controller

A number of approaches have been proposed in the past to overcome the problems of PID controllers (Brehm, 1994; Tzafestas and Papanikolopoulos, 1990). Kwok *et al.* (1990, 1991) have considered a novel means of decomposing a PID controller into a fuzzy PD controller in parallel with various types of fuzzy gains, fuzzy integrators, fuzzy PI controller and deterministic integral control. For a process whose steady-state gain is known or can be measured easily as k_p, integral action is not necessary. If k_p is not known, integral action is required which can be achieved by placing a conventional integral controller in parallel with the fuzzy PD controller. These hybrid types of PID controllers are strictly not true fuzzy PID controllers as they include deterministic terms. A detailed description of these kinds of decompositions can be found in Harris *et al.* (1993).

A typical method for rule reduction in a fuzzy PID-type controller is to divide the three-term PID controller into separate fuzzy PD and fuzzy PI parts (Chen and Linkens, 1998; Kwok *et al.*, 1990). This hybrid PD and PI controller, with n linguistic labels on each input variable, requires only $n \times n + n \times n = 2n^2$ rules. Thus, for $n = 5$ there will be 50 rules which is significantly smaller than the 125 for a generic PID controller. The number of rules processed during control execution defines the amount of processing time and memory space requirements. A further reduction is possible if the controller is switched from PD- to absolute PI-type after a certain period of time. In that case only one set of rules (n^2 rules for each type of controller) will be executed at a time. The executed rules in the rule-base will be reduced to only 25, with 5 linguistic labels on each input variable. A switching-type FLC scheme was developed for the flexible-link manipulator where a PD-type FLC was executed first and then switched to a PI-type FLC. The block diagram of this switching PD–PI-type controller is shown in Figure 15.20.

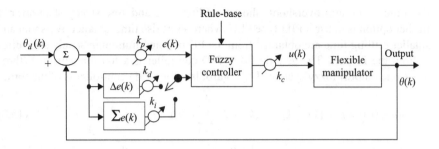

Figure 15.20   Block diagram of a PD-PI-type FLC system

The state variables used in the PD–PI-type FLC were the same as in the PD- and PI-type in Sections 15.4.1 and 15.4.2, respectively. Triangular membership functions were again chosen for both the inputs and the output. The membership functions for hub-angle error, change of hub-angle error, sum of hub-angle error and torque input were the same as used earlier for the PD- and PI-type controllers shown in Figures 15.12 and 15.16, respectively. The rule-bases for both PD-type and PI-type controllers were also the same as described in Sections 15.4.1 and 15.4.2, respectively.

Determination of switching point is important and can result in a frustration by bad performance if chosen incorrectly. If the controller is switched at the point of maximum overshoot of the PD-type FLC, it can yield the best performance but, surprisingly, will not give a good result. Experimental investigations showed that a switching point just before, or after, maximum overshoot will give better results than at the point of maximum overshoot.

15.4.3.2   Experimental results

The switching PD–PI-type FLC was implemented and tested on the single-link flexible manipulator. The performance of the PD–PI-type FLC was verified for different switching points, control output scaling factors (k_c) and different integral summations $\sum e(k)$. Figure 15.21 shows the performance of the controller for a demanded hub-angle of 36° with control output scaling factor $k_c = 86$ and switching point at 2.8 s.

A higher value of k_c yielded a shorter rise time of 1.96 s causing larger overshoot of 53.63°. An earlier switching point than 2.8 s caused a negative overshoot, showing the dominance of the PI-type controller over PD-type. It was concluded from this experiment (Figure 15.21) that a switching time earlier than 2.8 s will cause the maximum overshoot to increase and produce more oscillations around the set point.

Figure 15.22 shows the hub-angle with different switching times, namely, at 3.22, 3.36 and 3.5 s, with a scaling factor $k_c = 76$ for a set point of 36°. A switching time at 3.22s, indicated by solid line with dots, showed a slightly larger overshoot and settling time and caused the steady-state error to increase. The switching time at 3.36 and 3.5 s yielded better control performances. These two curves have the same rise time of 2.1s, the same overshoot of 52.16° and negligible steady-state error. In all these experiments, the scaling factors k_p, k_d and k_i were assumed to be unity.

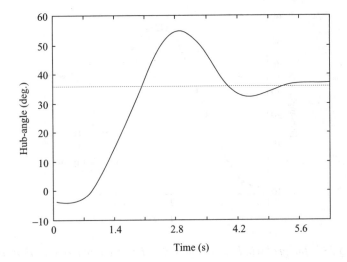

Figure 15.21   Hub-angle with a switching time 2.8 s and scaling factor $k_c = 86$.

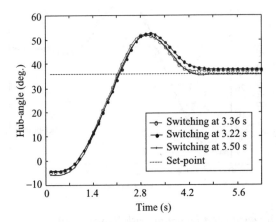

Figure 15.22   Hub-angle at different switching times with scaling factor $k_c = 76$.

The effect of integral wind-up, described in Section 15.4.2.1, on the PD–PI-type FLC was also verified for a demanded hub-angle of 36°, a switching time at 2.38 s and different values of N in the calculation of integral summation. This switching point was chosen to see the effect of integral action on the overshoot, observed earlier in Figure 15.22. Figure 15.23 shows the integral wind-up action for $N = 15$ and for $N = 5$. As can be seen from the figure, in both cases, there are larger positive overshoots and a smaller negative overshoot. The overshoot is a little larger (56.18°) and the settling time is longer (5.74s) for $N = 15$ than for $N = 5$.

The integral wind-up actions, shown in Figure 15.24, are for values of $N = 10$ (line with circles) and $N = 11$. The overshoot and settling time for $N = 10$ were

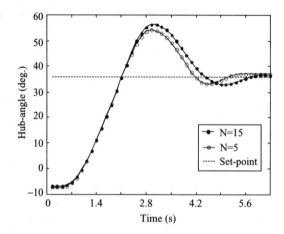

Figure 15.23   Integral wind-up action with N less than and greater than 10

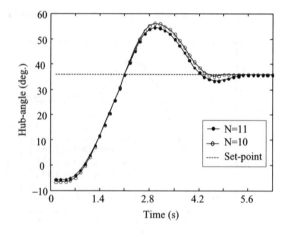

Figure 15.24   Integral wind-up action with N around 10

55.23° and 4.34 s respectively, whereas these for $N = 11$ were 55.17° and 5.18 s, respectively. The experiment confirmed that $N = 10$ gave the best result in terms of overshoot and settling time. The control output-scaling factor k_c was set to 76 in these experiments.

Although, the hybrid PD–PI fuzzy controller has separate PD and PI components, the combined effect is that of a PID controller. The PD controller provides a response with a short rise time and minimal peak overshoot and the PI controller results in good steady-state performance. However, a hybrid PD–PI switching-type controller also encounters a number of problems. These include designing the membership functions for error and change of error and the rule-base for both parts of the FL controller. The switching point of the PD–PI fuzzy controller can be determined within a few trials.

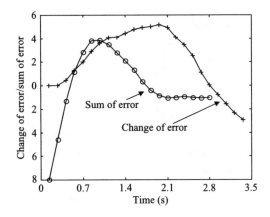

Figure 15.25 Change and sum of error within a common universe of discourse

Some effort has been made to automate the construction of rule-bases and to define the membership functions in various ways using NNs and GAs (Lin and Lee, 1991, 1992, 1993, 1995; Nauck and Kruse, 1992, 1993). In most of these cases, either the rule-base is fixed and the parameters of the membership functions are adjusted or the membership functions are fixed and the rule-base is optimised with GAs.

The next section focuses on the use of GAs where MF parameters are optimised while using a common rule-base for both PD- and PI-type controllers.

15.4.4   GA optimisation of fuzzy controller

The main idea here is to represent the complete set of membership functions by an individual chromosome in the GA and to evolve the optimum shape and location of the triangular membership functions. The change of error before the switching time and the sum of errors after the switching time are plotted in Figure 15.25. As can be seen, the range of change of error and sum of error fall within such an interval that they can be brought together within a common universe of discourse. In FLC design, the actual values of the inputs do not matter, rather the fuzzy sets for each linguistic variable are important. Therefore, a common rule-base was used for both the PD- and PI-parts and the GA only then needed to find the optimum shape of the membership functions for inputs, output and their common rule-base. The initial universes of discourse for change of error and sum of error were chosen within the same interval $[-25, +25]$. The block diagram of the GA-fuzzy controller is shown in Figure 15.26.

15.4.4.1   Genetic representation for membership functions

A real-valued GA was used to optimise parameters of the triangular membership functions of the input and output fuzzy sets and the rule-base. The input variables were partitioned into overlapping sets. The triangular membership functions used here were defined by three parameters: left position, peak and the right position.

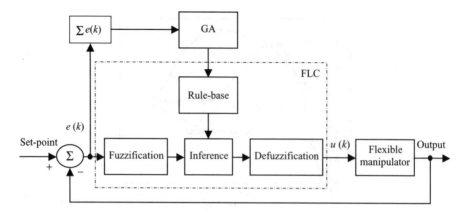

Figure 15.26   GA-based optimisation of MF parameters and rules

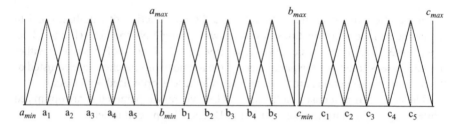

Figure 15.27   Fixed upper and lower limits of the membership functions

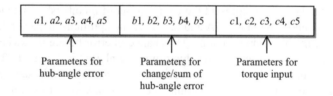

Figure 15.28   Chromosome representation for membership functions

Overlapping of the fuzzy sets (of not more than 50 per cent) was required to ensure good performance of the FLC system. The left and peak positions of successive fuzzy sets were therefore the same as shown in Figure 15.27.

Five parameters were needed to define five fuzzy sets with fixed upper and lower limit, as shown in Figure 15.27. Thus, a total of 15 parameters were required to represent the membership functions in a chromosome. Figure 15.28 shows such a representation, where the a_is are parameters for the hub-angle error, b_is are parameters for the change of error and sum of error and c_is are parameters for the torque input.

A GA was used to find optimal parameters of the membership functions through minimisation of a cost function J expressed as the sum of absolute hub-angle error

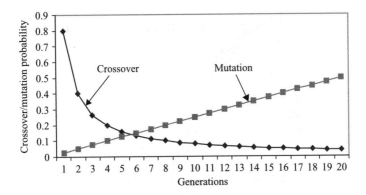

Figure 15.29   Crossover and mutation over generation

and defined as

$$J = \sum_{k=1}^{n} |e\,(k)|$$
(15.9)

where, n is some reasonable time by which the system can be assumed to have settled.

Grefenstette (1986) has shown that the parameter settings for best online performance are a population size 20–30, a crossover probability 0.75–0.95 and a mutation rate 0.0005–0.01. Lee and Takagi (1993) used a population size of 13, crossover and mutation rates of 0.9 and 0.08, respectively. Because of the time taken in the evaluation, a population size of 10 was considered appropriate in this study. A steady-state GA (Michalewicz, 1994), with one-point crossover between two selected parents and mutation with random replacement of selected real values, was utilized for evaluation of the population. For a small population size, normal generating operators do not help to achieve much improvement over successive generations. Therefore, dynamic crossover and mutation operators with probabilities p_{cd} and p_{md}, respectively, were used and defined as

$$p_{cd} = \frac{p_c}{g}, \qquad p_{md} = p_m \left(\frac{g}{g_{max}} \right)$$
(15.10)

where p_c and p_m are the crossover and mutation probabilities and g and g_{max} are the number of generations and maximum generations in the genetic evaluations, respectively. The initial values of p_c, p_m and g_{max} were chosen as 0.8, 0.2 and 20, respectively.

Figure 15.29 shows the relationship between p_{cd}, p_{md} and g. It can be seen that the dynamic crossover probability decreases whereas the mutation probability increases over successive generations. This sort of arrangement is required to keep a balance between uniformity and diversity in a small size population.

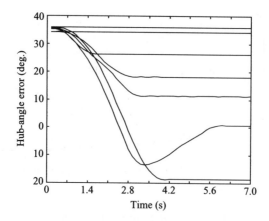

Figure 15.30   Performance of the FLC with first population

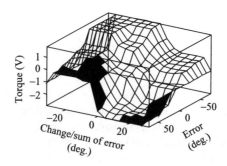

Figure 15.31   Control surface of the best individuals showing dominant rules

15.4.4.2   Experimental results

The GA scheme was applied to the fuzzy controller for the single-link flexible manipulator. The generated chromosomes were used to define fuzzy membership functions in the PD–PI-type controller and then applied to the manipulator system. The performance of the controller was then determined in terms of the fitness function defined earlier in equation (15.9). The smaller the sum of absolute error, the higher is the fitness of the chromosomes. After coding each chromosome as membership functions into the controller program, it was applied to the single-link manipulator system for a demanded hub-angle of 36°. The value of the hub-angle position was monitored and the sum of absolute error calculated.

Figure 15.30 shows the performance achieved with the initial population, which was applied for a demanded hub-angle of 36°. It was observed that some individuals did not respond well, while others achieved good performance with a very small steady-state error. Figure 15.31 shows the control surface of the dominant rules in the first generation.

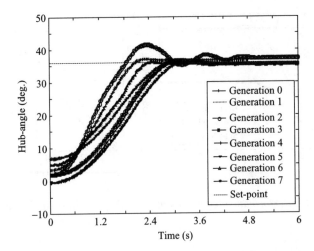

Figure 15.32 Best individuals from 0th generation to 7th generation

Table 15.3 Rise time, settling time, overshoot and steady-state error for different generations

Generation	Rise-time (s)	Overshoot (deg.)	Settling time (s)	Steady-state error (deg.)
0	2.04	37.01	2.88	−0.635
1	2.88	36.95	3.84	0.95
2	1.80	41.55	4.82	−1.32
3	2.88	36.84	3.84	0.63
4	2.88	37.05	4.20	0.42
5	3.08	36.90	4.08	0.58
6	2.40	35.94	2.76	0.52
7	2.88	36.58	3.84	0.47

The performance of the best individuals from the initial population to generation 7 is shown in Figure 15.32. The corresponding performance improvement in terms of rise-time, overshoot, settling time and steady-state error are given in Table 15.3.

Earlier generations show better performance with respect to rise-time but have significant oscillation around the set-point. The performance of later generations is better with respect to settling time and steady-state error. Generations 5, 6 and 7 all achieved an improvement in steady-state error and in the total time. The overall performance of the system did not improve much after generation 4 but rather the difference between the performance of the best individuals and worst individuals was minimised. This results in an improvement in the average fitness of the

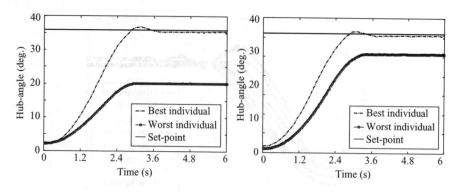

Figure 15.33   Performance improvement from 3rd generation to 7th generation

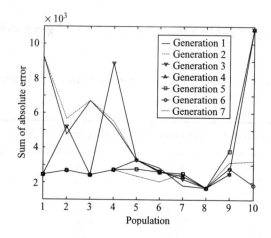

Figure 15.34   Sum of absolute error of population from generation 1 to 7

population over successive generations. Figure 15.33(a) shows a difference of 15°
between the best and worst performances whereas Figure 15.33(b) shows a difference
of 5° between the best and worst performances in generation 3 and 7, respectively.

The characteristics of the learning profile were given by the fitness values of the
individuals of the population. Here fitness was calculated as the sum of absolute error,
as shown in Figure 15.34. Figure 15.35 shows the sum of squared error as a measure of
performance improvement. Another way of expressing the learning profile was as the
mean fitness of the population. The mean fitness, calculated as the mean of the sum
of absolute error and the mean of the sum of squared error is shown in Figures 15.36
and 15.37, respectively. These figures confirm the considerable improvement in the
performance achieved from generations 1 to 7.

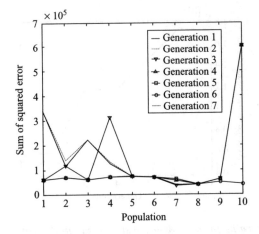

Figure 15.35   Sum of squared-error of population from generation 1 to 7

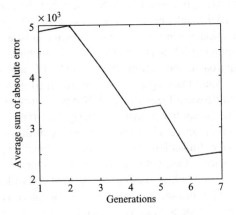

Figure 15.36   Learning profile - Average sum of absolute error

15.5   Summary

This chapter has described two SC approaches to controller design for trajectory control of a single-link flexible manipulator. In the first, offline construction of a modular neural controller with genetic learning was investigated. Although the GA took more training time compared to backpropagation learning, it produced a more general and flexible approach to the construction of a MNN controller. It has been shown that it is relatively straightforward to design a NN in modular form and that the scheme requires an insignificant level of human intervention. Offline construction of a modular NN controller for trajectory control of a flexible manipulator has also been demonstrated and shown to produce faster GA convergence. The modular aspect of the

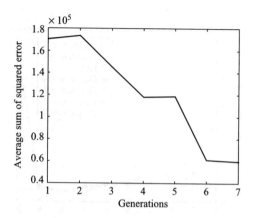

Figure 15.37   Learning profile - Average sum of squared-error

NN controller was analysed in detail. It has been shown that dominant controller input factors can easily be incorporated into a MNN, and that the corresponding control effort can be optimised through genetic evolution so as to achieve the desired control performance. The dominant effect of any control term can be easily accommodated by just adding (removing) a NN to (from) the control structure.

In the second approach, the performance of PD-, PI- and PID-type FLC was studied for online set-point tracking of the single-link flexible manipulator. A novel switching mechanism between fuzzy PD- and PI-type controllers was proposed and shown to have similar behaviour to that of a fuzzy PID-type controller, with considerable improvement in terms of number of rules. A further simplification of the common rule-base and optimisation of the membership function parameters was also discussed and it was shown that, with modification of some of the classical genetic operators, GAs can be used online to design an optimal controller.

This chapter provides insight into designing an adaptive controller from both offline and online perspective and can be applied to a wide range of applications.

Chapter 16

Modelling and control of smart material flexible manipulators

Z.P. Wang, S.S. Ge and T.H. Lee

In this chapter, dynamic modelling and control of smart material robots are investigated. First, dynamic modelling of a single-link smart material robot is introduced. Then, model-free control is presented for the system without employing the dynamics of the system explicitly. The non-model-based controllers are independent of system parameters and thus possess stability robustness to system parameter uncertainties. Finally, tracking problem of smart material robots is solved by employing singular perturbation techniques. Through the active stabilisation of the fast vibration related variables by smart material actuators, direct control of link deflection is possible and better control performances are expected. Simulation results are provided to show the effectiveness of the presented approaches.

16.1 Introduction

In recent years, with the promised advent of lightweight high-strength composite materials, much attention has been given to modelling and control of flexible-link manipulators (Book, 1990; Feliu *et al.*, 1990; Ge *et al.*, 1996, 1998a,c; Korolov and Chen, 1988; Kotnik *et al.*, 1988; Krishnan, 1988). The light-weight flexible-link robot needs less power consumption for high-speed operation, which makes it more feasible than the conventional rigid-link robots in many applications, such as in deep sea and space applications. These advantages greatly motivate the research in modelling and control of flexible robots.

To explicitly consider the effects of structural-link flexibility and deal with active and/or passive vibration control, it is highly desirable to have an explicit, complete and accurate dynamic model of the system available.

The most commonly used modelling methods include the assumed-modes method (AMM) and finite element (FE) method. In the AMM, the elastic deflection of the beam is represented by an infinite number of separable modes. Since the low-frequency modes are dominant in the system's dynamics, and controllers/actuators generally act as low-pass filters (Kanoh *et al.*,1986), the modes are truncated to a finite number to obtain a system with a finite dimensionality. The FE modelling of flexible manipulators can be found in Bayo (1987), Menq and Chen (1988), Usoro *et al.* (1984), among others. In this method, the flexible beam is divided into a finite number of elements, and elastic deformation of the link is represented in the form of a linear combination of admissible functions and generalized coordinates (displacements and rotations of nodes). There are many kinds of admissible functions to be selected. Actually, any function that meets boundary conditions of certain nodes can be applied (Meirovitch, 1975). Most commonly used admissible functions are solutions of the differential equation, which govern the static bending of the considered beam (Meirovitch, 1975). An alternative choice introduced in Matsuno and Yamamoto (1994) is the use of B-spline functions. In the FE method, all the generalized coordinates are physically meaningful; however, the concept of natural frequencies is not explicit.

The control problem for flexible robot manipulators belongs to the class of mechanical systems where the number of control variables is strictly less than the number of mechanical degrees of freedom. It is not easy to control both the rigid-body and elastic deformations precisely because of the undesired residual vibrations. Traditional controllers, for example, the proportional, derivative (PD) feedback, are no longer sufficient for fast and accurate positioning. The system, described by partial differential equations (PDEs), is actually a distributed-parameter system of infinite dimensions (Luo, 1993; Shifman, 1993; Spector and Flashner, 1990). The non-minimum phase behaviour from the base input to the end-point output makes it very difficult to achieve high-level performance and robustness simultaneously. The approaches established for control of rigid-link manipulators are theoretically no longer applicable to flexible ones.

The control targets are usually classified as

- rapid positioning of the flexible arm
- smooth trajectory motion tracking.

The tracking task can be assigned at the joint level, as if the manipulator were rigid. Satisfactory results may be obtained also at the end-effector level, provided that link deformation is kept limited.

On the basis of a truncated (finite-dimensional) model obtained from either the FE method or AMM (Bayo, 1987; Hastings and Book, 1987; Kanoh *et al.*, 1986; Wang and Vidyasagar, 1989*a*), various control approaches can be applied (Cannon and Schmitz, 1984; Siciliano *et al.*, 1986; Tzes and Yurkovich, 1993; Zuo and Wang, 1992). Some problems associated with the truncated model-based methods are highlighted in the literature. These include

- control and observation spillovers due to the ignored high-frequency dynamics

- computing burden caused by using a relatively high-order controller
- difficulty in implementing the controller from the engineering point of view since full state measurements/observations are often required.

In an attempt to overcome these shortcomings, an alternative method has been developed for the control of flexible robot systems in recent years, where, dynamic analysis and controller design are carried out directly based on the original PDE dynamics, named 'PDEs-based approach' (Luo, 1993; Shifman, 1993; Yigit, 1994a). Since the approach avoids the undesirable model truncation, it is more attractive in practical applications. In general, this approach needs more complicated derivations than the truncated model-based methods, because the PDEs of the original flexible robot systems are quite complicated, especially for multi-link flexible robots. This is also a reason that the references above have only considered the single flexible-link case.

In a traditional flexible-link robotic system, there is usually only one motor for each link acting as an actuator. However, the system is infinite dimension, as mentioned above, and this leads further to the difficulty in controller design. Therefore, an alternative way to solve the control problem of flexible-link robots has to be proposed.

Since piezoelectric materials can be bonded/embedded along a beam to achieve transformation between mechanical deformation and electrical field, they can be used to serve as actuators and sensors. Hence, an alternative approach to the control of flexible robots is the use of 'smart materials', that is, materials embedded or bonded with a network of distributed actuators, sensors, and processors. While the resulting system retains all the advantages of a conventional flexible-link robot, such as lightweight, high operational speed, it has additional sensing and control capabilities. It is expected to obtain better control performance for flexible robots easily with the application of smart materials.

Some theoretical and experimental work has been carried out on both modelling and controller design of smart material beams (Bailey and Hubbard, 1985; Crawley and Luis, 1987; Cundari and Abedian, 1991; Varadan *et al.*, 1990; Vukovich and Yousefi-Koma, 1996). All these controllers were designed based on the assumption that the system was exactly modelled. However, this assumption is too ideal to be true in practice. At the same time, problems associated with the truncated model-based methods remains.

For a point-to-point position control or regulation, the main task is to suppress system vibration. Although point-to-point position control is sufficient for certain applications, such as the automatic manufacturing assembly, it is more desirable to be able to drive the robot along a predefined trajectory. The tracking control problem of flexible robots is often solved by converting the problem into two sub-problems:

1. Tracking of the joint motion (the dynamic behaviour of the system is minimum phase, if the output is the joint angular position and the input is the torque at the joint).
2. Suppression of vibration of the flexible links.

Such a consideration directly leads to a singular perturbation treatment (Siciliano and Book, 1988). The attractive feature of this strategy is that the slow control can be designed based on the well-established control schemes for rigid-body manipulators, such as computed torque control (Sciavicco and Siciliano, 1996), robust control (Tang *et al.*, 1996) and adaptive control (Spong and Ortega, 1990).

In this chapter, dynamic modelling and controller design of smart material robots are investigated. Dynamic modelling of a single-link smart material robot is discussed first using the AMM and the FE method. Then, model-free control is presented to regulate the robot. Rather than using the dynamics of the links, the main stability results are obtained using the total energy and the energy–work relationship of the whole system. As the controllers are designed without invoking the dynamics of the system, they are model free and thus robust to uncertainties in the system. In addition, the non-model-based nature also makes the controller free of the problems associated with truncated model-based methods such as the control/observation spillover. Furthermore, adaptive composite tracking controller is presented for smart material robots. Active stabilisation of the fast vibration related variables is proposed through smart material actuators. Direct control of link deflection leads to improved control performance. As the controllers are designed based on signals that are easy to measure, they can be easily implemented in practice.

The rest of the chapter is organised as follows. Section 16.2 presents the dynamic modelling of a single-link smart material robot and some properties of the system. In Section 16.3, model-free robust control is presented for the system without employing the dynamics of the system explicitly. Finally, tracking control is discussed in Section 16.4, and Section 16.5 concludes the chapter.

16.2   Dynamic modelling of a single-link smart material robot

Dynamic modelling of manipulators deals with the mathematical formulation of the dynamic equations describing/characterising robotic arm motion. Model building is justified as (Ge *et al.*, 1998*b*)

- Simulation of the dynamic response of robots provides a means of designing prototype manipulators and testing control strategies without the expense of working with the actual robots.
- Any good regulator of a system must include, explicitly or implicitly, a model of that system, that is, success in the regulation of the system implies that a sufficiently similar model must first be built.

Without losing generality, it is assumed that (a) the deflection of the flexible link is small compared with the length of the link; (b) the payload attached to the free end-point of the flexible robot is a concentrated mass; (c) the base end of the robot is clamped to the rotor of a motor; (d) the effects of any damping are neglected; (e) the flexible robot is only operated in the horizontal X_0OY_0 plane (fixed-base frame), which implies that the dynamic models derived are also valid in space applications; and (f) electrical displacement $\mathbf{D}(x, t)$ is perpendicular to the beam in the X_1OY_1 plane which is the local reference frame with axis OX_1 that is tangent to the beam at the base. Thus, its z component $D_z = 0$. Moreover, due to the small deflection,

$D_x \ll D_y$, therefore, it is assumed that $D_x = 0$. The magnetic field intensity **H** is perpendicular to the plane X_1OY_1; consequently, $H_x = H_y = 0$. Thus, only D_y and H_z will be considered in the following discussion (Ge *et al.*, 1998a, 2001a).

In order to find the dominant physical properties of the proposed system and simplify the system model, it is assumed that the whole beam is very thin, which means that the deflection *w* is only a function of the distance *x* and is independent of the thickness. Furthermore, by choosing the polarisation directions of the upper and lower layer of smart materials opposite to each other, D_y in the upper layer is equal to that of the lower layer when the beam is under small deflection. Therefore, different notations are not used here.

Some fundamental relationships are listed here (Ge *et al.*, 1998a, 2001a).

Piezoelectric effects:

$$\mathbf{F}^S = \mathbf{c}^S \mathbf{S} - \mathbf{h}\mathbf{D}$$

where $\mathbf{F}^S \in R^6$ denotes the simplified stress vector of the smart material, $\mathbf{c}^S \in R^{6\times6}$ is the symmetric matrix of elastic stiffness coefficients of the smart material, $\mathbf{S} \in R^6$ is the simplified strain vector, $\mathbf{h} \in R^{6\times3}$ is the coupling coefficients matrix, and $\mathbf{D}(x,t) \in R^3$ denotes the electrical displacement of the smart material at location *x* and time *t*.

$$\mathbf{E} = -\mathbf{h}^T\mathbf{S} + \beta D$$

where $\mathbf{E} \in R^3$ is the electrical field intensity vector of the smart material, and $\beta \in R^{3\times3}$ denotes the symmetric matrix of impermittivity coefficients.

Magnetic properties neglecting the piezomagnetic effects:

$$\mathbf{B} = \mu\mathbf{H}$$

where $\mathbf{B} \in R^3$ is the magnetic flux density vector, $\mu \in R^{3\times3}$ is the permeability coefficients matrix, and $\mathbf{H}(x,t) \in R^3$ denotes the magnetic field intensity at location *x* and time *t*.

Mechanical properties of the pure beam:

$$\mathbf{F}^M = \mathbf{c}^M \mathbf{S}$$

where $\mathbf{F}^M \in R^6$ denotes the simplified stress vector of the pure beam, and $\mathbf{c}^M \in R^{6\times6}$ is the symmetric matrix of elastic stiffness coefficients of the pure beam.

Without loss of generality, it is assumed that there are *m* pairs of smart material bonded on the beam from l_{1i} to l_{2i}, $0 \leq l_{1i} < l_{2i} \leq L$, *L* is the length of the beam. The system geometry is shown in Figure 16.1, where r is the position vector of point *P* on the beam, *x* is the position of point *P* on the un-deformed beam in the local frame X_1OY_1, $w(x,t)$ denotes the beam deflection at location *x* and time *t*, θ is the joint angle at the hub, and m_3 is the end-point payload.

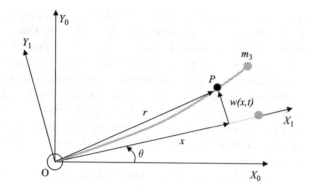

Figure 16.1   Description of a single-link smart material robot

Kinetic energy

The kinetic energy includes mechanical kinetic energy and electrical kinetic energy. The position vector r of a point P on the beam can be expressed in the fixed base frame as:

$$\mathbf{r} = \begin{bmatrix} \cos\theta & -\sin\theta \\ \sin\theta & \cos\theta \end{bmatrix} \begin{bmatrix} x \\ w(x,t) \end{bmatrix}$$

thus,

$$\dot{\mathbf{r}} = \frac{d\mathbf{r}}{dt} = \begin{bmatrix} \cos\theta & -\sin\theta \\ \sin\theta & \cos\theta \end{bmatrix} \begin{bmatrix} -w(x,t)\dot{\theta} \\ \dot{\theta}x + \dot{w}(x,t) \end{bmatrix}$$

where $\dot{x} = 0$ because the length of the beam is assumed to be constant.

The mechanical kinetic energy is given by

$$E_K^M = \frac{1}{2}I_h\dot{\theta}^2 + \int_0^L \frac{1}{2}m_1(x)\dot{\mathbf{r}}^T(x,t)\dot{\mathbf{r}}(x,t)dx + \sum_{i=1}^{m}\int_{l_{1i}}^{l_{2i}} \frac{1}{2}m_2(x)\dot{\mathbf{r}}^T(x,t)\dot{\mathbf{r}}(x,t)dx$$

(16.1)

where I_h is the inertia of the hub and

$$m_1(x) = \begin{cases} ab\rho_1, & 0 \le x < L \\ ab\rho_1 + m_3, & x = L \end{cases}$$
$$m_2(x) = (c_1 + c_2)b\rho_2$$

with a the thickness of the beam, b the width of the beam and that of the smart material, c_1 the thickness of upper surface smart material patch, c_2 the thickness of lower surface smart material patch, ρ_1 the mass per unit volume of the pure beam and ρ_2 the mass per unit volume of the smart material.

The first term on the right-hand side of equation (16.1) is the kinetic energy of the hub, the second term is that of the pure beam, and the last term is that of the smart material. The electrical kinetic energy, that is, magnetic energy, is derived as follows.

According to Maxwell equation,

$$\nabla \times \mathbf{H} = \frac{\partial \mathbf{D}}{\partial t} \tag{16.2}$$

and recalling that $D_x = D_z = 0$, $H_x = H_y = 0$, H_z can be expressed as

$$H_z(x,t) = -\int_0^x \dot{D}_y(\xi,t)d\xi$$

where it is assumed that $H_z(0,t) = 0$ because of the continuity of magnetic field intensity. It can be seen that $H_z(x,t)$ is a function of the time derivative of $D_y(x,t)$. Thus, the electrical kinetic energy E_K^E can be derived as

$$\begin{aligned} E_K^E &= \frac{b(c_1+c_2)\mu_{33}}{2} \sum_{i=1}^m \int_{l_{1i}}^{l_{2i}} \left[\int_0^x \dot{D}_y(\xi,t)d\xi\right]^2 dx \\ &= \frac{\mu_L}{2} \sum_{i=1}^m \int_{l_{1i}}^{l_{2i}} \left[\int_0^x \dot{D}_y(\xi,t)d\xi\right]^2 dx \end{aligned} \tag{16.3}$$

where μ_{33} represents the permeability coefficients matrix of the smart material, and $\mu_L = b(c_1 + c_2)\mu_{22}$ is the permeability per unit length of the smart material robot.

The total kinetic energy of the system E_K is the summation of mechanical kinetic energy and electrical kinetic energy, that is,

$$E_K = E_K^M + E_K^E \tag{16.4}$$

Potential energy

Potential energy includes three parts: the mechanical, the electrical and the coupled parts. The strain $\mathbf{S}$ is, in order, given as

$$S_1 = -y\frac{\partial^2 w(x,t)}{\partial x^2}$$
$$S_2 = S_3 = S_4 = S_5 = S_6 = 0$$

Hence, the total potential energy E_P is given as

$$\begin{aligned} E_P &= \frac{1}{2}b \sum_{i=1}^m \int_{l_{1i}}^{l_{2i}} \int_{-a/2-c_2}^{-a/2} [\mathbf{S}^T\mathbf{F}\mathbf{S} + \mathbf{E}^T\mathbf{D}]dydx + \frac{1}{2}b \int_0^L \int_{-a/2}^{a/2} \mathbf{S}^T\mathbf{F}\mathbf{S}^M dydx \\ &\quad + \frac{1}{2}b \sum_{i=1}^m \int_{l_{1i}}^{l_{2i}} \int_{a/2}^{a/2+c_1} [\mathbf{S}^T\mathbf{F}\mathbf{S} + \mathbf{E}^T\mathbf{D}]dydx \\ &= \frac{c_{L1}}{2} \int_0^L [\frac{\partial^2 w(x,t)}{\partial x^2}]^2 dx + \frac{c_{L2}}{2} \sum_{i=1}^m \int_{l_{1i}}^{l_{2i}} [\frac{\partial^2 w(x,t)}{\partial x^2}]^2 dx \\ &\quad + \frac{\beta_L}{2} \sum_{i=1}^m \int_{l_{1i}}^{l_{2i}} D_y^2(x,t)dx + h_L \sum_{i=1}^m \int_{l_{1i}}^{l_{2i}} D_y(x,t)\frac{\partial^2 w(x,t)}{\partial x^2}dx \end{aligned} \tag{16.5}$$

It is evident from equation (16.5) that the potential energy includes three parts: the mechanical potential energy, the electrical potential energy and the coupled potential energy.

Virtual work

Virtual work also includes mechanical and electrical components. The mechanical virtual work done by the applied force is in the form

$$\delta W^{\mathrm{M}} = \tau \delta \frac{\partial w(0, t)}{\partial x} + \tau \delta \theta$$

The electrical virtual work done by the applied voltage is

$$\delta W^{\mathrm{E}} = \sum_{i=1}^{m} \int_{l_{1i}}^{l_{2i}} b v_i(x, t) \delta D_y(x, t) dx$$

The total virtual work is thus given by

$$\delta W = \delta W^{\mathrm{M}} + \delta W^{\mathrm{E}}$$

16.2.1   *AMM modelling*

As the main purpose of smart materials is to make a highly integrated smart material robot with minor changes in the physical properties, according to AMM (Book, 1990; Ge *et al.*, 1998*a*; Krishnan, 1988), the elastic vibration of the smart material robot can be expressed as

$$w(x, t) = \sum_{i=0}^{\infty} \psi_i(x) q_i(t) \tag{16.6}$$

where $\psi_i(x)$ are the flexible mode functions, and $q_i(t)$ are the generalized coordinates. The electrical displacement is of the form

$$D_y(x, t) = \sum_{i=1}^{\infty} -\frac{h_L}{\beta_L} \psi_i''(x) q_i(t) \tag{16.7}$$

where $h_L = \frac{1}{2} h_{12} b (c_1 - c_2)(c_1 + c_2 + a)$ is the coupling parameter per unit length of the smart material robot with h_{12} the coupling parameter per unit volume of the piezo-electric material, and $\beta_L = b(c_1 + c_2)\beta_{22}$ is the impermittivity per unit length of the smart material robot with β_{22} the impermittivity per unit volume of the piezoelectric material.

As is well known that the first few modes are dominant in describing the system dynamics, the infinite series can be truncated into a finite set, that is,

$$w(x, t) = \sum_{i=0}^{n_{\mathrm{f}}} \psi_i(x) q_i(t) \quad , \qquad 0 \le x \le L$$

$$D_y(x, t) = \sum_{i=1}^{n_{\mathrm{f}}} -\frac{h_L}{\beta_L} \psi_i''(x) q_i(t), \qquad 0 \le x \le L$$

Defining the vector of generalized coordinates as

$$\mathbf{q} = [\, \theta \quad q_1 \quad \cdots \quad q_{n_{\mathrm{f}}} \,]^T$$

The kinetic energy of the system, from equations (16.1), (16.3), (16.4), (16.6) and (16.7) can be written as

$$E_K = E_K^M + E_K^E = \tfrac{1}{2}\dot{\mathbf{q}}^T \mathbf{M} \dot{\mathbf{q}}$$

where $M \in R^{(n_f+1) \times (n_f+1)}$ is a symmetric and positive definite inertial matrix.

Using equations (16.5), (16.6) and (16.7), the potential energy of the system can be rewritten as

$$E_P = \tfrac{1}{2}\mathbf{q}^T \mathbf{K} \mathbf{q}$$

where $\mathbf{K} = \text{diag}[0, \mathbf{K}_{\!f\!f}] \in R^{(n_f+1) \times (n_f+1)}$ is the stiffness matrix.

Considering the boundary condition $\frac{\partial}{\partial x}\delta w(0, t) = 0$, the mechanical virtual work can be obtained as

$$\delta W^M = \mathbf{F}_1^T \delta\mathbf{q}$$

where

$$\mathbf{F}_1 = [\tau, 0, \ldots, 0]^T \in R^{(n_f+1)} \tag{16.8}$$

Assuming that the voltage $V_i(x, t)$ does not depend on x_i, the electrical virtual work can be expressed as

$$\delta W^E = \mathbf{F}_2^T \delta\mathbf{q}$$

where

$$\mathbf{F}_2 = \begin{bmatrix} 0 \\ \mathbf{F}_f \mathbf{v} \end{bmatrix} \in R^{(n_f+1)} \tag{16.9}$$

$$\mathbf{F}_f = \begin{bmatrix} -\frac{bh_L}{\beta_L}\int_{l_{11}}^{l_{21}} \psi_1''(x)dx & \cdots & -\frac{bh_L}{\beta_L}\int_{l_{1m}}^{l_{2m}} \psi_1''(x)dx \\ -\frac{bh_L}{\beta_L}\int_{l_{11}}^{l_{21}} \psi_2''(x)dx & \cdots & -\frac{bh_L}{\beta_L}\int_{l_{1m}}^{l_{2m}} \psi_2''(x)dx \\ \vdots & \ddots & \vdots \\ -\frac{bh_L}{\beta_L}\int_{l_{11}}^{l_{21}} \psi_N''(x)dx & \cdots & -\frac{bh_L}{\beta_L}\int_{l_{1m}}^{l_{2m}} \psi_N''(x)dx \end{bmatrix} \in R^{n_f \times m}$$

$$\mathbf{v} = \begin{bmatrix} v_1 & v_2 & \cdots & v_m \end{bmatrix}^T \in R^m$$

is the vector of control voltages, and $\mathbf{F}_f$ is full row rank that is assured by the configuration of the smart material robot.

Applying the Euler–Lagrange equation

$$\frac{d}{dt}\frac{\partial L}{\partial \dot{q}_i} - \frac{\partial L}{\partial q_i} = F_i$$

the dynamic equation of the whole system can be obtained as

$$\mathbf{M(q)}\ddot{\mathbf{q}} + \mathbf{C(q, \dot{q})}\dot{\mathbf{q}} + \mathbf{Kq} = \mathbf{F}_1 + \mathbf{F}_2 \tag{16.10}$$

It follows from equations (16.8) and (16.9) that

$$\mathbf{F}_1 + \mathbf{F}_2 = \mathbf{Fu}$$

where

$$\mathbf{F} = \begin{bmatrix} \mathbf{I} & \mathbf{0} \\ \mathbf{0} & \mathbf{F}_f \end{bmatrix}, \quad \mathbf{u} = \begin{bmatrix} \tau & \mathbf{v} \end{bmatrix}^T$$

It can be seen that the generalized force u includes two parts, one part is the joint control torque τ, which enters the system through only rigid subsystem, another part is the control voltage $\mathbf{v}$, which enters the system through only the fast subsystem. Subsequently, the dynamic model of a smart material robot can be expressed as (Ge *et al.*, 2001*a*)

$$\mathbf{M}(\mathbf{q})\ddot{\mathbf{q}} + \mathbf{C}(\mathbf{q}, \dot{\mathbf{q}})\dot{\mathbf{q}} + \mathbf{Kq} = \mathbf{Fu}$$

where $\mathbf{M}(\mathbf{q})$ is the symmetric positive definite inertia matrix, $\mathbf{C}(\mathbf{q}, \dot{\mathbf{q}})\dot{\mathbf{q}}$ represents the Coriolis and Centrifugal forces, and $\mathbf{K}$ is the stiffness matrix of the smart material robot.

16.2.2   FE modelling

In this section, under the same assumptions as stated before, the model of smart material robot is derived using the FE method associated with Lagrangian approach. Accordingly, without loss of generality, it is assumed that the beam is divided into N parts of same length $h = L/N$. Thus, the system geometry changes to the form shown in Figure 16.2. It is assumed that the smart material bonded on each element is from s_{1i} to s_{2i}, where $0 \leq s_{1i} < s_{2i} \leq h$. When $s_{1i} = 0$, $s_{2i} = h$, the smart material covers the whole element.

16.2.2.1   FE analysis

Assuming the deflection $w_i(x_i, t), 0 \leq x_i \leq h$ can be represented by a weighted sum of v_{ia} and v_{ib}. In accordance with the boundary conditions, say, $w_i(0, t) = v_{ia}$,

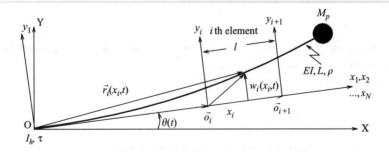

Figure 16.2   Description of the single-link smart material robot

$w_i(h, t) = v_{ib}$, the weights can be chosen as third-order polynomials. Therefore,

$$w_i(x_i) = \mathbf{P}_w^T(x_i)\alpha_i$$

where

$$\mathbf{P}_w(x_i) = \begin{bmatrix} 1 - \frac{3x_i^2}{h^2} + \frac{2x_i^3}{h^3} \\ \frac{3x_i^2}{h^2} - \frac{2x_i^3}{h^3} \end{bmatrix}, \qquad \alpha_i = \begin{bmatrix} v_{ia} \\ v_{ib} \end{bmatrix}$$

Moreover, the electrical displacement can be expressed as

$$D_{yi}(x_i) = \mathbf{P}_D^T(x_i)\mathbf{q}_{2i}$$

where

$$\mathbf{P}_D(x_i) = \begin{bmatrix} 1 - \frac{x_i}{h} & \frac{x_i}{h} \end{bmatrix}^T, \qquad \mathbf{q}_{2i} = \begin{bmatrix} d_{ia} & d_{ib} \end{bmatrix}^T$$

For clarity, Maxwell equation (16.2) is rewritten here

$$\nabla \times \mathbf{H} = \frac{\partial \mathbf{D}}{\partial t}$$

yielding

$$H_{zi}(x_i) = \mathbf{P}_H^T(x_i)\dot{\mathbf{q}}_{2i} \tag{16.11}$$

where

$$\mathbf{P}_H(x_i) = \begin{bmatrix} -x_i + \frac{x_i^2}{2h} & -\frac{x_i^2}{2h} \end{bmatrix}^T$$

It can be seen from equation (16.11) that the magnetic field intensity is related to the 'velocity' of the coordinates. Consequence, magnetic energy will be treated as electrical kinetic energy.

Kinetic energy

The position vector of point P in the ith element is given as

$$\mathbf{r}_i = \mathbf{o}_i + \mathbf{p}_i$$

where $\mathbf{o}_i$ and $\mathbf{p}_i$ are expressed in the fixed base frame as

$$\mathbf{o}_i = \mathbf{a}(\theta) \begin{bmatrix} (i-1)h \\ 0 \\ 0 \end{bmatrix}, \qquad \mathbf{p}_i = \mathbf{a}(\theta) \begin{bmatrix} x_i \\ w_i(x_i) \\ 0 \end{bmatrix}$$

with

$$\mathbf{a}(\theta) = \begin{bmatrix} \cos\theta & -\sin\theta & 0 \\ \sin\theta & \cos\theta & 0 \\ 0 & 0 & 1 \end{bmatrix}$$

representing the transformation matrix from the local reference frame to the fixed base frame. Thus, the mechanical kinetic energy of the ith element is

$$E_{K,i}^M = \int_0^h \frac{1}{2} m_1(x_i) \dot{\mathbf{r}}_i^T \dot{\mathbf{r}}_i dx_i + \int_{s_{1i}}^{s_{2i}} \frac{1}{2} m_2(x_i) \dot{\mathbf{r}}_i^T \dot{\mathbf{r}}_i dx_i$$

$$= \frac{1}{2} \dot{\mathbf{q}}_{1i}^T \mathbf{M}_i^M \dot{\mathbf{q}}_{1i}$$

where $q_{1i} = \begin{bmatrix} \theta & \alpha_i \end{bmatrix}^T$, and

$$m_1(x_i) = \begin{cases} ab\rho_1, & i = 1, \ldots, N-1 \\ ab\rho_1 + m_3\delta(x_i - h), & i = N \end{cases}, \qquad m_2(x_i) = (c_1 + c_2)b\rho_2$$

The electrical kinetic energy of the ith element is given as

$$E_{K,i}^E = \frac{1}{2} b \int_{s_{1i}}^{s_{2i}} \int_{-a/2-c_2}^{-a/2} \left[\mathbf{B}^T \mathbf{H} \right] dy dx_i + \frac{1}{2} b \int_{s_{1i}}^{s_{2i}} \int_{a/2}^{a/2+c_1} \left[\mathbf{B}^T \mathbf{H} \right] dy dx_i$$

$$= \frac{1}{2} \dot{\mathbf{q}}_{2i}^T \mathbf{M}_i^E \dot{\mathbf{q}}_{2i}$$

where $\mathbf{q}_{2i} = \begin{bmatrix} d_{ia} & d_{ib} \end{bmatrix}^T$. The total kinetic energy is thus given as

$$E_{K,i} = E_{K,i}^M + E_{K,i}^E = \frac{1}{2} \dot{\mathbf{q}}_{1i}^T \mathbf{M}_i^M \dot{\mathbf{q}}_{1i} + \frac{1}{2} \dot{\mathbf{q}}_{2i}^T \mathbf{M}_i^E \dot{\mathbf{q}}_{2i}$$

Potential energy

Similar to the above, the strain is in order given as

$$S_1 = -y \frac{\partial^2 w(x,t)}{\partial x^2} = -y \mathbf{P}_s^T(x_i) \mathbf{q}_{1i}$$

$$S_2 = S_3 = S_4 = S_5 = S_6 = 0$$

where

$$\mathbf{P}_s(x_i) = \begin{bmatrix} 0, & -\frac{6}{h^2} + \frac{12x_i}{h^3}, & \frac{6}{h^2} - \frac{12x_i}{h^3} \end{bmatrix}^T$$

The total potential energy for the ith element is thus given as

$$E_{P,i} = \frac{1}{2} b \int_{s_{1i}}^{s_{2i}} \int_{-a/2-c_2}^{-a/2} [\mathbf{S}^T \mathbf{F}^S + \mathbf{E}^T \mathbf{D}] dy dx_i + \frac{1}{2} b \int_0^h \int_{-a/2}^{a/2} \mathbf{S}^T \mathbf{F}^M dy dx_i$$

$$+ \frac{1}{2} b \int_{s_{1i}}^{s_{2i}} \int_{a/2}^{a/2+c_1} [\mathbf{S}^T \mathbf{F}^S + \mathbf{E}^T \mathbf{D}] dy dx_i$$

$$= \frac{1}{2} \mathbf{q}_{1i}^T \mathbf{K}_i^M \mathbf{q}_{1i} + \frac{1}{2} \mathbf{q}_{2i}^T \mathbf{K}_i^E \mathbf{q}_{2i} + \frac{1}{2} \mathbf{q}_{2i}^T \mathbf{K}_i^C \mathbf{q}_{1i}$$

Virtual work

The mechanical virtual work is obtained as

$$\delta W_i^M = \delta \mathbf{q}_{1i}^T \mathbf{F}_i^M$$

where

$$\mathbf{F}_1^M = \begin{bmatrix} \tau(t), & 0, & 0, & 0, & 0 \end{bmatrix}^T, \qquad \mathbf{F}_i^M = \begin{bmatrix} 0, & 0, & 0, & 0, & 0 \end{bmatrix}^T, \quad i \neq 1$$

The electrical virtual work done by the voltage is given as

$$\delta W_i^E = \int_{s_{1i}}^{s_{2i}} b v_i(x_i, t) \delta D_i(x_i, t) dx_i$$

Assuming the voltage $v_i(x_i, t)$ does not depend on x_i, the above yields

$$\delta W_i^E = \delta \mathbf{q}_{2i}^T \mathbf{F}_i^E$$

where

$$\mathbf{F}_i^E = \left[\begin{array}{cc} \dfrac{(s_{2i} - s_{1i})(2h - s_{2i} - s_{1i})}{2h} v_i & \dfrac{s_{2i}^2 - s_{1i}^2}{2h} v_i \end{array} \right]^T$$

16.2.2.2   Dynamic equations

The dynamic equation of the whole system can be obtained using the Euler–Lagrange equation,

$$\frac{d}{dt} \frac{\partial L}{\partial \dot{Q}_{m,j}} - \frac{\partial L}{\partial Q_{m,j}} = F_{m,j}$$

where $Q_{m,j}$ represents components of combination of vectors q_{1i} and $q_{2i}, i = 1, \ldots, N$, which lead to the generalized coordinates

$$\mathbf{Q}_1 = \left[\begin{array}{ccccc} \theta, & v_{1a}, & v_{2a}, & \ldots, & v_{Na}, & v_{Nb} \end{array} \right]^T$$

$$\mathbf{Q}_2 = \left[\begin{array}{cccc} d_{1a}, & d_{2a}, & \ldots, & d_{Na}, & d_{Nb} \end{array} \right]^T$$

$$L = \sum_{i=1}^{N} \left[E_{K,i} - E_{P,i} \right]$$

$$= \sum_{i=1}^{N} \left[\frac{1}{2} \left(\dot{\mathbf{q}}_{1i}^T \mathbf{M}_i^M \dot{\mathbf{q}}_{1i} + \dot{\mathbf{q}}_{1i}^T \mathbf{M}_i^M \dot{\mathbf{q}}_{1i} \right) - \left(\frac{1}{2} \mathbf{q}_{1i}^T \mathbf{K}_i^M \mathbf{q}_{1i} + \frac{1}{2} \mathbf{q}_{2i}^T \mathbf{K}_i^E \mathbf{q}_{2i} \right) - \mathbf{q}_{2i}^T \mathbf{K}_i^C \mathbf{q}_{1i} \right]$$

$$= \frac{1}{2} \dot{\mathbf{Q}}_1^T \mathbf{M}^M \mathbf{Q}_1 + \frac{1}{2} \dot{\mathbf{Q}}_2^T \mathbf{M}^E \dot{\mathbf{Q}}_2 - \left(\frac{1}{2} \mathbf{Q}_1^T \mathbf{K}^M \mathbf{Q}_1 + \frac{1}{2} \mathbf{Q}_2^T \mathbf{K}^E \mathbf{Q}_2 \right) - \frac{1}{2} \mathbf{Q}_2^T \mathbf{K}^C \mathbf{Q}_1$$

and

$$\mathbf{F}_1^T \delta \mathbf{Q}_1 = \sum_{i=1}^{N} \delta \mathbf{q}_{1i}^T \mathbf{F}_i^M = \left(\sum_{i=1}^{N} \mathbf{F}_{i,ext}^M \right)^T \delta \mathbf{Q}_1$$

$$\mathbf{F}_2^T \delta \mathbf{Q}_2 = \sum_{i=1}^{N} \delta \mathbf{q}_{2i}^T \mathbf{F}_i^E = \left(\sum_{i=1}^{N} \mathbf{F}_{i,ext}^E \right)^T \delta \mathbf{Q}_2$$

$$\mathbf{M}^M = \sum_{i=1}^{N} \mathbf{M}_{i,\text{ext}}^M, \mathbf{M}^E = \sum_{i=1}^{N} \mathbf{M}_{i,\text{ext}}^E$$

$$\mathbf{K}^M = \sum_{i=1}^{N} \mathbf{K}_{i,\text{ext}}^M, \mathbf{K}^E = \sum_{i=1}^{N} \mathbf{K}_{i,\text{ext}}^E, \mathbf{K}^C = \sum_{i=1}^{N} \mathbf{K}_{i,\text{ext}}^C$$

where $*_{i,\text{ext}}$ represents the extended form of matrix $*_i$ in accordance with $\mathbf{Q}_1$ and $\mathbf{Q}_2$, while $*_i$ is in accordance with $\mathbf{q}_{1i}$ and $\mathbf{q}_{2i}$. I_h is added to $\mathbf{M}_{1,11}^M$ to include the kinetic energy of the hub, that is, $\frac{1}{2}I_h\dot{\theta}^2$. With regard to the boundary conditions, $v_{1a} = 0$, the corresponding rows and/or columns of extended matrices should be removed.

In this manner, the dynamic equations of the whole system are obtained as

$$\mathbf{M}^M\ddot{\mathbf{Q}}_1 + \mathbf{C}^M(\mathbf{Q}_1,\dot{\mathbf{Q}}_1)\dot{\mathbf{Q}}_1 + \mathbf{K}^M\mathbf{Q}_1 + \mathbf{K}^{C^T}\mathbf{Q}_2 = \mathbf{F}_1$$

$$\mathbf{M}^E\ddot{\mathbf{Q}}_2 + \mathbf{K}^E\mathbf{Q}_2 + \mathbf{K}^C\mathbf{Q}_1 = \mathbf{F}_2 \qquad (16.12)$$

The jkth elements of centripetal/Coriolis matrix $\mathbf{C}^M$ have the form

$$C_{jk}^M = \frac{1}{2}\sum_{l=1}^{N}\left[\frac{\partial M_{jk}^M}{\partial Q_{1,l}} + \frac{\partial M_{jl}^M}{\partial Q_{1,k}} - \frac{\partial M_{kl}^M}{\partial Q_{1,j}}\right]\dot{Q}_{1,l}$$

If $\mathbf{Q}$ is defined as $\mathbf{Q} = \begin{bmatrix} \mathbf{Q}_1^T & \mathbf{Q}_2^T \end{bmatrix}^T$, equations (16.10) and (16.12) can be combined into

$$\mathbf{M}\ddot{\mathbf{Q}} + \mathbf{C}(\mathbf{Q},\dot{\mathbf{Q}})\dot{\mathbf{Q}} + \mathbf{K}\mathbf{Q} = \mathbf{F}$$

where

$$\mathbf{M} = \begin{bmatrix} \mathbf{M}^M & \mathbf{0} \\ \mathbf{0} & \mathbf{M}^E \end{bmatrix}, \quad \mathbf{C} = \begin{bmatrix} \mathbf{C}^M & \mathbf{0} \\ \mathbf{0} & \mathbf{0} \end{bmatrix},$$

$$\mathbf{K} = \begin{bmatrix} \mathbf{K}^M & \mathbf{K}^{C^T} \\ \mathbf{K}^C & \mathbf{K}^E \end{bmatrix}, \quad \mathbf{F} = \begin{bmatrix} \mathbf{F}_1 \\ \mathbf{F}_2 \end{bmatrix}$$

Thus, a full FE model is obtained for the considered smart material robotic system.

Since matrix $\mathbf{M}^E$ in equation (16.12) is very small, the dynamics in equation (16.12) are usually omitted. A reduced model can thus be obtained as

$$\mathbf{M}^M\ddot{\mathbf{Q}}_1 + \mathbf{C}^M(\mathbf{Q}_1,\dot{\mathbf{Q}}_1)\dot{\mathbf{Q}}_1 + (\mathbf{K}^M - \mathbf{K}^{C^T}\mathbf{K}^{E^{-1}}\mathbf{K}^C)\mathbf{Q}_1 = \mathbf{F}_1 - \mathbf{K}^{C^T}\mathbf{K}^{E^{-1}}\mathbf{F}_2$$

The above model can also be derived by removing the electrical kinetic energy from the system kinetic energy.

16.3   Model-free regulation of smart material robots

In this section, model-free robust regulation is investigated for the single-link smart material robot described in Section 16.2.

16.3.1   System description

The system considered here can be deployed either in space, or configured in the horizontal plane. In both cases, the effects of the gravity are neglected. There are n pairs of piezoelectric sensors and actuators collocated at n points along the beam. θ is the joint angular position, $\dot{\theta}$ is the joint angular velocity, θ_d is the desired joint angular position, v_i is the voltage applied to the ith piezoelectric actuator and τ is the torque applied to the hub. Because $\int_0^t I_i dt$ defines the electrical charge induced by the ith piezoelectric sensor, which is proportional to the strain, I_i therefore gives the electrical current induced by the ith piezoelectric sensor and is proportional to the change of strain (Ge *et al.*, 2000; Won, 1995). To facilitate further discussion, define the following vectors:

$$\mathbf{I} = \begin{bmatrix} I_1 & I_2 & \cdots & I_n \end{bmatrix}^T \in R^n$$

$$\mathbf{v} = \begin{bmatrix} v_1 & v_2 & \cdots & v_n \end{bmatrix}^T \in R^n$$

It follows from the above that there are two kinds of independent controls: torque control and voltage control, for the system under study. The motor is primarily used to move the smart material robot about the base, and piezoelectric patches are used to suppress the residual vibration of the robot.

For the system described above, the total work done by external inputs is

$$W = \int_0^t \tau(t)\dot{\theta}(t)dt + \int_0^t \mathbf{v}^T \mathbf{I} dt$$

which consists of the work done by the torque applied to the hub and the work contributed by the voltage applied to piezoelectric actuators. The energy–work relationship, that is, the increment in system energy equal to the work done by external inputs, yields (Ge *et al.*, 2000)

$$[E_K(t) + E_P(t)] - [E_K(0) + E_P(0)] = \int_0^t \dot{\theta}(t)\tau(t)dt + \int_0^t \mathbf{I}^T \mathbf{v} dt \qquad (16.13)$$

where $E_K(t)$ and $E_P(t)$ are the total kinetic energy and total potential energy of system at time t, respectively, $E_K(0)$ and $E_P(0)$ are constants representing the kinetic and potential energies at time 0, respectively. Therefore, taking time derivatives of both sides of equation (16.13) yields

$$\dot{E}_K(t) + \dot{E}_P(t) = \dot{\theta}(t)\tau(t) + \mathbf{I}^T \mathbf{v} \qquad (16.14)$$

which will be used for the controller design in the following section.

16.3.2   *Model-free controller design*

In this section, two classes of model-free controllers are presented by further extending the results in Ge *et al.* (1996, 1998c). The control objective here is to rotate the robot to a desired angular position and simultaneously suppress the residual vibrations effectively.

Strain feedback was introduced into traditional joint PD controller to improve the control performance of a single-link flexible robot in Ge *et al.* (1998c). But, the resulting performance is limited because there is only one actuator for each link. In this chapter, the idea of strain feedback for controlling flexible-link robots is extended to smart material robots for better performance by fully exploiting the fact that more control signals are available.

16.3.2.1   Decentralised model-free control

Decentralised controllers are desirable in certain cases because of the technical difficulties in implementing a centralised controller, and the inherent robustness of a decentralised control system (Ge *et al.*, 2000).

The decentralised model-free controller (DMFC) is given by

$$
\mathbf{u} = \begin{bmatrix} \tau \\ \mathbf{v} \end{bmatrix} = \begin{bmatrix} -k_{p\theta}[\theta(t) - \theta_d] - k_{v\theta}\dot{\theta}(t) \\ -\mathbf{k}_{ps}\int_0^t \mathbf{I}dt - \mathbf{k}_{vs}\mathbf{I} \end{bmatrix}
\tag{16.15}
$$

where $\mathbf{k}_{ps} = \text{diag}[k_{ps_i}] \in R^{n \times n}$, $\mathbf{k}_{vs} = \text{diag}[k_{vs_i}] \in R^{n \times n}$, and $k_{p\theta}$, $k_{v\theta}$, $k_{vs_i} > 0$, $k_{ps_i} \geq 0$. It is clear that the controller in equation (16.15) is decentralised in the sense that only local signals are used in constructing the local feedback control. The decentralised nature of the control scheme is also in line with the original objectives of smart materials that sensing and control are done locally for different physical properties. Furthermore, the controller can be easily implemented due to its simplicity. The stability of the system is summarised in the following theorem.

Theorem 16.1: Consider the smart material robot described in Section 16.3.1. Let the control objective be to rotate the robot to the desired angular position, while suppress the residual vibration simultaneously. Then, the proposed DMFC in equation (16.15) can guarantee the stability of the closed-loop smart material robot system.

Proof: Consider the following Lyapunov function candidate:

$$
V_1(t) = E_K(t) + E_P(t) + \frac{1}{2}k_{p\theta}[\theta(t) - \theta_d]^2 + \frac{1}{2}\left(\int_0^t \mathbf{I}dt\right)^T \mathbf{K}_{ps}\left(\int_0^t \mathbf{I}dt\right)
\tag{16.16}
$$

By virtue of equation (16.14), the time derivative of V_1 is given by

$$
\dot{V}_1(t) = \dot{\theta}(t)\tau(t) + \mathbf{I}^T\mathbf{v} + k_{p\theta}[\theta(t) - \theta_d]\dot{\theta}(t) + \mathbf{I}^T\mathbf{K}_{ps}\left(\int_0^t \mathbf{I}dt\right)
\tag{16.17}
$$

Substituting the control input from equation (16.15) into equation (16.17) yields

$$
\dot{V}_1(t) = -k_{p\theta}\dot{\theta}(t)^2 - \mathbf{I}^T\mathbf{K}_{ps}\mathbf{I} \leq 0
$$

Therefore, the closed-loop system is stable in the Lyapunov sense.

Remark 16.1: The decentralised controller is desirable for the control of a large-scale space structure, where communication among different parts of the structure is neither easy to implement nor cost effective. The decentralised controller is fault tolerant and robust to sensor/actuator failures. For example, if some of the sensors and actuators become faulty, the corresponding local control loops are broken with no input to the system. Furthermore, the overall closed-loop stability is not affected as can easily be seen from the proof of the theorem in Ge *et al.* (2000). The idea was further extended to control a closed-loop chain mechanism as a special case in Matsuno *et al.* (2002).

16.3.2.2 Centralised model-free controller

In order to further improve the control performance and consider those situations in which centralised controllers can be implemented, the centralised model-free control design, which fully utilizes the potential of the extra sensing and control capabilities of the smart material robots, can be studied (Ge *et al.*, 2000).

The centralised model-free controller (CMFC) is give by

$$\mathbf{u} = \begin{bmatrix} \tau \\ \mathbf{v} \end{bmatrix} = \begin{bmatrix} -k_{p\theta}[\theta(t) - \theta_d] - k_{v\theta}\dot{\theta}(t) - \tau_c \\ -\mathbf{k}_{ps} \int_0^t \mathbf{I} dt - \mathbf{k}_{vs}\mathbf{I} - \mathbf{v}_c \end{bmatrix} \tag{16.18}$$

where $\tau_c = \sum_{i=1}^{n_0} k_{i\theta} f_{i\theta}(t) \int_0^t \dot{\theta} f_{i\theta}(s)\, ds$, and $\mathbf{v}_c$ is a column vector defined as

$$\mathbf{v}_c = \begin{bmatrix} \sum_{j=1}^{n_1} k_{1j} f_{1j}(t) \int_0^t I_1(s) f_{1j}(s)\, ds \\ \sum_{j=1}^{n_{21}} k_{2j} f_{2j}(t) \int_0^t I_2(s) f_{2j}(s)\, ds \\ \vdots \\ \sum_{j=1}^{n_n} k_{nj} f_{nj}(t) \int_0^t I_n(s) f_{nj}(s)\, ds \end{bmatrix} \in R^n$$

where $f_{i\theta}, f'_{ij}$s are time integrable signals, $\mathbf{k}_{ps} = \mathrm{diag}[k_{ps_i}] \in R^{n \times n}$, $\mathbf{k}_{vs} = \mathrm{diag}[k_{vs_i}] \in R^{n \times n}$, and $k_{p\theta}, k_{v\theta}, k_{vs_i} > 0$, $k_{ps_i} \geq 0$, $k_{i\theta}, k_{ij} \geq 0$.

In comparison with the decentralised controller in equation (16.15), it is the presence of τ_c and $\mathbf{v}_c$ that makes equation (16.18) a centralised controller since $f_{i\theta}$ and f_{ij} can be signals from other locations on the link. The closed-loop stability of the system is summarised in the following theorem.

Theorem 16.2: Consider the smart material robot described in Section 16.3.1. Let the control objective be to rotate the robot to the desired angular position, while suppressing the residual vibration simultaneously. Then, the proposed CMFC in equation (16.18) can guarantee the stability of the closed-loop smart material robot system.

Proof: Consider the following Lyapunov function candidate:

$$V(t) = V_1(t) + \frac{1}{2}\sum_{i=1}^{n_0} k_{i\theta}\left[\int_0^t \dot{\theta}(s)f_{i\theta}(s)ds\right]^2 + \frac{1}{2}\sum_{i=1}^{n}\sum_{j=1}^{n_n} k_{ij}\left[\int_0^t I_i(s)f_{ij}(s)ds\right]^2$$

(16.19)

where $V_1(t)$ is as defined in equation (16.16).

By virtue of equation (16.14), the time derivative of V is given by

$$\dot{V}(t) = \dot{\theta}(t)\tau(t) + \sum_{i=1}^{n} v_i(t)I_i(t) + k_{p\theta}[\theta(t) - \theta_d]\dot{\theta}(t)$$

$$+ \sum_{i=1}^{n} k_{ps_i}\int_0^t I_i(s)dsI_i(t) + \sum_{i=1}^{n_0} k_{i\theta}\dot{\theta}(t)f_{i\theta}(t)\int_0^t \dot{\theta}(s)f_{i\theta}(s)ds$$

(16.20)

$$\sum_{i=1}^{n}\sum_{j=1}^{n_n} k_{ij}I_i(t)f_{ij}(t)\int_0^t I_i(s)f_{ij}(s)ds$$

Substituting equation (16.18) into equation (16.20) yields

$$\dot{V}(t) = -k_{v\theta}\dot{\theta}(t)^2 - \sum_{i=1}^{n} K_{vs_i}I_i^2(t) \leq 0$$

which means that the closed-loop system is stable in the Lyapunov sense.

Remark 16.2: Theoretically, item $\mathbf{k}_{vs}\mathbf{I}$ in equations (16.15) and (16.18) can be replaced by $\mathbf{k}_{vs}\mathbf{P}(\mathbf{I})$, where $\mathbf{P}(\mathbf{I})$ is a column vector with its ith element as a function of I_j. It can be found that as long as $p_i(I_i)I_i \geq 0, i = 1,\ldots,n,,$ the stability of the closed loop can be guaranteed. Some examples of $p_i(I_i)$ are listed as follows:

$$p_i(I_i) = \begin{cases} I_i \\ \text{sgn}(I_i) \\ I_i^3 \end{cases}$$

$p_i(I_i)$ can be selected in the implementation of the controller according to different constructions of the system.

Remark 16.3: The stability results are obtained by using the energy–work relationship of the whole system and the closed-loop stability proofs are independent of the system dynamics, Thus, drawbacks/problems associated with truncated model-based approaches are essentially avoided.

Remark 16.4: The control parameters in equations (16.15) and (16.18) are independent of the system parameters, and thus both controllers are robust to system parameter uncertainties. In fact, the closed-loop systems are stable as long as $k_{p\theta}$, $k_{v\theta}, k_{vs_i} > 0, k_{ps_i} \geq 0, k_{i\theta}, k_{ij} \geq 0$.

Remark 16.5: From the construction of DMFC controller, it can be seen that besides the conventional joint torque control, there is an additional voltage control applied to the flexible links. The voltage control is constructed from electrical current induced in the piezoelectric sensor. As the electric current induced in the piezoelectric sensor is proportional to the change of strain (Won, 1995), the voltage control is a kind of direct control of the vibration of the link. For DMFC in equation (16.15), with the aid of local voltage control, the vibration of the flexible link can be controlled directly and the control performance can be improved. In an attempt to include explicit evaluation of the vibration into the controller, non-linear terms τ_c and v_c are introduced into CMFC in equation (16.18) to improve the control performance without destabilising the system. At this point, it is desired to improve the control performance without destabilising the system. Theoretically, the stability of the system will not be affected for any time-integrable function $f_{i\theta}, f_{ij}$, but it is preferable to select $f_{i\theta}, f_{ij}$ to be associated with the vibration of the flexible link. Depending on the actual instrumentation, $f_{i\theta}, f_{ij}$ can be chosen as any variable or any combination of variables related to the vibration of the flexible link at any place along the beam. The model-free controllers in equations (16.15) and (16.18) constructed here can be easily implemented because $\theta(t)$, $\dot{\theta}(t)$ and $I_i(t)$ are easily obtainable, while $f_{i\theta}$ and f_{ij} can be chosen as any other measurable signals according to different system configurations.

Remark 16.6: It is known that collocated actuators/sensors lead to an energy dissipative system in which the stability and a certain degree of robustness are guaranteed (Spector and Flashner, 1990). The simplest collocated controller is the joint PD controller, which uses only joint angle and joint velocity feedback and controls the flexible robot as a rigid one. The proposed DMFC belongs to collocated control. Thus, it can guarantee the stability of the whole system with random sensor locations. As stated in Spector and Flashner (1990), non-collocated sensor feedback may lead to instability. In Luo (1993), the extra feedback selected was the base strain of the flexible link. This feedback is also collocated since the base strain actually represents the bending moment of the beam at the joint. However, for the CMFC proposed in this work, it is shown that extra non-collocated feedback, if introduced in a certain way, can also be utilized without destroying the stability.

Remark 16.7: When the system is subjected to bounded disturbances at the collocated positions of the control inputs, that is, τ_{dis} and v_{disi}, $i = 1, \ldots, n$, the proposed controllers can be modified to compensate for the effects of external disturbances.

When the system is subjected to bounded disturbances at collocated positions of the control inputs, the increment of the system energy is given by

$$[E_K(t)+E_P(t)]-[E_K(0)+E_P(0)] = \int_0^t \dot{\theta}(t)(\tau(t) + \tau_{dis})dt + \sum_{i=1}^n \int_0^t I_i(v_i + v_{disi})dt$$

where τ_{dis} and v_{dis} are upper bounded by τ_{dism} and v_{dism}, respectively. Accordingly, this yields

$$\dot{E}_K(t) + \dot{E}_P(t) = \dot{\theta}(t)(\tau(t) + \tau_{dis}) + \sum_{i=1}^n I_i(v_i + v_{disi}) \tag{16.21}$$

Consider the following modified model-free controller:

$$\mathbf{u} = \begin{bmatrix} \tau \\ \mathbf{v} \end{bmatrix} = \begin{bmatrix} -k_{p\theta}[\theta(t) - \theta_d] - k_{v\theta}\dot\theta(t) - \tau_c \\ -\mathbf{k}_{ps} \int_0^t \mathbf{I}dt - \mathbf{k}_{vs}\mathbf{I} - \mathbf{v}_c \end{bmatrix} \tag{16.22}$$

where $\tau_c = -\tau_{\text{dism}}\text{sgn}(\dot\theta)$, and $\mathbf{v}_c$ is a column vector defined as

$$\mathbf{v}_c = \begin{bmatrix} -v_{\text{dism}}\text{sgn}(I_1) \\ -v_{\text{dism}}\text{sgn}(I_2) \\ \vdots \\ -v_{\text{dism}}\text{sgn}(I_n) \end{bmatrix} \in R^n$$

$\mathbf{k}_{ps} = \text{diag}[k_{ps_i}] \in R^{n \times n}$, $\mathbf{k}_{vs} = \text{diag}[k_{vs_i}] \in R^{n \times n}$, and $k_{p\theta}, k_{v\theta}, k_{vs_i} > 0$, and $k_{ps_i} \geq 0$.

The closed-loop system is stable by choosing the following Lyapunov function candidate:

$$V_m(t) = E_K(t) + E_P(t) + \frac{1}{2}k_{p\theta}[\theta(t) - \theta_d]^2 + \frac{1}{2}\left(\int_0^t \mathbf{I}dt\right)^{\mathrm{T}} \mathbf{K}_{ps} \left(\int_0^t \mathbf{I}dt\right)$$

By virtue of equation (16.21), the time derivative of V_m is given by

$$\dot V_m(t) = \dot\theta\tau + \dot\theta\tau_{\text{dis}} + \sum_{i=1}^n v_i(t)I_i(t) + \sum_{i=1}^n v_{\text{disi}}I_i$$

$$+ k_{p\theta}[\theta(t) - \theta_d]\dot\theta + \sum_{i=1}^n k_{ps_i} \int_0^t I_i(s)ds I_i$$

Substituting the control input from equation (16.22) into the above equation yields

$$\dot V_m(t) = -k_{v\theta}\dot\theta(t)^2 - \sum_{i=1}^n K_{vsi}I_i^2(t) - |\dot\theta|\tau_{\text{dism}} + \dot\theta\tau_{\text{dis}} - \sum_{i=1}^n |I_i|v_{\text{dism}} + \sum_{i=1}^n I_i v_{\text{disi}}$$

$$\leq 0$$

Therefore, the closed-loop system is stable in the Lyapunov sense.

Remark 16.8: In both Theorems 16.1 and 16.2, only closed-loop stability is claimed. It is not easy to prove the asymptotic stability of the system because of the infinite dimensionality of the system. Asymptotic tracking control of an Euler–Bernoulli beam has been achieved in Shifman (1993) under the unrealistic assumption of zero hub inertia. In the following, before giving rigorous proof based on a truncated model, it is shown that in practice the smart material robot can only possibly stop at $\theta = \theta_d$ without vibration. Assume that the link stops at the position $\theta = \theta_1$ (hence $\dot\theta = 0$) with $\theta_1 \neq \theta_d$. In practice, due to the existence of internal structural damping in the smart material robot (structural damping is neglected in the proofs of Theorem 16.1 and Theorem 16.2), the link must tend to stop vibrating and finally be static at the un-deformed position. As I_i is proportional to the change of strain, $\mathbf{I} = \mathbf{0}$. Hence, there is no energy input to the system since $\dot\theta\tau + \mathbf{I}^T\mathbf{v} = 0$ and the robot under study is in the horizontal plane. Furthermore, $\int_0^t I_i dt$ is proportional to the strain, $\int_0^t \mathbf{I} dt = 0$ when the link is un-deformed. As signals $f_{i\theta}, f_{ij}$, are chosen to be zero when the link is at rest because these functions are associated with the vibration of the link, consequently, all the other terms except the first one in τ equal to zero and $\mathbf{v} = \mathbf{0}$. Therefore, τ is a non-zero constant and thus $\theta = \theta_1$ cannot hold. The only possibility is that $\theta = \theta_d$, which implies that end-point regulation can be achieved in practice.

Theorem 16.3: The controller in equation (16.15) can guarantee asymptotic stability of the damped truncated system where the flexible deflection is described arbitrarily by any finite number of flexible modes. Furthermore, the controller in equation (16.18) can also guarantee asymptotic stability of the same truncated system if $f_{i\theta}$ and f_{ij} are selected as functions equal to zero when the flexible smart material robot is un-deformed.

Proof: Using the same Lyapunov functions in equations (16.16) and (16.19), the motion of the system is considered in the largest invariant set in $\dot V_1 = 0 (\dot V = 0)$. In both cases, $\dot\theta \equiv 0$ and $\mathbf{I} \equiv \mathbf{0}$. Subsequently, $\ddot\theta = 0$, $\tau = -k_{p\theta}(\theta - \theta_d)$. Furthermore, according to Remark 16.8, $\mathbf{v} = \mathbf{0}$.

Considering the motion of system in $\dot V_1 = 0 (\dot V = 0)$, the following PDEs and boundary conditions of the system hold:

$$c_e \frac{\partial^2 y(0,t)}{\partial x^2} - k_{p\theta}(\theta - \theta_d) = 0 \tag{16.23}$$

$$\rho L \ddot y(x,t) = -c_e \frac{\partial^4 y(x,t)}{\partial x^4}$$

$$y(0,t) = 0, \quad \frac{\partial y(0,t)}{\partial x} = 0, \quad \frac{\partial^2 y(0,t)}{\partial x^2} = 0$$

$$c_e \frac{\partial^3 y(L,t)}{\partial x^3} = m_3 \ddot y(L,t)$$

where $y(x,t)$ is the deflection of the link, c_e is the electro-mechanic stiffness per length of the smart material beam, m_3 is the mass of payload at the end-point.

The solution $y(x,t)$ has already been given in Section 16.2 as

$$y(x,t) = \sum_{i=1}^{\infty} \phi_i(x) q_i(t)$$

where

$$\phi_i(x) = A_i \left[\cosh \frac{\beta_i x}{L} - \cos \frac{\beta_i x}{L} - \gamma_i \left(\sinh \frac{\beta_i x}{L} - \sin \frac{\beta_i x}{L} \right) \right]$$

$$q_i(t) = B_i \cos \omega_i t + C_i \sin \omega_i t$$

$$\omega_i = \frac{\beta_i^2}{L^2}\sqrt{\frac{c_e}{\rho_L}}$$

Note that when A_i, B_i and C_i are determined using the method given in Section 16.2, the '*initial*' moment should denote the moment when the system motion enters the invariant set, rather than the initial operating moment since the motion of the system is considered in the largest invariant set in $\dot{V} = 0$.

From modal analysis, those modes with comparatively low frequencies are dominant. The following reasonable approximation can be made

$$y(x, t) = \sum_{i=1}^{N} \phi_i(x) q_i(t) \tag{16.24}$$

where N represents the number of flexible modes used to approximate the deflection.
Substituting equation (16.24) into equation (16.23) yields

$$\sum_{i=1}^{N} a_i q_i(t) = 0 \tag{16.25}$$

where

$$a_0 = 1, \quad q_0 = k_{p\theta}(\theta - \theta_d), \quad a_i = c_e \phi^{''}(0) = 2\frac{A_i \beta_i^2}{L^2} \neq 0$$

Note that q_0 is a constant. For the summation in equation (16.25), the inner product defined in Shifman (1993) is applicable, that is,

$$\langle q_i, \quad q_j \rangle := \lim_{T \to \infty} \int_0^T q_i(t) q_j(t) dt = \begin{cases} = 0, & i \neq j \\ \neq 0, & i = j \end{cases}$$

Applying this inner product to the summation in equation (16.25) leads to the fact that each q_i, $(i = 0, 1, \ldots, N)$ must be zero, and subsequently $\theta = \theta_d$ and $w(x, t) = 0$.

Invoking the truncation assumption, the elastic deflection of the link is assumed to be described by a finite number of flexible modes, and subsequently the system is of only finite dimension. For this truncated system, because it has been proven that the largest invariant set $\dot{V}_1 = 0(\dot{V} = 0)$ is the final equilibrium position, the asymptotic stability directly follows LaSalle's theorem.

Remark 16.9: Dynamic modelling and control of multi-link flexible robots are, in general, much more difficult and complex. If the truncated model of finite dimension is used for control design, the complexity and order of the corresponding controller increases dramatically as the order of the model increases. The extension from single link to multi-link is usually not that straightforward. However, the approach presented based on the system energy–work relationship applies to systems of any number of links, and the results obtained above can be easily extended to multi-link cases without the influence of gravity (Ge *et al.*, 2001*b*).

The effectiveness of the proposed controllers are shown through simulation studies. In the simulations, the system parameters are given in Table 16.1. The system is

Table 16.1   *System parameters of a single-link smart material robot*

Parameter	Symbol (unit)	Value
Beam thickness	a(m)	0.008
Piezoelectric actuator thickness	c_1(m)	0.0008
Piezoelectric sensor thickness	c_2(m)	0.0004
Width of the beam	b(m)	0.01
Length of the piezoelectric material	l(m)	0.01
Length of the beam	L(m)	1
Density of the beam	ρ_1(kg/m^3)	500
Density of the piezoelectric material	ρ_2(kg/m^3)	100
Moment of inertia of the hub	I_h(kgm^2)	3.0
Payload at the end-point	m_3(kg)	0.001
Stiffness of the beam	c_{11}^m (N/m^2)	6×10^8
Stiffness of the piezoelectric material	c_{11}^s (N/m^2)	8×10^6
Permeability of the piezoelectric material	μ_{33} (H/m)	1.2×10^6
Coupling parameter	h_{12}(V/m)	5×10^{11}
Impermittivity	β_{22}(m/F)	4×10^{14}

simulated using an AMM model in which the first four mode shape functions are considered. A fourth-order Runge–Kutta programme with adaptive step-size is used to numerically solve the ordinary differential equations (Ge *et al.*, 1998*b*). The sampling interval is set to be 0.005 s. For simplicity, only one pair of piezoelectric actuator and sensor is used in the simulation because it is enough to show the effectiveness of the model-free controllers. The pair of piezoelectric patches are located at $x = 0.1L$ of the beam with L representing the full length of the link. The initial and final desired positions of the smart material robot are $\theta(0) = 0°$ and $\theta_d = 60°$ respectively. The robot is assumed to be initially at rest without any deformation.

First consider the closed-loop performance under pure PD control;

$$\tau(t) = -k_{p\theta}[\theta(t) - \theta_d] - k_{v\theta}\dot{\theta}$$

while the voltage input to the piezoelectric actuator is set to zero, which means that no explicit effort is made on the residual vibration suppression.

To avoid possible overshoot of the angular position, assume that the equivalent system is lightly overdamped with a damping factor of $\xi = 1.0195$. Letting $\omega = 2.5487$, yields $k_{p\theta} = 19.5813$ and $k_{v\theta} = 15.6650$, which are used for the pure joint PD controller and the controllers listed below.

The energy-based robust controller (EBRC) using end-point deflection feedback is given by

$$\text{EBRC}: \tau(t) = -k_{p\theta}[\theta(t) - \theta_d] - k_{v\theta}\dot{\theta} - k_{1\theta}w(L,t)\int_0^t \dot{\theta}(\xi)w(L,t)d\xi$$

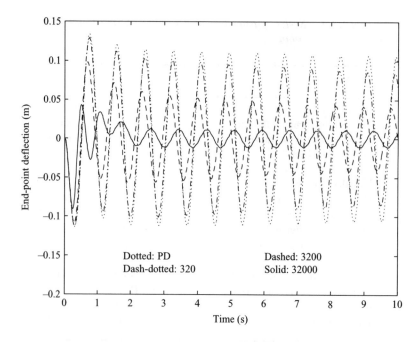

Figure 16.3   End-point deflections with PD control and EBRC

where $w(L, t)$ represents the end-point deflection of the robot.

The DMFC and CMFC to be compared with are given by

$$\text{DMFC} : \mathbf{u} = \begin{bmatrix} \tau \\ \mathbf{v} \end{bmatrix} = \begin{bmatrix} -k_{p\theta}[\theta(t) - \theta_d] - k_{v\theta}\dot{\theta}(t) \\ -\mathbf{k}_{vs}\mathbf{I} \end{bmatrix}$$

and

$$\text{CMFC} : \mathbf{u} = \begin{bmatrix} \tau \\ \mathbf{v} \end{bmatrix} = \begin{bmatrix} -k_{p\theta}[\theta(t) - \theta_d] - k_{v\theta}\dot{\theta}(t) - k_{1\theta}w(L, t) \int_0^t \dot{\theta}w(L, \xi)d\xi \\ -\mathbf{k}_{vs}\mathbf{I} \end{bmatrix}$$

PD control versus EBRC

The end-point deflections under PD control and EBRC with different $k_{1\theta}$s are shown in Figure 16.3. It can be seen that PD control gives the worst performance while EBRC can suppress the vibration effectively when $k_{1\theta}$ is large enough, say $k_{1\theta} = 32\,000$.

EBRC versus DMFC

Figure 16.4 shows the end-point deflection with DMFC with different feedback gains ($k_{vs1} = 3.2 \times 10^4, 3.2 \times 10^5$ and 3.2×10^6) in comparison with the relatively good EBRC with gain $k_{1\theta} = 3.2 \times 10^4$. It can be seen that when the feedback gain k_{vs1} is large enough, DMFC is much more effective in residual vibration suppression in comparison with EBRC. While there is no residual vibration at steady state with DMFC,

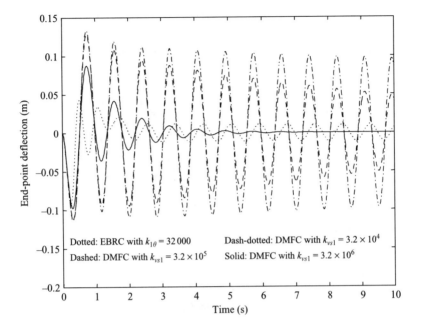

Figure 16.4   End-point deflections with EBRC and DMFC

there is residual vibration of small magnitude with EBRC. For better understanding, and a clear presentation, other signals in the closed-loop are compared between EBRC and DMFC with $k_{vs1} = 3.2 \times 10^6$. Figure 16.5 shows the bounded joint angle responses of the corresponding EBRC and DMFC. Though not much conclusion can be drawn from the joint angle responses, the torque control with DMFC is much smoother during the transient period in comparison with that with EBRC as shown in Figure 16.6. The voltage acting on the piezoelectric actuator in DMFC is shown in Figure 16.7.

DMFC versus CMFC

Figure 16.8 shows the end-point deflection with the DMFC and CMFC with voltage feedback gains $k_{vs1} = 3.2 \times 10^5$ and 3.2×10^6. It can be seen that the performance of CMFC is better than that of DMFC in residual vibration suppression for the same feedback gain. The joint angle with DMFC and CMFC is shown in Figure 16.9, where there is not much difference between the two responses.

If more precise control performance is required, more pairs of piezoelectric actuators and sensors can be bonded, and appropriate $f_{i\theta}$s, f_{ij}s can be chosen to improve the performance. When properly constructed, both DMFC and CMFC can achieve good control performance. Through the simulation study, it can be seen that both DMFC and CMFC are very effective in residual vibration control despite their low authority in control action.

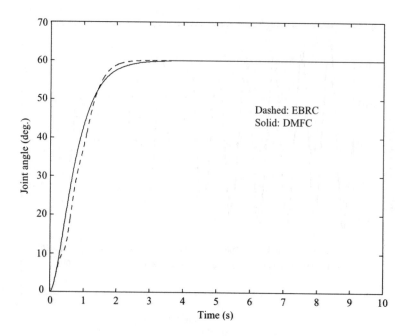

Figure 16.5   Joint angles with EBRC and DMFC

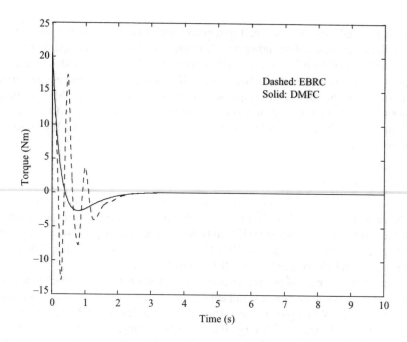

Figure 16.6   Variation in torque control with EBRC and DMFC

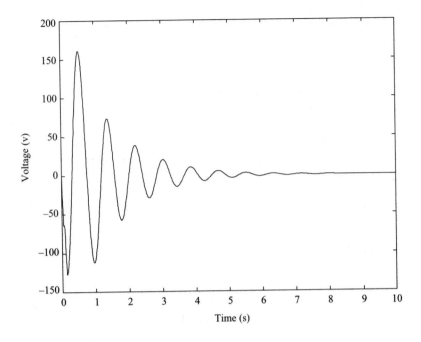

Figure 16.7   Voltage input of DMFC

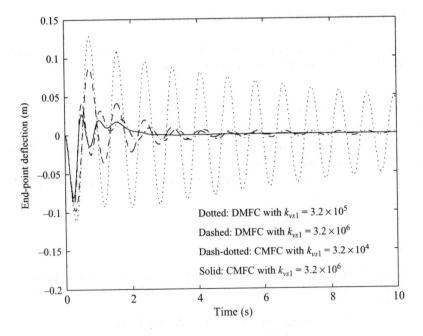

Figure 16.8   End-point deflections with DMFC and CMFC

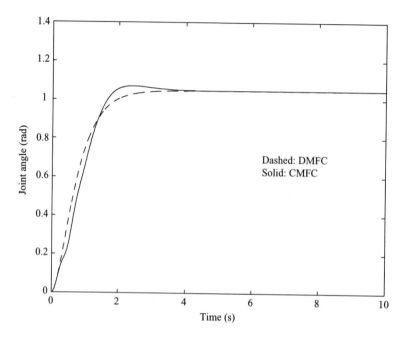

Figure 16.9   Joint angles with DMFC and CMFC

16.4   Tracking control of smart material robots

In this section, tracking control of the single-link smart material robot described in Section 16.2 is investigated using singular perturbation.

16.4.1   Singular perturbed smart material robots

Smart material robots retain the benefit of flexible-link robots and at the same time have additional sensing and control abilities via piezoelectric material bonded/embedded along the links.

The dynamics of the considered smart material robot is given by Ge *et al.* (2001*a*); and Wang *et al.* (2001)

$$\mathbf{M(q)\ddot{q}} + \mathbf{C(q, \dot{q})\dot{q}} + \mathbf{Kq} = \mathbf{Fu} \tag{16.26}$$

where

1. $\mathbf{q} = \begin{bmatrix} \mathbf{q}_r^T & \mathbf{q}_f^T \end{bmatrix}^T \in R^n, n = n_f + n_r$, with $\mathbf{q}_r \in R^{n_r}$ the vector of the rigid variables and $\mathbf{q}_f \in R^{n_f}$ the vector of flexible variables. Here, $n_r = 1$ for the considered system.
2. $\mathbf{M(q)} \in R^{n \times n}$ is the symmetric positive definite inertia matrix.
3. $\mathbf{C(q, \dot{q})} \in R^n$ represents the Coriolis and Centrifugal forces.
4. $\mathbf{K} \in R^{n \times n}$ is the stiffness matrix of the smart material robot.

5. $\mathbf{u} = \begin{bmatrix} \boldsymbol{\tau} & \mathbf{v}^T \end{bmatrix}^T$ is the vector of generalized torque with $\boldsymbol{\tau} \in R^{n_r}$ the vector of joint control torques, $\mathbf{v} \in R^m$ the vector of the control voltages to the piezoelectric material where m is the number of the piezoelectric material actuators.

6. $\mathbf{F} = \text{diag}[\mathbf{I}, \mathbf{F}_f]$ with $\mathbf{I} \in R^{n_r \times n_r}$ is an identity matrix and $\mathbf{F}_f \in R^{n_f \times m}$ is of full row rank.

Note that the model obtained using AMM can guarantee that $\mathbf{F}_f$ is of full row rank. On the other hand, it is not necessarily true when FE method is used in dynamic modelling.

Exploiting the natural timescale separation between the faster flexible mode dynamics and the slower desired rigid mode dynamics, the singular perturbation theory is used to formulate a boundary layer correction that stabilises non-minimum phase internal dynamics. The dynamic equation (16.26) can be rewritten as

$$\begin{bmatrix} \mathbf{M}_{rr} & \mathbf{M}_{rf} \\ \mathbf{M}_{fr} & \mathbf{M}_{ff} \end{bmatrix} \begin{bmatrix} \ddot{\mathbf{q}}_r \\ \ddot{\mathbf{q}}_f \end{bmatrix} + \begin{bmatrix} \mathbf{H}_r \\ \mathbf{H}_f \end{bmatrix} + \begin{bmatrix} \mathbf{0} \\ \mathbf{K}_{ff}\mathbf{q}_f \end{bmatrix} = \begin{bmatrix} \boldsymbol{\tau} \\ \mathbf{F}_f \mathbf{v} \end{bmatrix} \tag{16.27}$$

where

$$\mathbf{H}_r = \mathbf{C}_{rr}\dot{\mathbf{q}}_r + \mathbf{C}_{rf}\dot{\mathbf{q}}_f$$
$$\mathbf{H}_f = \mathbf{C}_{fr}\dot{\mathbf{q}}_r + \mathbf{C}_{ff}\dot{\mathbf{q}}_f$$

It should be noted that $\dot{\mathbf{M}} - 2\mathbf{C}$ is skew-symmetric as in the rigid robot case. Correspondingly, $\dot{\mathbf{M}}_{rr} - 2\mathbf{C}_{rr}$ is also skew-symmetric. Since inertia matrix $\mathbf{M}$ is positive definite, its inverse exists and is denoted by $\mathbf{D}$ as

$$\mathbf{M}^{-1} = \mathbf{D} = \begin{bmatrix} \mathbf{D}_{rr} & \mathbf{D}_{rf} \\ \mathbf{D}_{fr} & \mathbf{D}_{ff} \end{bmatrix}$$

where

$$\mathbf{D}_{rr} = (\mathbf{M}_{rr} - \mathbf{M}_{rf}\mathbf{M}_{ff}^{-1}\mathbf{M}_{fr})^{-1} \tag{16.28}$$

$$\mathbf{D}_{rf} = -\mathbf{M}_{rr}^{-1}\mathbf{M}_{rf}(\mathbf{M}_{ff} - \mathbf{M}_{fr}\mathbf{M}_{rr}^{-1}\mathbf{M}_{rf})^{-1} \tag{16.29}$$

$$\mathbf{D}_{fr} = -\mathbf{M}_{ff}^{-1}\mathbf{M}_{fr}(\mathbf{M}_{rr} - \mathbf{M}_{rf}\mathbf{M}_{ff}^{-1}\mathbf{M}_{fr})^{-1} \tag{16.30}$$

$$\mathbf{D}_{ff} = (\mathbf{M}_{ff} - \mathbf{M}_{fr}\mathbf{M}_{rr}^{-1}\mathbf{M}_{rf})^{-1} \tag{16.31}$$

Solving equation (16.26) for $\ddot{\mathbf{q}}_r$ and $\ddot{\mathbf{q}}_f$ yields

$$\ddot{\mathbf{q}}_r = -\mathbf{D}_{rr}\mathbf{H}_r - \mathbf{D}_{rf}\mathbf{H}_f - \mathbf{D}_{rf}\mathbf{K}_{ff}\mathbf{q}_f + \mathbf{D}_{rr}\boldsymbol{\tau} + \mathbf{D}_{rf}\mathbf{F}_f\mathbf{v} \tag{16.32}$$

$$\ddot{\mathbf{q}}_f = -\mathbf{D}_{fr}\mathbf{H}_r - \mathbf{D}_{ff}\mathbf{H}_f - \mathbf{D}_{ff}\mathbf{K}_{ff}\mathbf{q}_f + \mathbf{D}_{fr}\boldsymbol{\tau} + \mathbf{D}_{ff}\mathbf{F}_f\mathbf{v} \tag{16.33}$$

Introducing an appropriate scaling factor k such that

$$\mathbf{K}_{\!f\!f} = k\tilde{\mathbf{K}}$$

The following new variables can be defined as $\xi := k\tilde{\mathbf{K}}\mathbf{q}_f$, and defining $\varepsilon^2 = 1/k$, equations (16.32) and (16.33) become

$$\ddot{\mathbf{q}}_r = -\mathbf{D}_{rr}(\mathbf{q}_r, \varepsilon^2\xi)\mathbf{H}_r(\mathbf{q}_r, \dot{\mathbf{q}}_r, \varepsilon^2\xi, \varepsilon^2\dot{\xi}) - \mathbf{D}_{rf}(\mathbf{q}_r, \varepsilon^2\xi)\mathbf{H}_f(\mathbf{q}_r, \dot{\mathbf{q}}_r, \varepsilon^2\xi, \varepsilon^2\dot{\xi})$$
$$- \mathbf{D}_{rf}(\mathbf{q}_r, \varepsilon^2\xi)\xi + \mathbf{D}_{rr}(\mathbf{q}_r, \varepsilon^2\xi)\tau + \mathbf{D}_{rf}(\mathbf{q}_r, \varepsilon^2\xi)\mathbf{F}_f\mathbf{v} \tag{16.34}$$

$$\varepsilon^2\ddot{\xi} = -\mathbf{D}_{fr}(\mathbf{q}_r, \varepsilon^2\xi)\mathbf{H}_r(\mathbf{q}_r, \dot{\mathbf{q}}_r, \varepsilon^2\xi, \varepsilon^2\dot{\xi}) - \mathbf{D}_{ff}(\mathbf{q}_r, \varepsilon^2\xi)\mathbf{H}_f(\mathbf{q}_r, \dot{\mathbf{q}}_r, \varepsilon^2\xi, \varepsilon^2\dot{\xi})$$
$$- \mathbf{D}_{ff}(\mathbf{q}_r, \varepsilon^2\xi)\xi + \mathbf{D}_{fr}(\mathbf{q}_r, \varepsilon^2\xi)\tau + \mathbf{D}_{ff}(\mathbf{q}_r, \varepsilon^2\xi)\mathbf{F}_f\mathbf{v} \tag{16.35}$$

which is a singular perturbed model of the smart material robot arm. Notice that all variables on the right hand side of equation (16.35) have been scaled by $\tilde{\mathbf{K}}$. Formally, setting $\varepsilon = 0$ and solving for ξ in equation (16.35), yields

$$\bar{\xi} = \mathbf{D}_{ff}^{-1}(\bar{\mathbf{q}}_r, 0)[-\mathbf{D}_{fr}(\bar{\mathbf{q}}_r, 0)\mathbf{H}_r(\bar{\mathbf{q}}_r, \dot{\bar{\mathbf{q}}}_r, 0, 0) + \mathbf{D}_{fr}(\bar{\mathbf{q}}_r, 0)\bar{\tau}] - \mathbf{H}_f(\bar{\mathbf{q}}_r, 0) + \mathbf{F}_f\bar{\mathbf{v}} \tag{16.36}$$

where the bars are used to indicate that the system is considered with $\varepsilon = 0$. Substituting equation (16.36) into equation (16.34) with $\varepsilon = 0$ yields

$$\ddot{\bar{\mathbf{q}}}_r = [\mathbf{D}_{rr}(\bar{\mathbf{q}}_r, 0) - \mathbf{D}_{rf}(\bar{\mathbf{q}}_r, 0)\mathbf{D}_{ff}^{-1}(\bar{\mathbf{q}}_r, 0)\mathbf{D}_{fr}(\bar{\mathbf{q}}_r, 0)][\mathbf{H}_r(\bar{\mathbf{q}}_r, \dot{\bar{\mathbf{q}}}_r, 0, 0) + \bar{\tau}] \tag{16.37}$$

Using equations (16.28) to (16.31) yields

$$\mathbf{D}_{rr}(\bar{\mathbf{q}}_r, 0) - \mathbf{D}_{rf}(\bar{\mathbf{q}}_r, 0)\mathbf{D}_{ff}^{-1}(\bar{\mathbf{q}}_r, 0)\mathbf{D}_{fr}(\bar{\mathbf{q}}_r, 0) = \mathbf{M}_{rr}^{-1}(\bar{\mathbf{q}}_r)$$

Thus, equation (16.37) becomes

$$\mathbf{M}_{rr}(\bar{\mathbf{q}}_r)\ddot{\bar{\mathbf{q}}}_r + \mathbf{C}_{rr}(\bar{\mathbf{q}}_r, \dot{\bar{\mathbf{q}}}_r)\dot{\bar{\mathbf{q}}}_r = \bar{\tau} \tag{16.38}$$

which corresponds to the rigid-body robot dynamic model.

Remark 16.10: It can be seen that $\bar{\mathbf{v}}$ does not appear in equation (16.37), which means that the voltage control $\mathbf{v}$ has no influence on the slow subsystem, thus, $\bar{\mathbf{v}} = 0$. This coincides with the physical function of the voltage control of smart material robots (Ge *et al.*, 2001*a*; Wang *et al.*, 2001).

To identify the fast subsystem, define a fast timescale $\tau_t = t/\varepsilon$ and boundary layer correction variables $\bar{\mathbf{v}}$ and $\mathbf{z}_2 = \varepsilon\dot{\xi}$. It follows from equation (16.35) and the fact that $d\bar{\xi}/d\tau_t = \varepsilon\dot{\bar{\xi}} = \mathbf{0}$, that

$$\frac{d\mathbf{z}_1}{d\tau_t} = \mathbf{z}_2$$

$$\frac{d\mathbf{z}_2}{d\tau_t} = -\mathbf{D}_{fr}(\mathbf{q}_r, \varepsilon^2(\mathbf{z}_1 + \overline{\xi}))\mathbf{H}_r(\mathbf{q}_r, \dot{\mathbf{q}}_r, \varepsilon^2(\mathbf{z}_1 + \overline{\xi}), \varepsilon \mathbf{z}_2)$$

$$- \mathbf{D}_{ff}(\mathbf{q}_r, \varepsilon^2(\mathbf{z}_1 + \overline{\xi}))\mathbf{H}_f(\mathbf{q}_r, \dot{\mathbf{q}}_r, \varepsilon^2(\mathbf{z}_1 + \overline{\xi}), \varepsilon \mathbf{z}_2)$$

$$- \mathbf{D}_{ff}(\mathbf{q}_r, \varepsilon^2(\mathbf{z}_1 + \overline{\xi}))(\mathbf{z}_1 + \overline{\xi}) + \mathbf{D}_{fr}(\mathbf{q}_r, \varepsilon^2(\mathbf{z}_1 + \overline{\xi}))\tau$$

$$+ \mathbf{D}_{ff}(\mathbf{q}_r, \varepsilon^2(\mathbf{z}_1 + \overline{\xi}))\mathbf{F}_f \mathbf{v}$$

Setting $\varepsilon = 0$, and substituting for $\overline{\xi}$ from equation (16.36) results in

$$\frac{d\mathbf{z}_1}{d\tau_t} = \mathbf{z}_2$$

$$\frac{d\mathbf{z}_2}{d\tau_t} = -\mathbf{D}_{ff}(\overline{\mathbf{q}}_r, 0)\mathbf{z}_1 + \mathbf{D}_{fr}(\overline{\mathbf{q}}_r, 0)\tau_f + \mathbf{D}_{ff}(\overline{\mathbf{q}}_r, 0)\mathbf{F}_f \mathbf{v}$$

which is a linear system parameterised in the slow variable $\overline{\mathbf{q}}_r$ and $\tau_f = \tau - \overline{\tau}$. Thus, the fast subsystem can be obtained as a linear system with the slow variables $\overline{\mathbf{q}}_r$ as parameters, that is,

$$\frac{d\mathbf{z}}{d\tau_t} = \mathbf{A}\mathbf{z} + \mathbf{B}_1 \mathbf{u}_f \qquad (16.39)$$

where

$$\mathbf{z} = \begin{bmatrix} \mathbf{z}_1^T & \mathbf{z}_2^T \end{bmatrix}^T, \qquad \mathbf{u}_f = \begin{bmatrix} \tau_f^T & \mathbf{v}^T \end{bmatrix}^T$$

and

$$\mathbf{A} = \begin{bmatrix} \mathbf{0} & \mathbf{I} \\ -\mathbf{D}_{ff}(\overline{\mathbf{q}}_r, 0) & \mathbf{0} \end{bmatrix}, \qquad \mathbf{B}_1 = \begin{bmatrix} \mathbf{0} & \mathbf{0} \\ \mathbf{D}_{fr}(\overline{\mathbf{q}}_r, 0) & \mathbf{D}_{ff}(\overline{\mathbf{q}}_r, 0)\mathbf{F}_f \end{bmatrix}$$

By examining equation (16.39), it can be found that there are two kinds of control inputs τ_f and $\mathbf{v}$ that can be used to suppress the vibration, $\mathbf{z}$.

If $\mathbf{v}$ is unavailable or not activated, the fast subsystem becomes

$$\frac{d\mathbf{z}}{d\tau_t} = \mathbf{A}\mathbf{z} + \mathbf{B}_2 \tau_f$$

where $\mathbf{B}_2 = \begin{bmatrix} \mathbf{0}^T & \mathbf{D}_{fr}^T(\overline{\mathbf{q}}_r, 0) \end{bmatrix}^T$.

If τ_f is not used, the fast subsystem becomes

$$\frac{d\mathbf{z}}{d\tau_t} = \mathbf{A}\mathbf{z} + \mathbf{B}_3 \mathbf{v}$$

where $\mathbf{B}_2 = \begin{bmatrix} \mathbf{0}^T & \mathbf{F}_f^T \mathbf{D}_{ff}^T(\overline{\mathbf{q}}_r, 0) \end{bmatrix}^T$.

16.4.2 Adaptive composite controller design

In practice, it is hard to get the exact model of the system. Thus, to eliminate the need for an exact dynamic model, an adaptive composite control approach can be adopted.

16.4.2.1   Adaptive control of the slow subsystem

To facilitate controller design, the following property of the dynamic model in equation (16.38) is given.

Property 16.1: The dynamics described by equation (16.38) are linear in the parameters, that is

$$\mathbf{M}_{rr}(\overline{\mathbf{q}}_r)\ddot{\chi} + \mathbf{C}_{rr}(\overline{\mathbf{q}}_r, \dot{\overline{\mathbf{q}}}_r)\dot{\chi} = \boldsymbol{\psi}\mathbf{P}$$

where $\mathbf{P} \in R^l$ are the parameters of interest, $\boldsymbol{\psi} = \boldsymbol{\psi}(\overline{\mathbf{q}}_r, \dot{\overline{\mathbf{q}}}_r, \dot{\chi}, \ddot{\chi}) \in R^{n \times l}$ is the regressor matrix, and $\dot{\chi}, \ddot{\chi} \in R^n$.

Consider a controller of the form

$$\overline{\boldsymbol{\tau}} = \boldsymbol{\psi}_r \hat{\mathbf{P}} + \mathbf{K}_p \mathbf{r} \tag{16.40}$$

where $(\hat{*})$ represents the estimate of $(*)$, the estimation error is given as $(\tilde{*}) = (*) - (\hat{*})$, $\mathbf{K}_p$ is positive definite, and

$$\boldsymbol{\psi}_r = \boldsymbol{\psi}(\overline{\mathbf{q}}_r, \dot{\overline{\mathbf{q}}}_r, \dot{\mathbf{q}}_v, \ddot{\mathbf{q}}_v)$$

Applying the control law in equation (16.40) to the system in equation (16.38) yields the closed-loop system error equation:

$$\mathbf{M}_{rr}\dot{\mathbf{r}} + \mathbf{C}_{rr}\mathbf{r} + \mathbf{K}_p \mathbf{r} = \boldsymbol{\psi}_r \tilde{\mathbf{P}} \tag{16.41}$$

Theorem 16.4: For the closed-loop system given in equation (16.41), asymptotic stability, that is, $\mathbf{r} \rightarrow \mathbf{0}$ as $t \rightarrow \infty$, is achieved if $\mathbf{K}_p > 0$, and the parameter adaptation laws are given by

$$\dot{\hat{\mathbf{P}}} = \Gamma \boldsymbol{\psi}_r^T \mathbf{r} \tag{16.42}$$

where Γ is dimensional compatible symmetric positive definite matrix, then $\hat{\mathbf{P}} \in L^\infty$ and $\mathbf{e} \in L^2_{n_r} \cap L^\infty_{n_r}$, $\dot{\mathbf{e}} \in L^2_{n_r}$, $\mathbf{e}$ is continuous and $\mathbf{e}, \dot{\mathbf{e}} \rightarrow \mathbf{0}$ as $t \rightarrow \infty$ (Wang *et al.*, 2001).

Proof: Consider the Lyapunov function candidate

$$V = \frac{1}{2}\mathbf{r}^T \mathbf{M}_{rr}(\overline{\mathbf{q}}_r)\mathbf{r} + \frac{1}{2}\tilde{\mathbf{P}}^T \Gamma^{-1} \tilde{\mathbf{P}}$$

The time derivative of V is given by

$$\dot{V} = \mathbf{r}^T \mathbf{M}_{rr}\dot{\mathbf{r}} + \frac{1}{2}\mathbf{r}^T \dot{\mathbf{M}}_{rr}\mathbf{r} + \tilde{\mathbf{P}}^T \Gamma^{-1}\dot{\tilde{\mathbf{P}}}$$

Substituting equations (16.40) and (16.41) into the above equation leads to:

$$\dot{V} = -\mathbf{r}^T \mathbf{K}_p \mathbf{r} \leq 0$$

thus $\mathbf{r} \in L^2_{n_r}$. Consequently, $\mathbf{e} \in L^2_{n_r} \cap L^\infty_{n_r}$; $\mathbf{e}$ is continuous and $\mathbf{e}, \rightarrow \mathbf{0}$ as $t \rightarrow \infty$; and $\dot{\mathbf{e}} \in L^2_{n_r}$. It is easy to show that $\mathbf{r} \rightarrow \mathbf{0}$ as $t \rightarrow \infty$, and hence, $\dot{\mathbf{e}} \rightarrow \mathbf{0}$ as $t \rightarrow \infty$.

16.4.2.2   Stabilisation of fast subsystem

Passive damping of flexible robot arm is not adequate due to additional mass and its inability to adjust to changing flexibility effects. Hence, some kind of active damping is desired to control the vibration. Although there are two kinds of control inputs for smart material robots, it is shown that design of τ_f to stabilise the fast subsystem is difficult because (a) the dynamics of the system are assumed to be unknown and (b) there is no nice property about $\mathbf{D}_{fr}^T(\bar{\mathbf{q}}_r, \mathbf{0})$ that can be used. Fortunately, for the design of voltage control $\mathbf{v}$, the positivity of $\mathbf{D}_{ff}^T(\bar{\mathbf{q}}_r, \mathbf{0})$ can be fully utilized to design a controller, which is independent of the unknown dynamics, and yet stabilises the fast subsystem. Thus, a scheme where the slow subsystem is controlled by $\bar{\tau}$, and the fast subsystem is controller by $\mathbf{v}$ only by choosing $\tau_f = \mathbf{0}$ is investigated in this work.

By letting $\tau_f = \mathbf{0}$, the fast subsystem reduces to

$$\frac{d^2\mathbf{z}_1}{d\tau_t^2} + \mathbf{D}_{ff}(\bar{\mathbf{q}}_r, \mathbf{0})\mathbf{z}_1 = \mathbf{D}_{ff}(\bar{\mathbf{q}}_r, \mathbf{0})\mathbf{F}_f\mathbf{v} \tag{16.43}$$

where $\mathbf{D}_{ff}(\bar{\mathbf{q}}_r, \mathbf{0})$ is positive definite and $\mathbf{F}_f$ is an known full row rank matrix.

Theorem 16.5: For the fast subsystem in equation (16.43), if $\mathbf{F}_f$ is a full row rank matrix and known, then the system in equation (16.39) is uniform exponential stable if the control is chosen as (Ge *et al.*, 2001a):

$$\mathbf{v} = -\varepsilon\kappa\mathbf{UF}_f^+\dot{\xi}, \quad \kappa > 0 \tag{16.44}$$

where $\mathbf{U}$ is the permutation matrix such that $\mathbf{F}_f\mathbf{U} = \begin{bmatrix} \mathbf{F}_{f1} & \mathbf{F}_{f2} \end{bmatrix}$ with $\mathbf{F}_{f1} \in R^{n_f \times n_f}$ being non-singular, and the right inverse $\mathbf{F}_f^+$ is, for any dimensionally compatible matrix $\mathbf{F}_{f3}$, given by (Ortega, 1987):

$$\mathbf{F}_f^+ = \begin{bmatrix} \mathbf{F}_{f1}^{-1} - \mathbf{F}_{f1}^{-1}\mathbf{F}_{f2}\mathbf{F}_{f3} \\ \mathbf{F}_{f3} \end{bmatrix}$$

Proof: Consider the Lyapunov function candidate

$$V = \frac{1}{2}\mathbf{z}_1'^T\mathbf{z}_1' + \frac{1}{2}\mathbf{z}_1^T\mathbf{D}_{ff}(\bar{\mathbf{q}}_r, \mathbf{0})\mathbf{z}_1$$

where $\mathbf{z}_1'$ denotes $d\mathbf{z}_1/d\tau_t$.

The control law in equation (16.44) can be rewritten in the fast timescale as

$$\mathbf{v} = -\varepsilon\kappa\mathbf{UF}_f^+\dot{\mathbf{z}}_1 = -\kappa\mathbf{UF}_f^+\mathbf{z}_1' \tag{16.45}$$

Substituting equation (16.45) into the fast subsystem, yields

$$\frac{d^2\mathbf{z}_1}{d\tau_t^2} + \kappa\mathbf{D}_{ff}(\bar{\mathbf{q}}_r, \mathbf{0})\mathbf{F}_f\mathbf{UF}_f^+\frac{d\mathbf{z}_1}{d\tau_t} + \mathbf{D}_{ff}(\bar{\mathbf{q}}_r, \mathbf{0})\mathbf{z}_1 = 0 \tag{16.46}$$

thus,

$$\frac{d^2\mathbf{z}_1}{d\tau_t^2} + \kappa\mathbf{D}_{ff}(\bar{\mathbf{q}}_r, \mathbf{0})\frac{d\mathbf{z}_1}{d\tau_t} + \mathbf{D}_{ff}(\bar{\mathbf{q}}_r, \mathbf{0})\mathbf{z}_1 = 0 \tag{16.47}$$

Accordingly, the following holds along the trajectory in equation (16.47):

$$\frac{dV}{d\tau_t} = -\kappa \mathbf{z}_1'^T \mathbf{D}_{ff}(\bar{\mathbf{q}}_r, 0)\mathbf{z}_1' \leq 0$$

Global asymptotic stability of the boundary layer system in equation (16.46) then follows from LaSalle's theorem. Since the system is linear and it is uniform in $\bar{\mathbf{q}}_r$, the fast subsystem is exponentially uniformly stable for fixed $\bar{\mathbf{q}}_r$. Consequently, it is concluded that if the slow subsystem in equation (16.38) has a unique solution defined on an interval $t \in [0, t_1]$ and if the fast subsystem in equation (16.39) is exponentially uniformly stable in $(t, \bar{\mathbf{q}}_r)$, there exists ε^* such that for all $\varepsilon < \varepsilon^*$

$$\xi(t) = \bar{\bar{\xi}}(t) + \mathbf{z}_1(\tau_t) + O(\varepsilon)$$

$$\mathbf{q}_r(t) = \bar{\mathbf{q}}_r(t) + O(\varepsilon)$$

hold uniformly for $t \in [0, t_1]$.

Remark 16.11: When the system model is known, then both τ_f and $\mathbf{v}$ can be used to control the fast subsystem. Under the assumption that pairs $(\mathbf{A}, \mathbf{B}_2)$, $(\mathbf{A}, \mathbf{B}_3)$ and hence $(\mathbf{A}, \mathbf{B}_1)$ are uniformly stabilisable for any slow trajectory $\bar{\mathbf{q}}_r(t)$, different linear control strategies such as LQR and pole placement methods can be used to design the fast feedback control law $\mathbf{u}_f = \mathbf{Kz}$ to make the fast subsystem uniformly exponentially stable (Ge *et al.*, 2001a; Wang *et al.*, 2001).

Remark 16.12: In practice, the dynamic models of robots based on the Lagrange-Euler formation are simple in its representation but are difficult, time consuming and error prone to obtain, except for simple cases. Moreover, it is desirable to have generic non-linear model structures for dynamic modelling. Neural networks are legitimate candidates substantiated by the fact that certain classes of feedforward neural networks are known to possess the so-called universal approximation property. The introduction of neural networks can remove the need for tedious dynamic modelling and time consuming computation of the regression matrix as in conventional adaptive techniques (Ge and Postlethwaite, 1995). Adaptive neural network composite controller was successfully proposed in Ge *et al.* (2001a) for smart material robots.

Numerical simulations are carried out below on the single-link smart material robot considered in Section 16.3 to verify the effectiveness of the proposed controller. The smart material robot is simulated using AMM.

Trajectory planning

The desired trajectory for rigid joint angle is expressed as a Hermite polynomial of the fifth degree in t with continuous bounded position, velocity and bounded acceleration. The general expression for the desired position trajectory is:

$$q_d(t, t_d) = q_0 + \left(6.0\frac{t^5}{t_d^5} - 15\frac{t^4}{t_d^4} + 10.0\frac{t^3}{t_d^3} \right)(q_f - q_0)$$

where t_d represents the time the desired arm trajectory reaches the desired final position q_f starting from the desired initial position q_0. Here $q_0 = 0.0$, $q_f = 1.0$ and $t_d = 2.0$ s used.

It has been shown that the simple joint PD control can make the closed-loop flexible robotic system stable (Yigit, 1994a). Therefore, the performance of PD controller is shown first. When the update law for the controller in equation (16.40) is not activated, the controller becomes the traditional PD controller, given as

$$\bar{\tau} = K_p r = K_p \dot{e} + K_e e$$

In the simulation, the adaptive composite controller (ACC) is given by

$$\mathbf{u} = \begin{bmatrix} \tau \\ w \end{bmatrix} = \begin{bmatrix} \bar{\tau} \\ w \end{bmatrix} = \begin{bmatrix} \psi_r^{\hat{P}} + K_p r \\ -\varepsilon \kappa K_w^{\xi} \end{bmatrix}$$

where

$$\dot{\hat{P}} = \Gamma \psi_r^T r$$

Since only one piezoelectric actuator and sensor were used in the simulation, only the first mode shape function was used in the fast subsystem controller design, and the control parameter was chosen as $\kappa K_w = -500$ to make $\kappa D_{ff}(\bar{q}_r, 0) F_f K_w$ positive definite.

Figure 16.10 shows the joint angle trajectory under PD and ACC control, while Figure 16.11 shows the end-point deflection. It can be seen that ACC can also achieve good performance in joint angle tracking, and simultaneously suppress the residual vibration effectively. There is no residual vibration at steady state with ACC. For completeness and clarity in presentation, other signals in the closed loop are included. Figure 16.12 shows the bounded joint control torque signals under both controllers, while Figure 16.13 shows the control voltages with ACC.

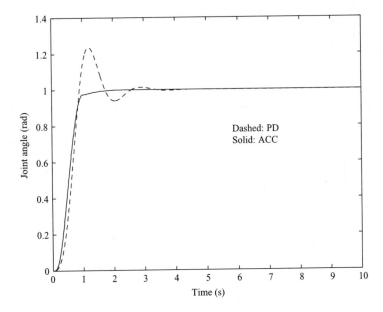

Figure 16.10   Joint angle trajectories with ACC

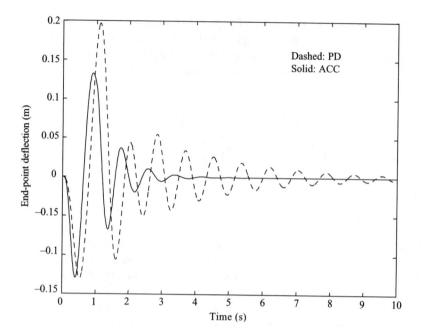

Figure 16.11   *End-point deflections with ACC*

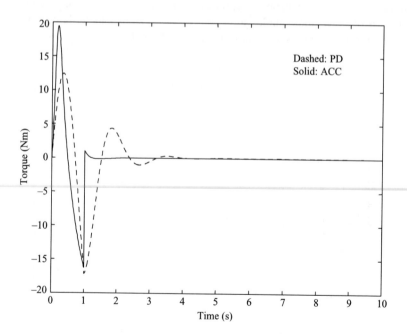

Figure 16.12   *Torque control with ACC*

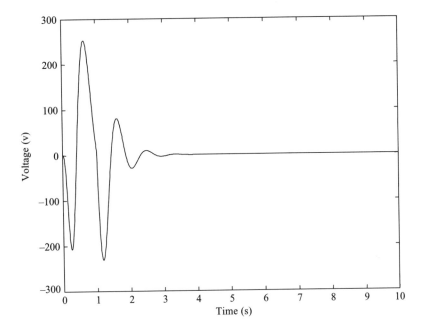

Figure 16.13   Voltage control with ACC

Though only a single-link smart material robot was used to simulate the effectiveness of the proposed controller, it can be used to control higher degrees of freedom robots as well.

16.5   Summary

In this chapter, dynamic modelling and controller design of smart material robots have been presented. Model-free control has been designed for the regulation of the robot. Theoretical proofs have shown that the closed-loop system is stable, and the model-free controllers are independent of system parameters and hence possess stability robustness to parameter variations. Furthermore, adaptive composite tracking controller has been presented. The smart material bonded along the links has been used to actively suppress the residual vibration. Numerical simulations have shown that the system response converges fast and the residual vibrations are effectively suppressed with the proposed controllers.

Chapter 17

Modelling and control of rigid–flexible manipulators

A.S. Yigit

Dynamic modelling and control of manipulators involving both rigid and flexible links are presented in this chapter. Various modelling approaches are discussed, and as an example, the equations of motion of a rigid–flexible two-link manipulator are derived using Hamilton's principle. Discretisation of the governing equations and the effects of coupling between rigid and flexible degrees of freedom are examined. The response to an impact with a rigid surface is studied to investigate the effect of link flexibility on the dynamic response. Various control strategies used in rigid–flexible manipulators are reviewed. As an example, the performance of a non-model-based independent joint proportional, derivative (PD) control is demonstrated. Trajectory tracking and various implementation issues are also discussed.

17.1   Introduction

Originally, manipulators were built to be rather massive requiring high power and operated at low speeds. Consequently, the modelling was based on the assumption that links behave as rigid bodies. With the increasing demand for precise high-speed operation, it was no longer adequate to treat certain links in a manipulator as rigid. Initially, the effect of link flexibility was studied by considering a single-link flexible arm. Clearly, the single-link case does not have sufficient kinematic complexity to generalize the results for practical multi-link manipulators. Despite its relative simplicity as compared to multi-link flexible manipulators, a rigid–flexible two-link manipulator model includes the non-linear, coupled dynamics of multi-link systems as well as the behaviour of flexible structures including non-minimum phase characteristics. Therefore, a rigid–flexible two-link manipulator can be considered as an ideal starting system for more complicated multi-link flexible manipulators. Besides this pedagogical significance, these manipulators can also be good models for a wide range of experimental or industrial implementations since the first link is generally built to be much more rigid than the second (Khorrami and Jain, 1993). Though most of the studies published involve theoretical and numerical work, few implementation examples have also been documented. Oakley and Cannon (1988) reported some experiments on an experimental two-link manipulator. In order to observe the effect of link flexibility, the forearm was designed to be very flexible. Lucibello and Ulivi (1993) used a similar experimental set-up to validate their analytical model as well as to test some simple controllers. Direct drive joint motors were used in both experimental set-ups to avoid gear backlash. Optical encoders and d.c. tachometers are generally used to measure joint angular motion. Optical transducers with CCD cameras, accelerometers or strain guages are typically used to measure the elastic deflection of the flexible link. This chapter provides a summary of various approaches to dynamic modelling and control of rigid–flexible manipulators.

Various dynamic models have been developed for purposes of simulation and control design. For rigid–flexible manipulators, the equations of motion are hybrid ordinary–partial differential equations (OPDEs). Therefore, research on manipulator dynamics has concentrated on developing adequate reduced-order models that capture all essential dynamics such as modal convergence, mutual coupling between rigid and flexible motions and geometric stiffening. On the other hand, research on control has been focused on designing effective controllers, which are robust to modelling errors and uncertainties. To this end, this chapter provides a summary of various approaches to dynamic modelling and control of rigid–flexible manipulators.

17.2   Dynamic modelling

Various approaches are available to derive the equations of motion for a manipulator with rigid and flexible links. One approach is to derive the equations of motion for the distributed parameter system using either a Newtonian or Lagrangian technique, and then discretising the resulting equations of motion. Another approach is to write discrete equations directly by employing an assumed-modes approach. In this case, the kinetic and potential energy expressions are expressed in finite-dimensional discrete form. Amirouche (1992) presents two separate formulations for multi-body systems with terminal flexible links, one based on motion overlapping and another based on the finite element (FE) method. A number of automated procedures for computer generation of these equations have also been proposed, see for example, Cetinkunt and Book (1989), Lin and Lewis (1994) and Lucibello *et al.* (1986).

While most of the studies involve manipulators with only revolute joints, few researchers have investigated the dynamics and control of rigid–flexible manipulators with prismatic joints. Chalhoub and Ulsoy (1987) demonstrated the feasibility of reducing arm vibrations through end-point motion feedback and high bandwidth joint actuators. Their set-up involved a rigid link attached to a flexible link driven by a lead screw. Pan *et al.* (1990) used the same system and experimentally demonstrated the effects of coupling between the rigid and elastic motion, as well as

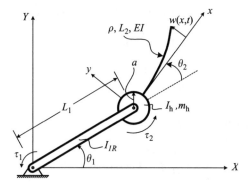

Figure 17.1   Sketch of a two-link manipulator

centrifugal stiffening. Buffinton (1992) presented a general formulation of dynamics of manipulators involving prismatic joints.

In this section, the equations of motion of a rigid–flexible manipulator with revolute joints are derived. The system to be considered consists of two links connected with a revolute joint as shown in Figure 17.1. The first link is assumed to be rigid and the second link is composed of a slender flexible beam cantilevered onto a rigid rotating hub. Longitudinal deformations are neglected and it is assumed that Euler–Bernoulli beam theory is adequate to describe flexural motion of the second link. No damping is assumed and the manipulator moves in the horizontal plane, thus gravity is not considered.

To describe the kinematics, a frame moving with the rigid hub is introduced. Inspection of Figure 17.1 shows that the position vector $\mathbf{r}$, relative to an inertial frame, of a point on the deformed centreline of the beam may be written as

$$
\begin{aligned}
\mathbf{r} = &[L_1 \cos \theta_1 + (a + x) \cos \theta - w(x, t) \sin \theta]\,\mathbf{i} \\
&+ [L_1 \sin \theta_1 + (a + x) \sin \theta + w(x, t) \cos \theta]\,\mathbf{j}
\end{aligned}
\tag{17.1}
$$

where L_1 is the length of the rigid link, a is the radius of the rigid hub, x gives the location of the un-deformed point on the flexible beam, w is the transverse elastic displacement, θ_1 and θ_2 represent the shoulder and elbow joint angular motions, respectively, and $\theta_1 + \theta_2 = \theta$. Note that axial deformations are not considered. For the beam considered here flexural motion does not induce significant axial vibration (Yigit *et al.*, 1988).

The total kinetic energy of the system can be written as

$$
C = \frac{1}{2}I_{1R}\dot{\theta}_1^2 + \frac{1}{2}I_h\dot{\theta}^2 + \frac{1}{2}m_h L_1^2\dot{\theta}_1^2 + \frac{1}{2}\int_0^{L_2} \rho A\,\dot{r}^2\,dx
\tag{17.2}
$$

where I_{1R} is the inertia of the first link with respect to the shoulder joint axis, m_h and I_h are the mass and inertia of the rigid hub with respect to the elbow axis of the second link, respectively. ρ, A and L_2 are the material density, cross-sectional area and the length of the flexible link, respectively, and $\dot{r}^2$ can be obtained from equation (17.1) as

$$\dot{r}^2 = L_1^2\dot{\theta}_1^2 + [(a+x)\,\dot{\theta} + \dot{w}]^2 + w^2\dot{\theta}^2$$
$$+ 2L_1\dot{\theta}_1[(a+x)\,\dot{\theta} + \dot{w}]\cos\theta_2 - 2L_1\dot{\theta}_1\dot{\theta}\,w\,\sin\theta_2$$

The potential energy can be written as

$$P = \frac{1}{2}\int_0^{L_2} EI(w'')^2 dx + \frac{1}{2}\int_0^{L_2} F_c(x,t)\,(w')^2 dx \tag{17.3}$$

where E is the Young modulus, I denotes the area moment of inertia of the beam, $(\ )'$ denotes partial derivative with respect to x. $F_c(x,t)$ is given for a uniform beam as (Yigit *et al.*, 1988)

$$F_c(x,t) = \frac{1}{2}\rho\,\dot{\theta}^2(L_2^2 - x^2) + \rho\,\dot{\theta}^2\,a(L_2 - x)$$

The first term in equation (17.3) represents the strain energy of the second link according to Euler–Bernoulli beam theory, while the second term represents the work done due to axial shortening that is responsible for the effect of centrifugal stiffening for rotating beams (Meirovitch, 1980). Standard Euler–Bernoulli beam theory without this correction would erroneously give a softening effect due to rotation. Though it is important to include this stiffening effect for large angular velocities, it may be neglected for most manipulators since during normal operation of a manipulator, angular velocities do not reach very large values.

The equations of motion can be derived using the extended Hamilton's principle;

$$\int_{t_1}^{t_2} \delta\,(C - P)\,dt + \int_{t_1}^{t_2} \delta\,Wdt = 0 \tag{17.4}$$

where δW denotes the virtual work done by the joint torques, τ_1 and τ_2, at the shoulder and the elbow joints, respectively. Substitution of equations (17.2) and (17.3) into equation (17.4), integration by parts, and some algebra yield the following equations of motion and the associated boundary conditions:

$$I_{1R}\ddot{\theta}_1 + I_h\ddot{\theta} + m_h L_1^2\ddot{\theta}_1 + \int_0^{L_2} \rho A \Big\{(a+x)^2\,\ddot{\theta} + \ddot{w}(a+x) + L_1^2\,\ddot{\theta}_1 + 2w\dot{w}\dot{\theta} + w^2\ddot{\theta}$$
$$+2L_1(a+x)\ddot{\theta}_1\cos\theta_2 + L_1\ddot{w}\cos\theta_2 - 2L_1(a+x)\dot{\theta}_1\dot{\theta}_2\sin\theta_2 - 2L_1\dot{\theta}_1\,w\sin\theta_2$$
$$-L_1(a+x)\dot{\theta}_2^2\sin\theta_2 - 2L_1\dot{w}\dot{\theta}\sin\theta_2 - 2L_1 w\dot{\theta}_1\,\dot{\theta}_2\cos\theta_2 - L_1\dot{\theta}_2 w\cos\theta_2$$
$$-\tfrac{1}{2}(L_2^2 - x^2 + 2aL_2 - 2ax)\,(w')^2\ddot{\theta}\ - (L_2^2 - x^2 + 2aL_2 - 2ax)\,w'\dot{w}'\,\dot{\theta}\Big\}\,dx = \tau_1$$
$$\tag{17.5}$$

$$I_h\ddot{\theta} + \int_0^{L_2} \rho A \left\{ (a+x)^2\,\ddot{\theta} + \ddot{w}(a+x) + 2w\dot{w}\dot{\theta} + w^2\ddot{\theta} + L_1(a+x)\ddot{\theta}_1\,\cos\,\theta_2 \right.$$

$$-L_1\ddot{\theta}_1 w\,\sin\,\theta_2 + L_1\dot{\theta}_1^2(a+x)\,\sin\,\theta_2 + L_1\dot{\theta}_1^2 w\,\cos\,\theta_2$$

$$\left. -\tfrac{1}{2}(L_2^2 - x^2 + 2aL_2 - 2ax)(w')^2\ddot{\theta} - (L_2^2 - x^2 + 2aL_2 - 2ax)\,w'\dot{w}'\,\dot{\theta} \right\}\,dx = \tau_2$$

$$(17.6)$$

$$\rho A(a+x)\,\ddot{\theta} + \rho\ddot{w} + \rho L_1\ddot{\theta}\,\cos\,\theta_2 - \rho w\dot{\theta}^2 + \rho L_1\dot{\theta}_1^2\,\sin\,\theta_2 + EIw''''$$

$$- \left\{ \tfrac{1}{2}\rho\,\dot{\theta}^2(L_2^2 - x^2) + \rho\,\dot{\theta}^2 a(L_2 - x) \right\}w'' + \rho\,\dot{\theta}^2(x+a)w' = 0$$

$$(17.7)$$

$$\begin{aligned} w(0,t) &= 0; & w'(0,t) &= 0 \\ w''(L_2,t) &= 0; & w'''(L_2,t) &= 0 \end{aligned}$$

$$(17.8)$$

17.2.1 Discrete equations of motion

Equations (17.5) to (17.8) constitute a highly non-linear, coupled set of hybrid differential equations. An analytical solution cannot be found and approximate means must be used for simulation. Commonly used methods are the FE method, assumed modes, and various forms of weighted residual methods such as Galerkin's method. Assume

$$w(x,t) = \sum_{i=1}^{N} \phi_i(x)\,q_i(t) \tag{17.9}$$

where $q_i(t)$ represents unknown generalised coordinates and $\phi_i(x)$ a set of admissible functions, that is, they satisfy the geometric boundary conditions (Meirovitch, 1980). Here the admissible functions $\phi_i(x)$ are chosen as the mode shapes for the non-rotating cantilever beam. The technique yields the following non-linear coupled set of ordinary differential equations:

$$\mathbf{M}(\mathbf{q})\ddot{\mathbf{q}} + \mathbf{h}(\mathbf{q},\dot{\mathbf{q}}) + \mathbf{K}(\mathbf{q}) = \mathbf{u}(t) \tag{17.10}$$

where $\mathbf{q}$ is the vector of generalised coordinates representing both the rigid-body and the elastic degrees of freedom, $\mathbf{M}(\mathbf{q})$ is configuration-dependent mass matrix, $\mathbf{h}(\mathbf{q},\dot{\mathbf{q}})$ is the vector of non-linear terms representing centrifugal and Coriolis terms, $\mathbf{K}(\mathbf{q})$ is the stiffness matrix and $\mathbf{u}(t)$ is the vector of joint torques and actuator forces. The elements of $\mathbf{M}, \mathbf{K}, \mathbf{h}$ and $\mathbf{u}$ are given below. Note that in these expressions the elastic deflections are assumed to be small enough so that the high-order terms in q_i and $\dot{q}_i$ are neglected. This is in agreement with the Euler–Bernoulli theory that neglects some terms of the same order of magnitude (e.g. rotary inertia and shear effects).

The elements of the symmetric mass matrix $\mathbf{M}$ are given as

$$\mathbf{M}(1,\ 1) = I_{2T} + b + 2d\cos\theta_2 + I_{1R} + m_h L_1^2 - 2L_1 \sin\theta_2 \left(\sum_{j=1}^{N} m_j^* q_j \right)$$

$$\mathbf{M}(1,\ 2) = I_{2T} + d\cos\theta_2 - L_1 \sin\theta_2 \left(\sum_{j=1}^{N} m_j^* q_j \right)$$

$$\mathbf{M}(1,\ 2+j) = S_j + L_1 m_j^* \cos\theta_2$$

$$\mathbf{M}(2,\ 2) = I_{2T}$$

$$\mathbf{M}(2,\ 2+j) = S_j$$

$$\mathbf{M}(2+i,\ 2+j) = m_{ij}$$

The elements of the symmetric stiffness matrix $\mathbf{K}$ are given as

$$\mathbf{K}(1,\ 1) = \mathbf{K}(1,2) = 0$$
$$\mathbf{K}(2+i,\ 2+j) = k_{ij}$$

The elements of vectors $\mathbf{h}$ and $\mathbf{u}$ are given as

$$\mathbf{h}(1) = - \left[d\sin\theta_2 + L_1 \cos\theta_2 \left(\sum_{j=1}^{N} m_j^* q_j \right) \right] \dot{\theta}_2^2$$

$$- \left[2d\sin\theta_2 + 2L_1 \cos\theta_2 \left(\sum_{j=1}^{N} m_j^* q_j \right) \right] \dot{\theta}_1 \dot{\theta}_2 - 2L_1 \sin\theta_2 \left(\sum_{j=1}^{N} m_j^* \dot{q}_j \right) \dot{\theta}$$

$$\mathbf{h}(2) = \left[d\sin\theta_2 + L_1 \cos\theta_2 \left(\sum_{j=1}^{N} m_j^* q_j \right) \right] \dot{\theta}_1^2$$

$$\mathbf{h}(2+i) = \sum_{j=1}^{N} (c_{ij} - m_{ij}) q_j \dot{\theta}^2 + (m_i^* L_1 \sin\theta_2)\dot{\theta}_1^2$$

$$\mathbf{u}(1) = \tau_1$$
$$\mathbf{u}(2) = \tau_2$$
$$\mathbf{u}(2+i) = 0$$

where, $i, j = 1, 2, \ldots, N,$

$$I_{2T} = I_h + \int_0^{L_2} \rho A(x+a)^2 dx$$

$$b = \rho A L_1^2 L_2$$

$$d = \frac{1}{2}\rho A L_1 L_2^2 + \rho A L_1 L_2 a$$

$$m_{ij} = \int_0^{L_2} \rho A \phi_i(x)\phi_j(x)dx$$

$$m_j^* = \int_0^{L_2} \rho A \phi_j(x)dx$$

$$S_j = \int_0^{L_2} \rho A (x+a)\phi_j(x)dx$$

$$k_{ij} = \int_0^{L_2} EI\phi_i''(x)\phi_j''(x)dx$$

$$c_{ij} = \int_0^{L_2} \left\{ \rho A (x+a)\phi_i'(x)\phi_j(x) - \frac{1}{2}\rho A (L_2^2 - x^2 + 2aL_2 - 2ax)\phi_i''(x)\phi_j(x) \right\} dx$$

17.2.2 Convergence of the solution

Generally, open-loop simulations are performed (a) to study the convergence of the solution, (b) to validate the dynamic model used, (c) to understand the dynamic characteristics of the system and (d) to establish reasonable performance specifications. Modal convergence of the solution depends on the excitation of the system. For smooth excitation, the first few modes generally suffice. Impact excitation is the most demanding in terms of number of modes to be included. In order to determine the number of modes that can sufficiently describe the dynamics of the system for a realistic physical input, a set of simulations are performed. Rectangular torque pulses are generally used due to their rich frequency content (Yigit, 1994a). Simulations with increasing number of modes are performed until no perceptible difference is observed.

17.3 Coupling between rigid and flexible motion

Identifying rigid and flexible degrees of freedom will allow partitioning the matrices and vectors in equation (17.10) as follows:

$$\mathbf{q} = \{\mathbf{q}_r, \mathbf{q}_f\}^T$$

$$\mathbf{M} = \begin{bmatrix} \mathbf{M}_{rr} & \mathbf{M}_{fr} \\ \mathbf{M}_{rf} & \mathbf{M}_{ff} \end{bmatrix}$$

where $\{\}^T$ denotes the matrix transpose, and subscripts r and f are used to distinguish between the rigid and flexible generalised coordinates respectively, and the partitions are to be understood accordingly.

In general, for manipulators that consist of rigid and flexible links, there is coupling through mass matrix and the non-linear terms. Since there is a two-way coupling between the rigid and elastic degrees of freedom, the equations are known as fully coupled equations. In this approach the rigid-body motion is not considered to be known *a priori*, and is solved simultaneously with the elastic motion. In some earlier studies, the effect of elastic link deflections on the rigid-body motion was assumed negligible and consequently the elastic link deflections were considered to be perturbations about the rigid-body motion (Kojima, 1986). The equations of motion obtained with this approach are referred to as uncoupled equations. Here, uncoupled equations refer to the situation where the elastic motion does not affect the rigid-body motion, while the rigid-body motion affects the elastic motion. If the rigid-body motion is prescribed, then the resulting equations will be uncoupled. In this case the inertial forces resulting from the rigid-body motion are considered to be forcing functions for the elastic motion equations. For this reason a linear strain model will result in linear elastic motion equations. The general form of the equations will be that of equation (17.10) with $\mathbf{M}_{fr} = 0$.

The importance of coupling between the elastic and rigid-body motion was discussed in Yigit *et al.* (1988), where a cantilever beam attached to a rotating rigid hub was studied. It was shown that when the rigid-body inertia is small with respect to the inertia of the flexible parts, the uncoupled equations lead to incorrect results particularly with regard to resonance frequencies. It was shown that there is a significant difference in dominant resonance frequencies obtained from simulations using coupled and uncoupled equations. The dominant resonance frequency obtained using uncoupled equations is basically that of the cantilever beam, since mode shapes of a non-rotating cantilever beam was used. The dominant resonance frequency of the coupled equations, however, is in considerable deviation from either the non-rotating cantilever or the pinned-free beam. In limiting cases, when the ratio of the rigid-body inertia to the flexible beam inertia is very large or very small, the dominant resonance frequency will approach that of a cantilever or pinned-free beam, respectively (Yigit *et al.*, 1988). Another alternative is to consider the flexible link as a hinged-free beam and to use eigenfunctions of hinged-free beam with hub inertia as the admissible functions in equation (17.9). The issue here is related to the choice of trial functions in the discretisation procedure. For an uncoupled model, for instance, when the rigid hub inertia becomes smaller, the configuration will be that of a hinged-free beam, and cantilever mode shapes will no longer be admissible, resulting in a poor approximation. Similarly, if the hub inertia becomes larger, using simple hinged-free mode shapes will not be adequate. In general, it is advisable to model the system with a separate generalised coordinate for the rigid-body rotation, and thus, obtain a fully coupled system. In this case, any choice of admissible functions yields reasonably good results for the dominant response frequencies.

Using uncoupled equations has advantages in controller design for they yield a set of uncoupled harmonic oscillators. One can then design a controller for each mode separately. On the other hand, this modelling approach results in an unobservable system with measurement of only rigid-body motion. This can be verified by noting that in this case $\mathbf{M}_{fr} = 0$. Therefore, the rigid-body equations are completely decoupled from the elastic motion equations. Thus, a controller design based on the uncoupled equations requires that the elastic motion be measured. A controller designed based on the uncoupled equations (e.g. a rigid-body controller), may give good results when evaluated using the fully coupled model (Yigit and Ulsoy, 1989). On the other hand, using uncoupled equations in simulations to check the performance of any partic- ular controller can be misleading since the simulations will not include the correct coupling between rigid and flexible degrees of freedom.

17.4 Impact response

During the operation of a manipulator, impacts may occur due to several reasons. One example may be a mass capture operation where the end-effector has to con- tact the payload with a non-zero velocity. Another example may be a nailing-type operation where the end-effector is required to hit a rigid surface with considerably large velocity. The impact response of flexible manipulators has received relatively little attention. Chapnik *et al.* (1991) have studied the impact response of a one-link flexible robotic arm. Their impact model was based on an assumed contact force profile. Shi *et al.* (1999) presented a dynamic model for a constrained rigid–flexible robot arm where the end-effector is assumed to contact a constraint surface with a controlled contact force.

The impact response of a two-link rigid–flexible manipulator shown in Figure 17.1 was investigated in Yigit (1994*b*). A spring-dashpot model based on the Hertzian contact law incorporated with a non-linear damping mechanism was used for impact modelling. A contact algorithm was developed to accurately capture the multiple impacts, which happen due to the excitation of higher structural modes.

Simulations were carried out to obtain the response as well as contact force his- tories during impact. The parameters of the system studied were taken as follows: $I_{1R} = 0.0834 \, \text{kgm}^2$, $m_1 = 1.00 \, \text{kg}$, $L_1 = 0.5 \, \text{m}$, $I_h = 0.0020 \, \text{kgm}^2$, $\rho A = 0.15 \, \text{kg/m}$, $L_2 = 0.5 \, \text{m}$, $EI = 1.00 \, \text{Nm}^2$, $m_h = 0.5 \, \text{kg}$ and $a = 0.04 \, \text{m}$.

In the following, the term 'rigid manipulator' refers to the case where the second link is also rigid. The rigid and flexible manipulators do not differ significantly with respect to the shoulder angle as seen in Figure 17.2. The difference becomes evident in the elbow angle response as it is seen in Figure 17.3. Before the impact (which occurs at around 0.18 s) the responses are very similar, soon after the impact, however,

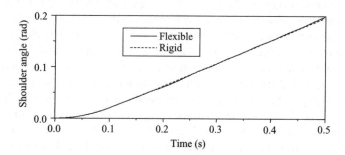

Figure 17.2   Shoulder angle responses

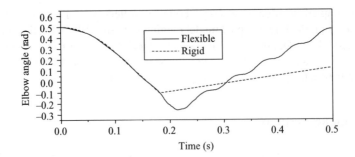

Figure 17.3   *Elbow angle responses*

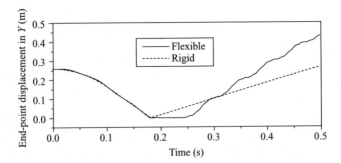

Figure 17.4   *End-point displacements in Y direction*

they differ significantly. Owing to the coupling between the flexible- and rigid-body motions, the response becomes oscillatory in the flexible case. It also appears that the energy loss due to impact is much less than that of the rigid manipulator. It is interesting to note that the end-point of the manipulator stays at the impact region for an extended period of time. This is confirmed by the end-point displacement history shown in Figure 17.4. This may be advantageous from an application viewpoint where one would like to perform some compliant tasks during this phase.

Figure 17.5 shows the end-point trajectory. The end-point trajectory for the flexible manipulator is very similar to that of the rigid manipulator up to the impact, but differs afterwards, rebounding to a much greater distance. During contact it moves along the contact surface (see the flat portion in Figure 17.4). The elastic end-point deflection for the flexible link is given in Figure 17.6. It appears that the first mode is dominant, with a small second mode component excited after the impact. As expected, there is a large peak due to impact. Typical non-minimum phase behaviour is observed at the beginning of the motion (i.e. the end-point moves in the opposite direction to the torque input). It is quite clear that the energy loss due to impact is much less for the flexible system than that of the rigid manipulator. The second link has a higher angular velocity after the impact. Inspection of the impact force histories reveals that the peak contact force for the rigid manipulator is much higher than that of the flexible manipulator (Yigit, 1994*b*). Also the duration of the contact is much smaller in the

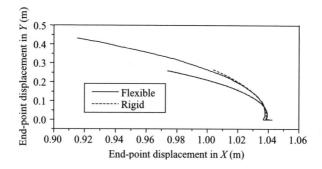

Figure 17.5   End-point trajectories

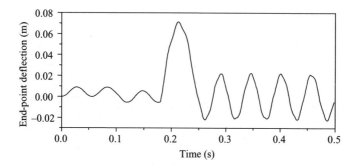

Figure 17.6   Elastic end-point deflections

flexible case. It should be noted though for the flexible case that there are a series of impacts rather than a single one as in the rigid case. These multiple impacts occur due to the excitation of higher modes of the flexible system. In fact, impacts, which appear single to the naked eye, have been shown to consist of several sub-impacts in quick succession (Yigit *et al.*, 1990). High-speed video recordings also confirmed the existence of multiple impacts (Yigit, 1988).

It is shown that the impulse due to impact is much higher for the rigid case, even though the impact conditions (e.g. impact velocity) did not differ significantly. It can be concluded that some of the energy is utilized to excite the flexible motion instead of being dissipated due to impact. Another interpretation is that the flexibility reduces the severity of the impact. This may be desirable for some applications if one wants to minimise the joint reaction forces due to impact. In any case, the flexibility of the links should not be neglected before a careful examination. For instance, simulations are carried out for flexural rigidity values of $EI = 64 \text{ Nm}^2$ and $EI = 128 \text{ Nm}^2$, which can be thought of as quite rigid compared to the previous simulations where the flexural rigidity was $EI = 4 \text{ Nm}^2$. Figure 17.7 shows that, as the beam becomes more rigid ($EI = 128 \text{ Nm}^2$) the end-point displacement tends to approach that of the rigid case. There is still, however, considerable difference in terms of the post-impact behaviour. The end-point elastic deflections for two different rigidity values are given

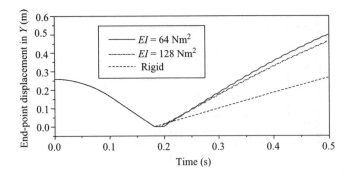

Figure 17.7   The effect of flexural rigidity on end-point displacement

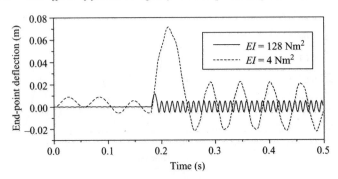

Figure 17.8   The effect of flexural rigidity on the elastic motion

in Figure 17.8. It is seen that the impact is able to excite some vibration modes of the stiff beam. Though relatively small in magnitude, this elastic motion carries some of the energy which otherwise will be dissipated due to impact.

In summary, it is clear that flexibility significantly affects the post-impact response, more specifically, it reduces the severity of impact, which can be characterised by smaller impulses. It has been found that as long as some elastic motion is excited due to impact, the response is significantly different than that of the rigid case even for relatively stiff links.

17.5   Control of rigid–flexible manipulators

The work by Book and co-workers (Book *et al.*, 1975) represents one of the earliest studies in control of flexible manipulators. They linearised the equations of motion about a nominal configuration and applied several linear control schemes to control a two-link flexible manipulator. A linear PD control was also applied by Oakley and Cannon (1988) and Fukuda (1985). In Oakley and Cannon (1989) a linear quadratic regulator (LQR) was designed based on a linearised model at a specific arm configuration. Linear designs based on a nominal configuration generally perform well when the links move at relatively low speeds and around a specified configuration at which the model is linearised. However, when angular velocities of the link are high and/or the manipulator moves away from the specified configuration, non-lineraties become significant and the performance degrades with a possibility of instability.

The computed torque method (also known as feedback linearisation) that was originally developed for rigid manipulators (Luh *et al.*, 1982) was also tried on flexible-link systems. The complexity of the inverse dynamics makes a straight-forward application of computed torque method impossible. In fact, Wang and Vidyasagar (1989*b*) showed that a multi-link manipulator with one flexible link is not input-state feedback linearisable. However, they demonstrated that for an appropriately chosen output variable, the input–output equations are feedback linearisable. Yigit and Ulsoy (1989) proposed a modified Corless–Leitman robust control scheme for a rigid–flexible two-link manipulator, based on partial feedback linearisation. This approach includes feeding back the flexible motion but utilises no control actuator for the degrees of freedom of the elastic motion. Later, this strategy was applied to force and motion control of a constrained rigid–flexible robot arm (Hu and Ulsoy, 1994). Experimental results were excellent in combined force and motion control.

Some approximate schemes of inverse dynamics have previously been proposed for open- and closed-loop control (Asada *et al.*, 1990; Bayo *et al.*, 1989; Pfeiffer, 1989). Lucibello and Di Benedetto (1993) developed an inversion-based control of a two-link rigid–flexible manipulator. They showed that for certain end-point trajectories (e.g. periodic) exact tracking with bounded elastic vibration can be achieved. Li (1994) proposed a non-linear control design based on decoupling the shoulder and elbow link dynamics through feedback linearisation. The method has been applied to trajectory tracking for a two-link rigid–flexible manipulator.

The main drawback of all model-based controllers is that one never has an exact model. Therefore, the robustness to parameter uncertainties has been a major concern in control design for manipulators (Asada and Slotine, 1986). Another difficulty with flexible systems is the so-called 'spillover' problem. Since the actual system is a distributed parameter system, any controller designed based on finite-dimensional models will generally suffer from the control and observation spillovers (Balas, 1978).

Independent joint PD controllers have been shown to be stable for rigid manipulators (Asada and Slotine, 1986). Oakley and Cannon (1988) performed some experiments on the performance of independent joint PD control of a two-link rigid–flexible manipulator, and they showed experimentally that the strategy works for flexible links. Since their design was based on a linearised model, the angles of manoeuvre were restricted to small values. Kelkar *et al.* (1992) studied the asymptotic stability of static dissipative compensators for flexible multi-body systems. However, they used a discretised model to prove the stability, making use of the skew-symmetric property of Coriolis and centrifugal terms. Lucibello and Ulivi (1993) implemented an independent joint PD control on a two-link rigid–flexible manipulator. Yigit (1994*a*) studied the stability robustness of independent joint PD control. Lyapunov stability theorem was used to show the global stability of the controller. Moreover, the control gains do not depend on the parameters of the system, the stability result is independent of any spatial discretisation, and linearisation is not required for design and implementation of the controller. Simulations are carried out to investigate the performance of independent joint PD control.

17.5.1   Stability of independent joint control for a two-link rigid–flexible manipulator

The independent joint PD control is given as

$$\tau_1 = -K_p(\theta_1 - \theta_{1f}) - K_d\dot{\theta}_1$$
$$\tau_2 = -G_p(\theta_2 - \theta_{2f}) - G_d\dot{\theta}_2 \tag{17.11}$$

where, θ_{1f} and θ_{2f} denote the desired final values of θ_1 and θ_2, respectively, and the gain values are to be positive. The control given in equation (17.11) stabilises the system given in equations (17.5) to (17.8). This can be proven as follows.

Considering the total energy in the system, the following function is taken as a candidate Lyapunov function:

$$V(t) = \frac{1}{2}I_{1R}\dot{\theta}_1^2 + \frac{1}{2}I_h\dot{\theta}^2 + \frac{1}{2}m_hL_1^2\dot{\theta}_1^2 + \frac{1}{2}\int_0^{L_2} \rho \dot{r}^2 dx + \frac{1}{2}\int_0^{L_2} EI(w'')^2 dx$$

$$+ \frac{1}{2}K_p(\theta_1 - \theta_{1f})^2 + \frac{1}{2}G_p(\theta_2 - \theta_{2f})^2 \tag{17.12}$$

The last two terms on the right-hand side of equation (17.12) are added to the total energy to make the final state be a global minimum of V. Clearly $V(t) \geq 0$ for all t, and $V = 0$ only at the desired state. These properties make V a candidate Lyapunov function.

Differentiation, integration by parts and substitution of equations of motion into equations (17.5)–(17.7) with centrifugal stiffening terms neglected and proper factorisation lead to

$$\dot{V}(t) = \dot{\theta}_1[\tau_1 + K_p(\theta_1 - \theta_{1f})] + \dot{\theta}_2[\tau_2 + G_p(\theta_2 - \theta_{2f})] \tag{17.13}$$

Substitution of equation (17.11) into equation (17.13) yields

$$\dot{V}(t) = -K_d\dot{\theta}_1^2 - G_d\dot{\theta}_2^2 \tag{17.14}$$

which is negative semi-definite. In order to show the asymptotic stability LaSalle's theorem has to be used (Vidyasagar, 1978).

Suppose $\dot{V}(t) \equiv 0$, then, equation (17.14) implies that

$$\dot{\theta}_1 \equiv 0 \quad \text{and} \quad \dot{\theta}_2 \equiv 0 \tag{17.15}$$

and hence

$$\ddot{\theta}_1 \equiv 0 \quad \text{and} \quad \ddot{\theta}_2 \equiv 0 \tag{17.16}$$

Substitution of equations (17.11), (17.15) and (17.16) into the equations of motion, equations (17.5)–(17.8), yields

$$\int_0^{L_2} \rho A[\ddot{w}(a + x) + L_1\ddot{w} \cos \theta_2] dx + K_p(\theta_1 - \theta_{1f}) = 0 \tag{17.17}$$

$$\int_0^{L_2} \rho A[\ddot{w}(a+x)]dx + G_p(\theta_2 - \theta_{2f}) = 0 \tag{17.18}$$

$$\ddot{w} + EIw'''' = 0 \tag{17.19}$$

$$w(0,t) = 0; \qquad w'(0,t) = 0$$
$$w''(L_2,t) = 0; \qquad w'''(L_2,t) = 0 \tag{17.20}$$

The solution to equation (17.19) with the boundary conditions in equation (17.20) is given as (Meirovitch, 1986)

$$w(x,t) = \sum_{r=1}^{\infty} Y_r(x) \left(q_{r0} \cos \omega_r t + \frac{\dot{q}_{r0}}{\omega_r} \sin \omega_r t \right) \tag{17.21}$$

where $Y_r(x)$ and ω_r are the eigenfunctions and eigenfrequencies of the regular cantilever beam and

$$q_{r0} = \int_0^{L_2} \rho A y_0(x) Y_r(x)dx \qquad r = 1, 2, \ldots$$

$$\dot{q}_{r0} = \int_0^{L_2} \rho A v_0(x) Y_r(x)dx \qquad r = 1, 2, \ldots$$

where $y_0(x)$ and $v_0(x)$ are the initial deflection and velocity profiles for the second link. Substituting equation (17.21) into equations (17.17) and (17.18) yields θ_1 and θ_2 as functions of time. On the other hand, equations (17.15) and (17.16) imply that θ_1 and θ_2 are constants. The only possibility that this would be true is when $w(x,t) \equiv 0$, which in turn yields $\theta_1 = \theta_{1f}$ and $\theta_2 = \theta_{2f}$, which is the null solution for the controlled system. Thus, it has been shown that the only solution of the system given by equations (17.5)–(17.8) satisfying $\dot{V}(t) \equiv 0$ is the null solution (desired states). Therefore, the controlled system is asymptotically stable, (Vidyasagar, 1978). The following remarks can be made regarding this control:

- This control does not require state estimation since joint angular positions and velocities can easily be measured.
- This control does not depend on any discretisation of the equations; therefore, there is no spillover effect, which is a general problem in most control strategies for flexible structures (Balas, 1978).
- This control has the stability robustness to parameter uncertainties since the gains are not computed using the parameters of the system. For all physically reasonable values of the parameters of the system, V remains to be a valid Lyapunov function.

17.5.2 Closed-loop simulations

In the simulations presented here, only step response (set-point tracking) is considered. The desired values for the shoulder and the elbow joint angles are $\theta_{1f} = 1.0$ rad,

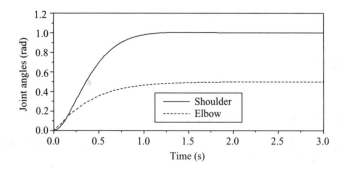

Figure 17.9   Step responses of joint angles

and $\theta_{2f} = 0.5$ rad, respectively. In order to have physically reasonable control inputs, the following procedure is applied to determine the control gains. As a very crude approximation, the second link is assumed to be rigid and fully extended with elbow joint locked. This represents the highest inertia configuration with respect to the shoulder axis. The equation of motion for this approximate system is

$$I_{eq}\ddot{\theta}_1 + K_d\dot{\theta}_1 + K_p\theta_1 = 0$$
$$\ddot{\theta}_1 + 2\zeta\omega_n\dot{\theta}_1 + \omega_n^2\theta_1 = 0$$

where I_{eq} is the total inertia of the approximate system with respect to the shoulder axis, and

$$2\zeta\omega_n = K_d/I_{eq}$$
$$\omega_n^2 = K_p/I_{eq}$$
(17.22)

ζ and ω_n can be selected based on the desired damping ratio and the control bandwidth, respectively. Then, K_p and K_d are calculated from equation (17.22). A similar procedure can also be applied to the elbow joint. For the simulations given here, the same gain values are used for both joints. The gain values used here ($K_p = 5$ and $K_d = 2$) correspond to $\zeta = 1.0$ and $\omega_n = 5.0$ rad/s.

Figures 17.9 and 17.10 show the rigid-body motion and the second-link elastic end-point deflection, respectively. The joint torques are given in Figure 17.11. It is seen that the performance of the controller is very good for both slewing and vibration suppression. There is a small overshoot in the rigid-body positions. The overshoot may not always be small and depends on the controller gains. The effect on elastic end-point deflection is not significant. It seems that in order to avoid overshoot one has to perform some simulations until a satisfactory performance is achieved. It is shown that as the position feedback gains increase the speed of response increases, at the expense of overshoot.

The effect of a possible actuator saturation was investigated in Yigit (1994a), where it is assumed that the maximum available torques for the shoulder and the

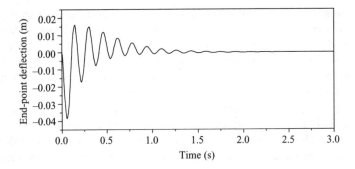

Figure 17.10   Elastic end-point deflections

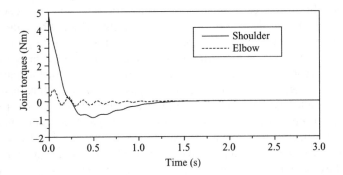

Figure 17.11   Joint torque inputs

elbow joint are 2 Nm and 1 Nm, respectively. The performance of the controller was found to be very good with a small delay due to saturation.

The robustness of the proposed controller was also investigated by perturbing the system parameters in the simulation (Yigit, 1994*a*). The performance of the controller designed for the nominal system was checked, when the inertia of the first link was increased by 100 per cent. The performance was very good in relation to the elastic motion. There was some overshoot in the shoulder angle. It can be concluded that the controller performed reasonably well in the case of uncertainties – if a small amount of overshoot can be tolerated, or a conservative value for the damping ratio is selected – thus yielding a slower response.

In cases where no overshoot is tolerated, this control strategy is not very efficient for carrying out both large-angle slewing and vibration suppression. Typically relatively larger gains are needed for vibration suppression near the desired state, which are generally not appropriate during the initial phase of a large-angle manoeuvre. This fact puts a limit on the achievable bandwidth for the joint PD controllers. The bandwidth of the controller can be increased if the control is modified to include either the end-point deflection or the base strain feedback (Ge *et al.*, 1996).

The manipulator here was assumed to move in the horizontal plane, so that the gravity was not considered. In cases where the manipulator moves in the vertical plane, the effect of gravity is to result in some offset in the final values. Though this can be compensated to some extent (De Luca and Panzieri, 1994), gravity will destroy nice robustness properties of the controller. The reason is that compensation for the gravity will depend on the parameters of the system.

In most applications of robotics, the main objective is to follow a desired trajectory. A good control law should have good tracking capability for a wide class of trajectories. The control law given in equation (17.11) has to be modified if one wishes to have the tracking property. A possible modification is given in Choura and Yigit (2001), which uses an open-loop trajectory. An inversion based end-effector tracking control is difficult due to the instability of zero-dynamics, and unconditional asymptotic stability is not easy to prove. If some bounded elastic vibration is tolerated, then it is possible to track certain trajectories with sufficiently slow rigid-body motion (Lucibello and Di Benedetto, 1993). Paden *et al.* (1993) showed both theoretically and experimentally that passive joint controllers, and the feedforward of nominal joint torques corresponding to a delayed nominal trajectory results in exponentially stable tracking control for sufficiently stiff and damped links (structural and/or joint damping). Rossi *et al.* (1997) showed that the passivity of multi-link manipulators with a last flexible link is determined by the passivity of the last link. They performed experiments on a five-bar linkage manipulator with a single flexible link, and showed that for large effective hub inertia, the manipulator can track a four-leaf clover trajectory. Bigras *et al.* (1998) proposed a tracking controller for manipulators made of rigid and flexible links. Provided there is an actuator for each rigid generalised coordinate, this controller can achieve exponential stability of tracking error in a virtual joint space as long as the flexibility effects are small and there is viscous friction. From the foregoing short review, it is clear that the issue of tracking controllers for rigid–flexible manipulators is still an open problem.

17.6   Summary

Rigid–flexible manipulators have come to existence because despite their relative simplicity as compared to multi-link flexible manipulators, they include the non-linear, coupled dynamics of multi-link systems as well as the behaviour of flexible structures including non-minimum phase characteristics in the control problem. This chapter reviewed the literature dealing with dynamics and control of manipulators consisting of rigid and flexible links.

Most of the manipulators are the so-called elbow robots with two links. Generally, the first link is sufficiently rigid, while the second link is designed to be quite flexible. Lagrangian dynamic formulations are generally preferred over Newton–Euler approach. This is not surprising as the former uses easily derived energy expressions while the latter requires a more detailed investigation of all forces involved. Although, in principle, the two formulations are equivalent, care must be exercised in using Lagrangian formulations with linear strain expressions since one may lose the so-called geometric or centrifugal stiffening term. The most favoured formulations appear to be those based on modal models although FE techniques are also used due to their generality. It has been demonstrated that the model should retain the full coupling between the rigid and elastic degrees of freedom for an accurate representation of the dynamic behaviour. It has also been shown that flexibility effects may be important even for relatively stiff links in case of impulsive excitations such as mass capture or impact with the environment.

Control design and implementation for rigid–flexible manipulators has also received considerable interest in the last decade. Although some linear optimal controllers have been designed based on a linearised model around a nominal trajectory, most favoured control designs are essentially modifications of controllers used for rigid-link robots. Since it is difficult to employ actuators for the elastic motion, most control strategies rely on joint actuators. Passive joint controllers, inversion based feedforward–feedback schemes and dynamic decoupling controllers are among the control strategies that are implemented on rigid–flexible manipulators. It has been shown that asymptotic tracking of the end-effector can be achieved if certain conditions are met in the design of the manipulator.

In summary, a great deal has been achieved in dynamic modelling and control of rigid–flexible manipulators. As our understanding matured, however, the need to model certain links as rigid bodies subsided since a rigid link can effectively be modelled as a sufficiently stiff flexible link. Therefore, most if not all of the results of the research on multi-link flexible manipulators are applicable to manipulators with rigid and flexible links.

Chapter 18

Analysis and design environment for flexible manipulators

O. Ravn and N.K. Poulsen

The focus in this chapter is on a design environment for modelling and control of flexible-link robots. The environment consists of physical equipment and a software environment for design and analysis. The physical equipment comprises two computer platforms, a design platform for analysis and design and, a real-time Linux platform for controlling a laboratory rig equipped with actuator and sensing systems. The rig has several configurations including a horizontal and a vertical mounting of the robot, and is controlled from a real-time Linux platform. The software environment for design, analysis and simulation is based on a Matlab/Simulink workbench extended with a mechatronic Simulink library (MSL). This allows combination of different principles of modelling and control resulting in various control designs and strategies. A simple switch can change the configuration from simulation to real-time application. The real-time code can be generated using real-time workshop (RTW) and transferred to the real-time platform in order to perform the experiments. The collected data can also be transmitted back to the design platform for further analysis.

18.1   Introduction

The desire for high-performance manipulators and the benefits offered by a lightweight flexible arm have lead to analysis in which flexibility is the essential issue. This is especially the case for manoeuvring of large payloads. The high-performance requirements will inevitably produce designs that during operation will excite vibration modes in the manipulator structure.

The flexibility leads to a severe problem in controlling the motion owing to the inevitable excitation of structural vibrations, which affect the positioning accuracy of the manipulator. Therefore, a successful controller implementation of a flexible

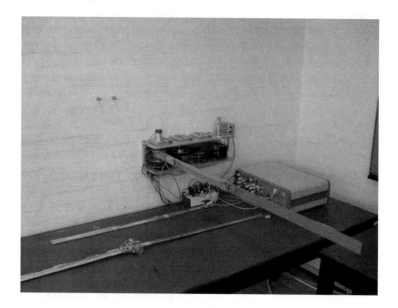

Figure 18.1   The flexible manipulator rig

manipulator system is contingent on achieving acceptable performance taking into account variations in, for example, payload and environmental disturbances.

The aim of the controller is to suppress the structural vibration while minimising the cycle time of the manipulator system. For flexible manipulator systems, it is necessary to use a model-based controller in order to mitigate the first harmonics. However, changes in payload degrade the model and consequently the performance of the control system, unless some sort of adaptation or gain scheduling is utilized to estimate such effects.

In order to investigate different aspects of control of flexible-link robot configurations an experimental set-up has been made. This experimental set-up forms the basis of the work described in this chapter. The set-up, shown in Figures 18.1 and 18.2, consists of the design platform, a real-time system and the actual experimental rig. The rig has several configurations and it can be mounted in a horizontal and a vertical position. In the horizontal position the gravity plays a negligible role. The rig has a 1 degree of freedom (DOF) and 2 DOF configuration. The rig consists (in its 2 DOF configuration) of two very flexible links with two actuators located at the joints. The geometry of the links makes the predominant bending to take place in this plane making it possible to ignore torsion. The actuators are d.c. motors with a sufficient gear ratio, and tachometers making an analogue velocity feedback feasible. This suppresses the friction and other non-linearities in the actuators. Apart from the tachometers there are two kinds of sensors on the set-up, a potentiometer in each joint for measurement of the (angular) position of the joint and a number of strain gauges located on each link for measurement of the bending of the link. When dealing with real laboratory rigs it is very important to have a Control System Design environment

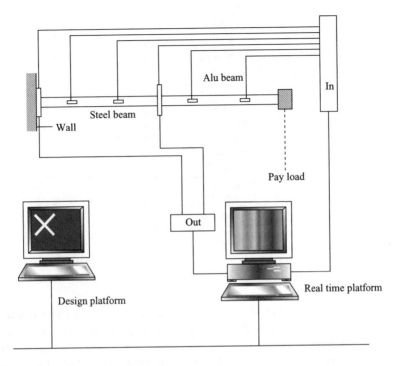

Figure 18.2   Schematic diagram of the flexible manipulator system

that supports the design process in order to be able to put emphasis on the controller design and not on practical details.

18.2   Computer aided control engineering design paradigm

In this section some background on computer aided control engineering (CACE) is given in order to motivate the design of the software for controlling the flexible robot rig. In the classical model of the control development process presented in (Ravn and Szymkat, 1992) shown in Figure 18.3 the following general phases are distinguished:

Goal generation. This phase initiates the design process. The problem and the desired features of the solution are determined. This phase is normally done in cooperation with the customer, engineers and other team members. Normally no formalised tools or methods are used in this phase.

Modelling. The modelling phase is used to determine a model of the system to be controlled. This is normally a mathematical model, which can be used in subsequent phases. Models of different complexity may be derived, such as linear plant models for the design of linear controllers and then non-linear plant models in the evaluation phase. Many CACE tools exist for assisting the user during this phase.

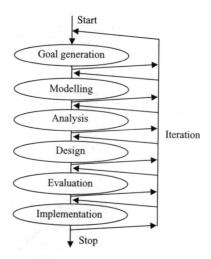

Figure 18.3   The classical design model

Analysis. The derived model is analysed in order to gain an understanding of the system and the potential problems. The analysis results are used as a basis for choosing a controller structure. Not just the normal numerical tools are applicable in this phase; the potential benefits of using symbolic manipulation tools are becoming more and more evident and many of the numerical packages have built-in symbolic tools or interface to them.

Design. A possible controller structure is selected and the parameters are chosen in order to match the design goals. It may be useful to consider more controller structures and compare their performances in parallel. Many tools for designing controllers such as standard linear quadratic (LQ), linear quadratic Gaussian (LQG), and so on exist.

Evaluation. The different controllers are considered in this phase and compared with respect to the features of the desired solution set up in the first phase of the design process. The degree of compliance with the goals is determined and the best controller selected. The evaluation phase may use simulation of the system or use partially the real-time interface in order to select the best controller. More models may be used in order to gain insight into what features of the system and the controller limit the performance.

Implementation. The chosen mathematical description of the controller is implemented. More and more tools are emerging in this field. Standard packages have C-code generation tools and offer hardware, which can be used for testing the controller in a laboratory environment. The main problem here is the balance between code efficiency, hardware dependency and the degree of automation.

Another element of the design process model is the iteration, which is its fundamental property. Iteration can be performed manually, semi-automatically or automatically. The iterative nature of the design process is also an important element. An overall evaluation of the design phases indicates that most CACE tools are

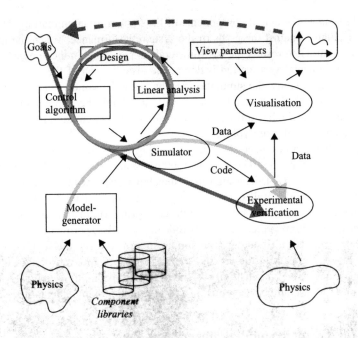

Figure 18.4   A simulation centred design process model seen from viewpoints of algorithm and application designers

available for the *modelling, analysis* and *design* phases. Some tools are also available for the *implementation* phase. However there is a lack of tools for the rest of the phases and the iteration. Many alternative design process models have been discussed in numerous papers, see Barker *et al.* (1993), MacFarlane *et al.* (1989) and Ravn and Szymkat (1992, 1994).

The proposed design models emphasise the importance of simulation most commonly becoming central and integrating phase of the design cycle. The simulation centred model of the design process presented in this section is based on the fact that different users have different approaches to the design problem. Two major categories are found: the application engineer, who is focused on getting the application controlled in the optimal way and the algorithm designer, whose main goal is to develop and validate the control algorithm on some object. The model is shown in Figure 18.4.

Consider the model in Figure 18.4, in the first instance, from the application engineer's point of view. The model of the design process is centred on the simulator; this is not a simulator in the limited sense of the term but includes the ability of making real-time code, simulation data and linearising the non-linear model of the system. The simulator takes a control algorithm and a system model as inputs and produces results in terms of data and real-time code. The system model is generated by a model generator based on the physics of the system components drawn from a component library. The control algorithm is found in an iterative way as described earlier. The results of the simulation are visualised in different ways, as plot of variables or as a visualisation of the actual components of the system in an animation. The result is then evaluated against the goals of the system performance. When the goals are met by the simulated results the real-time code is generated in the simulation block and the experimental verification is done. The visualisation of the results obtained here is validated against the simulation results using the same visualisation module and compared with the goals. Finally the design should be implemented for production purposes, which is quite different from the implementation for experimental verification and rapid prototyping.

The application engineer starts by modelling the system to be controlled and ends with the experiments. The starting point of the algorithms developer is different as shown in Figure 18.4. The algorithm developer starts with a candidate control algorithm and some overall performance goals for the system. Iterations are done modifying the parameters or structure of the control algorithm until the goals are met. The control algorithm is then possibly validated through experiments.

The two points of view are quite similar in structure but the focus is different in each case. The result of the process should not just be the control algorithms but also some measure on the sensitivity to imperfect initial conditions in the design. The validation results should be presented in a form suitable for documentation purposes.

18.3   Mechatronic Simulink library

In the design of mechatronic Simulink library (MSL) (Ravn and Szymkat, 1995; Ravn *et al.*, 1996; Szymkat *et al.*, 1995) a number of important observations were made. In the modelling phase it is very important to look at the following aspects:

- Component-based modelling so that the simulation model structure closely resembles the physical model structure to facilitate easy exchange of components.
- Consistent definition of the input/output of components.
- Handling of parameters through a database and/or tuneable through an interface.
- Tuneable granularity/complexity as described below.

The prototype implementation of MSL incorporates a number of components often used in the laboratory. The current groups of components are shown in Figure 18.5. It should be noted that MSL is designed in such a way that the addition of new component types is simple and the user should use this facility, as only the most basic component types are included from the beginning.

The d.c. motor model is shown in Figure 18.6, where the dialogue box can be used to enter type of d.c. motor, as well as if the parameters of friction, stiction and

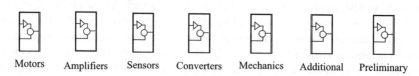

Motors   Amplifiers   Sensors   Converters   Mechanics   Additional   Preliminary

Figure 18.5   The mechatronic Simulink library

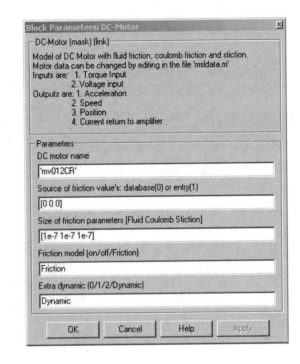

Figure 18.6   DC motor symbols and dialogue box

fluid friction should be taken from the database (default) or entered. Furthermore it can be chosen if friction should be simulated or not and the level of extra dynamics included, in this case the electrical time constant of the drive. The model uses a feature of Simulink called triggered blocks enabling the user to only simulate certain blocks in the diagram. This is utilized in MSL to enhance simulation performance of the simplest complexity level significantly.

To demonstrate the possibility of making simple changes in the complexity of the MSL models, a simple servomechanism is modelled and calibrated, see Figure 18.7. Figure 18.8 shows the shaft position of the servomechanism when the global variable 'Friction' is respectively 'on' and 'off'.

Another system example is a simple 1 DOF flexible robot arm. The arm is a single-joint robot where the arm is made from a very flexible material. The bending of arm is measured using two strain gauges. Data and more details relating to the flexible robot arm and its modelling in MSL is found in Rostgaard (1995).

Figure 18.9 shows the finished MSL model of the flexible robot. Figure 18.10 shows plots of the end-point angle, motor angle and the harmonic time functions q for a step input in motor position using a simple position control and without taking the flexibility into account. The simulation is shown with one and two modes, respectively. The change is done by inputting the number of modes into the dialogue box of the 'flex' block.

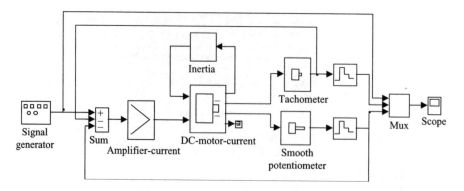

Figure 18.7   MSL position servomechanism

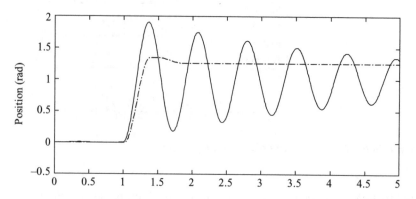

Figure 18.8   Step response for the simple servomechanism with and without friction

The MSL-based model is thus an excellent alternative to modelling in plain Simulink. More information on MSL can be found at http://www.oersted.dtu.dk/personal/or/MSL.

18.4   Design models

The experimental set-up is in horizon and vertical configurations, where gravity plays a negligible role in the horizontal configuration. The control design is based on physical models in this work. These models can be based on a number of different approaches including modal or eigenvalue method (Kruise, 1990) and finite element (FE) method (Sakawa *et al.*, 1985). Example on an FE model is given below for a 1 DOF configuration.

The flexible manipulator system studied here, see Figure 18.11, carries a payload, m_p at its tip (end-point) and moves in the horizontal or the vertical plane. For the 2 DOF flexible-link robot the active DOF are the two rotational angles θ_{b1} and θ_{b2}, see Figure 18.12.

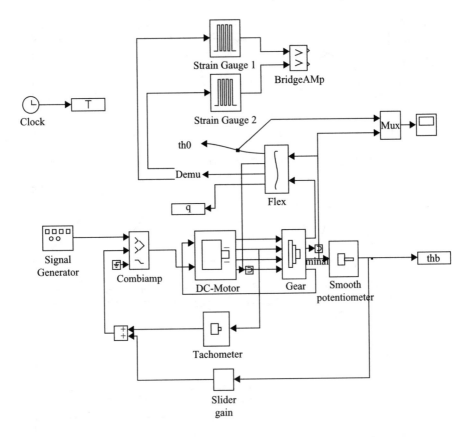

Figure 18.9   Mechatronic Simulink library model of the flexible robot

18.4.1   Dynamics of actuators

The actuators consist mainly of a d.c. motor supplied with a gear and a tacho feedback. Consider a d.c. motor, which can be described (to a reasonable degree) by a first-order model as depicted in Figure 18.13. Let V_a denote the voltage input, T_m the resulting output torque (i.e. torque not included in the model) and $\dot\theta$ the angular velocity of the motor shaft. Then using Newton's second law the model can be given as

$$J\ddot\theta_m = T_m - f\dot\theta_m + \frac{k_t}{R_a}(V_a - k_e\dot\theta_m)$$

where $k_e = k_t$ represent motor constants, f the total viscous friction of the motor, R_a the electrical resistance and J the total inertia of the motor.

In order to reduce the influence of disturbances, non-linearities and other imperfects, the d.c. motor is included in a tacho loop, in which the difference between the reference voltage u and the tacho voltage $V_{tg} = k_{tg}\dot\theta_m$ is amplified (gain k_g) and fed

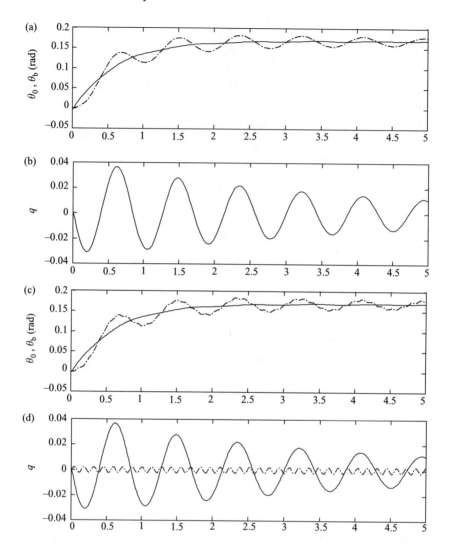

*Figure 18.10   Step responses for the flexible robot modelled with one and two modes:
(a) one mode (theta sub zero, theta sub b), (b) one mode (q), (c) two
modes (theta sub zero, theta sub b), (d) two modes (q)*

to the motor, that is,

$$V_a = k_p(u - k_{tg}\dot{\theta}_m)$$

A gear is introduced between the motor shaft (θ_m) and the manipulator (θ_b). The gear is considered stiff and is described as

$$\theta_b = \frac{1}{N}\theta_m$$

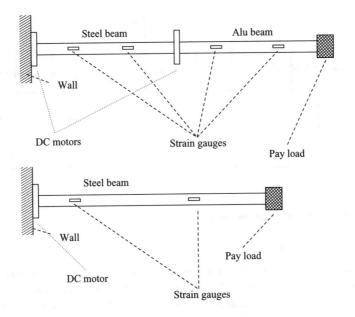

Figure 18.11 The experimental rig

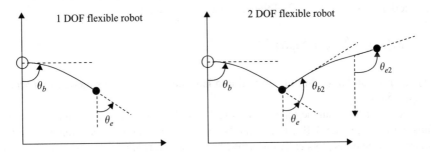

Figure 18.12 Definitions of angles for the 1DOF and 2DOF configuration of the flexible link robot

If the external momentum $T_b = NT_m$ is introduced the actuator can be given as

$$\ddot{\theta}_b = k_1 u + k_2 T_b + k_3 \dot{\theta}_b \qquad (18.1)$$

where k_1, k_2 and k_3 are constants. Notice that this is a second-order model.

Owing to the span in time constants for the actuator and for the rest of the robot the fast part of the actuator dynamics is often neglected. In that case the actuators are modelled as

$$k_1 u + k_2 T_b + k_3 \dot{\theta}_b = 0$$

which is a first-order model.

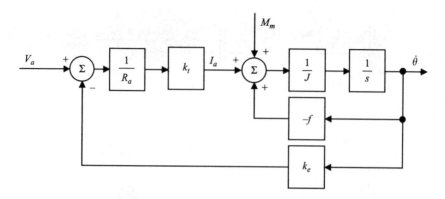

Figure 18.13   Block diagram for a DC-motor

18.4.2   Modal models

The model of the flexible-link robot consists of four parts, namely, models for the two actuators and the two arms. The dynamics of the flexible arms can be described by a partial differential equation (PDE), which can be transformed into an ordinary differential equation (ODE) by using the FE method or the method of separation of variables. In that case, the resulting model becomes a modal model in which the deflection, $w_j(x, t), j = 1, 2$ of the arms can be written as

$$w_j(x, t) = \sum_{i=1}^{\infty} \varphi_{ji}(x) q_{ji}(t)$$

where $\varphi_{ji(x)}$ and $q_{ji}(t)$ are the normal and harmonic functions of mode i and arm j, respectively.

Consider the beam segment in Figure 18.14. Let x denote the distance along the beam and $y(x, t)$ the deflection of the beam. Here a is the cross-section area of the beam. The beam is physically described by b, h and L representing the width, height and length.

Since transversal vibrations appear athwart to the beam the cross-section area and inertia are

$$a = bh, \quad I = \int_0^h (2 \int_0^{b/2} r^2 dr) dy = \frac{b^3 h}{12}$$

Taking the resulting shear forces and torque the following relations can be obtained:

$$\frac{\partial F}{\partial x} + \rho a \frac{\partial^2 w(x, t)}{\partial^2 t} = 0, \quad F = \frac{\partial M}{\partial x} \tag{18.2}$$

It follows from elementary flexural theory (Timoshenko *et al.*, 1974) that

$$M = EI \frac{\partial^2 w(x, t)}{\partial^2 x} \tag{18.3}$$

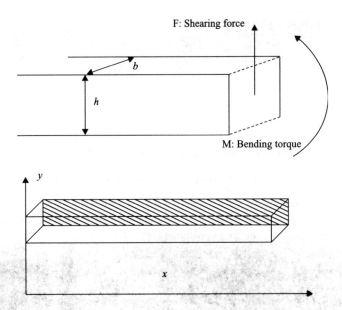

Figure 18.14   Forces and torque in a cross section of a beam

Equations (18.2) and (18.3) yield the Euler–Bernoulli equation for the beam as:

$$EI\frac{\partial^4 w(x,t)}{\partial x^4} = -\rho a\frac{\partial^2 w(x,t)}{\partial t^2} \qquad (18.4)$$

The solution to equation (18.4) with boundary conditions can be found using the separation of variables method, that is,

$$w(x,t) = \sum_{i=1}^{\infty} \varphi_i(x)q_i(t)$$

which is equivalent to expanding the deflection of the beam into modes. The functions $\varphi_i(x)$ define the shape of the natural modes of vibration and are called principal functions or normal functions. It can be shown (Rostgaard, 1995) that the normal functions are orthogonal and possess other interesting properties. The function $q_i(t)$ describes the time dependence of the modal deflection. The expansion thus yields

$$M = EI\sum_{i=0}^{\infty} q_i(t)\frac{\partial^2 \varphi_i(x)}{\partial x^2}, \qquad F = EI\sum_{i=0}^{\infty} q_i(t)\frac{\partial^3 \varphi_i(x)}{\partial x^3}$$

Each mode has the following time dependence

$$\frac{\partial^2 q_i(t)}{\partial t^2} + \omega_i^2 q_i(t) = 0 \quad \text{or} \quad \ddot{q}_i(t) + \omega_i^2 q_i(t) = 0 \qquad (18.5)$$

The modal function is given by

$$\frac{\partial^4 \varphi_i(x)}{\partial x^4} - \gamma_i^2 \varphi_i(x) = 0 \qquad\qquad (18.6)$$

where

$$\omega_i^2 = \frac{EI}{\rho a} \gamma_i^4$$

The general solution to the mode shape equation in equation (18.6) can be found as

$$\varphi_i(x) = c_{1i} \cosh(\gamma_i x) + c_{2i} \sinh(\gamma_i x) + c_{3i} \cos(\gamma_i x) + c_{4i} \sin(\gamma_i x)$$

where the constants in the above equation can be determined from the end-point constraints. Consider, for example, that for a clamped free beam the following four conditions (to determine the four constants) hold:

$$\varphi_i(0) = 0, \qquad \frac{\partial \varphi_i(0)}{\partial x} = 0, \qquad \frac{\partial^2 \varphi_i(L)}{\partial x^2} = 0, \qquad \frac{\partial^3 \varphi_i(L)}{\partial x^3} = 0$$

where, the first two conditions in the above equation are because of the clamped end ($x = 0$) and the last two are caused by the free end ($F(L) = 0$, $M(L) = 0$). In order to obtain non-trivial solution the 'frequency equation'

$$\cosh(\gamma_i L) \cos(\gamma_i L) = -1$$

has to be fulfilled, that is, determine γ_i (and ω_i). The first four normal functions (mode functions) are plotted in Figure 18.15.

A modal model of the flexible-link robot can now be developed. To demonstrate this only the 2 DOF configuration of the robot is considered and a summary is provided. The robot consists of two actuators, two flexible links, two tacho encoders

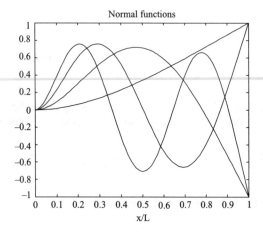

Figure 18.15   *The first four mode shape functions*

and four strain gauges. The four basic differential equations can be summarised as follows (Rostgaard, 1995).

For the shoulder-actuator the following holds:

$$k_{13}\dot{\theta}_{b1} + k_{11}u_1 + k_{12}E_1I_1 \sum_{i=0}^{\infty} \varphi_{1i}''(0)q_{1i}(t) = \ddot{\theta}_{b1} \tag{18.7}$$

where k_{11}, k_{12} and k_{13} are constants related to the shoulder actuator, I_1 is the beam inertia for the upper arm, E_1 is the Young modulus for the beam and u_1 is the input (control) voltage.

For the elbow actuator the relation in a similar manner is given as

$$k_{23}\dot{\theta}_2 + k_{21}u_2 + k_{22}E_2I_2 \sum_{i=0}^{\infty} \varphi_{12i}''(0)q_{2i}(t) = \ddot{\theta}_{b2} \tag{18.8}$$

For the lower arm the following ODE holds for each mode ($i = 1, 2, \ldots, n$):

$$\omega_{2i}^2 q_{2i}(t) + 2\zeta_{2i}\omega_{2i}\dot{q}_{2i}(t) + \ddot{q}_{2i}(t) = \sum_{j=1}^{\infty} \kappa_{2ij}^* \ddot{q}_{2j}(t) + \alpha_{2i}^* \left[\ddot{\theta}_{b2} + \ddot{\theta}_{b1} + \sum_{j=1}^{\infty} \varphi_{1j}'(L_1)\ddot{q}_{1j}(t) \right]$$

$$+ \beta_{2i}^* \left[L_1\ddot{\theta}_{b1} + \sum_{j=1}^{\infty} \varphi_{1j}(L_1)\ddot{q}_{1j} \right] \cos\left(\theta_{b2}\right)$$

Here ω_{2i} and ζ_{2i} are the harmonic frequency and damping for mode i (and the second or lower arm). L_1 is the length of arm and the modal parameters for the lower arm are linearly dependent on the payload, that is,

$$\alpha_{2i}^* = \alpha_{2i} - \frac{m_p}{\mu_2}L_2\varphi_{2i}(L_2)$$

$$\beta_{2i}^* = \beta_{2i} - \frac{m_p}{\mu_2}\varphi_{2i}(L_2) \tag{18.9}$$

$$\kappa_{2ij}^* = -\frac{m_p}{\mu_2}\varphi_{2i}(L_2)\varphi_{2j}(L_2)$$

where, the payload free modal parameters, α_{2i} and β_{2i} depend on the geometry and the normal functions. The parameter μ_2 is one quarter of the mass of the link, that is, $\mu_2 = \frac{1}{4}m_{l2}$.

For the upper arm the situation becomes a little more complicated. This is because of the coupling between elbow-actuator and the deflection of the upper arm. Here the modal equations are

$$\omega_{1i}^2 q_{1i}(t) + 2\zeta_{1i}\dot{q}_{1i}(t) + \ddot{q}_{1i}(t) = \sum_{j=1}^{\infty} \kappa_{1ij}^* \ddot{q}_{1j} + \ddot{\theta}_{b1}\left[\alpha_{1i}^* + \frac{J_h \varphi_{1i}'(L_1)}{\mu_1}\right]$$

$$+ \frac{J_h \varphi_{1i}'(L_1)}{\mu_1} \sum_{j=1}^{\infty} \varphi_{1i}'(L_1)\ddot{q}_{1j} - \frac{F_{ye}^{(1)}}{\mu_1}\varphi_{1i}'(L_1) + \frac{J_2\varphi_{1i}'(L_1)}{\mu_1}\left[k_{21}u_2 + k_{23}\dot{\theta}_{b2}\right]$$

where J_h, J_2 are hub and rotor inertia of actuator 2. Furthermore,

$$F_{ye}^{(1)} = F_{b2}\cos(\theta_{b2}) + F_{x2}\sin(\theta_{b2})$$

$$F_{b2} = EI_2 \sum_{j=1}^{\infty} \varphi_{2j}'''(0)\, q_{2j}(t) \tag{18.10}$$

$$F_{x2} = (m_{l2} + m_p)\sin(\theta_{b2})\left[L_1\ddot{\theta}_{b1} + \sum_{j=1}^{\infty}\varphi_{1j}(L_1)\ddot{q}_{1j}(t)\right]$$

where m_{l2} is the mass of arm 2 and the modal parameters α_{1i}^* and β_{1i}^* are independent of the payload mass (but depend on the mass of actuator 2). Notice the linear dependence on the payload mass, m_p in equation (18.10).

If the actuator equations (18.6) and (18.7) are used for obtaining the angular accelerations in equations (18.8) and (18.9) the four main equations can be written in a more compact form. Truncating the sums involved and introducing the notation

$$z = \begin{bmatrix} \theta_{b1} \\ \theta_{b2} \\ q_1 \\ q_2 \end{bmatrix}, \qquad u = \begin{bmatrix} u_1 \\ u_2 \end{bmatrix}$$

where

$$q_1 = \begin{bmatrix} q_{11} \\ \vdots \\ q_{1n} \end{bmatrix}, \qquad q_2 = \begin{bmatrix} q_{21} \\ \vdots \\ q_{2n} \end{bmatrix}$$

the description of the flexibility, equations (18.8) and (18.9), can be linearised and brought into the following compact form:

$$\ddot{z} = M_1 z + M_2 \dot{z} + M_3 \ddot{z} + M_4 u \tag{18.11}$$

where, the matrices, M_k, $k = 1, \dots, 4$ are affine in m_p. Notice, the matrices depend on the point of linearisation. In this case the matrices depend only on θ_{b2}. Also, notice the angular acceleration, $\ddot{z}$ appears on both sides of the equation.

The compact description in equation (18.11) can be transformed into a state-space description. The algebraic loop, related to $\ddot{z}$ in equation (18.11), can be solved if the following matrix inverse exists

$$\Delta = (I - M_3)^{-1}$$

Note that M_3 depends linearly on the payload mass m_p. Using the state vector definition

$$x = \begin{bmatrix} z \\ \dot{z} \end{bmatrix}$$

the following state-space description is obtained:

$$\dot{x} = \underline{\underline{A}}x + \underline{\underline{B}}u$$

where

$$\underline{\underline{A}} = \begin{bmatrix} 0 & I \\ \Delta M_1 & \Delta M_2 \end{bmatrix}, \quad \underline{\underline{B}} = \begin{bmatrix} 0 \\ \Delta M_4 \end{bmatrix}$$

where the state transition matrix is of order $2 + 4n$ with n as the number of considered modes. For an arbitrary linearisation angle, this representation is a linear approximation to the dynamics, that is, M_k, $k = 1, \dots, 4$ are functions of the linearisation angle. Thus, one can use the measurements of θ_2 to obtain a running linear description around the actual orientation.

The measurement system consists of two tachometers and four strain gauges. The tachometers give measurements of the link angles θ_{b1} and θ_{b2} whereas the strain gauges are located tactically on the links in order to give measurements of the deflections $\bar{q}$.

The measurements are connected to the state of the description through

$$y_m(t) = C_m x(t), \quad C_m = \begin{bmatrix} C_{tg} & 0 & 0 & 0 \\ 0 & C_{sg1} & 0 & 0 \\ 0 & 0 & C_{sg2} & 0 \end{bmatrix}$$

where C_{tg}, C_{sg1} and C_{sg2} are observation matrices for the two tachometers and the strain gauges located on the two links. These are

$$C_{tg} = \begin{bmatrix} k_{tg1} & 0 \\ 0 & k_{tg2} \end{bmatrix}, \quad C_{sg1} = \begin{bmatrix} k_{sg11}\varphi_{11}''(l_{11}) & k_{sg11}\varphi_{12}''(l_{11}) & \cdots \\ k_{sg12}\varphi_{11}''(l_{12}) & k_{sg12}\varphi_{12}''(l_{12}) & \cdots \end{bmatrix}$$

The output matrix, C_{sg2}, is defined in a similar manner. k_{tgj} and k_{sgji} are constants characterising the tachometers and the strain gauge, whereas l_{ji} represents location i on link j.

The control objective is to control the end-point position and velocity. The controlled quantities are related to the system state according to

$$y_c(t) = C_c x(t), \quad C_c = \begin{bmatrix} C_p & 0 \\ 0 & C_p \end{bmatrix},$$

$$C_p = \begin{bmatrix} 1 & 1 & \dfrac{\varphi_{11}(L_1)}{L_1} & \cdots & \dfrac{\varphi_{1n}(L_1)}{L_1} & \dfrac{\varphi_{21}(L_2)}{L_2} & \cdots & \dfrac{\varphi_{2n}(L_2)}{L_2} \end{bmatrix}$$

This clearly shows the dependency of the link angles, θ_{b1} and θ_{b2}, the beam deflections and velocities.

18.4.3   FE model

In the previous section a modal representation of the flexible robot arm was presented. The displacement of the link was described by a weighted infinite sum series of mode shape functions defined on the entire special domain and subsequently, the system was modally truncated to obtain a finite-order model. Instead of this modal discretisation and subsequent reduction to a certain number of modes, the FE method makes use of a spatial discrete model by dividing the robot arm into a finite number of elements and assuming that the arm displacement for each element can be represented by a certain specified well-known type of function. The model can be made arbitrarily accurate by choosing a suitable number of elements and suitable placements of the corresponding nodal points.

The robot arm is formulated in spatial discrete model by dividing it into $n - 1$ elements resulting in n nodal points as illustrated in Figure 18.16. The n nodal points are arbitrarily placed along the link. However, the first nodal point is fixed because it has to separate the link from the hub and the final point is fixed too at the end-point to separate a possible payload from the link.

The displacement function of the kth element $w_k(x, t)$ defined in the interval $x_k \leq x < x_{k+1}$ is an approximate cubic spline function, which is the most commonly used function for 1 DOF flexible structures. Furthermore, the degree of the chosen polynomial is the lowest possible to choose for the purpose of describing the vibrations

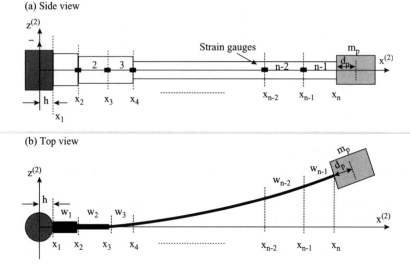

Figure 18.16   Segmentation of the flexible robot link into $n-1$ elements [Baungaard, 1996]

in the link while these are described by the fourth-order Euler–Bernoulli equation

$$w_k(x, t) = a_k(x - x_k)^3 + b_k(x - x_k)^2 + c_k(x - x_k) + d_k \quad \text{for} \quad x_k \le x < x_{k+1}$$

where x_k is the position of the kth nodal point. The time dependence is embedded in the coefficients, which are related to the curvature, w_k'', and the displacement, w_k in the nodal points through

$$a_k = \frac{1}{6}\beta_k(w_{k+1}'' - w_k'')$$

$$b_k = \frac{1}{2}w_k''$$

$$c_k = \beta_k(w_{k+1} - w_k) - \frac{1}{6}\alpha_k(w_{k+1}'' - w_k'')$$

$$d_k = w_k$$

Let $\overline{w} = (w_1, \ \ldots \ w_k, \ \ldots \ w_n)^T$ and $\overline{w}'' = (w_1'', \ \ldots \ w_k'', \ \ldots \ w_n'')^T$. Owing to the internal smoothness it is possible to express the displacement $\overline{w}$ as a simple function of the curvature, $\overline{w}''$ as

$$\overline{w} = N\overline{w}''$$

Consequently, it is a natural choice, among several possibilities, to use the curvature $\overline{w}''$ and the angular position θ_{b1} as state variables, that is,

$$q = \begin{bmatrix} \theta_{b1} \\ \overline{w}'' \end{bmatrix}$$

Accordingly, the following system description can be derived:

$$T = M\ddot{q} + D\dot{q} + Kq \tag{18.12}$$

where

$$M = \begin{bmatrix} J_t & \gamma^T \\ \gamma & M_b \end{bmatrix}, \quad K = \begin{bmatrix} 0 & 0 \\ 0 & K_b \end{bmatrix}, \quad D = \begin{bmatrix} 0 & 0 \\ 0 & \zeta I \end{bmatrix}$$

where M, D and K represent the system mass/inertia, damping and stiffness matrices, respectively. The external forces are because of the actuator and

$$T = \begin{bmatrix} T_{\text{ext}} \\ 0 \end{bmatrix}$$

In this section only the 1 DOF version of the robot is considered. Combining the FE model for the flexible link described in the previous section with the actuator model in section 18.4.1, a state-space model for the flexible-link robot can be obtained.

Here the full actuator model is used. This can be written as

$$T_{\text{act}} = -T_b = \frac{k_1}{k_2}u + \frac{k_3}{k_2}\dot{\theta}_b - \frac{1}{k_2}\ddot{\theta} \tag{18.13}$$

Combining equation (18.13) with the FE model, equation (18.12), yields

$$0 = M^* \ddot{q} + D^* \dot{q} + Kq + Vu + G(q)$$

where

$$M^* = \begin{bmatrix} J_t + 1/k_2 & \gamma^T \\ \gamma & M_b \end{bmatrix}, \quad D^* = \begin{bmatrix} -k_3/k_2 & 0 \\ 0 & K_b \end{bmatrix}, \quad V = \begin{bmatrix} -k_1/k_2 \\ 0 \end{bmatrix}$$

Introducing the state vector, x,

$$x = \begin{bmatrix} q \\ \dot{q} \end{bmatrix}$$

the linear state-space model

$$\dot{x} = Ax + Bu$$

is obtained, where

$$A = \begin{bmatrix} 0 & I \\ -(M^*)^{-1}K & -(M^*)^{-1}D^* \end{bmatrix}, \quad B = \begin{bmatrix} 0 \\ -(M^*)^{-1}V \end{bmatrix}$$

For the FE model the measured quantities are tacho measurement and the strain gauge voltage, that is,

$$y_m(t) = C_m x(t), \quad C_m = \begin{bmatrix} K_{tg} & 0 & 0 \\ 0 & K_{sg} & 0 \end{bmatrix}$$

The position and the velocity of the end-point, which are the controlled variables, are given as

$$y_c(t) = C_c x(t), \quad C_c = \begin{bmatrix} C_p & 0 \\ 0 & C_p \end{bmatrix}, \quad C_p = \begin{bmatrix} 1 & 0 & \dfrac{1}{L+h}N \end{bmatrix}$$

These are influenced by the link angle and the curvature of the beam.

18.5   Control design

The controllers used in this workbench are standard discrete time LQG controllers eventually augmented with an integral state. The design of the controllers is based on a linear model (linearised at a specific operating point). The measurement system (tacho and strain gauge sensors) is introduced to obtain reasonable values for the system states. In order to reduce the effect of measurement noise the states are estimated via a Kalman filter. The estimated states are used in the controller and are fed into the actuator(s).

The state estimator is based on a stationary version of the Kalman filter for either the FE model or the modal model. Assume that A_d and B_d are system matrices in the discrete time description of the system. The stationary Kalman filter consists of two sets of recursions, namely the time update

$$\hat{x}(t+1|t) = A_d \hat{x}(t|t) + B_d u(t)$$

and the data update

$$\hat{x}(t|t) = \hat{x}(t|t-1) + K_{est}\left[y(t) - C_m\hat{x}(t|t-1)\right]$$

where, K_{est} is determined as the optimal gain, from system information such as system matrices and variance of process and measurement noise.

The controllers implemented in the workbench, are variants of LQG controllers. They are in standard version and in a form that includes an integral action for cancelling stationary errors that might occur owing to model mismatch or non-zero d.c. components in the disturbances. In this case the state vector is augmented with an integral state.

Let $\tilde{x}_i = x_i - x_0$ be the deviation from the stationary point, equivalent to reference or set point. In order to introduce the integral action the state vector is augmented with the integral state z_i. The total design model can then be stated as

$$\begin{bmatrix} \tilde{x} \\ z \end{bmatrix}_{i+1} = \begin{bmatrix} A_d & 0 \\ -C_c & 1 \end{bmatrix}\begin{bmatrix} \tilde{x} \\ z \end{bmatrix}_i + \begin{bmatrix} B_d \\ 0 \end{bmatrix}u_i, \quad \tilde{y}_i = \begin{bmatrix} C_c & 0 \\ 0 & 1 \end{bmatrix}\begin{bmatrix} \tilde{x} \\ z \end{bmatrix}_i$$

The objective is to control the end-point position and velocity with due consideration of the control effort. In the LQG framework this is formulated as a minimisation of the performance index:

$$J = \sum_{i=0}^{\infty} \|\tilde{y}\|_Q^2 + \|u_i\|_R^2$$

where the weight matrices Q and R should compromise between performance and control effort. In the work bench the performance weight matrix is given as

$$Q = \begin{bmatrix} q_p & 0 & 0 \\ 0 & q_v & 0 \\ 0 & 0 & q_i \end{bmatrix}$$

which contains the weights q_p, q_v and q_i accounting for the error in position, velocity and integral state, respectively. This results in a well-known feedback law

$$u_i = -K_c\tilde{x}_i - K_z z$$

for the deviation from the stationary point and the integral state.

18.6 CACE environment

The main panel of the CACE environment used for controlling the flexible-link robot is shown in Figure 18.17. This is designed using the graphical interface in Matlab and uses the Matlab Control System Toolbox functions for computing the controllers from the corresponding model description. The main panel has five sections, of which two are related to modelling and design. One section is related to specification of the reference signal. In the last two sections it is possible to impose different plotting and other types of commands (Caspersen, 2000).

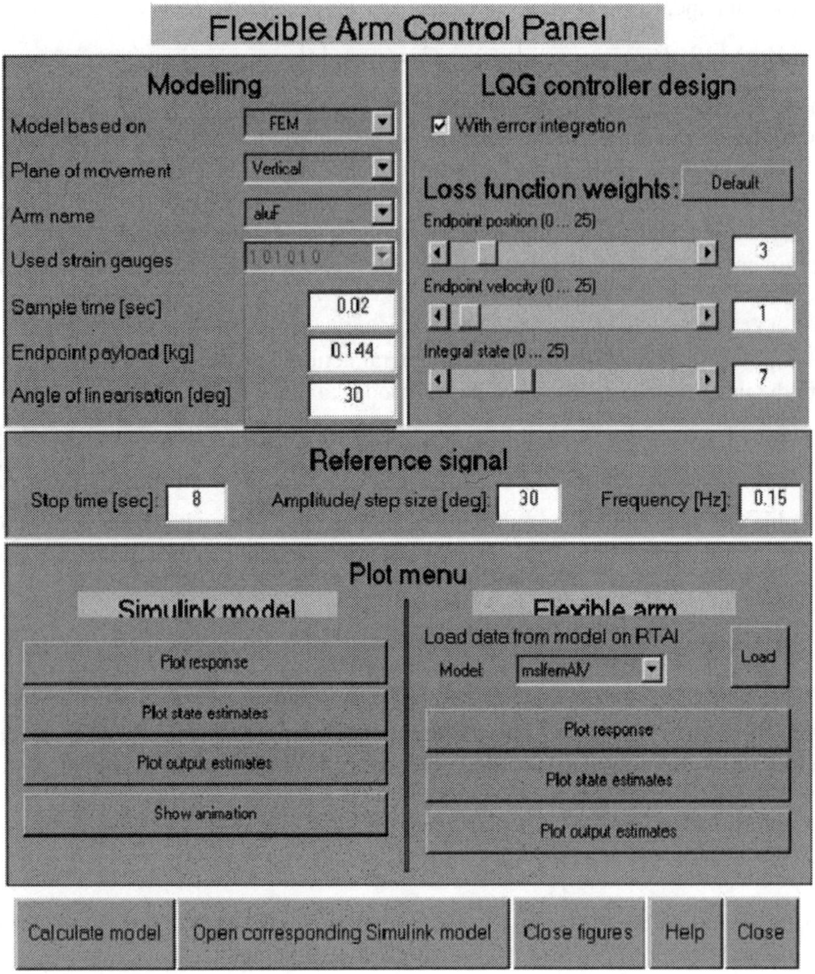

Figure 18.17   Main panel of the software system

The models can, as previously mentioned, be based on a modal description (Section 18.4.2) or on a FE description (Section 18.4.3). The positions of the strain gauges can be specified and it can be determined which one is currently in use. The payload and sampling interval is also specified in this section of the main panel. In the design part of the main panel the weights q_p, q_v and q_i are entered as well as indication of a switch as required by the integral action. The reference signal is specified and the corresponding Simulink model is generated and opened. A plot menu allows specifying which simulation results be plotted. The unified plot menu ensures that the resulting plots from the simulation and the experiments are easily comparable.

The corresponding Simulink model is shown in Figure 18.18. Notice the two blocks realising the flexible-link system. One block, *mlsflexo,* is the simulation model,

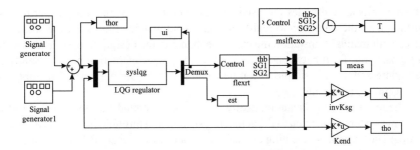

Figure 18.18   Simulink diagram of the flexible-link system

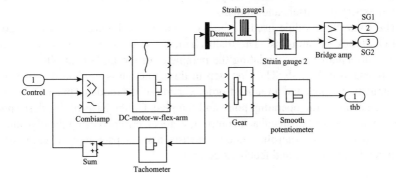

Figure 18.19   The block realising simulation of the flexible-link system

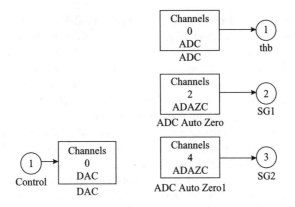

Figure 18.20   The block realising real-time version of the flexible-link system

shown in Figure 18.19, and another block, *flexrt,* is the real-time system interface, shown in Figure 18.20. Switching the two blocks convert the Simulink system from a simulation model to a system from which real-time code can be generated using the Matlab real-time workshop (RTW). The main advantage of this approach is that the same blocks and codes are used for simulation and for real-time experiments. Furthermore, the transition to real-time code is automated eliminating the source of errors with manual translation means and enhancing the design process, as it is easy to go back and forth between the simulation and experimentation phases. Safety is also addressed, as it is simple to test new controllers by simulation before applying to real-time experiments.

The real-time code is automatically generated, compiled and downloaded to the real-time system connected to the actual physical model through the process interface boards. The computer executing the real-time code is an ordinary PC running the real-time application interface (RTAI) for Linux (Quaranta and Mantegazza, 2001). The data is automatically collected during the experiment and can be plotted from the main window menu. The system has proven quite flexible and easy to use especially for non-experienced users and students.

The controller module used in the workbench (see Figure 18.18) is based on the LQG block of Simulink. The main challenge is to convert the system model into a form suitable for calculating the parameter of the LQG controller using the *dlqr* command in Matlab. The first step in this direction is to choose between the full simulation model and a reduced-order version of the model. This model is then linearised at a specific operating point and sampled. On the basis of this description, the Kalman filter and the feedback control law are determined using the *dlqe* and the *dlqr* commands. The controller module realises a state-space representation of the combined Kalman filter and feedback control law.

18.7   Summary

This chapter has focused on software design for modelling and control using an experimental platform. The modelling is carried out for both the horizontal and the vertical case. An LQG control design that suits both types of models is designed and analysed. The aim of the control is to dampen the vibration of the arm as well as position the end-point of the flexible link according to the reference signal. With regards to vertical movement the controller must compensate for the static deflection. At the same time the controller must dampen the vibration and position the end-point accurately. The model and the control strategy depend on the mass of a payload at the end-point. Methods for estimation of the payload are examined. These methods are incorporated into the control algorithm.

The requirements for a computer aided control system design environment have been outlined and a model of the design process described. Furthermore, a block library for Simulink has been described enabling simulations and experiments with scalable granularity in the simulations.

The controllers have been implemented and illustrated within simulations and experimental set-ups, using an environment based on Matlab/Simulink and RTW to generate real-time code for a real-time RTAI Linux based platform.

Chapter 19

SCEFMAS – An environment for simulation and control of flexible manipulator systems

M.O. Tokhi, A.K.M. Azad, M.H. Shaheed and H. Poerwanto

This chapter presents the development of SCEFMAS (Simulation and Control Environment for Flexible MAnipulator Systems) software package. This is a user-friendly interactive software environment based on Matlab and associated tool-boxes. The environment incorporates a finite difference (FD) simulation algorithm of a constrained planar single-link flexible manipulator system for analysis, simulation, modelling of dynamic behaviour of the system. The package also incorporates a range of control techniques, including open-loop control such as filtered command, Gaussian-shaped and command shaping, collocated and non-collocated closed-loop control methods of fixed and adaptive types, and intelligent soft computing control techniques. The environment allows the user to set-up the system by providing its physical parameters and to select the controller type through an interactive graphical user interface (GUI). Data analyses can be performed in time and frequency domains on the controller and system input and outputs signals. The environment is suitable as an education package as well as for research purposes for investigating and developing various simulations, modelling and controller designs for flexible manipulator systems.

19.1 Introduction

During the last decade a considerable amount of research work has been accomplished on the development of software environments for flexible manipulators. SCEFMAS (Simulation and Control Environment for Flexible MAnipulator Systems) as an example of such efforts provides an environment for modelling, simulation and control for flexible manipulators, where students can study the behaviour of flexible manipulators and their controllers without going into the details of programming and data analysis. The first version of SCEFMAS was provided with a finite difference (FD) simulation of a flexible manipulator system along with a number of open-loop and closed-loop classical control strategies. The initial development was implemented with Matlab 4.2 and Simulink 1.3 and includes FD simulation along with classical control methods (open-loop and closed-loop methods) (Tokhi *et al.*, 1999). With the first version, the major part of SCEFMAS implementation was keyboard driven.

Considering the new developments in flexible manipulators and the introduction of newer versions of Matlab with various powerful features, a major task was undertaken to upgrade SCEFMAS during 2002. The aim was to incorporate new research findings within the SCEFMAS and also to upgrade the package to the new version of Matlab and its associated toolboxes. As a result, the second version of SCEFMAS was produced. This version of SCEFMAS is enhanced in four areas (Azad *et al.*, 2004): (a) inclusion of various advanced control schemes; (b) inclusion of intelligent modelling techniques based on neural networks (NNs) and genetic algorithms (GAs); (c) development of highly interactive graphical user interface (GUI) using Matlab Guide toolbox; (d) upgrading of the package from Matlab 4.2 and Simulink 1.3 to Matlab 6.5 and Simulink 5.0 respectively.

This chapter describes the development of SCEFMAS with associated features. The chapter is organised as follows. Section 19.2 briefly describes an experimental flexible manipulator system used as a basis of development of SCEFMAS. Section 19.3 provides an outline of the theoretical basis for the simulation, modelling and control facilities provided within SCEFMAS. Section 19.4 presents a number of case studies utilizing some of the features provided in SCEFMAS. Section 19.5 concludes the chapter by highlighting the main features of SCEFMAS.

19.2   The flexible manipulator system

A schematic model of the flexible manipulator system considered for the development of SCEFMAS is shown in Figure 19.1, where X_0OY_0 and XOY represent the stationary and moving co-ordinate frames respectively. The axis OX coincides with the neutral line of the link in its un-deformed configuration, and is tangent to it at the clamped end in a deformed configuration. τ represents the applied torque at the hub. E, I, ρ, A, I_h and M_p represent the Young modulus, area moment of inertia, mass density per unit volume, cross-sectional area, hub inertia and payload of the manipulator respectively. $\theta(t)$ denotes angular displacement (hub-angle) of the manipulator and $w(x, t)$ denotes elastic deflection (deformation) of a point along the manipulator at a distance x from the hub. In this work, the motion of the manipulator is confined to the X_0OY_0 plane. Since the manipulator is long and slender, the shear deformation and rotary inertia effects are neglected. This allows the use of the Bernoulli–Euler beam theory to model the elastic behaviour of the manipulator. The manipulator is assumed to be stiff in vertical bending and torsion, allowing it to vibrate dominantly in the horizontal direction and thus, gravity effects are neglected. Moreover, the manipulator is considered to have constant cross section and uniform material properties.

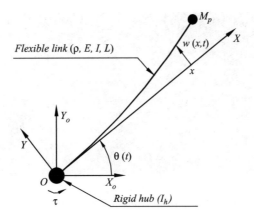

Figure 19.1   Schematic model of a flexible manipulator system

The dynamic equation describing the motion of the flexible manipulator, presented in terms of $y(x, t)$, is given as (Azad, 1994; Tokhi *et al.*, 1995a)

$$EI \frac{\partial^4 y(x,t)}{\partial x^4} + \rho \frac{\partial^2 y(x,t)}{\partial t^2} - D_S \frac{\partial^3 y(x,t)}{\partial x^2 \partial t} = \tau(t) \tag{19.1}$$

where D_S represents the damping constant, with the corresponding boundary conditions as

$$y(0, t) = 0$$

$$I_h \frac{\partial^3 y(0,t)}{\partial t^2 \partial x} - EI \frac{\partial^2 y(0,t)}{\partial x^2} = \tau(t)$$

$$M_p \frac{\partial^2 y(l,t)}{\partial x^2} - EI \frac{\partial^3 y(l,t)}{\partial x^3} = 0 \tag{19.2}$$

$$EI \frac{\partial^2 y(l,t)}{\partial x^2} = 0$$

and initial conditions as

$$y(x, 0) = 0, \qquad \frac{\partial y(x,0)}{\partial x} = 0 \tag{19.3}$$

For experimental verification of the simulation environment, an experimental flexible manipulator rig, shown in Figure 19.2, is considered. This is the Sheffield Manipulator described in Chapter 1 of this book.

19.3   Structure of SCEFMAS

The development of SCEFMAS involves the use of Matlab, a number of Matlab toolboxes, Simulink and Guide environment (Marchand and Holland, 2003). There are a number of facilities provided within the package, involving a single-link flexible

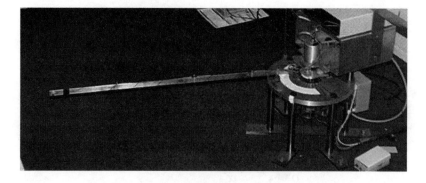

Figure 19.2   The experimental flexible manipulator system

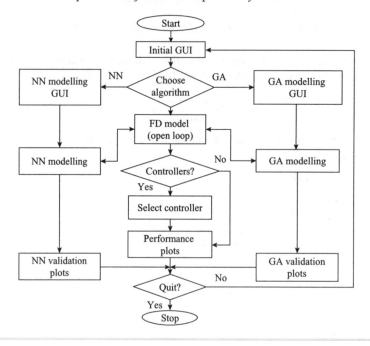

Figure 19.3   Structure of the SCEFMAS environment

manipulator system. Although most of the models and data are provided for a specific system but the users can introduce their own specifications. An execution flowchart for SCEFMAS is shown in Figure 19.3. The facilities provided within the package can be classified into following groups:

• FD simulation and control
• Intelligent modelling and validation
• GUI.

FD simulation and control. This part provides the FD simulation and position and vibration control of single-link flexible manipulator systems. The user can run an open-loop simulation (for observing open-loop response) of the flexible manipulator or simulation with a specific control scheme. The control part involves both open-loop and closed-loop control methods. These include classical, advanced, and intelligent control techniques.

Intelligent modelling and validation. This part includes NN and GA modelling and subsequent model verification. The training data used for NN and GA modelling is obtained from an open-loop FD simulation using random or composite PRBS torque inputs. For NN and GA modelling, the user is provided with GUI driven facilities where desired model parameters can be entered. The progress in modelling will also be reported through graph windows within the GUI. Following the modelling process, there is provision for model validation, so that the user can understand and assess the appropriateness of the chosen model structure.

Graphical user interface. This part constitutes one of the major strengths of SCEF-MAS. There are a number of GUIs provided within the package so that the user can perform almost all the tasks using a mouse. This makes the package much more user friendly and easy to follow. The GUIs provided within the package are: initial GUI, results GUI, intelligent modelling GUIs, and model validation GUIs. The initial GUI allows the user to provide specifications of a single-link flexible manipulator and set the FD simulation parameters. All the simulation, modelling, and control activities within the environment are linked with the parameters provided within this GUI. The results-GUI allows the user to monitor results of the simulation and control exercises through the display of input and various outputs both in time and frequency domains. This GUI also displays a 3D motion profile of a flexible manipulator for a given simulation run. The intelligent modelling GUIs allow the user to choose their desired NN and GA modelling structures and also to monitor the modelling process. The model validation GUIs allow the user to verify the quality of the developed models. Time and frequency domain validations including correlation tests are provided.

19.3.1 FD simulation and control

The FD simulation and control is one of the main parts of SCEFMAS. Through this route the user is able to perform open-loop simulation of a single-link flexible manipulator system as well as implement various control strategies. The implementation flowchart for the FD simulation and control is shown in Figure 19.4.

For the open-loop simulation the user can observe the open-loop responses of a flexible manipulator with various input choices. The input torque choices provided are random, composite PRBS, and bang-bang. The choice of input is valid only for observing the open-loop system behaviour. The control part involves the implementation of various open- and closed-loop classical control methods, including a number of adaptive controllers, and a neuro-adaptive inverse controller. For the open-loop controller design the input torque is either developed from scratch or by modifying a bang-bang input by removing the natural vibration frequencies of the flexible manipulator. The closed-loop controllers are provided with a reference angle input. However, the user can incorporate any other input torque patterns of his/her choice.

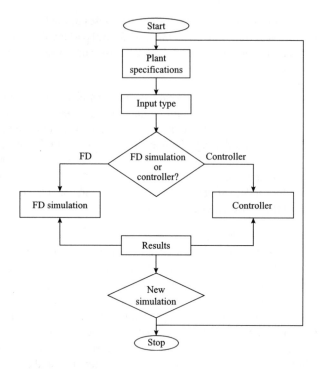

Figure 19.4   Flowchart of the FD simulation and controller implementation

19.3.1.1   FD simulation algorithm

The development of flexible manipulator simulation environment involves the solution of equation (19.1) using the FD method. With the FD method, the length of the flexible manipulator is divided into a suitable number of sections. For a given input torque, the displacement of each of these sections is calculated using an appropriate boundary condition. The FD method uses a set of equivalent difference equations defined by the central FD quotients of the FD method, obtained by discretising the partial differential equation (PDE) in equation (19.1) with its associated boundary and initial conditions in equations (19.2) and (19.3). The process involves dividing the manipulator into n sections each of length Δx and considering the deflection of each section at sample times Δt. In this manner, a solution of the PDE is obtained by generating the central difference formulae for the partial derivative terms of the response $y(x, t)$ of the manipulator at points $x = i\Delta x$, $t = j\Delta t$, $i = 1, 2, \ldots, n$, $j = 0, 1, \ldots, m$. For details of the algorithm development see Chapter 5 and (Tokhi et al., 1995a).

19.3.1.2   Controller designs

The controller implementation in SCEFMAS is divided into two main types: open-loop control and closed-loop control. The open-loop controller implementation involves the design of suitable input torque that does not excite the resonance modes of the system. The resulting waveform is then used as the input torque profile for the flexible manipulator system. The main problem with open-loop control is that the torque profile needs to be changed with any change in the manipulator system that causes a change in natural frequency of the system. The open-loop controller options provided within SCEFMAS are: (a) Gausian-shaped torque command; (b) low-pass filtered and (c) band-stop filtered.

The Gausian torque command is generated in such a way that all its frequency components lie below the system's first resonance frequency. While with the low-pass filtered torque input, the input signal amplitude (energy) at all the resonance frequencies of the system are reduced using a low-pass filter. This method reduces the input signal energy at all the frequency components starting from just below the system's first resonance frequency. Due to the removal of energy at all the higher frequency modes, both of these methods make the system slower. To include contribution of the input at frequencies near system resonance frequencies one can remove input energy over short frequency bands around each dominant mode of vibration of the system. This can help to achieve faster system response. The method is achieved through the use of band-stop filters. The band-stop filters are tuned to absorb input energy at frequency bands around the dominant resonance modes. For both, low-pass and band-stop filtered cases, Butterworth and Elliptic type filters are used (Tokhi and Azad, 1995*b*). Further details can be found in Chapter 8 of this book.

The closed-loop control strategies are of two main types: classical methods and advanced controller design. The classical controllers are implemented in two ways, namely joint-based collocated feedback control and joint and end-point feedback based hybrid (collocated and non-collocated) control. These controller implementations involve the use of proportional, derivative (PD) controllers, proportional, integral, derivative (PID) controllers, and combination of PD and PID (Tokhi and Azad, 1996*a,b*), see Chapter 12 for further details. The advanced controller implementation involves various adaptive control strategies using both the hub and end-point feedback. One of these controllers also involves the use of NN as a part of the controller feedback loop and is called adaptive neuro-inverse controller. The details of these designs are provided earlier in Chapter 12. The controller designs provided within the SCEFMAS environment are as follows:

- Fixed controllers
    Joint based controllers
    Hybrid controllers
- Adaptive controllers
    Adaptive joint based controller
    Adaptive hybrid controller
- Adaptive neuro-inverse controller.

19.3.2 *Intelligent modelling and model validation*

This section describes the NN and GA modelling strategies used in SCEFMAS. This involves NN and GA modelling and validation techniques, where the user can specify

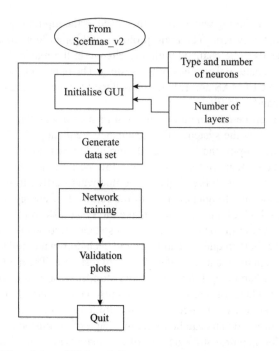

Figure 19.5   Flowchart of NN modelling process

the desired model structure, perform a modelling exercise and validated the model to assess its effectiveness.

19.3.2.1   NN modelling

The development of NN model for a flexible manipulator system is implemented using multi-layered perception (MLP) networks (Luo and Unbehauen, 1997; Shaheed, 2000; Sze, 1995). The flowchart for the NN model implementation is shown in Figure 19.5. The environment includes the simultaneous development of three models. These are: input torque to hub-angle, hub-velocity and end-point acceleration.

In this implementation, the user will have an option of choosing the number of layers, number of neurons in each layer and the type of neurons. The input and outputs from an open-loop FD simulation model are used as the training data for the NN modelling exercise. The developed NN models can also be validated both in time and frequency domains along with appropriate correlation tests.

19.3.2.2   GA modelling

This part involves the development of a GA model of a user specified flexible manipulator system. A flowchart describing the GA modelling process is shown in Figure 19.6. In this process the user will have the choice of specifying GA model parameters, such as number of individuals, maximum number of generations,

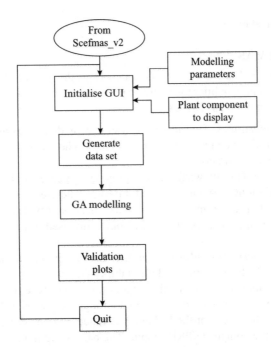

Figure 19.6   Flowchart of GA modelling process

generation gap, binary precision, and the order of the GA model (Chipperfield *et al.*, 1994; Karr, 1991). Similar to NN modelling, the specification of the desired flexible manipulator, which needs to be modelled, can also be provided. The input and outputs obtained from open-loop simulation for the specified system will be used as for the model development process. The quality of the developed model can be verified both in time and frequency domains along with suitable correlation tests.

19.3.3   Graphical user interfaces

The interactive GUIs are one of the key features of SCEFMAS. The provision of GUI allows the user to concentrate on the model structure, control algorithm and verification of implementation results rather than spending time on developing programs for algorithm implementation and data analysis. This feature makes the environment very attractive in an educational context as well as in research, where there is very little time to spare. This section will discuss the features of the GUIs contained in SCEFMAS. These are

- SCEFMAS_V2
- Results
- NN modelling and validation
- GA modelling and validation.

These are described below.

19.3.3.1   SCEFMAS_V2 GUI

This is the main GUI and serves as the gateway to the SCEFMAS environment. According to the functionalities the facilities provided within this GUI can be divided into five major components. These are

Manipulator specifications. This allows the user to provide the physical dimensions, hub inertia, and payload condition of the manipulator. The user can change the values by using the respective sliders.

Material properties. This allows the user to provide properties of the intended material which is used for the desired manipulator design. The user can choose the values using respective drop-down menus. The environment is provided with some readily available material properties, which are commonly used in flexible manipulator designs.

Simulation parameters. This allows the user to set the simulation time, number of segments for the FD discretisation and the stability factor.

Input types. This allows the user to choose a desired input type for open-loop simulation. The input choices provided are: random input, composite PRBS input and bang-bang input. However, for the NN and GA modelling one can choose only the random input or the composite PRBS input. The bang-bang input is not suitable in these two cases.

Algorithm types. There are three option buttons of algorithm types. These are

- FD simulation and control button, which will direct the user to the FD simulation and control applications;
- NN modelling button; this is used for performing NN modelling and validation exercises; and
- GA modelling button, which is used for GA modelling and validation exercises.

For NN and GA modelling, the input–output data from an open-loop FD simulation of the manipulator are used.

19.3.3.2   Results GUI

This GUI is used for displaying the input and outputs obtained from an FD simulation and control run. The input and outputs that can be viewed through this GUI are input torque, hub-angle, hub-velocity, hub-acceleration, end-point displacement, end-point velocity, end-point acceleration and end-point residual motion. There are three graph windows, two of them are for time and frequency domain plots of a selected input or output, while the third graph displays a 3D plot showing complete motion of the manipulator for the whole period of simulation. The user can rotate the 3D plot to view the profile from different angles.

19.3.3.3   NN modelling and validation GUIs

There are two GUIs used for NN modelling and validation process. The 'NN modelling GUI' is used to carry out NN modelling of a flexible manipulator system. The user can choose two, three, or four layers of neurons. The types and number of neurons in each layer can also be selected. After specifying the desired NN structure the user needs to generate the training data using open-loop simulation of the system. After the generation of necessary training data the user can initiate the training process. The progress through the training can be monitored through a graph window within the GUI. The validation process is performed by the 'NN validation GUI', which can be activated through the 'NN Modelling GUI'. This GUI will allow the user to examine the quality of a trained NN model. The models can be validated both in time and frequency domains as well as through correlation tests.

19.3.3.4 GA modelling and validation GUI

Similar to the NN process the GA modelling and validation process is implemented through two GUIs. The 'GA modelling GUI' is used to carry out GA modelling of a flexible manipulator system. This GUI can be invoked by clicking the 'GA modelling' button within the 'SCEFMAS_V2 GUI'. The user can choose specifications such as, number of individuals, maximum number of generations, generation gap, binary precision and order of the genetic model along with the type of model required. The model options provided are input torque to hub-angle, input torque to hub-velocity and input torque to end-point acceleration. Only one of these options can be selected for a given modelling run.

After specifying the desired GA structure the user needs to generate the training data and run the modelling process. Progress of the training can be monitored through the graph window provided within the GUI. This graph window will provide a real-time adaptation plot. The GA model validation is performed using the 'GA validation GUI'. These validations are done in time and frequency domains.

19.4 Case studies

This section will provide four case studies demonstrating some of the implementation features of the SCEFMAS package. The case studies are provided for open-loop simulation (natural response), adaptive inverse dynamics active control, NN modelling and validation, and GA modelling and validation.

19.4.1 Open-loop FD simulation

This case study will demonstrate the steps required to perform simulation of a single-link flexible manipulator system in an open-loop configuration. The simulation will be followed by visualisation of the input and outputs. To perform this exercise the user can use the open-loop model provided within SCEFMAS with their desired system specification.

Step 1. Within the Matlab editor set the path where the SCEFMAS folder has been placed. Type SCEFMAS_V2 within the Matlab command window. This will open the initial GUI as shown in Figure 19.7. Within the GUI, the user needs to provide

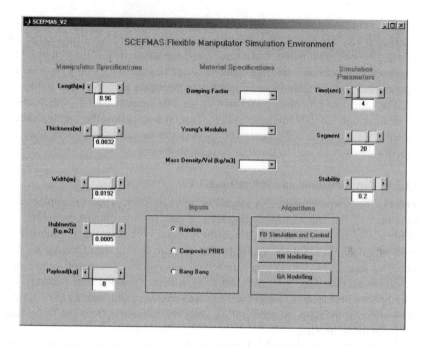

Figure 19.7   SCEFMAS_V2 GUI as the gateway for the SCEFMAS environment

the manipulator specification, material properties, simulation parameters, and input torque options. There are three input torque options to choose from. These are 'random input', 'composite PRBS input' and 'bang-bang input'. In addition to these, the users can develop their own torque profiles. For this case study a bang-bang torque input is going to be used by selecting the radio button against the 'bang-bang'. The stability factor needs to be between zero and 0.25.

Step 2. A model for open-loop simulation has been provided within SCEFMAS. This model can be opened by following the corresponding sequence of steps. A click on the 'FD Simulation and Control' button within the 'SCEFMAS_V2 GUI' will open the Simulink part of the SCEFMAS environment and the user can view a Simulink window as shown in Figure 19.8. The GUI is named 'SCEFMAS LIBRARY AND CASE STUDIES'. The Simulink drives the components within this window. This window consists of three main components. The first one is the 'SCEFMAS library', the second is 'open-loop case studies' and the third is the 'closed-loop case studies'. The user can open any of these components by clicking on an appropriate block shown in Figure 19.8.

The Simulink model for open-loop FD simulation is provided within the 'open-loop case studies' block. A click on this block will open a new window with the 'OPEN-LOOP CASE STUDIES' (Figure 19.9). There is a block called open-loop response in this window (top left corner) that contains Simulink model of an open-loop flexible manipulator system. A double click on this block will open the model (Figure 19.10).

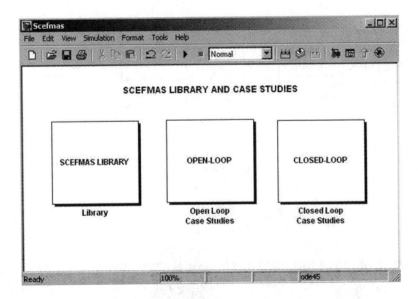

Figure 19.8   SCEFMAS library components

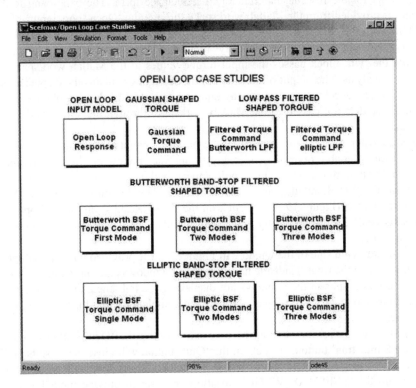

Figure 19.9   Open-loop case studies

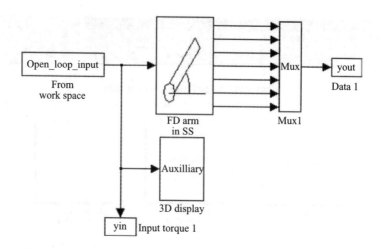

Figure 19.10   Open-loop model with bang-bang torque input

The model consists of a 'FD arm in SS' block, which implements the FD algorithm for the flexible manipulator in state-space form. The bang-bang torque input is provided from the Matlab workspace through the 'Open-loop_input' block. Along with the 'FD arm in SS' block this torque input is also passed to the 'auxiliary' and 'yin' blocks. The 'auxiliary' block produces data for 3D displacement, while the 'yin' block passes the input torque values to the Matlab workspace for further analysis. The output of the 'FD arm in SS' block contains the displacement, velocity, and acceleration data for the hub and end-point of the flexible manipulator. These data are passed to the Matlab workspace through the 'Mux' and 'yout' blocks.

Step 3. After a simulation run all the input and outputs data will be available within the Matlab workspace and ready to analyse. Following the completion of the simulation run a new button will appear within the 'initial GUI' window, next to the 'FD Simulation and Control' button. This button is called the 'View Results' button (Figure 19.11). A click on the 'View Results' button will open the 'results GUI' for displaying the input and outputs. The new window is shown in Figure 19.12.

The left-hand side of the window in Figure 19.12 is provided with all the option buttons while the right-hand side has three graph windows and four control buttons. The current graphs in the windows are displaying the hub-angle and its frequency spectrum, along with a 3D figure showing the complete motion for the given input torque.

Step 4. After a viewing session the user can click on the 'Quit' button or on the 'New Simulation' button. A click on the 'Quit' button will close the 'results GUI'. A click on the 'New Simulation' button will close the 'results GUI' window and bring up the 'SCEFMAS_V2 GUI' to allow the user to provide input for a new simulation run.

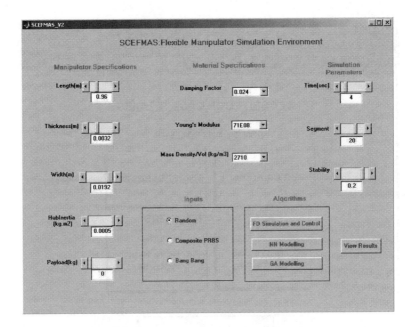

Figure 19.11   Initial GUI with the 'view results' button activated

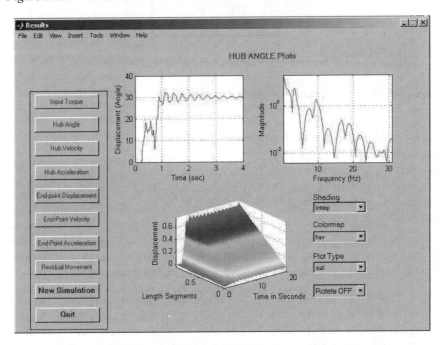

*Figure 19.12   Results GUI displaying hub angle output from an open-loop simula-
tion run with a bang-bang input*

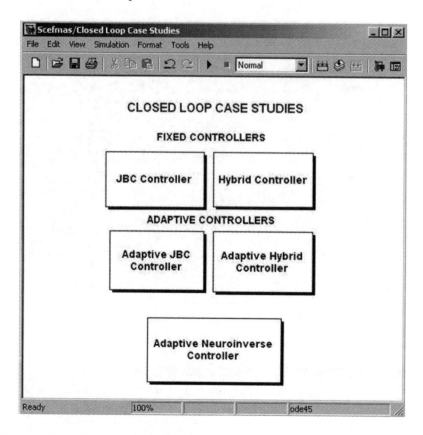

Figure 19.13   Closed-loop case studies

19.4.2   Adaptive inverse dynamic active control

The model for this controller design is provided within the closed-loop case studies block shown in Figure 19.13. The main theme of the controller is to drive the flexible manipulator with an additional signal from a controller whose transfer function is the inverse of that of the plant itself. The adaptive inverse controller is active when the flexible manipulator is in motion, so that the computed torque is used to force the end-point vibration to a minimum level. Since the plant is generally unknown, it is necessary to adapt or adjust the parameters of the controller in order to create the true plant inverse. Steps involved in this process are described below.

Step 1. Similar to the open-loop case studies the user needs to set the Matlab editor path where the SCEFMAS folder has been placed. Typing 'scefmas_v2' within the Matlab command window will open the 'SCEFMAS_V2 GUI' as shown in Figure 19.7. Within the GUI the user needs to provide all details of the manipulator and simulation. A hub-angle reference input is provided at the input of the corresponding Simulink model. However, the user can also develop their desired angle profile.

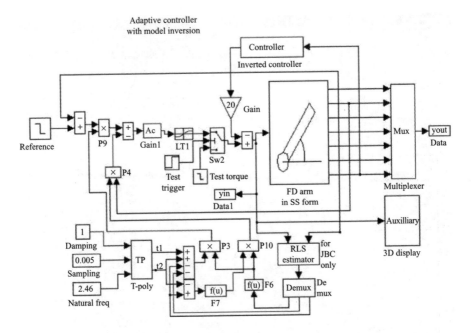

Figure 19.14   Simulink model of a flexible manipulator with adaptive inverse-dynamic active control

Step 2. A Simulink model with an 'adaptive inverse dynamic active controller' has been provided within the 'Closed-loop Case Studies' block (Figure 19.8). A click on this block will open a new Simulink window, which includes all the closed-loop controller designs for a single-link flexible manipulator (Figure 19.13).

To open the 'adaptive inverse dynamic active controller' model one needs to click on the 'Adaptive hybrid controller block' within the Simulink window. The Simulink model for a flexible manipulator system with 'adaptive inverse dynamic active controller' will then appear as shown in Figure 19.14. A click on the Simulink run button will initiate the simulation process.

Step 3. After the completion of the simulation run all the input and outputs data will be available within the Matlab workspace. A new button called 'View Results' will appear within the 'SCEFMAS_V2 GUI', shown in Figure 19.7. A click on this button will open the 'results GUI'. According to the description in Section 19.3.3.2, the user can use this GUI to view the input and outputs obtained from the simulation run. An image of the GUI with the end-point acceleration profile of the manipulator with the controller is shown in Figure 19.15.

19.4.3   NN modelling and validation

The NN modelling and validation process involves specifying the system and simulation parameters, NN model structure, execution of modelling process, and model validation. Steps involved in this process are described below.

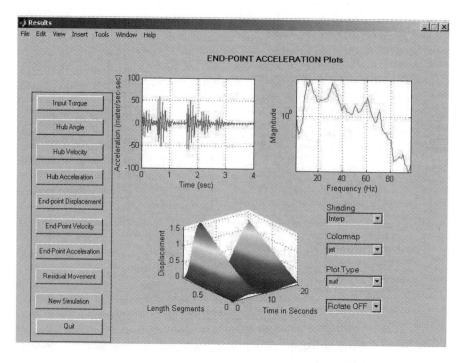

Figure 19.15   GUI displaying the end-point acceleration output

Step 1. Similar to the previous two case studies the user needs to set the path and open the 'SCEFMAS_V2 GUI' by typing 'scefmas_v2' within the Matlab command window. Within the GUI the user needs to provide the manipulator specification, material properties, and simulation parameters. The input type option is valid for this case and the selected input will be used for the data generation process through FD simulation of the flexible manipulator in an open-loop configuration. In this case, the user can choose either 'random input' or 'composite PRBS input'. Only one of these inputs can ensure the excitation of all the vibration modes of the flexible manipulator system.

Step 2. Once all the options have been selected within the 'SCEFMAS_V2 GUI', the user needs to click on the 'NN Modelling' button and this will open the 'NN modelling GUI', which will be used to carry out the NN modelling process (Figure 19.16).

The user can choose two, three, or four layers of neurons. In this case, a three-layer structure has been chosen with purelin as the activation function for each of the neurons. The types of neuron along with the number of neurons in each layer can also be selected. After specifying the desired NN structure the user needs to generate the training data by clicking on the 'Generate Data' button within the 'NN modelling GUI' (Figure 19.16). This will open a Simulink model with open-loop input as shown in Figure 19.10. The input will be either 'random input' or 'composite PRBS input', as selected within the 'SCEFMAS_V2 GUI'. Subsequent run of the open-loop FD

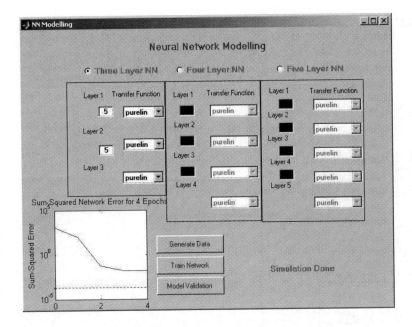

Figure 19.16   The GUI used for neural network modelling

simulation model will produce the input–output data necessary for NN modelling. The data will automatically pass to the Matlab environment for subsequent use in the NN training process.

Step 3. At this point, to complete the training the user needs to click on the 'Train Network' button. This will start the NN training process (Figure 19.16). The progress through the training can be monitored with the graph window shown on the left bottom side of the GUI. The graph will plot the sum-squared error. At the completion of the training process the 'simulation done' message will appear within the 'NN modelling GUI'. This will be followed by the appearance of another button called the 'Model Validation' button (Figure 19.16).

Step 4. A double click on the 'Model Validation' button will open the 'NN validation GUI' (Figure 19.17). As mentioned before, there are two validation tests for each developed model. One test will be in time and frequency domains and the other is through correlation tests. The time and frequency domain comparison between the actual (data used for training) and predicted (obtained through the developed model) outputs is shown in Figure 19.17. The correlation test results for the same model are shown in Figure 19.18.

19.4.4   GA modelling and validation

The GA modelling and validation process involves specifying the system and simulation parameters, desired GA model structure, running the modelling process, and model validation. Steps involved in this process are described below.

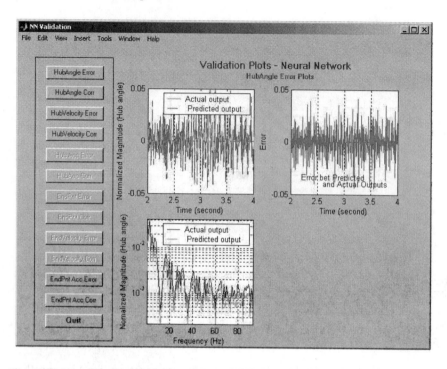

Figure 19.17   Time and frequency domain validation plots for a developed NN model

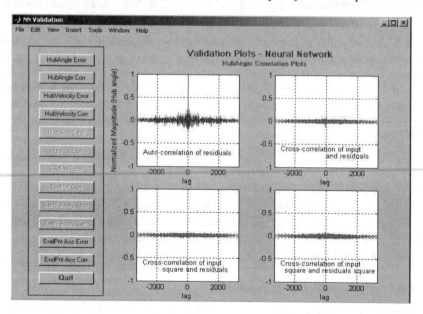

Figure 19.18   Validation plots for a developed hub angle model through correlation tests

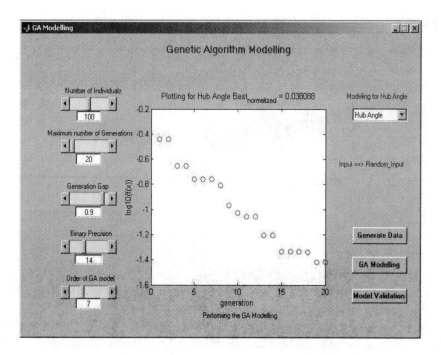

Figure 19.19   GUI used for providing GA model structure

Step 1. Similar to the previous case studies the user needs to set the path and open the initial GUI by typing 'scefmas_v2' within the Matlab command window. Within the GUI the user needs to provide the manipulator specification, material properties and simulation parameters. The input type option is valid for this case and the selected input will be used for the data generation process. The generated data will subsequently be used for the GA model development.

Step 2. Once all the options have been selected within the initial GUI, the user needs to click on the 'GA Modelling' button within the 'SCEFMAS_V2 GUI'. A click on the 'GA Modelling' button will open the 'GA modelling GUI' (Figure 19.19). On the left hand side of the GUI, the user enters the number of individual as (100, say), maximum number of generation (500, say), generation gap (0.9, say), binary precision (14, say), and order of GA model (7, say). On the right-hand side of the GUI the model option is chosen (hub-angle, say). For data generation process choice of the input appears as a message within the GUI, which has been selected in this case study as the random input.

Step 3. After setting all the desired model parameters and model type options the user needs to generate reference data by clicking on the 'Generate Data' button on the right bottom part of the GUI. This will open a Simulink model with open-loop input (Figure 19.17) and subsequent simulation run will generate the reference data within the Matlab environment and will be available for GA modelling process.

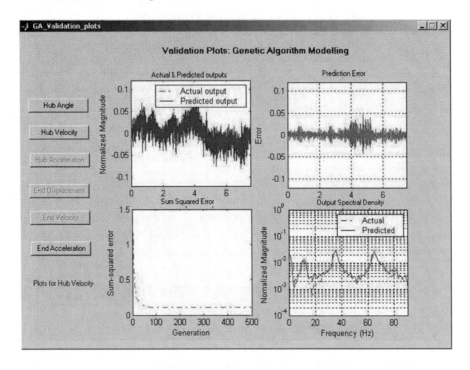

Figure 19.20   GUI for GA model validation

Step 4. After generating the training data, the user can proceed with the GA modelling process by clicking on 'GA Modelling' button within the 'GA Modelling GUI'. The adaptation plot displays the generation error (Figure 19.19). The error reduces as the adaptation process proceeds to higher number of generations. At the completion of a GA modelling process, a 'Model Validation' button will be activated. The user can use this button to initiate the validation process.

Step 5. The validation process involves a GUI, known as the 'GA validation' and is shown in Figure 19.20, where the GA modelling option was selected for input torque to hub-velocity model. This is done by clicking on the hub-velocity button within the GUI in Figure 19.20. The validation plots for the GA hub-velocity model thus obtained are shown in Figure 19.20.

19.5   Summary

The development of an interactive software environment for performing modelling, simulation, and control of single-link flexible manipulator systems has been presented. The user has freedom to select physical specifications and material properties of the manipulator. For the simulation of the system the FD method has used as the simulation algorithm. A number of controller designs have been provided within the package so that the user can try them before developing their own. The simulation and controller designs have been provided through the Simulink environment. The NN and GA modelling approach and subsequent model verification facilities have been provided under intelligent modelling techniques. All the facilities are supported with highly interactive user-friendly GUI where the user can perform almost all the operations with only using a mouse. Matlab Guide toolbox has been used for the GUI development process.

The environment has proven to be a valuable education tool for understanding the behaviour of flexible manipulator systems and development of various controller designs for position and vibration control of such systems. The environment can also be used as a test-bed for newly designed controllers, where control engineers and researchers could test their controllers for vibration and positioning control of a flexible manipulator system and study their performance. Moreover, students can test the effectiveness of their controller designs without spending much time on system simulation.

The first version of SCEFMAS is in use in the Department of Automatic Control and Systems Engineering, University of Sheffield (UK), as a support tool in the delivery of a module in the MSc programme of the department. Although, there is no formal assessment of the effectiveness of this environment as a leaning tool, but students opinions gathered through end-of-module questionnaires have indicated that the software environment enables the students to understand the behaviour of a flexible manipulator system and also the effect of parameter variations without much difficulty. The learning process for simulation, modelling and control of a flexible manipulator system could be much difficult without this package.

References

AGHILI, F., DUPUIS, E., MARTIN, E. and PIEDBOEUF, J.-C. (2001). Force/moment accommodation control for tele-operated manipulators performing contact tasks in stiff environment. *Proceedings of the IEEE/RSJ International Conference on Intelligent Robotics and Systems*, Maui, Hawaii, USA, 29 October–30 November 2001, 2227–2233.

ALBERTS, T. E. and COLVIN, J. A. (1991). Observations on the nature of transfer functions for control of piezoelectric laminates. *Journal of Intelligent Material Systems & Structures*, **2** (4), 528–541.

ALBERTS, T. E., DuBOIS, T. V. and POTA, H. R. (1995). Experimental verification of transfer functions for a slewing piezoelectric laminate beam. *Control Engineering Practice*, **3** (2), 163–170.

AMIROUCHE, F. M. L. (1992). *Computational methods in multibody dynamics*. Prentice Hall, Englewood Cliffs, New Jersey, USA.

ARMADA, M. (1987). Análisis y diseño de sistemas de control de robots industriales en el dominio multivariable de la frecuencia. *Automática e Instrumentación*, **175**, 147–153.

ASADA, H., MA, Z. D. and TOKUMARU, H. (1990). Inverse dynamics of flexible robot arms: Modeling and computation for trajectory control. *Transactions of ASME: Journal of Dynamic Systems, Measurement, and Control*. **112**, 177–185.

ASADA, H. and SLOTINE J. J. E. (1986). *Robot analysis and control*, John Wiley and Sons, Inc., New York, USA.

ASCHER, U., PAI, D. K. and CLOUTIER, B. (1997). Forward dynamics, elimination methods and formulation stiffness in robot simulation. *International Journal of Robotics Research*, **16** (2), 749–758.

ASPINWALL, D. M. (1980). Acceleration profiles for minimising residual response. *Transactions of ASME: Journal of Dynamic Systems, Measurement, and Control*, **102** (1), 3–6.

ÅSTRÖM, K. J. and HÄGGLUND, T. (1984). Automatic tuning of simple regulators with specifications on phase and amplitude margins. *Automatica*, **20** (5), 645–651.

ÅSTRÖM, K. J. and HÄGGLUND, T. (1995). *PID controllers: Theory, design, and tuning*, second edition. Instrument Society of America (ISA), North Carolina, USA.

ÅSTRÖM, K. J. and WITTENMARK, B. (1997). *Computer-controlled systems: Theory and design*, third edition. Prentice-Hall Inc., Englewood Cliffs, New Jersey, USA.

ATHERTON, D. P. (1999). PID controller tuning. *Computing and Control Engineering Journal*, **10** (2), 44–50.

ATHERTON, D. P. and BOZ, A. F. (1998). Using standard forms for controller design. *Proceedings of the UKACC International Conference on Control*, Swansea, UK, 1–4 September 1998, 1066–1071.

ATHERTON, D. P. and MAJHI, S. (1998). Tuning of optimum PIPD controllers. *Proceedings of the 3rd Portuguese Conference on Automatic Control (CONTROLO '98)*, Coimbra, Portugal, 9–11 September 1998, 549–554.

ATHERTON, D. P. and MAJHI, S. (1999). Limitations of PID controllers. *Proceedings of the American Control Conference*, San Diego, California, USA, 2–4 June 1999, 3843–3847.

AUBRUN, J.-N. (1980). Theory of the control structures by low-authority controllers. *Journal of Guidance and Control*, **3**, 444–451.

AZAD, A. K. M. (1994). Analysis and design of control mechanisms for flexible manipulator systems. PhD thesis, Department of Automatic Control and Systems Engineering, The University of Sheffield, UK.

AZAD, A. K. M., TOKHI, M. O., PATHANIA, A. and SHAHEED, M. H. (2004). A Matlab/Simulink based environment for intelligent modelling and simulation of flexible manipulator systems. *Proceedings of American Society for Engineering Education Annual Conference & Exposition*, Utah, USA, 20–23 June 2004, Document-826.

AZAM, M., SINGH, S. N., IYER, A. and KAKAD, Y. P. (1992). Detumbling and reorientation maneuvers and stabilization of NASA SCOLE system. *IEEE Transactions on Aerospace and Electronics Systems*, **28** (1), 80–92.

BAI, M., ZHOU, D.-H. and SCHWARZ, H. (1998). Adaptive augmented state feedback control for an experimental planar two-link flexible manipulator. *IEEE Transactions on Robotics and Automation*, **14** (6), 940–950.

BAILEY, T. and HUBBARD, J. E. (1985). Distributed piezoelectric-polymer active vibration control of a cantilever beam. *Journal of Guidance, Control, and Dynamics*, **8** (5), 605–611.

BALAS, M. J. (1978). Feedback control of flexible systems. *IEEE Transactions on Automatic Control*, **AC-23** (4), 673–679.

BALAS, M. J. (1979). Direct velocity feedback control of large space structures. *Journal of Guidance and Control*, **2** (3), 252–253.

BANKS, S. (1990). *Signal processing image processing and pattern recognition.* Prentice Hall International, Englewood Cliffs, New Jersey, USA.

BANERJEE, A., PEDREIRO, N. and SINGHOSE, W. (2001). Vibration reduction for flexible spacecraft following momentum dumping with/without slewing. *AIAA Journal of Guidance, Control, and Dynamics*, **24** (May–June), 417–428.

BAR-COHEN, Y., XUE, T, SHAHINPOOR, M., SIMPSON, J. and SMITH, J. (1998). Flexible, low-mass robotic arm actuated by electroactive polymers and operated equivalently to human arm and hand. *Rototics 98: The 3rd Conference and Exposition/Demonstration on Robotics for Challenging Environments*, New Mexico, USA, 26–30 April 1998, 15–21.

BARKER, H. A., JOBLING, C. P., RAVN, O., and SZYMKAT, M. (1993). A requirements analysis of future environments for computer aided control engineering. *The 12th IFAC World Congress*, Sydney, Australia, 13–18 July 1993, **8**, 373–376.

BATTITI, R. (1992). First and second order methods for learning: Between steepest descent and Newton's method. *Neural Computation*, **4** (2), 141–166.

BAUNGAARD, J. R. (1996). Modelling and control of flexible robot links. Ph.D. *thesis*, Department of Automation, Technical University of Denmark.

BAYO, E. (1987). A finite-element approach to control the end-point motion of a single-link flexible robot. *Journal of Robotic Systems*, **4** (1), 63–75.

BAYO, E. (1988). Computed torque for the position control of open-loop flexible robots, *Proceedings of IEEE International Conference on Robotics and Automation*, Philadelphia, USA, 25–29 April 1988, 316–321.

BAYO, E. (1989). Timoshenko versus Bernoulli–Euler beam theories for the inverse dynamics of flexible robots. *International Journal of Robotics and Automation*, **4** (1), 53–56.

BAYO, E. and MOULIN, H. (1989). An efficient computation of the inverse dynamics of flexible manipulators in the time domain. *Proceedings of the IEEE International Conference on Robotics and Automation*, Scottsdale, Arizona, USA, 14–19 May 1989, 710–715.

BAYO, E., SERNA, M. A., PAPADOPOULOS, P. and STUBBE, J. (1989). Inverse dynamics and kinematics of multi-link elastic robots: An iterative frequency domain approach. *The International Journal of Robotics Research*, **8** (6), 49–62.

BAZ, A. and POH, S. (1988). Performance of an active control system with piezoelectric actuators. *Journal of Sound and Vibrations*, **126** (2), 327–343.

BAZ, A., POH, S. and FEDOR, J. (1992). Independent modal space control with positive position feedback. *Transactions of ASME: Journal of Dynamic Systems, Measurement, and Control*, **114**, 96–103.

BENDAT, J. S. and PIERSOL, A. G. (1986). *Random data: Analysis and measurement procedures*. Wiley Interscience, New York, USA.

BENNETT, S. (1994). *Real-time computer control: An introduction*, second edition. Prentice-Hall, London, UK.

BENOSMAN, M. and LE VEY, G. (2002). Joint trajectory tracking for planar multi-link flexible manipulator: Simulation and experiment for a two-link flexible manipulator. *Proceedings of the IEEE International Conference on Robotics and Automation*, Washington DC, USA, 11–15 May 2002, 2461–2466.

BHAT, S. P. and MIU, D. K. (1990). Precise point-to-point positioning control of flexible structures. *Transactions of ASME: Journal of Dynamic Systems, Measurement, and Control*, **112** (4), 667–674.

BHAT, S. P., TANAKA, M. and MIU, D. K. (1991). Experiments on point-to-point position control of a flexible beam using Laplace transform technique-Part 1: Open-loop. *Transactions of ASME: Journal of Dynamic Systems, Measurement, and Control*, **113**, 432–437.

BIGRAS, P., SAAD, M. and O'SHEA, J. (1998). Exponential trajectory tracking in the workspace of a class of flexible robots. *Journal of Robotic Systems*, **15** (9), 487–504.

BILLINGS, S. A. and VOON, W. S. F. (1986). Correlation based model validity tests for non-linear systems. *International Journal of Control*, **15** (6), 601–615.

BISHOP, C. M. (1995). *Neural networks for pattern recognition*. Clarendon Press, Oxford, UK.

BLUM, E. K. and LI, L. K. (1991). Approximation theory and feedforward theory. *Neural Networks*, **4**, 511–515.

BOLZ, R. E. and TUVE, G. L. (1973). *CRC Handbook of tables for applied engineering science*. CRC Press, Inc., Boca Raton, Florida, USA.

BOOK, W. J. (1984). Recursive Lagrangian dynamics of flexible manipulator arms. *International Journal of Robotics Research*, **3** (3), 87–101.

BOOK, W. J. (1990). Modeling, design and control flexible manipulator arms: A tutorial review. *Proceedings of 29th Conference on Decision and Control*, Honolulu, Hawaii, USA, 5–7 December 1990, 500–506.

BOOK, W. J. (1993). Controlled motion in an elastic world. *Transactions of ASME:Journal of Dynamic Systems, Measurement, and Control*, **115**, 252–261.

BOOK, W. J., ALBERTS, T. E. and HASTINGS, G. G. (1986). Design strategies for high-speed lightweight robots. *Computers in Mechanical Engineering*, **5** (2), 26–33.

BOOK, W. J., MAIZZA-NETO, O. and WHITNEY, D. E. (1975). Feedback control of two beam, two joint systems with distributed flexibility. *Transactions of ASME: Journal of Dynamic Systems, Measurement, and Control*, **97** (4), 424–431.

BOOK, W. J. and MAJETTE, M. (1983). Controller design for flexible distributed parameter mechanical arms via combined state-space and frequency domain techniques. *Transactions of ASME: Journal of Dynamic Systems, Measurement and Control*, **105** (4), 245–254.

BOROWIEC, J. and TZES, A. (1996). Frequency-shaped explicit output feedback for flexible link manipulators. *Proceedings of the 35th Conference on Decision and Control*, Kobe, Japan, 11–13 December 1996, 4106–4111.

BOSSERT, D., LY U.-L. and VAGNERS, J. (1996). Experimental evaluation of robust reduced-order hybrid position/force control on a two-link manipulator. *Proceedings of the IEEE International Conference on Robotics and Automation*, Minneapolis, Minnesota USA, 22–28 April 1996, 2573–2578.

BOYER, F. and COIFFET, P. (1996). Generalization of Newton–Euler model for flexible manipulators. *Journal of Robotic Systems*, **13** (1), 11–24.

BREAKWELL, J. A. (1980). Control of flexible spacecraft. PhD thesis, Department of Aeronautics and Astronautics, Stanford University, USA.

BREHM, T. (1994). Hybrid fuzzy logic PID controller. *Proceedings of 3rd IEEE Conference on Fuzzy Systems*, **3**, 1682–1687.

BREMER, H. and PFEIFFER, F. (1992). *Elastic multibody systems*. Teubner, Stuttgard, Germany. In German.

BUENO, S. S., DE KEYSER, R. M. C. and FAVIER, G. (1991). Auto-tuning and adaptive tuning of PID controllers. *Journal of Acoustics*, **32** (1), 28–34.

BUFFINTON, K. W. (1992). Dynamics of elastic manipulators with prismatic joints. *Transactions of ASME: Journal of Dynamic Systems, Measurement, and Control*, **114**, 41–49.

BUFFINTON, K. W. and LAM, J. (1992). A comparative study of simple dynamic models and control schemes for elastic manipulators. *Proceedings of the American Control Conference*, Chicago, Illinois, USA, 24–26 June 1992, 3334–3339.

BURDEN, R. L. and FAIRES, J. D. (1989). *Numerical analysis*. PWS-KENT Publishing Company, Boston, Massachusetts, USA.

CALVERT, J. F. and GIMPEL, D. J. (1957). Method and apparatus for control of system output response to system input. US Patent No. 2,801,351.

CANNON, R. H. and SCHMITZ, E. (1984). Initial experiments on the end-point control of a flexible one-link robot. *International Journal of Robotics Research*, **3** (3), 62–75.

CANUDAS DE WIT, C., SICILIANO, B. and BASTIN, G. (Eds) (1996). *Theory of robot control*. Springer-Verlag, London.

CAPONETTO, R. C., FORTUNA, L., MANGANARO, G. and XIBILIA M. G. (1995). Chaotic system identification via genetic algorithm. *Proceedings of*

GALESIA-95: 1st IEE/IEEE International Conference on GAs in Engineering Systems: Innovations and Applications, Sheffield, UK, 12–14 September 1995, 170–174.

CASDAGLI, M. (1989). Non-linear prediction of chaotic time series. *Physics D*, **35**, 335–356.

CASPERSEN, M. K. (2000). Control of a flexible link robot. Master thesis, Department of Automation, The Technical University of Denmark, Denmark.

CETINKUNT, S. and BOOK, W. J. (1989). Symbolic modeling and dynamic simulation of robotic manipulators with compliant links and joints. *Robotics and Computer-Integrated Manufacturing*, **5** (4), 301–310.

CETINKUNT, S. and ITTOP, B. (1992). Computer-automated symbolic modeling of dynamics of robotic manipulators with flexible links.*IEEE Transactions on Robotics and Automation*, **8** (1), 94–105.

CETINKUNT, S. and YU, W.-L. (1991). Closed-loop behavior of a feedback-controlled flexible arm: A comparative study. *The International Journal of Robotics Research*, **10** (3), 263–275.

CHALHOUB, N. G. and ULSOY, A. G. (1987). Control of a flexible robot arm: Experimental and theoretical results. *Transactions of ASME: Journal of Dynamic Systems, Measurement, and Control*, **109** (4), 299–309.

CHAO, C.-T. and TENG, C.-C. (1997). A PD-like self-tuning fuzzy controller without steady-state error. *Fuzzy Sets and Systems*, **87**, 141–154.

CHAPNIK, B. V., HEPPLER, G. R., and APLEVICH, J. D. (1991). Modelling impact on a one-link flexible robotic arm. *IEEE Transactions on Robotics and Automation*, **7** (4), 479–488.

CHAPRA, S. and CANALE, R. (1988). *Numerical methods for engineers*, second edition. McGraw-Hill, New York, USA.

CHEN, J.-S. and MENQ, C.-H. (1990). Experiments on the payload-adaptation of a flexible one-link manipulator with unknown payload. *IEEE International Conference on Robotics and Automation*, Cincinnati, Ohio, USA, 13–18 May 1990, 1614–1619.

CHEN, M. and LINKENS, D.A. (1998). A hybrid neuro-fuzzy PID controller. *Fuzzy Sets and Systems*, **99**, 27–36.

CHEN, S., COWAN, C. F. N. and GRANT, P. M. (1991). Orthogonal least squares learning algorithm for radial basis function networks. The *IEEE Transactions on Neural Networks*, **2** (2), 302–309.

CHENG, W. and WEN, J. T. (1993). A neural controller for the tracking control of flexible arms. *Proceedings of the IEEE International Conference on Neural Networks*, San Francisco, USA, 28 March-01 April 1993, 749–754.

CHEONG, J., CHUNG, W. and YOUM, Y. (2001). Fast suppression of vibration for multi-link flexible robots using parameter adaptive control. *Proceedings of the IEEE/RSJ International Conference on Intelligent Robots and Systems*, Maui, Hawaii, USA, 29 October–03 November 2001, 913–918.

CHIANG, W.-W., KRAFT, R. and CANNON, Jr., R. H. (1991). Design and experimental demonstration of rapid precise end-point control of a wrist carried by a very flexible manipulator. *The International Journal of Robotics Research*, **10** (1), 30–40.

CHIAVERINI, S. and SCIAVICCO, L. (1993). The parallel approach to force/position control of robotic manipulators. *IEEE Transactions on Robotics and Automation*, **9**, 361–373.

CHIAVERINI, S., SICILIANO, B. and VILLANI, L. (1994). Force/position regulation of compliant robot manipulators. *IEEE Transactions on Automatic Control*, **39**, 647–652.

CHIOU, B. C. and SHAHINPOOR, M. (1990). Dynamic stability analysis of a two-link force-controlled flexible manipulator. *Transactions of ASME: Journal of Dynamic Systems, Measurement, and Control*, **112**, 661–666.

CHIPPERFIELD, A. J., FLEMING, P. J., POHLHEIM, H. and FONSECA, C. (1994). A genetic algorithm toolbox for MATLAB. *Proceedings of the International Conference on Systems Engineering*, Coventry, UK, 6–8 September 1994, 200–207.

CHO, K., HORI, N. and ANGELES, J. (1991). On the controllability and observability of flexible beams under rigid-body motion. *Proceedings of the 1991 International Conference on Industrial Electronics, Control and Instrumentation*, Kobe, 28 October–1 November 1991, 455–460.

CHOI, S. B., GANDHI, M. V. and THOMPSON, B. S. (1988). An experimental investigation of an articulating robotic manipulator with a graphite epoxy arm. *Journal of Robotic Systems*, **5**, 73–79.

CHOURA, S. and YIGIT, A. S. (2001). Control of a two-link rigid-flexible manipulator with a moving payload mass. *Journal of Sound and Vibration*, **243**, 883–897.

CHOW, J. H. and KOKOTOVIC, P. V. (1978). Two-time-scale feedback design of a class of non-linear systems. *IEEE Transactions on Automatic Control*, **23** (3), 438–443.

CHUNG, H.-Y., CHEN, B.-C. and LIN, J.-J. (1998). A PI-type fuzzy controller with self-tuning scaling factors. *Fuzzy Sets and Systems*, **93**, 23–28.

CLOUGH, R. W. and PENZIEN, J. (1975). *Dynamics of structures*. McGraw-Hill Book Company, New York, USA.

COOK, R. D. (1981). *Concepts and applications of finite element analysis*. Wiley, New York, USA.

CRANDALL, S. H., KARNOPP, D. C., KURTZ, E. F. and PRIDMORE-BROWN, D. C. (1968). *Dynamics of mechanical and electromechanical systems*. McGraw-Hill Inc., New York, USA.

CRAWLEY, E. F. and ANDERSON, E. H. (1990). Detailed models of piezoelectric actuation of beams. *Journal of Intelligent Material Systems and Structures*, 1, 4–25.

CRAWLEY, E. F. and LUIS, J. (1987). Use of piezoelectric actuators as elements of intelligent structures. *AIAA Journal*, 25 (10), 1373–1385.

CUNDARI, M. and ABEDIAN, B. (1991). The dynamic behavior of a polyvinylidene fluoride piezoelectric motional device. *Proceedings of Smart Structures and Materials, Winter Annual Meeting of the American Society of Mechanical Engineers*, Atlanta, Georgia, USA, 1–6 December 1991, 25–31.

DADO, M. and SONI, A. H. (1986). A generalized approach for forward and inverse dynamics of elastic manipulators. *Proceedings of IEEE Conference on Robotics and Automation*, San Francisco, California, USA, 7–10 April 1986, 359–364.

DALEY, S. and LIU, G. P. (1999). Optimal PID tuning using direct search algorithms. *Computing and Control Engineering Journal*, 10 (2), 51–56.

DANCOSE, S. and ANGELES, J. (1989). Modelling and simulation of flexible beams using cubic splines and zero-order holds. *International Symposium on the Mathematical Theory of Networks and Systems*, Amsterdam, The Netherlands, 19–23 June 1989, 553–564.

DANCOSE, S., ANGELES, J. and HORI, N. (1989). Optimal vibration control of a rotating flexible beam. *Proceedings of the 12th Biennial ASME Conference on Mechanical Vibration and Noise, Diagnostics, Vehicle Dynamics and Special Topics*, Montreal, Canada, 17–20 September 1989, DE-Vol. 18-5, 259–264.

DAVIS, J. H. and HIRSCHORN, R. M. (1988). Tracking control of flexible robot link. *IEEE Transactions on Automatic Control*, 33 (3), 238–248.

DE CARUFEL, J., MARTIN, E. and PIEDBOEUF, J.-C. (2000). Control strategies for hardware-in-the-loop simulation of flexible space robots. *IEE Proceedings – D: Control Theory and Applications*, 147 (6), 569–579.

D'ELEUTERIO, G. M. T. (1992). Dynamics of an elastic multibody chain: Part C-Recursive dynamics. *Dynamics and Stability of Systems*, 7 (2), 61–89.

DE LUCA, A., ISIDORI, A. and NICOLO, F. (1985*a*). An application of nonlinear model matching to the dynamic control of robot arms with elastic joints. *Proceedings of Syroco '85: IFAC Symposium on Robotic Control*, Barcelona, Spain, 6–8 November 1985, 219–225.

DE LUCA, A., ISIDORI, A. and NICOLO, F. (1985*b*). Control of robot arm with elastic joints via nonlinear dynamic feedback. *Proceedings of the 24th IEEE Conference on Decision and Control*, Fort Lauderdale, Florida, USA, 11–13 December 1985, 1671–1679.

DE LUCA, A., LUCIBELLO, P. and NICOLO, F. (1988). Automatic symbolic modelling and non-linear control of robots with flexible links. *Proceedings of IEE Seminar on Robotics and Control*, Oxford, UK, 13–15 April 1988, 62–70.

DE LUCA, A. and PANZIERI, S. (1994). An iterative scheme for learning gravity compensation in flexible robot arms. *Automatica*, **30** (6), 993–1002.

DEWCA, A. and SICILIANO, B. (1989). Trajectory control of a non-linear one-line flexible arm. *International Journal of Control*, **50**, 1699–1715.

DE LUCA, A. and SICILIANO, B. (1991). Closed-form dynamic model of planar multilink lightweight robots. *IEEE Transactions on Systems, Man, and Cybernetics*, **21** (4), 826–839.

DE LUCA, A. and SICILIANO, B. (1993). Regulation of flexible arms under gravity. *IEEE Transactions on Robotics and Automation*, **9**, 463–467.

DE SCHUTTER, J. and VAN BRUSSEL, H. (1988). Compliant robot motion II. A control approach based on external control loops. *International Journal of Robotics Research*, **7** (4), 18–33.

DELLMAN, R., GLICKSBER, I. and GROSS, O. (1956). On the bang-bang control problem. *Quarterly of Applied Mechanics*, **14**, 11–18.

DEROOVER, D., BOSGRA, O. H., SPERLING, F. B. and STEINBUCH, M. (1996). High-performance motion control of a wafer stage. *Philips Conference on Applications of Control Technology*, Epe, The Netherlands, 29–30 October 1996.

DEROOVER, D., SPERLING, F. B. and BOSGRA, O. H. (1998). Point-to-point control of a MIMO servomechanism. *Proceedings of the American Control Conference*, Philadelphia, Pennsylvania, USA, June 1998, 2648–2651.

DOETSCH, K. H. and MIDDLETON, J. A. (1987). Canada's space program. *Canadian Aeronautics and Space Journal*, **33** (4), 61–89.

DONNE, J. D. and ÖZGÜNER, U. (1994). Neural control of a flexible-link manipulator. *Proceedings of the IEEE International Conference on Neural Networks*, Orlando, Florida, USA, 27 June–2 July 1994, 2327–2332.

DORATO, P. (1987). A historical review of robust control. *IEEE Control Systems Magazine*, **7**, 44–47.

DOYON, M., PIEDBOEUF, J.-C., AGHILI, F., GONTHIER, Y. and MARTIN, E. (2003). The SPDM task verification facility: On the dynamic emulation in one-g environment using hardware-in-the-loop simulation. *The 7th International Symposium on Artificial Intelligence and Robotics & Automation in Space: i-SAIRAS 2003*, Nara, Japan, 19–23 May 2003.

DRAPEAU, V. and WANG, D. (1993). Verification of a closed-loop shaped-input controller for a five-bar-linkage manipulator. *Proceedings of IEEE International Conference on Robotics and Automation*, Atlanta, Georgia, USA, 2–6 May 1993, 216–221.

DUBOWSKY, S. and PAPADOPOULOS, E. (1993). The kinematics, dynamics, and control of free-flying and free-floating space robotic systems. *IEEE Transactions on Robotics and Automation*, **9** (5), 531–543.

DUTTON, K., THOMPSON, S. and BARRACLOUGH, B. (1997). *The Art of Control Engineering*. Addison-Wesley, Harlow, VK.

DWIGHT, H. B. (1957). *Tables and integrals and other mathematical data*, third edition. Macmillan, New York, USA.

ELANAYAR, S. V. T. and YUNG, C. S. (1994). Radial basis function neural network for approximation and estimation of non-linear stochastic dynamic systems. The *IEEE Transactions on Neural Networks*, **5** (4), 594–603.

EPPINGER, S. and SEERING, W. (1988). Modeling robot flexibility for endpoint force control. A.I. Memo 1046. MIT Artificial Intelligence Lab., May 1988.

FEATHERSTONE, R. (1987). *Robot dynamics algorithms*. Kluwer Academic Publishers, London.

FEDDEMA, J. T. (1993). Digital filter control of remotely operated flexible robotic structures. *Proceedings of the American Control Conference*, San Francisco, California, USA, 2–4 June 1993, 2710–2715.

FEDDEMA, J., DOHRMANN, C., PARKER, G., ROBINETT, R., ROMERO, V. and SCHMITT, D. (1997). Control for slosh-free motion of an open container. *IEEE Control Systems Magazine*, **17** (1), 29–36.

FEDDEMA, J., PETTERSON, B. and ROBINETT, R. (1998). Operator control systems and methods for swing-free Gantry-style cranes, US Patent No. 5, 785, 191.

FELIÚ, V., RATTAN, K. S. and BROWN, Jr, H. (1990). Adaptive control of a single-link flexible manipulator. *IEEE Control Systems Magazine*, **10**, 29–33.

FELIÚ, V., RATTAN, K. S. and BROWN Jr, H. B. (1992). Modelling and control of single link flexible arms with lumped masses. *Transactions of ASME: Journal of Dynamic Systems, Measurement, and Control*, **114**, 59–69.

FELIÚ, V., RATTAN, K. and BROWN Jr, H. (1995). Control of a two-degree-of-freedom lightweight flexible arm with friction in the joints. *Journal of Robotic Systems*, **12** (1), 1391–1395.

FERNÁNDEZ, G. (1997). *Control multivariable en el dominio de la frecuencia de robots rígidos y flexibles*. PhD thesis, University of Valladolid, Valladolid, Spain.

FERNÁNDEZ, G., ARMADA, M. and GRIECO, J. C. (1993). Estudio del acoplamiento en modelos de robots industriales. *3er Congreso de la Asociación Española de Robótica*. 17–19 November 1993, Zaragoza, Spain, **1**, 75–84.

FERNÁNDEZ, G., GRIECO, J. C. and ARMADA, M. (1994a). Decoupling control for robot manipulators using multivariable frequency domain techniques.

5th International Symposium on Application of Multivariable System Techniques. Mechanical Engineering Publishers Ltd, UK, London, 235–245.

FERNÁNDEZ, G., GRIECO, J. C. and ARMADA, M. (1994*b*). Achieving diagonal dominance for robot control. *Studies in Informatics and Control*, **3** (2), 241–252.

FERNÁNDEZ, G., GRIECO, J. C., ARMADA, M. and GONZÁLEZ de SANTOS, P. (2002*a*). Following strategy for deflection avoiding on a flexible manipulator. *CLCA 2002: IFAC Latin-American Conference on Automatic Control*. Guadalajara, México, 3–5 December, 2002.

FERNÁNDEZ, G., GRIECO, J. C. and ARMADA, M. (2002*b*). Pre-filter design for decoupling control of a flexible manipulator. *CLCA 2002: IFAC Latin-American Conference on Automatic Control*. Guadalajara, México, 3–5 December, 2002.

FLIESS, M. (1989). Nonlinear control theory and differential algebra. In *Modelling and Adaptive Control*. Lecture Notes in Control and Information Sciences, **15**, Springer-Verlag, Berlin, 135–145.

FONSECA, C. M., MENDES, E. M., FLEMING, P. J. and BILLINGS, S. A. (1993). Non-linear model term selection with genetic algorithms. *Proceedings of Workshop on Natural Algorithms in Signal Processing*, Essex, UK, 15–16 November 1993, 27/1–27/8.

FORD, M. P. and DALY, K. C. (1979). Dominance improvement by pseudodecoupling. *Proceedings of the IEE*, **126**, 1316–1320.

FRASER, A. R. and DANIEL, R. W. (1991). *Perturbation techniques for flexible manipulators*. Kluwer Academic Publishers, Boston, Massachusetts, USA.

FRIEDLAND, B. (1986). *Control system design – An introduction to state-space methods*. McGraw-Hill, New York, USA.

FRIMAN, M. and WALLER, K. V. (1994). Autotuning of multiloop control systems. *Journal of Industrial and Engineering Chemistry Research*, **33** (7), 1708–1717.

FUKUDA, T. (1985). Flexibility control of elastic robot arms. *Journal of Robotic Systems*, **2** (1), 73–88.

GE, S. S., LEE, T. H. and GONG, J. Q. (1998*a*). Dynamic modeling of a smart materials robot. *AIAA Journal*, **36** (8), 1466–1478.

GE, S. S., LEE, T. H., GONG, J. Q. and WANG, Z. P. (2000). Model-free controller design for a single-link flexible smart materials robot. *International Journal of Control*, **73** (6), 531–544.

GE, S. S., LEE, T. H. and HARRIS, C. J. (1998*b*). *Adaptive neural network control of robot manipulators*. World Scientific, New Jersey, USA.

GE, S. S., LEE, T. H. and TAN, E. G. (1997). Adaptive neural network control of flexible link robots based on singular perturbation. *Proceedings of the IEEE International*

Conference on Control Applications, Hardford, Maryland, USA, 5–7 October 1997, 365–369.

GE, S. S., LEE, T. H. and WANG, Z. P. (2001*a*). Adaptive neural network control for smart materials robots using singular perturbation technique. *Asian Journal of Control*, **3** (2), 143–155.

GE, S. S., LEE, T. H. and WANG, Z. P. (2001*b*). Model-free regulation of multi-link flexible smart materials robots. *IEEE/ASME Transactions on Mechatronics*, **6** (3), 346–351.

GE, S. S., LEE, T. H. and ZHU, G. (1996). Energy-based robust controller design for multi-link flexible robots. *Mechatronics*, **6** (7), 779–798.

GE, S. S., LEE, T. H. and ZHU, G. (1998*c*). Improving regulation of a single-link flexible manipulator with strain feedback. *IEEE Transactions on Robotics and Automation*, **14** (1), 179–185.

GE, S. S. and POSTLETHWAITE, I. (1995). Adaptive neural network controller design for flexible joint robots using singular perturbation technique. *Transactions of the Institute of Measurement and Control*, **17** (2), 120–131.

GENIELE, H., PATEL, R.V. and KHORASANI, K. (1992). Control of flexible manipulator. *Proceedings of the 4th ASME International Symposium on Robotics and Manufacturing*, New Mexico, USA, 11–13 November 1992, 567–572.

GÉRADIN, M. and CARDONA, A. (2001). *Flexible multibody dynamics: A finite element approach*. John Whiley and Sons, Chichester, UK.

GERALD, C. F. and WHEATLEY, P. O. (1985). *Applied Numerical Analysis*. Addison-Wesley Publishing Company, Reading, Massachusetts, USA.

GEVARTER, W. B. (1970). Basic relations for control of flexible vehicles. *AIAA Journal*, **8**, 666–672.

GIOVAGNONI, M. (1994). A numerical and experimental analysis of a chain of flexible bodies. *Transactions of ASME: Journal of Dynamic Systems, Measurement, and Control*, **106** (1), 73–80.

GOH, S. P. (2001). Investigation and implementation of advanced control techniques for flexible manipulators. PhD thesis, Department of Mechanical Engineering, the University of Leeds, UK.

GOH, S. P., PLUMMER, A. R. and BROWN, M. D. (2000*a*). Decoupling multivariable pole-placement controller of a flexible manipulator. *EUREL European Advanced Robotics Systems Masterclass and Conference – Robotics 2000*, Salford, UK, 12–14 April 2000.

GOH, S. P., PLUMMER, A. R. and BROWN, M. D. (2000*b*). Digital control of a flexible manipulator. *Proceedings of the American Control Conference*, Chicago, Illinois, USA, 28–30 June 2000, 2205–2209.

GOLDBERG, D. E. (1989). *Genetic algorithms in search, optimisation and machine learning*. Addison-Wesley, Reading, Massachusetts, USA.

GONTHIER, Y., MCPHEE, J., LANGE, C. and PIEDBOEUF, J.-C. (2004). A regularized contact model with asymmetric damping and dwell-time dependent friction. *Multibody System Dynamics*, **11** (3), 209–233.

GORINEVSKI, D. M., FORMALSKY, A. M. and SCHNEIDER. A. Y. (1997). *Force control of robotics systems*. CRC Press, Boca Raton, Florida, USA.

GREFENSTETTE, J. J. (1986). Optimisation of control parameters for genetic algorithms. *IEEE Transactions on Systems, Man and Cybernetics*, **16** (1), 122–128.

GU, M. (2002). Statics of flexible manipulator: Modeling and validation. Technical Report CSA-TR-2002-0003, Canadian Space Agency, Canada.

GU, M. and PIEDBOEUF, J.-C. (2002). Three-dimensional kinematic analysis and verification for a flexible robot arm. *ASME Design Engineering Technical Conferences and Computers and Information in Engineering Conference*, Montreal, Canada, 29 September–2 October 2002, Paper No. MECH-34260.

GUTIÉRREZ, L. B., LEWIS, F. L. and LOWE J. A. (1998). Implementation of a neural network tracking controller for a single flexible link: Comparison with PD and PID controllers. The *IEEE Transactions on Industrial Electronics*, **45** (2), 307–318.

GUYAN, R. J. (1965). Reduction of stiffness and mass matrices. *AIAA Journal*, **3** (2), 380.

HAGAN, M. T. and MENHAJ, M. B. (1994). Training feedforward networks with Marquardt algorithm. *IEEE Transactions on Neural Networks*, **5** (6), 989–993.

HANSON, L. and SALAMON, P. (1990). Neural network ensembles. *IEEE Transactions on Pattern Analysis and Machine Intelligence*, **12**, 993–1001.

HAPPEL, B. M. and MURRE, J. J. (1994). Design and evolution of modular neural network architectures. *Neural Networks*, **7**, 985–1004.

HARA, S. and YOSHIDA, K. (1994). Simultaneous optimization of positioning and vibration controls using time-varying criterion function. *Journal of Robotics and Mechatronics*, **6** (4), 278–284.

HARASHIMA, F. and UESHIBA, T. (1986). Adaptive control of flexible arm using the end-point position sensing. *Proceedings of Japan-USA Symposium on Flexible Automation*, Osaka, 15–20 July, 225–229.

HARDT, M. W. (1999). Multibody dynamical algorithms, numerical optimal control, with detailed studies in the control of jet engine compressors and biped walking. PhD Thesis, Electrical Engineering, University of California, San Diego, California, USA.

HARRIS, C. J., MOORE, C. G. and BROWN, M. (1993). *Intelligent control: Aspects of fuzzy logic and neural nets*. World Scientific, London.

HASHTRUDI-ZAAD, K. and KHORASANI, K. (1996). Control of non-minimum phase singularly perturbed systems with applications to flexible manipulators. *International Journal of Control*, **63**, 679–701.

HASTINGS, G. G. and BOOK, W. J. (1987). A linear dynamic model for flexible robotic manipulator. *IEEE Control Systems Magazine*, **7** (1), 61–64.

HE, S. Z., TAN, S. H., XU, F. L. and WANG, P. Z. (1993). PID self-tuning control using a fuzzy adaptive mechanism. *Proceedings of the 2nd IEEE International Conference on Fuzzy Systems*, San Francisco, California, USA, 28 March–01 April 1993, **2**, 708–713.

HE, X. and LAPEDES, A. (1993). Non-linear modelling and prediction by successive approximation using radial basis functions. *Physics D*, **70**, 289–301.

HILLSLEY, K. L. and YURKOVICH, S. (1993). Vibration control of a two-link flexible robot arm. *Dynamics and Control*, **3**, 261–280.

HODGE, L., AUDA, G. and KAMEL, M. (1999). Learning decision fusion in cooperative modular neural networks. *International Joint Conference on Neural Networks: IJCNN'99*, Piscataway, New Jersey, USA, 10–16 July 1999, **4**, 2777–2781.

HOGAN, N. (1985). Impedance control: An approach to manipulation: Parts I–III. *Transactions of ASME: Journal of Dynamic Systems, Measurement, and Control*, **107**, 1–24.

HOGAN, N. (1987). Stable execution of contact task using impedance control. *Proceedings of IEEE International Conference on Robotics and Automation*, Raleigh, North Carolina, USA, 31 March–3 April 1987, 1047–1054.

HOLLAND, J. (1975). *Adaptation in natural and artificial systems*, University of Michigan Press, Michigan, USA.

HOLLAND, J. (1992). *Adaptation in natural and artificial systems: An introductory analysis with applications to biology*. MIT Press, Massachusetts, USA.

HOSSAIN, M. A., TOKHI, M. O., CHIPPERFIELD, A. J., FONSECA, C. M. and DAKEV, N. V. (1995). Adaptive active vibration control using genetic algorithms. *Proceedings of GALESIA-95: 1st IEE/IEEE International Conference on GAs in Engineering Systems: Innovations and Applications*, Sheffield, UK, 12–14 September 1995, 175–180.

HU, A. (1993). A survey of experiments for modeling verification and control of flexible robotic manipulators. *Proceedings of The 1st IEEE Regional Conference on Aerospace Control Systems*, Westlake, California, USA, 25–27 May 1993, 344–353.

HU, F. L. and ULSOY, A. G. (1994). Force and motion control of a constrained flexible robot arm. *Transactions of ASME: Journal of Dynamic Systems, Measurement, and Control*, **116**, 336–343.

HUGHES P. C. (1987). Space structure vibration modes: How many exists? Which are important. *IEEE Control Systems Magazine*, **7** (1), 22–28.

HURTY, W. C. (1965). Dynamic analysis of structural systems using component modes. *AIAA Journal*, **3** (4), 678–685.

HYDE, J. M. and SEERING, W. P. (1991). Using input command pre-shaping to suppress multiple mode vibration. *Proceedings of IEEE International Conference on Robotics and Automation*, Sacramento, California, USA, 9–11 April 1991, 2604–2609.

ISERMANN, R. (1989). *Digital control systems – Vol. 1: Fundamentals, deterministic control*, second revised edition. Springer-Verlag, Berlin Heidelberg, Germany.

IWAMOTO, K., KOIKE, Y., NONAMI, K., TANIDA, K. and IWASAKI, I. (2002). Output feedback sliding mode control for bending and torsional vibration control of 6-story flexible structure. *JSME International Journal*, Series C. **45** (1), 150–158.

JACKSON, L. B. (1989). *Digital filters and signal processing*. Kluwer Academic Publishers, London.

JACOBS, R. A. and JORDAN, M. I. (1993). Learning piecewise control strategies in a modular neural network architecture. *IEEE Transactions on Systems, Man, and Cybernetics*, **23** (2), 337–345.

JAIN, A. and RODRIGUEZ, G. (1992). Recursive flexible multibody system dynamics using spatial operators. *AIAA Journal of Guidance, Control, and Dynamics*, **15** (6), 1453–1466.

JAIN, A. K. and MOHIUDDIN K. M. (1996). Artificial neural networks: A tutorial. *Computer*, **29** (3), 31–44.

JANSEN, J. F. (1992). Control and analysis of a single-link flexible beam with experimental verification. ORNL/TM-12198, Oak Ridge National Laboratory, Oak Ridge, Tennessee, USA.

JAYASURIYA, S. and CHOURA, S. (1991). On the finite settling time and residual vibration control of flexible structures. *Journal of Sound and Vibration*, **148**, 117–136.

JOHANSSON, K. H., JAMES, B., BRYAN, G. F. and ÅSTRÖM, K. J. (1998). Multivariable controller tuning. *Proceedings of the American Control Conference*, Philadelphia, Pennsylvania, USA, 24–26 June 1998, **6**, 3514–3518.

JONES, J. F., PETTERSON, B. J. and STRIP, D. R. (1991). Methods of and system for swing damping movement of suspended objects. US Patent No. 4, 997, 095.

JONES, S. and ULSOY, A. G. (1999). An approach to control input shaping with application to coordinate measuring machines. *Transactions of ASME: Journal of Dynamic Systems, Measurement, and Control*, **121** (June), 242–247.

JORDAN, S. (2002). Eliminating vibration in the nano-world. *Photonics Spectra*, **36** (July), 60–62.

JUNKINS, J. L. and BANG, H. (1993). Maneuver and vibration control of hybrid coordinate systems using Lyapunov stability theory. *AIAA Journal of Guidance, Control, and Dynamics*, **16** (4), 668–676.

KANE, T. R., RYAN, R. R. and BANERJEE, A. K. (1987). Dynamics of a cantilever beam attached to a moving base. *AIAA Journal of Guidance, Control and Dynamics*, **10** (12), 139–151.

KANOH, H., TZAFESTAS, S., LEE, H. G. and KALAL, J. (1986). Modelling and control of flexible robot arms. *Proceedings of 25th Conference on Decision & Control*, Athens, Greece, 10–12 December 1986, 1866–1870.

KARAKASOGLU, A., SUDHARSANAN, S. I. and SUNDARESHAN, M. K. (1993). Identification and decentralised adaptive control using dynamical neural networks with application to robotic manipulators. *IEEE Transactions on Neural Networks*, **4** (6), 919–930.

KARGUPTA, H. and SMITH, R. E. (1991). System identification with evolving polynomial networks. *Proceedings of the 4th International Conference on Genetic Algorithms*, San Diego, California, USA, 14–17 July 1991, 370–376.

KARR, C. L. (1991). Design of an adaptive fuzzy logic controller using genetic algorithms. *Proceedings of the 4th International Conference on Genetic Algorithms*, San Diego, California, USA, 13–16 July 1991, 450–457.

KARRAY, F. and MODI, V. (1995). On the pointing robustness issue of a class of new generation spacecraft. *IEEE Transactions on Automatic Control*, **40** (12), 2132–2137.

KECMAN, V. (1996). System identification using modular neural network with improved learning. *Proceedings of the International Workshop on Neural Networks for Identification, Control, Robotics, and Signal/Image Processing*, Los Alamitos, California, USA, 21–23 August 1996, 40–48.

KELKAR, A. G., ALBERTS, T. E. and JOSHI, S. M. (1992). Asymptotic stability of dissipative compensators for flexible multibody systems in the presence of actuator dynamics and actuator/sensor nonlinearities. *Proceedings of ASME 1992 Winter Annual Meeting*, Anaheim, California, USA, 8–13 November 1992, **AMD-141**, 35–38.

KENISON, M. and SINGHOSE, W. (1999). Input shaper design for double-pendulum planar gantry cranes. *Proceedings of IEEE Conference on Control Applications*, Hawaii, USA, 22–27 August 1999, 539–544.

KENISON, M. and SINGHOSE, W. (2002). Concurrent design of input shaping and proportional plus derivative feedback control. *Transactions of ASME: Journal of Dynamic Systems, Measurement, and Control*, **124** (3), 398–405.

KERWIN Jr, E. M. (1959). Damping of flexural waves by a constrained viscoelastic layer. *Journal of the Acoustical Society of America*, **31**, 952–962.

KHORRAMI, F. and JAIN, S. (1993). Nonlinear control with end-point acceleration feedback for a two-link flexible manipulator: Experimental results. *Journal of Robotic Systems*, **10** (4), 505–530.

KHORRAMI, F., JAIN, S. and TZES, A. (1993). Adaptive nonlinear control and input preshaping for flexible-link manipulators. *Proceedings of the American Control Conference*, San Francisco, California, USA, 2–4 June 1993, 2705–2709.

KHORRAMI, F., JAIN, S. and TZES, A. (1994). Experiments of rigid body-based controllers with input preshaping for a two-link flexible manipulator. *IEEE Transactions on Robotics and Automation*, **10** (1), 55–65.

KHORRAMI, F. and ÖZGÜNER, Ü. (1988). Perturbation methods in control of flexible link manipulators, *Proceedings of the IEEE International Conference on Robotics and Automation*, Philadelphia, Pennsylvania, USA, 25–29 April 1988, 310–315.

KIM J.-S., SUZUKI, K., KONNO A. and UCHIYAMA M. (1996). Force control of constrained flexible manipulators. *Proceedings of the 1996 IEEE International Conference on Robotics and Automation*, Minneapolis, Minnesota, USA, 22–28 April 1996, 635–640.

KINO, M., GODEN, T., MURAKAMI, T. and OHNISHI, K. (1998). Reaction torque feedback based vibration control in multi-degrees of freedom motion system. *Proceedings of the 24th Annual Conference of the Industrial Electronics Society of the IEEE*, Aachen, Germany, 31 August–4 September 1998, **3**, 1807–1811.

KOJIMA, H. (1986). Transient vibrations of a beam/mass system fixed to a rotating body. *Journal of Sound and Vibration*, **107**, 149–154.

KOJIMA, H. and NAKAJIMA, N. (2003). Multi-objective trajectory optimization by a hierarchical gradient algorithm with fuzzy decision logic. *AIAA Guidance, Navigation, and Control Conference*, Austin, Texas, USA, 11–14 August 2003, AIAA2003–5568.

KOKOTOVIC, P. V. (1984). Applications of singular perturbation techniques to control problems. *SIAM Review*, **26** (4), 502–550.

KOKOTOVIC, P., KHALIL, H. K. and O'REILLY, J. (1986). *Singular perturbation methods in control: Analysis and design*. Academic Press, New York, USA.

KOMATSU, T., UEYAMA, M., IIKURA, S., MIURA, H. and SHIMOYAMA, I. (1990). Autonomous satellite robot testbed. *Proceedings of International Symposium on Artificial Intelligence, Robotics and Automation in Space*, Kobe, Japan, 18–20 November 1990, 113–116.

KOROLOV, V. V. and CHEN, Y. H. (1988). Robust control of a flexible manipulator arm. *Proceedings of 1988 IEEE International Conference on Robotics and Automation*, Philadelphia, Pennsylvania, USA, 24–29 April 1988, 159–164.

KOROLOV, V. V. and CHEN, Y. H. (1989). Controller design robust to frequency variation in a one-link flexible robot arm. *Transactions of ASME: Journal of Dynamic Systems, Measurement, and Control*, **111** (1), 9–14.

KOTNIK, P. T., YORKOVICH, S. and ÖZGÜNER, Ü. (1988). Acceleration feedback for control of a flexible manipulator arm. *Journal of Robotic Systems*, **5** (3), 181–196.

KOURMOULIS, P. K. (1990). Parallel processing in the simulation and control of flexible beam structures. PhD thesis, Department of Automatic Control and Systems Engineering, The University of Sheffield, UK.

KRESS, R. L., JANSEN, J. F. and NOAKES, M. W. (1994). Experimental implementation of a robust damped-oscillation control algorithm on a full sized, two-DOF, AC induction motor-driven crane. *Proceedings of 5th International Symposium on Robotics and Manufacturing*, Maui, Hawaii, USA, 15–17 August 1994, 585–592.

KRESS, R. L., LOVE, L., DUBEY, R. and GIZELAR. A (1997). Waste tank cleanup manipulator modeling and control. *IEEE International Conference on Robotics and Automation*, Albuquerque, New Mexico, USA, 20–25 April 1997, 662–668.

KRISHNAN, H. (1988). *Bounded input discrete-time control of a single-link flexible beam*. Master thesis, Department of Electrical Engineering, University of Waterloo, Canada.

KRISTINSSON, K. and DUMONT, G. (1992). System identification and control using genetic algorithms. *IEEE Transactions on Systems, Man, and Cybernetics*, **22** (5), 1033–1046.

KRUISE, L. (1990). Modelling and control of a flexible beam and robot arm. PhD thesis, University of Twente, The Netherlands.

KWAK, H. J., SUNG, S. W. and LEE, I.-B. (1997). On-line process identification and autotuning for integrating processes. *Journal of Industrial and Engineering Chemistry Research*, **36** (12), 5329–5338.

KWOK, D. P., TAM, D., LI, C. K. and WANG, P. (1990). Linguistic PID controllers. *Proceedings of 11th IFAC World Congress*, Tallin, USSR, 13–17 August 1990, 192–197.

KWOK, D. P., TAM, D., LI, C. K. and WANG, P. (1991). Analysis and design of fuzzy PID control systems, *Proceedings of IEE Control '91 Conference*, Edinburgh, UK, 25–28 March 1991, **2**, 955–960.

KWON, D.-S. and BOOK, W. J. (1990). An inverse dynamic method yielding flexible manipulator state trajectories. *Proceedings of the American Control Conference*, San Diego, California, USA, 23 – 25 May 1990, 186–193.

LAMBERT, M., MOORE, B. and AHMADI, M. (2001). Essential real-time and modeling tools for robot rapid prototyping. *The 6th International Symposium on Artificial Intelligence and Robotics & Automation in Space: i-SAIRAS 2001*, Montreal, Canada, 18–22 June 2001.

LANDAU, I. D. (1998). The R–S–T digital controller design and applications. *Control EngineeringPractice,* **6** (2), 155–165.

LANDAU, I., LANGER, J., REY, D. and BARNIER, J. (1996). Robust control of a 360° flexible arm using the combined pole placement/sensitivity function shaping method. *IEEE Transactions on Control Systems Technology,* **4** (4), 369–383.

LAPIDUS, L. (1982). *Numerical solution of partial differential equations in science and engineering.* John Wiley and Sons, New York, USA.

L'ARCHEVÊQUE, R., DOYON, M., PIEDBOEUF, J.-C. and GONTHIER, Y. (2000). SYMOFROS: Software architecture and real time issues. *Proceedings of DASIA 2000 – Data Systems in Aerospace,* ESA, Montreal, Canada, 22–26 May 2000, **SP-457**, 41–46.

LARCOMBE, P. J. and BROWN, I. C. (1997). Computer algebra: A brief overview and application to dynamic modelling. *Journal of Computing and Control Engineering,* **8** (2), 53–57.

LA SALLE, J. and LEFSCHETZ, S. (1961). *Stability by Liapunov's direct method.* Academic Press, San Diego, California, USA.

LAU, M. and PAO, L. (2002). Characteristics of time-optimal commands for flexible structures with limited fuel usage. *Journal of Guidance, Control, and Dynamics,* **25** (1), 222–231.

LEE, J. (1993). On methods for improving performance of PI-type fuzzy logic controllers. *IEEE Transactions on Fuzzy Systems,* **1** (1), 298–301.

LEE, J. X., VUKOVICH, G. and SASAIDEK, J. Z. (1994). Fuzzy control of a flexible link manipulator. *Proceedings of the American Control Conference,* Baltimore, Maryland, USA, 29 June–1 July 1994, 568–574.

LEE, M. A. and TAKAGI, H. (1993). Dynamic control of genetic algorithms using fuzzy logic techniques. *Proceedings of 5th International Conference on Genetic Algorithms,* Illinois, USA, 17–21 July 1993, 76–83.

LESHNO, M. (1993). Multilayer feedforward networks with a nonpolynomial activation function can approximate any function. *Neural Networks,* **6**, 861–867.

LEW, J. Y. and BOOK, W. J. (1993). Hybrid control of flexible manipulators with multiple contact. *Proceedings of the IEEE International Conference on Robotics and Automation,* Atlanta, Georgia, USA, 2–6 May 1993, **2**, 242–247.

LEWIS, D., PARKER, G. G., DRIESSEN, B. and ROBINETT, R. D. (1998). Command shaping control of an operator-in-the-loop boom crane. *Proceedings of the American Control Conference,* Philadelphia, Pennsylvania, USA, 24–26, June 1998, 2643–2647.

LEWIS, H. W. (1997). *The foundation of fuzzy control.* Plenum Press, New York, USA.

LI, D. (1994). Nonlinear control design for tip position tracking of a flexible manipulator arm. *International Journal of Control*, **69** (6), 1065–1082.

LI, Y. F. and CHEN, X. B. (2001). End-Point sensing and state observation of a flexible-link robot. *IEEE/ASME Transactions on Mechatronics*, **6** (3), 351–356.

LIN, C.-T. and LEE, C. S. G. (1991). Neural-network-based fuzzy logic control and decision system. *IEEE Transactions on Computer*, **40**, 1320–1336.

LIN, C.-T. and LEE, C. S. G. (1992). Real-time supervised structure-parameter learning for fuzzy neural network. *Proceeding of the IEEE International Conference on Fuzzy Systems*, San Diego, California, USA, 8–12 March 1992, 1283–1290.

LIN, C.-T. and LEE, C. S. G. (1993). Reinforced structure-parameter learning for neural-network-based fuzzy logic control systems. *Proceeding of the IEEE International Conference on Fuzzy Systems*, San Francisco, California, USA, 28 March–1 April 1993, 88–93.

LIN, C.-T. and LEE, C. S. G. (1995). A neural fuzzy control system with structure and parameter learning. *Fuzzy Sets and Systems*, **70**, 183–212.

LIN, I.-C. and FU, L.-C. (1998). Adaptive hybrid force/position control of a flexible manipulator for automated deburring with on-line cutting trajectory modification. *Proceedings of the IEEE International Conference on Robotics and Automation*, Leuven, Belgium, 16–20 May 1998, 818–825.

LIN J. (2003). Hierarchical fuzzy logic controller for a flexible link robot arm performing constrained motion tasks. *IEE Proceedings – D: Control Theory and Applications*, **150** (4), 355–364.

LIN, J. and LEWIS, F. L. (1994). A symbolic formulation of dynamic equations for a manipulator with rigid and flexible links, *International Journal of Robotics Research*, **13** (5), 454–466.

LITTLE, J. N. and SHURE, L. (1988). *Signal processing toolbox for MATLAB*. The MathWorks, Inc., Natick, Massachusetts, USA.

LIU, Q. and WIE, B. (1992). Robust time-optimal control of uncertain flexible spacecraft. *Journal of Guidance, Control, and Dynamics*, **15** (3), 597–604.

LJUNG, L. and SJÖBERG, J. (1992). A system identification perspective on neural networks. *Neural Networks for Signal Processing II. Proceedings of the IEEE-SP Workshop*, Helsingoer, Denmark, 31 August–2 September 1992, 423–435.

LOH, A. P. and VASNANI, V. U. (1992). Multiloop controller design for multivariable plants. *Proceedings of the 31st Conference on Decision and Control*, Tucson, Arizona, USA, 16–18, December 1992, 181–182.

LONGMAN, R. W. (1990). Attitude tumbling due to flexibility in satellite-mounted robots. *The Journal of Astronautical Sciences*, **38** (4), 487–509.

LONGMAN, R. W. and LINDBERG, R. E. (Eds) (1990). Special issue on robotics in space. *The Journal of the Astronautical Sciences*, **38** (4), 395–396.

LÓPEZ-LINARES, S. (1993). Control de robots flexibles. Tesis Doctoral. Escuela Superior de Ingenieros Industriales. Universidad de Navarra. San Sebastián.

LORON, L. (1997). Tuning of PID controllers by the non-symmetrical optimum method. *Automatica*, **33** (1), 103–107.

LOW, K. H. and VIDYASAGAR, M. (1988). A Lagragian formulation of dynamic model for flexible manipulator systems. *Transactions of ASME: Journal of Dynamic Systems, Measurement, and Control*, **110**, 175–181.

LUBICH, C., NOWAK, U., POHLE, U. and ENGSTLER, C. (1992). MEXX – numerical software for the integration of constrained mechanical multibody systems. Tech Report SC92-12, Konrad-Zuse-Zentrum für Informationstechnik, Berlin.

LUCIBELLO, P. and DI BENEDETTO, M. D. (1993). Output tracking for a nonlinear flexible arm. *Transactions of ASME: Journal of Dynamic Systems, Measurement, and Control*, **115**, 78–85.

LUCIBELLO, P., NICOLO, F. and PIMPINELLI, R. (1986). Automatic symbolic modelling of robots with a deformable link. *Proceedings of IFAC International Symposium on Theory of Robots*, Vienna, 3–5 December 1986, 183–187.

LUCIBELLO, P. and ULIVI, G. (1993). Design and realization of a two-link direct-drive robot with a very flexible forearm. *International Journal of Robotics and Automation*, **8** (3), 113–128.

LUH, J. Y. S., WALKER, M. W. and PAUL, R. P. (1982). Resolved motion force control of robot manipulators, *Transactions of ASME: Journal of Dynamic Systems, Measurement, and Control*, **102**, 126–133.

LUO, F-L. and UNBEHAUEN, R. (1997). *Applied neural networks for signal processing*. Cambridge University Press, Cambridge, UK.

LUO, Z. H. (1993). Direct strain feedback control of flexible robot arms: new theoretical and experimental results. *IEEE Transactions on Automatic Control*, **38** (11), 1610–1622.

MA, O. (1995). Contact dynamics modelling for the simulation of the space station manipulators handling payloads. *Proceedings of IEEE International Conference on Robotics and Automation*, Nagoya, Japan, 21–27 May, 1995, 1252–1258.

MA, O. (2000). CDT – A general contact dynamics toolkit. *Proceedings of the 31st International Symposium on Robotics*, Montreal, Canada, 14–17 May 2000, 468–473.

MA, O. (2002). *Contact dynamics toolkit (cdt) user's guide*. MacDonald Dettwiler Space and Advanced Robotics Ltd., Research and Development. Version CDT_2000, Release 3.

MA, O., BUHARIWALA, K., ROGER, N., MACLEAN, J. and CARR, R. (1997). MDSF – a generic development and simulation facility for flexible, complex robotic systems. *Robotica*, **15**, 49–62.

MACE, B. R. (1991). The effects of transducer inertia on beam vibration measurements. *Journal of Sound and Vibration*, **145**, 365–379.

MACFARLANE, A. G. J., GRÜBEL, G. and ACKERMANN, J. (1989). Future design environments for control engineering.*Automatica*, **25** (2), 165–176.

MACIEJOWSKI, J. M. (1991). *Multivariable feedback design*. Addison-Wesley, Workingham, UK.

MADHAVAN, S. K. and SINGH, S. N. (1991). Inverse trajectory control and dynamic sensitivity of an elastic manipulator. *International Journal of Robotics and Automation*, **6** (4), 179–191.

MAGEE, D. P. and BOOK, W. J. (1993). Eliminating multiple modes of vibration in a flexible manipulator. *Proceedings of the IEEE International Conference on Robotics and Automation*, Atlanta, Georgia, USA, 2–6 May 1993, 474–479.

MAGEE, D. P. and BOOK, W. J. (1995). Filtering micro-manipulator wrist commands to prevent flexible base motion. *Proceedings of the American Control Conference*, Seattle, Washington, USA, 21–23 June 1995, 924–928.

MAGEE, D. P., CANNON, D. W. and BOOK, W. J. (1997). Combined command shaping and inertial damping for flexure control. *Proceedings of the American Control Conference*, Albuquerque, New Mexico, USA, 4–6 June 1997, 1330–1334.

MAHIL, S. S. (1982). On the application of Lagrange's method to the description of dynamic systems. *IEEE Transactions on System, Man, and Cybernetics*, **SMC-12** (6), 877–889.

MARCHAND, P. and HOLLAND, T. (2003). *Graphics and GUI's with MATLAB*, third edition. CRC Press, Boca Raton, Florida, USA.

MARQUARDT, D. (1963). An algorithm for least squares estimation of non-linear parameters. *Journal of the Society for Industrial and Applied Mathematics*, **11** (2), 431–441.

MARTIN, G. (1978). On the control of flexible mechanical system. PhD thesis, Department of Aeronautics and Astronautics, Stanford University, Stanford, California, USA.

MARTINS, J. M., BOTTO, M. A. and SÁ DA COSTA, J. (2002). Modelling of flexible beams for robotic manipulators. *Multibody System Dynamics*, **7** (1), 79–100.

MARTINS, J. M., MOHAMED, Z., TOKHI, M. O., SÁ DA COSTA, J. and BOTTO, M. A. (2003). Approaches for dynamic modelling of flexible manipulator systems. *IEE Proceedings – D: Control Theory and Applications*, **150** (4), 401–411.

MASUTANI, Y., MIYAZAKI, F. and ARIMOTO, S. (1989). Sensory feedback control for space manipulators. *Proceedings of IEEE International Conference on Robotics and Automation*, Scottsdale, Arizona, USA, 14–19 May, 1346–1351.

MATSUNO, F., ASANO, T. and SAKAWA, Y. (1994). Modeling and quasi-static hybrid position/force control of constrained planar two-link flexible manipulators. *IEEE Transactions on Robotics and Automation*, **10**, 287–297.

MATSUNO, F., OHNO, T. and ORLOV, Y. V. (2002). Proportional derivative and strain (PDS) boundary feedback control of a flexible space structure with a closed-loop chain mechanism. *Automatica*, **38** (7), 1201–1211.

MATSUNO, F. and YAMAMOTO, K. (1994). Dynamic hybrid position/force control of a two degree-of-freedom flexible manipulator. *Journal of Robotic Systems*, **11** (5), 355–366.

MCCLAMROCH, N. H. and WANG, D. (1988). Feedback stabilization and tracking of constrained robots. *IEEE Transactions on Automatic Control*, **33**, 419–426.

MECKL, P. H. and SEERING, W. P. (1990). Experimental evaluation of shaped inputs to reduce vibration of a cartesian robot. *Transactions of ASME: Journal of Dynamic Systems, Measurement, and Control*, **112** (6), 159–165.

MEIROVITCH, L. (1967). *Analytical methods in vibrations*. Macmillan, New York, USA.

MEIROVITCH, L. (1970). *Methods of analytical dynamics*. McGraw-Hill, New York, USA.

MEIROVITCH, L. (1975). *Elements of vibration analysis*. McGraw-Hill, Inc., New York, USA.

MEIROVITCH, L. (1980). *Computational methods in structural dynamics*. Sijthhoff and Noordhoff, Amsterdam.

MEIROVITCH, L. (1986). *Elements of vibration analysis*. McGraw-Hill, New York, USA.

MEIROVITCH, L., BARUH, H. and OZ, H. (1983). A comparison of control techniques for large flexible systems.*Journal of Guidance, Control, and Dynamics*, **6**, 302–310.

MENQ, C. and CHEN, J. (1988). Dynamic modeling and payload-adaptive control of a flexible manipulator. *Proceedings of the IEEE International Conference on Robotics and Automation*, Philadelphia, Pennsylvania, USA, 24–29 April 1988, 488–493.

MEYER, G. E. (1971). *Analytical methods in conduction heat transfer*. McGraw-Hill, New York, USA.

MEYER, J. L. and SILVERBERG, L. (1996). Fuel optimal propulsive maneuver of an experimental structure exhibiting spacelike dynamics. *Journal of Guidance, Control, and Dynamics*, **19** (1), 141–149.

MICHALEWICZ, Z. (1994). *Genetic algorithms + Data structures = Evolution programs.* Springer-Verlag, New York, USA.

MILLS, J. K. (1992). Stability and control aspects of flexible link robot manipulators during constrained motion tasks. *Journal of Robotic Systems,* **9,** 933–953.

MINSKY, M. and PAPERT, S. (1969). *Perceptrons: An Introduction to Computational Geometry,* MIT Press, Cambridge, USA.

MISIR, D., MALKI, H. A. and CHEN, G. (1996). Design and analysis of a fuzzy proportional-integral-derivative controller. *Fuzzy Sets and Systems,* **79** (3), 297–314.

MIYABE, T., YAMANO, M., KONNO, A. and UCHIYAMA, M. (2001). An approach toward a robust object recovery with flexible manipulators. *Proceedings of the IEEE/RSJ International Conference on Intelligent Robots and Systems,* Maui Hawaii, USA, 29 October–3 November 2001, 907–912.

MOALLEM, M., KHORASANI, K. and PATEL, R. V. (1997). An integral manifold approach for tip-position tracking of flexible multi-link manipulators. *IEEE Transactions on Robotics and Automation,* **13** (6), 823–837.

MOHAMMED, Z. and TOKHI, M. O. (2002). A symbolic manipulation approach for modelling and performance analysis of flexible manipulator systems. *International Journal of Acoustics and Vibration,* **7** (1), 27–37.

MOHAMED, Z. and TOKHI, M. O. (2004). Command shaping techniques for vibration control of a flexible robot manipulators. *Mechatronics,* **14** (1), 69–90.

MOORE, B., KÖVECSES, J. and PIEDBOEUF, J.-C. (2003). Symbolic model formulation for dynamic parameters identification. In J. A. Ambrósio (Ed.), *Proceedings of Multibody Dynamics 2003,* Lisbon, Portugal, 1–4 July 2003, 1–11. MB2003-058.

MOORE, B., PIEDBOEUF, J.-C. and BERNARDIN, L. (2002). Maple as an automatic code generator? *Maple Summer Workshop,* Ontario, Canada, 28–30 July 2002.

MOULIN, H. and BAYO, E. (1991). On the accuracy of end-point trajectory tracking for flexible arms by noncausal inverse dynamic solution. *Transactions of ASME: Journal of Dynamic Systems, Measurement, and Control,* **113,** 320–324.

MUDI, R. K. and PAL, N. R. (1999). A robust self-tuning scheme for PI- and PD-type fuzzy controllers. *IEEE Transactions on Fuzzy Systems,* **7** (1), 2–16.

MUENCHHOF, M. and SINGH, T. (2002). Concurrent feedforward/feedback controller design using time-delay filters. *AIAA Guidance, Navigation and Control Conference,* Monterey, California, USA, 5–8 August 2002.

MUROTSU, Y., SENDA, K., MITSUYA, A., YAMANE, K., HAYASHI, M. and NUNOHARA, T. (1993). Theoretical and experimental studies for continuous path control of flexible manipulators mounted on a free-flying space robot. *Proceedings of AIAA Guidance, Navigation and Control Conference,* Monterey, California, USA, 17–19 August 1993, **3,** 1458–1471.

MUROTSU, Y., TSUJIO, S., SENDA, K. and HAYASHI, M. (1990). Dynamics and positioning control of space robot with flexible manipulators. *Proceedings of AIAA Guidance, Navigation and Control Conference*, Portland, Oregon, USA, 20–22 August 1990, **1**, 735–742.

MUROTSU, Y., TSUJIO, S., SENDA, K. and HAVASHI, M. (1992). Trajectory control of flexible manipulators on a free-flying space robot. *IEEE Control Systems*, **12** (3), 51–57.

NAGARKATTI, S. P., RAHN, C. D., DAWSON, D. M. and ZERGEROGLU, E., (2001). Observer-based modal control of flexible systems using distributed sensing. *Proceedings of the 40th IEEE Conference on Decision and Control*, Orlando, Florida, USA, 4–7 December 2001, 4268–4273.

NAGATHAN, G. and SONI, A. H. (1986). Non-linear flexibility studies for spatial manipulators. *Proceedings of the IEEE International Conference on Robotics and Automation*, San Francisco, California, USA, 7–10 April 1986, 373–378.

NARENDRA, K. S. and PARTHASARATHY, K. (1990). Identification and control of dynamical systems using neural networks. *IEEE Transactions on Neural Networks*, **1** (1), 4–27.

NAUCK, D. and KRUSE, R. (1992). Interpreting changes in the fuzzy sets of a self-adaptive neural fuzzy controller, *Proceeding of IFIS'92: 2nd Workshop on Industrial Applications of Fuzzy Control and Intelligent Systems*, College Station, Texas, USA, 2–4 December 1992, 146–152.

NAUCK, D. and KRUSE, R. (1993). A fuzzy neural network learning fuzzy control rules and membership functions by fuzzy error backpropagation. *Proceeding of IEEE International Conference on Neural Networks*, San Francisco, California, USA, 28 March–1 April 1993, 1022–1027.

NERRAND, O., ROUSSEL-RAGOT, P., URBANI, D., PERSONNAZ, L. and DREYFUS, G. (1994). Training recurrent neural networks: why and how? An illustration in dynamical process modelling. *IEEE Transactions on Neural Networks*, **5** (2), 178–184.

NESLINE, F. W. and ZARCHAN, P. (1984). Why modern controllers can go unstable in practice. *AIAA Journal of Guidance, Control, and Dynamics*, **7**, 495–500.

NEWLAND, D. E. (1996). *Introduction to random vibrations, spectral and wavelet analysis*. Addison-Wesley Longman, California, USA.

NEWTON, R. T. and XU, Y. (1993). Neural network control of a single-link flexible manipulators, *IEEE Control Systems Magazine*, **12**, 14–22.

NOVOZHILOV, V. V. (1961). *Theory of elasticity*. Pergamon Press Ltd, London, UK.

OAKLEY, C. M., and CANNON, R. H. (1988). Initial experiments on the end-point control of a two-link manipulator with a very flexible forearm, *Proceedings of the American Control Conference*, Atlanta, Georgia, USA, 15–17 June 1988, 996–1002.

OAKLEY, C. M. and CANNON, R. H. (1989). End-point control of a two-link manipulator with a very flexible forearm: Issues and experiments. *Proceedings of the American Control Conference*, Pittsburgh, Pennsylvania, USA, 21–23 June 1989, 1381–1388.

OAKLEY, C. M. and CANNON, R. (1990). Anotomy of an experimental two-link flexible manipulator under end-point control. *Proceedings of the IEEE 29th Conference on Decision and Control*, Honolulu, Hawaii, USA, 5–7 December 1990, 507–513.

OGATA, K. (2001). *Modern control engineering*. Prentice-Hall, Inc., Englewood Cliffs, New Jersey, USA.

OLUROTIMI, O. (1994). Recurrent neural network training with feedforward complexity. *IEEE Transactions on Neural Networks*, **5** (2), 185–197.

OMATU, S., KHALID, M. and YUSOF, R. (1996). *Neuro-control and its applications*. Springer, London.

OMATU, S. and SEINFELD, J. H. (1986). Optimal sensor actuator locations for linear distributed parameter systems. *Proceedings of the 4th IFAC Symposium on Control of Distributed Parameter Systems*, Los Angeles, California, USA, 30 June–2 July 1986, 215–220.

ONSAY, T. and AKAY, A. (1991). Vibration reduction of a flexible arm by time optimal open-loop control. *Journal of Sound and Vibration*, **142** (2), 283–300.

OOSTING, K. and DICKERSON, S. L. (1988). Simulation of a high-speed lightweight arm. *Proceedings of the IEEE International Conference on Robotics and Automation*, Philadelphia, Pennsylvania, USA, 24–29 April 1988, 494–496.

ORTEGA, J. M. (1987). *Matrix theory – A second course*. Plenum Press, New York, USA.

OUSSAMA, K., CRAIG, J. and LOZANO-PÉREZ, T. (1989). *The robotics review*. MIT Press, Boston, Massachusetts, USA.

OWER, J. C. and VAN de VEGTE, J. (1987). Classical control design for a flexible manipulator: modeling and control system design. *IEEE Journal of Robotics and Automation*, **RA-3** (5), 485–489.

PADEN, B., CHEN, D., LEDESMA, R. and BAYO, E. (1993). Exponentially stable tracking control for multijoint flexible-link manipulators. *Transactions of ASME: Journal of Dynamic Systems, Measurement, and Control*, **115**, 53–59.

PAI, D. K., ASCHER, U. M. and KRY, P. G. (2000). Forward dynamics algorithms for multibody chains and contact. *Proceedings of IEEE International Conference on Robotics and Automation*, San Francisco, California, USA, 24–28 April 2000, 857–863.

PALMOR, Z. J., HALEVI, Y. and KRASNEY, N. (1995). Automatic tuning of decentralized PID controllers for TITO processes. *Automatica*, **31** (7), 1001–1010.

PAN, Y. C., ULSOY, A. G. and SCOTT, R. A. (1990). Experimental model validation for a flexible robot with a prismatic joint. *Journal of Mechanical Design*, **112**,
315–323.

PAO, L. Y. (1996). Minimum-time control characteristics of flexible structures.
Journal of Guidance, Control, and Dynamics, **19** (1), 123–129.

PAO, L. Y. (1999). Multi-input shaping design for vibration reduction. *Automatica*,
35 (1), 81–89.

PAO, L. Y. and SINGHOSE, W. E. (1997). Verifying robust time-optimal commands
for multi-mode flexible spacecraft. *AIAA Journal of Guidance, Control, and
Dynamics*, **20** (4), 831–833.

PAO, L. Y. and SINGHOSE, W. E. (1998). Robust minimum time control of flexible
structures. *Automatica*, **34** (2), 229–236.

PAPOULIS, A. (1962). *The Fourier integral and its applications*. McGraw-Hill Book
Company, Inc., New York, USA.

PASSINO, K. M. and YURKOVICH, S. (1998). *Fuzzy control*. Addison-Wesley
Longman Inc., Massachusetts, USA.

PFEIFFER, F. (1989). A feedforward decoupling concept for the control of elastic
robots. *Journal of Robotic Systems*, **6** (4), 407–416.

PIEDBOEUF, J.-C. (1993). Kane's equations or Jourdain's principle. *36th Midwest
Symposium on Circuits and Systems*, Detroit, Michigan, USA, 16–18 August 1993.

PIEDBOEUF, J.-C. (1996). Modelling flexible robots with maple. *Maple Tech: The
Maple Technical Newsletter*, **3** (1), 38–47.

PIEDBOEUF, J.-C. (1998). Recursive modelling of serial flexible manipulators. *The
Journal of the Austronautical Sciences*, **46** (1), 1–24.

PIEDBOEUF, J.-C., AGHILI, F., DOYON, M., GONTHIER, Y., MARTIN, E. and
ZHU, W.-H. (2001). Emulation of space robot through hardware-in-the-loop simulation. *The 6th International Symposium on Artificial Intelligence and Robotics &
Automation in Space: i-SAIRAS 2001*, Canadian Space Agency, St-Hubert, Canada,
18–22 June 2001.

PIEDBOEUF, J.-C. and MOORE, B. (2002). On the foreshortening effects of a rotating flexible beam using different modeling methods. *Mechanics of Structure and
Machines*, **30** (1), 83–102.

PLUMMER, A. R. and VAUGHAN, N. O. (1997). Decoupling pole-placement
control, with application to a multi-channel electro-hydraulic servosystem. *Control
Engineering Practice*, **5** (3), 313–323.

PLUNKELL, R. and LEE, C. T. (1970). Length optimization for constrained
viscoelastic layer damping. *Journal of the Acoustical Society of America*, **48**,
150–161.

POERWANTO, H. (1998). Dynamic simulation and control of flexible manipulator systems. PhD thesis, Department of Automatic Control and Systems Engineering, The University of Sheffield, Sheffield, UK.

POTA, H. R. and ALBERTS, T. E. (1995). Multivariable transfer functions for a slewing piezoelectric laminate beam. *Transactions of ASME: Journal of Dynamic Systems, Measurement, and Control*, **117** (3), 352–359.

PREUMONT, A. (1997). *Vibration control of active structures: An introduction.* Kluwer Academic, Dordrecht, The Netherlands.

QIAN, W. T. and MA, C. C. H. (1992). A new controller design for a flexible one-link manipulator. *IEEE Transactions on Automatic Control*, **37** (1), 133–137.

QIN, S.-Z., SU, H.-T., and MCAVOY, T. J. (1992). Comparison of four neural net learning methods for dynamic system identification. *IEEE Transactions on Neural Networks*, **3** (1), 122–130.

QUARANTA, G. and MANTEGAZZA, P. (2001). Using MATLAB-Simulink RTW to build real time control applications in user space with RTAI-LXRT. *Real-Time Linux Workshop*, Milano, Italy, 19–26 November 2001.

RAD, A. B., LO, W. L. and TSANG, K. M. (1997). Self-tuning PID controller using Newton–Raphson search method. *IEEE Transactions on Industrial Electronics*, **44** (5), 717–725.

RAIBERT, M. H. and CRAIG, J. J. (1981). Hybrid position/force control of manipulators. *Transactions of ASME: Journal of Dynamic Systems, Measurement, and Control*, **103**, 126–133.

RAKSHA, F. and GOLDENBERG, A. A. (1986). Dynamic modelling of a single-link flexible robot. *The IEEE International Conference on Robotics and Automation*, San Francisco, California, USA, 7–10 April 1986, 820–828.

RAO, S. S. (1989). *The finite element method in engineering*. Paragon Press, Oxford, UK.

RAPPOLE, B. W., SINGER, N. C. and SEERING, W. P. (1994). Multiple-mode impulse shaping sequences for reducing residual vibrations. *Proceedings of 23rd Biennial Mechanisms Conference*, Minneapolis, Minnesota, USA, 11–14 September 1989, 11–16.

RAVN, O. and SZYMKAT, M. (1992). The evolution of CACSD tools – a software engineering perspective. *Proceedings of IEEE Control Systems Society Symposium on Computer-Aided Control System Design*, Napa, California, USA, 17–19 March 1992, 225–231.

RAVN, O. and SZYMKAT, M. (1994). Requirements for user interaction support in future CACE environments. *Proceedings of IEEE Control Systems Society Symposium on Computer-Aided Control System Design*, Tucson, Arizona, USA, 7–9 March 1994, 381–386.

RAVN, O. and SZYMKAT, M. (1995). Mechatronics approach to robotic modelling – software engineering perspective. *Proceeding of MMAR'95: 2nd International Symposium on Methods and Models in Automation and Robotics*, Miedzyzdroje, Poland, 30 August–2 September 1995, 455–466.

RAVN, O., SZYMKAT, M., UHL, T., BETEMPS, M., PJETURSSON, A. and ROD, J. (1996). Mechatronic blockset for simulink – concept and implementation. *Proceedings of CACSD'96*, Dearborn, Michigan, USA, 15–18 September 1996, 530–535.

REDDY, J. N. (1993). *An introduction to the finite element method*, second edition. McGraw-Hill, New York, USA.

ROBINETT, R., PARKER, G., FEDDEMA, J., DOHRMANN C. and PETTERSON, B. (1999). Sway control method and system for rotary cranes, US Patent No. 5,908,122.

ROCCO, P. and BOOK, W. J. (1996). Modelling for two-time scale force/position control of flexible robots. *Proceedings of the 1996 IEEE International Conference on Robotics and Automation*, Minneapolis, Minnesota, USA, 22–28 April 1996, 1941–1946.

RODRIGUEZ, G., JAIN, A. and KREUTZ-DELGADO, K. (1992). Spatial operator algebra for multibody system dynamics. *The Journal of the Austronautical Sciences*, **40**, 27–50.

ROMANO, M., AGRAWAL, B. and BERNELLI-ZAZZERA, F. (2002). Experiments on command shaping control of a manipulator with flexible links. *Journal of Guidance, Control and Dynamics*, **25** (2), 232–239.

ROSENBROCK, H. H. (1974). *Computer-aided control system design*. Academic Press, London.

ROSS, C. T. F. (1996). *Finite element techniques in structural mechanics*. Albion Publishing Limited, West Sussex, UK.

ROSSI, M., WANG, D. and ZUO, K. (1997). Issues in the design of passive controllers for flexible-link robots. *The International Journal of Robotics Research*, **16** (4), 577–588.

ROSTGAARD, M. (1995): Modelling, estimation and control of fast sampled dynamic systems. PhD theses. Department of Mathematical Modelling, The Technical University of Denmark, Denmark.

RUMELHART, D. E., HINTON, G. E. and WILLIAMS, R. J. (1986). Learning internal representations by error propagation. *Parallel Distributed Processing: Exploration in the Microstructure of Cognition*, MIT Press, Cambridge, Massachusetts, USA, 318–362.

SAAD, M., PIEDBOEUF, J.-C., AKHRIF, O. and SAYDY, L. (2000*a*). A comparison of different shape functions in assumed-mode models of a flexible slewing beam. Technical Report, Canadian Space Agency, Canada.

SAAD, M., PIEDBOEUF, J.-C., AKHRIF, O. and SAYDY, L. (2000*b*). Shape function comparison in a model of a flexible slewing beam. *Proceedings of International Symposium on Robotics and Automation, ISRA'2000*, Monterey, Mexico, 10–12 November 2000, 237–243.

SADEGH, N. (1995). Synthesis of a stable discrete-time repetitive controller for MIMO systems. *Transactions of ASME: Journal of Dynamic Systems, Measurement, and Control*, **117** (March), 92–97.

SAKAWA, Y., MATSUNO, F. and FUKUSHIMA, S. (1985). Modelling and feedback control of a flexible arm. *Journal of Robotic Systems*, **2**, 453–472.

SALISBURY, J. K. (1980). Active stiffness control of a manipulator in Cartesian coordinates. *Proceedings of 19th IEEE Conference on Decision and Control*, Albuquerque, New Mexico, USA, 10–12 December 1980, 95–100.

SANDIA NATIONAL LABORATORIES (2004). Flexible robotics, Sandia National Laboratories, USA (http://endo.sandia.gov/9234/FlexRobot).

SCHMITZ, E. (1985). Experiments on the end-point position control of a very flexible one-link manipulator. PhD thesis, Stanford University, Stanford, California, USA.

SCHOENWALD, D. A. and Özgüner, U. (1990). On combining slewing vibration control in flexible manipulator via singular perturbation. *Proceedings of the 29th IEEE Conference on Decision and Control*, Honolulu, Hawaii, USA, 5–7 December 1990, 533–538.

SCHULZ, G. and HEIMBOLD, G. (1983). Dislocated actuator/sensor positioning and feedback design of flexible structures. *Journal of Guidance, Control, and Dynamics*, **6**, 361–367.

SCHWERTASSEK, R. and WALLRAPP, O. (1999). *Dynamic flexible multibody systems*. Vieweg, Braunschweig, Germany. In German.

SCIAVICCO, L. and SICILIANO, B. (1996). *Modelling and control of robot manipulators*. McGraw-Hill, Inc., New York, USA.

SELCOM SELECTIVE ELECTRONICS (1994). *SELSPOT hardware users manual and software users manual*. Selcom Selective Electronics Co AB, Partille, Sweden.

SENDA, K. (1993). Dynamics and control of rigid/flexible space manipulators. PhD thesis, Osaka Prefecture University, Osaka, Japan.

SENDA, K. and MUROTSU, Y. (1993). Controlability and observability of flexible manipulators and a stability condition of PD-control. *Proceedings of the Asia–Pacific Vibration Conference*, Kitakyushu, Japan, 14–18 November 1993, **3**, 1343–1348.

SENDA, K. and MUROTSU, Y. (1994a). Stable manipulation variable feedback control for flexible manipulators. *Proceedings of Japan–USA Symposium on Flexible Automation*, Kobe, Japan, **II**, 397–404.

SENDA, K. and MUROTSU, Y. (1994b). Manipulation varible feedback control of flexible manipulators by using virtual rigid manipulator concept. *Proceedings of the AIAA Guidance, Navigation and Control Conference*, Scottsdale, Arizona, USA, 1–3 August 1994, **3**, 1030–1039.

SETH, N., RATTAN, K. and BRANDSTETTER, R. (1993). Vibration control of a coordinate measuring machine. *Proceeding of the IEEE Conference on Control Applications*, Dayton, Ohio, USA, 13–16 September 1993, 368–373.

SHAHEED, M. H. (2000). Neural and genetic modelling, control and real-time finite element simulation of flexible manipulators. PhD thesis, Department of Automatic Control and Systems Engineering, The University of Sheffield, Sheffield, UK.

SHARF, I. (1996). Geometrically non-linear beam element for dynamics simulation of multibody systems. *International Journal for Numerical Methods in Engineering*, **39**, 763–786.

SHARMA, S. K. (2000). Soft computing for modelling and control of fynamic dystems with spplication to flexible manipulators. PhD thesis. Department of Automatic Control and System Engineering, The University of Sheffield, UK.

SHARMA, S. K., IRWIN, G. W., TOKHI, M. O. and MCLOONE, S. F. (2003). Learning soft computing control strategies in a modular neural network architecture. *Engineering Applications of Artificial Intelligence*, **16** (5–6), 395–405.

SHARMA, S. K. and TOKHI, M. O. (2000). Neural network optimisation using genetic algorithm: A hierarchical fuzzy method. *Proceedings of AARTC00: 6th IFAC Workshop on Algorithms and Architectures for Real-Time Control*, Palma de Mallorca, Spain, 15–17 May 2000, 71–76.

SHCHUKA, A. and GOLDENBERG, A. A. (1989). Tip control of a single-link flexible arm using feedforward technique. *Mechanical Machines Theory*, **24**, 439–455.

SHI, Z. X., FUNG, E. H. K. and LI, Y. C. (1999). Dynamic modelling of a rigid-flexible manipulator for constrained motion task control. *Applied Mathematical Modelling*, **23**, 509–525.

SHIFMAN, J. J. (1993). Lyapunov functions and the control of Euler-Bernoulli beam. *International Journal of Control*, **57**, 971–992.

SICILIANO, B. (1990). A closed-loop inverse kinematic scheme for on-line joint based robot control. *Robotica*, **8**, 231–243.

SICILIANO, B. (1998). An inverse kinematics scheme for a flexible arm in contact with a compliant surface. *Proceedings of 37th IEEE Conference on Decision and Control*, Tampa, Florida, USA, 16–18 December 1998, 3617–3622.

SICILIANO, B. (1999). Closed-loop inverse kinematics algorithm for constrained flexible manipulators under gravity. *Journal of Robotic Systems*, **16**, 353–362.

SICILIANO, B. and BOOK, W. J. (1988). A singular perturbation approach control of lightweight flexible manipulators. *International Journal of Robotics Research*, **7** (4), 79–90.

SICILIANO, B., PRASAD, J. V. R. and CALISE, A. J. (1992). Output feedback two-time scale control of multi-link flexible arms. *Transactions of ASME: Journal of Dynamic Systems, Measurement, and Control*, **114**, 70–77.

SICILIANO, B. and VILLANI, L. (1999). *Robot force control*. Kluwer Academic Publishers, Boston, Massachusetts, USA.

SICILIANO, B. and VILLANI, L. (2000). Parallel force and position control of flexible manipulators. *IEE Proceedings–D: Control Theory and Applications*, **147** (6), 605–612.

SICILIANO, B. and VILLANI, L. (2001). An inverse kinematics algorithm for interaction control of a flexible arm with a compliant surface. *Control Engineering Practice*, **9**, 191–198.

SICILIANO, B., YUAN, B. and BOOK, W. J. (1986). Model reference adaptive control of a one link flexible arm. *Proceedings of 25th Conference on Decision and Control*, Athens, Greece, 10–12 December 1986, 91–95.

SILVA, J. M. M. and MAIA, N. M. M. (1988). Single mode identification techniques for use with small microcomputers. *Journal of Sound and Vibration*, **124** (1), 13–26.

SIMO, J. C. (1985). A finite strain beam formulation. The three-dimensional dynamic problem. Part I. *Computer Methods in Applied Mechanics and Engineering*, **49**, 55–70.

SIMO, J. C. and VU-QUOC, L. (1986). On the dynamics of flexible beams under large overall motions – The plane case: Part I. *ASME Journal of Applied Mechanics*, **53** (4), 849–854.

SIMO, J. C. and VU-QUOC, L. (1987). The role of non-linear theories in transient dynamic analysis of flexible structures. *Journal of Sound and Vibration*, **119** (3), 487–508.

SINCARSIN, G. B. and D'ELEUTERIO, G. M. T. and HUGHES, P. C. (1993). Dynamics of an elastic multibody chain: Part D-Modelling of joints. *Dynamics and Stability of Systems*, **8** (2), 127–146.

SINGER, N. C. (1989). Residual vibration reduction in computer controlled machines, AITR-1030, MIT Artificial Intelligence Lab, Cambridge, Massachusetts, USA.

SINGER, N. C. and SEERING, W. P. (1990). Preshaping command inputs to reduce system vibration. *Transactions of ASME: Journal of Dynamic Systems, Measurement, and Control*, **112** (1), 76–82.

SINGER, N. C., SEERING, W. P. and PASCH, K. A. (1990). Shaping command inputs to minimize unwanted dynamics. US Patent No. 4,916,635.

SINGER, N., SINGHOSE, W. and KRIIKKU, E. (1997). An input shaping controller enabling cranes to move without sway. *Proceedings of ANS 7th Topical Meeting on Robotics and Remote Systems*, Augusta, Georgia, USA, 27 April–1 May 1997, 225–231.

SINGER, N. C., SINGHOSE, W. E. and SEERING, W. P. (1999). Comparison of filtering methods for reducing residual vibration. *European Journal of Control*, **5**, 208–218.

SINGH, T. and HEPPLER, G. R. (1993). Shaped input control of a system with multiple modes. *Transactions of ASME: Journal of Dynamic Systems, Measurement, and Control*, **115** (September), 341–347.

SINGH, G., KABAMBA, P. T. and McCLAMROCH, N. H. (1989). Planner time-optimal rest-to-rest slewing manoeuvre of flexible spacecraft. *Journal of Guidance, Control, and Dynamics*, **12**, 71–81.

SINGH, T. and VADALI, S. R. (1993). Robust time-delay control. *Transactions of ASME: Journal of Dynamic Systems, Measurement, and Control*, **115** (June), 303–306.

SINGH, T. and VADALI, S. R. (1994). Robust time-optimal control: A frequency domain approach. *Journal of Guidance, Control, and Dynamics*, **17** (2), 346–353.

SINGHOSE, W. E., BANERJEE, A. and SEERING, W. (1997*a*). Slewing flexible spacecraft with deflection-limiting input shaping. *AIAA Journal of Guidance, Control, and Dynamics*, **20** (2), 291–298.

SINGHOSE, W. E., BOHLKE, K. and SEERING, W. (1996*a*). Fuel-efficient pulse command profiles for flexible spacecraft. *AIAA Journal of Guidance, Control, and Dynamics*, **19** (4), 954–960.

SINGHOSE, W. E., CRAIN, E. A. and SEERING, W. P. (1997*b*). Convolved and simultaneous two-mode input shapers. *IEE Proceedings – D: Control Theory and Applications*, **144** (November), 515–520.

SINGHOSE, W. E., HUEY, J., LAWRENCE, J. and FRAKES, D. (2003). Input shaping curriculum: Integrating interactive simulations and experimental setups. *Proceedings of 11th Mediterranean Conference on Control Automation*, Rhodes, Greece, 18–20 June 2003, **IV**, 05–11.

SINGHOSE, W. E., PORTER, L., KENISON, M. and KRIIKKU, E. (2000). Effects of hoisting on the input shaping control of gantry cranes. *Control Engineering Practice*, **8** (10), 1159–1165.

SINGHOSE, W. E., PORTER, L. J., TUTTLE, T. D. and SINGER, N. C. (1997*c*). Vibration reduction using multi-hump input shapers. *Transactions of ASME: Journal of Dynamic Systems, Measurement, and Control*, **119** (June), 320–326.

SINGHOSE, W. E., SEERING, W. and SINGER, N. (1994). Residual vibration reduction using vector diagrams to generate shaped inputs. *ASME Journal of Mechanical Design*, **116** (June), 654–659.

SINGHOSE, W. E., SEERING, W. P. and SINGER, N. C. (1996*b*). Input shaping for vibration reduction with specified insensitivity to modeling errors. *Proceedings of the Japan-USA Symposium on Flexible Automation*, Boston, Massachusetts, USA, 8–10 June 1996, 307–313.

SINGHOSE, W. E. and SINGER, N. (1996). Effects of input shaping on two-dimensional trajectory following. *IEEE Transactions on Robotics and Automation*, **12** (6), 881–887.

SINGHOSE, W. E., SINGER, N., RAPPOLE, W., DEREZINSKI, S. and PASCH, K. (1997*d*). Methods and apparatus for minimizing unwanted dynamics in a physical system, US Patent No. 5,638,267.

SINGHOSE, W. E., SINGER, N. and SEERING, W. (1996*c*). Improving repeatability of coordinate measuring machines with shaped command signals. *Precision Engineering*, **18** (April), 138–146.

SINGHOSE, W. E., SINGER, N. and SEERING, W. (1997*e*). Time-optimal negative input shapers. *Transactions of ASME: Journal of Dynamic Systems, Measurement, and Control*, **119** (June), 198–205.

SINGHOSE, W., SINGH, T. and SEERING, W. (1999). On-off control with specified fuel usage. *Journal of Dynamic Systems, Measurement and Control*, **121** (2), 206–212.

SINHA, A. and KAO, C. K. (1991). Independent modal sliding mode control of vibration in flexible manipulator. *Journal of Sound and Vibration*, **147**, 352–358.

SIRA-RAMIREZ, H., AHMAD, S., and ZRIBI, M. (1992). Dynamical feedback control of robotic manipulators with joint flexibility. *IEEE Transactions on Systems, Man, and Cybernetics*, **22** (4), 736–747.

SMITH, C. and BARUH, H. (1991). Dominance of stiffening effects for rotating flexible beams. *AIAA Journal of Guidance, Control, and Dynamics*, **14** (5), 1072–1074.

SMITH, O. J. M. (1957). Posicast control of damped oscillatory systems. *Proceedings of the IRE*, **45** (September), 1249–1255.

SMITH, O. J. M. (1958). *Feedback control systems*. McGraw-Hill Book Co., Inc, New York, USA.

SONG, B. and KOIVO, A. J. (1998). Neural network model based control of a flexible link manipulator. *Proceedings of the IEEE International Conference on Robotics & Automation,* Leuven, Belgium, 16–20 May 1998, 812–817.

SPACE MACHINES LABORATORY (2004*a*). *Aerospace dual arm manipulator*, Tohoko University, Japan (http://www.space.mech.tohoku.ac.jp/research/adam/adam-e.html).

SPACE MACHINES LABORATORY (2004*b*). *The control of a 3D flexible manipulator*, Tohoko University, Japan (http://www.space.mech.tohoku.ac.jp/research/flebot2/flebot2-e.html).

SPECTOR, V. A. and FLASHNER, H. (1990). Modeling and design implications of noncollocated control in flexible systems. *Transactions of ASME: Journal of Dynamic Systems, Measurement and Control*, **112**, 186–193.

SPONG, M. V., KHORASANI, K. and KOKOTOVIC, P. V. (1987). An integral manifold approach to feedback control of flexible joint robots. *IEEE Journal of Robotics and Automation*, **3**, 291–300.

SPONG, M. W. and ORTEGA, R. (1990). On adaptive inverse dynamics control of rigid robots. *IEEE Transactions on Automatic Control*, **35** (1), 92–95.

SRINIVASAN, B., PRASAD, U. R., and RAO, N. J. (1994). Backpropagation through adjoints for the identification of non-linear dynamic systems using recurrent neural models. *IEEE Transactions on Neural Networks*, **5** (2), 213–228.

STADENNY, J. and BELANGER, P. (1986). Robot manipulator control by acceleration feedback: Stability, design and performance issues. *Proceedings of IEEE Conference on Decision and Control*, Athens, 10–12 December 1986, 80–85.

STARR, G. P. (1985). Swing-free transport of suspended objects with a path-controlled robot manipulator. *Transactions of ASME: Journal of Dynamic Systems, Measurement, and Control*, **107**, 97–100.

STEVENSAND, H. D. and HOW, J. P. (1996). The limitations of independent controller design for a multiple-link flexible macro-manipulator carrying a rigid mini-manipulator. *Proceedings of the 2nd ASCE Conference on Robotics for Challenging Environments*, Albuquerque, New Mexico, USA, 13–15 June 1996, 93–99.

STIEBER, M. E., McKAY, M., VUKOVICH, G. and PETRIU, (1999). Vision-based sensing and control for space robotics applications. *IEEE Transactions on Instrumentation and Measurement*, **48** (4), 807–812.

STRANG, G. and FIX, G. J. (1973). *An analysis of the finite element method*. Prentice-Hall, New York, USA.

SU, Z. and KHORASANI, K. (2001). A neural-network-based controller for a single-link flexible manipulator using the inverse dynamics approach. *IEEE Transactions on Industrial Electronics*, **48** (6), 1074–1086.

SUPINO, L. and ROMANO, P.M. (1997). Chatter reduction in sliding mode control of a disk drive actuator. US Patent No. 5, 699, 207.

SUR, S. and MURRAY, R. M. (1997). Simultaneous force-position control for grasping using flexible link manipulators. *Proceedings of the American Control Conference*, Albuquerque, New Mexico, USA, 4–6 June 1997, 1402–1406.

SUZUKI, E., YUH, J. and CHOI, B. (1993). On modeling and control of a 2-D.O.F. flexible robot having a prismatic joint. *Proceedings of the IEEE International Conference on Systems, Man and Cybernetics*, Le Touquet (France), 17–20 October 1993, **III**, 316–318.

SUZUKI, M. (1981). Composite controls for singularly perturbed systems. *IEEE Transactions on Automatic Control*, **26** (2), 438–443.

SWIGERT, J. C. (1980). Shaped torque techniques. *Journal of Guidance and Control*, **3**, 460–467.

SZE, T. L. (1995). System identification using radial basis neural networks, PhD thesis, Department of Automatic Control and Systems Engineering, The University of Sheffield, Sheffield, UK.

SZYMKAT, M., RAVN, O., TURNAU, A., KOLEK, K. and PJETURSSON, A. (1995). Integrated mechatronic modelling environments. *Proceedings of the International Conference on Recent Advances in Mechatronics*, Istanbul, Turkey, 14–16 August 1995, **II**, 767–772.

TABRIZI, M., JAMALUDDIN, H., BILLINGS, S. A. and SKAGGS, R. (1990). Use of identification techniques to develop a water-table prediction model. *Transactions of ASME: Journal of Dynamic Systems, Measurement, and Control*, **33** (6), 1913–1918.

TALEBI, H. A., KHORASANI, K. and PATEL, R. V. (1998*a*). Neural network based control schemes for flexible-link manipulator: simulations and experiments. *Neural Networks* (special issue), **11**, 1357–1377.

TALEBI, H. A., PATEL, R.V. and KHORASANI, K. (1997). Experimental evaluation of neural network based controllers for tracking the tip position of a flexible-link manipulator. *Proceedings of the IEEE International Conference on Robotics and Automation*, Albuquerque, New Mexico, USA, 20–25 April 1997, 3300–3305.

TALEBI, H. A., PATEL, R. V. and KHORASANI, K. (1998*b*). Inverse dynamics control of flexible-link manipulators using neural networks. *Proceedings of the IEEE International Conference on Robotics and Automation*, Leuven, Belgium, 16–20 May 1998, 806–811.

TALLMAN, G. H. and SMITH, O. J. M. (1958). Analog study of dead-beat Posicast control. *IRE Transactions on Automatic Control*, **4** (1), 14–21.

TANG, Y., TOMIZUKA, M. and GUERRERO, G. (1996). Robust control of rigid robots. *Proceedings of 36th IEEE Conference on Decision and Control*, Kobe, Japan, 11–13 December 1996, 791–796.

THEODORE, R. J. and GHOSAL, A. (1995). Comparison of the assumed modes and finite element models for flexible multi-link manipulators. *International Journal of Robotics Research*, **14** (2), 91–111.

THOMAS, S. and BANDYOPADHYAY, B. (1997). Comments on A new controller design for a flexible one-link manipulator. *IEEE Transactions on Automatic Control,* **42** (3), 425–429.

THOMPSON, B. S. and SUNG, C. K. (1986). A variational formulation for the dynamic viscoelastic finite element analysis of robotic manipulators constructed from composite materials. *ASME Journal of Mechanisms, Transmissions, and Automation Design,* **106**, 183–190.

TIMOSHENKO, S. and MACCULLOUGH, G. H. (1949). *Elements of strength of materials,* third edition. D. Van Nostrand Company, Inc., New Jersey, USA.

TIMOSHENKO, S., YOUNG, D. H. and WEAVER, W. (1974). *Vibration problems in Engineering,* fourth edition. John Wiley, New York, USA.

TOKHI, M. O. and AZAD, A. K. M. (1995a). Real-time finite difference simulation of a single-link flexible manipulator system incorporating hub inertia and payload. *Proceedings of IMechE-I: Journal of Systems and Control Engineering,* **209** (I1), 21–33.

TOKHI, M. O. and AZAD, A. K. M. (1995b). Active vibration suppression of flexible manipulator systems – open-loop control methods. *International Journal of Active Control,* **1** (1), 15–43.

TOKHI, M. O. and AZAD, A. K. M. (1996a). Control of flexible manipulator systems. *Proceedings of IMechE-I: Journal of Systems and Control Engineering,* **210** (12), 113–130.

TOKHI, M. O. and AZAD, A. K. M. (1996b). Collocated and non-collocated feedback control of flexible manipulator systems.*Machine Vibration,* **5**, 170–178.

TOKHI, M. O. and AZAD, A. K. M. (1997). Design and development of an experimental flexible manipulator system. *Robotica,* **15** (3), 283–292.

TOKHI, M. O., AZAD, A. K. M. and POERWANTO, H. (1999). SCEFMAS: A Simulink environment for dynamic characterisation and control of flexible manipulators. *International Journal of Engineering Education,* **15** (3), 213–226.

TOKHI, M. O. and LEITCH, R. R. (1991). Design and implementation of self tuning active noise control systems. *IEE Proceedings–D: Control Theory and Applications,* **138** (5), 421–430.

TOKHI, M. O. and MOHAMED, Z. (2003). Combined input shaping and feedback control of a flexible manipulator. *Proceedings of 10th International Congress on Sound and Vibration,* Stockholm, Sweden, 7–10 July 2003, 299–306.

TOKHI, M.O., MOHAMED, Z. and AZAD, A. K. M. (1997). Finite difference and finite element approaches to dynamic modelling of a flexible manipulator. *Proceedings of IMechE-I: Journal of Systems and Control Engineering,* **211** (I2), 145–156.

TOKHI, M. O., POWERWANTO, H. and AZAD, A. K. M. (1995). Dynamic simulation of flexible manipulator systems incorporating hub inertia, payload and structural damping. *Machine Vibration,* **4**, 106–124.

TOKHI, M. O. and VERES, S. M. (2002). *Active sound and vibration control – Theory and applications*. The Institution of Electrical Engineers, London, UK.

TOMIZUKA, M. (1987). Zero phase error tracking algorithm for digital control. *Transactions of ASME: Journal of Dynamic Systems, Measurement, and Control*, **109** (March), 65–68.

TRAUTMAN, C. and WANG, D. (1996). Noncollocated passive control of a flexible link manipulator. *Proceedings of the IEEE International Conference on Robotics and Automation*, Minneapolis, Minnesota, 22–28 April 1996, 1107–1114.

TRUCKENBRODT, A. (1980). *Bewegungsverhalten und Regelung hybrider Mehrkörpersysteme mit Anwendung auf Roboter*, Fortschrittsberichte der VDI-Z. Reihe 8, Nr. 33.

TSE, F. S., MORSE, I. E. and HINKEL, R. T. (1978). *Mechanical vibrations: Theory and applications*. Allyn and Bacon Inc, Boston, USA.

TUNG, E. D. and TOMIZUKA, M. (1993). Feedforward tracking controller design based on the identification of low frequency dynamics. *Transactions of ASME: Journal of Dynamic Systems, Measurement, and Control*, **115** (September), 348–356.

TUTTLE, T. and SEERING, W. (1997). Experimental verification of vibration reduction in flexible spacecraft using input shaping. *Journal of Guidance, Control, and Dynamics*, **20** (4), 658–664.

TUTTLE, T. and SEERING, W. (1999). Creating time optimal commands with practical constraints. *Journal of Guidance, Control, and Dynamics*, **22** (March–April), 241–250.

TZAFESTAS, S. and PAPANIKOLOPOULOS, N. P. (1990). Incremental fuzzy expert PID control. *IEEE Transactions on Industrial Electronics*, **37**, 365–371.

TZES, A. and YURKOVICH, S. (1993). An adaptive input shaping control scheme for vibration suppression in slewing flexible structures. *IEEE Transactions on Control Systems Technology*, **1** (2), 114–121.

TZES, A. P., YURKOVICH, S. and LANGER, F. D. (1989). A method for solution of the Euler-Bernoulli beam equation in flexible-link robotic systems. *Proceedings of the IEEE International Conference on Robotics and Automation*, Scottsdale, Arizona, USA, 14–19 May 1989, 557–560.

TZOU, H. S. (1988). Dynamic analysis and passive control of viscoelastically damped nonlinear dynamic contacts. *Journal of Finite Elements in Analysis and Design*, **4**, 209–224.

UMETANI, Y. and YOSHIDA, K. (1989). Resolved motion rate control of space manipulators with generalized Jacobian matrix. *IEEE Transactions on Robotics and Automation*, **5** (3), 303–314.

USORO, P. B., NADIRA, R. and MAHIL, S. S. (1984). A finite element/Lagrange approach to modelling lightweight flexible manipulator. *Transactions of ASME: Journal of Dynamic Systems, Measurement, and Control*, **108** (3), 198–205.

USORO, P. B., NADIRA, R., MAHIL, S. S. and MEHRA, R. K. (1983). *Advanced control of flexible manipulators*. Final report, NSF Award No. BCS8260419.

VANDEGRIFT, M. W., LEWIS, F. L. and ZHU, S. Q. (1994). Flexible-link robot arm control by a feedback linearization/singular perturbation approach. *Journal of Robotic Systems*, **11**, 591–603.

VARADAN, V. K., HONG, S. Y. and VARADAN, V. V. (1990). Piezoelectric sensors and actuators for active vibration damping using digital control. *Proceedings of IEEE Symposium on Ultrasonics*, Honolulu, Hawaii, USA, 4–7 December 1990, 1211–1214.

VIDYASAGAR, M. (1978). *Nonlinear systems analysis*, Prentice-Hall, Englewood Cliffs, New Jersey, USA.

VON SCHWERIN, R. (1999). *Multibody System Simulation: numerical methods, algorithms and software*. Springer-Verlag Berlin Heidelberg, Germany.

VUKOVICH, G. and YOUSEFI-KOMA, A. A. (1996). Non-collocated active traveling wave control of smart structures using distributed transducers. *Proceedings of the IEEE International Conference on Control Applications*, Michigan, USA, 15–18 September 1996, 297–302.

WANG, D. and VIDYASAGAR, M. (1989*a*). Transfer function of a single flexible link. *Proceedings of the IEEE International Conference on Robotics and Automation*, Scottsdale, Arizona, USA, 14–19 May 1989, 1042–1047.

WANG, D. and VIDYASAGAR, M. (1989*b*). Feedback linearizability of multi-link manipulators with one flexible link. *Proceedings of the 28th IEEE Conference on Decision and Control*, Florida, USA, 13–15 December 1989, 2072–2077.

WANG, Q.-G., ZOU, B., LEE, T. H. and BI, Q. (1997). Auto-tuning of multivariable PID controllers from decentralized relay feedback. *Automatica*, **33** (3), 319–330.

WANG, S. (1986). Open-loop control of a flexible robot manipulator. *International Journal of Robotics and Automation*, **1**, 54–57.

WANG, W.-J., LU, S.-S. and HSU, C.-F. (1989). Experiments on the position control of a one-link flexible arm. *IEEE Transactions on Robotics and Automation*, **5**, 373–377.

WANG, Z., ZENG, H., HO, D. and UNBENAUEN, H. (2002). Multiobjective control of a four-link flexible manipulator: A robust H_∞ approach. *IEEE Transactions on Control Systems Technology*, **10** (6), 866–875.

WANG, Z. P., GE, S. S. and LEE, T. H. (2001). Model-based and adaptive composite control of smart materials robots. *Proceedings of the 40th IEEE Conference on Decision and Control*, Orlando, Florida, USA, 4–7 December 2001, 4938–4943.

WASSERMAN, P. D. (1993). *Advanced methods in neural computing.* Van Nostrand Reinhold, New York, USA.

WEBER, M., MA, O. and SHARF, I. (2002). Identification of contact dynamics model parameters from contstrained robotic operations. *Proceedings of the DETC'02 ASME Design Engineering Technical Conferences and Computer and Information in Engineering Conference.* Montreal, Canada, 29 September–2 October 2002, DETC2002/MECH-34357.

WELLS, R. L. and SCHUELLER, J. K. (1990). Feedforward and feedback control of a flexible robotic arm. *IEEE Control Systems Magazine,* **10**, 9–15.

WELLSTEAD, P. E. and ZARROP, M. B. (1991). *Self-tuning systems: Control and signal processing.* John Wiley & Sons, Chichester, UK.

WHITNEY, D. E. (1969). Resolved motion rate control of manipulators and human prostheses. *IEEE Transactions on Man-Machine Systems,* **10**, 47–53.

WIDROW, B., GLOVER, J. R., MCCOOL, J. M. et al. (1975). Adaptive noise cancelling: Principles and applications. *Proceedings of the IEEE,* **63**, 1662–1676.

WIE, B., SINHA, R., SUNKEL, J. and COX, K. (1993). Robust fuel- and time-optimal control of uncertain flexible space structures. *Proceedings of the AIAA Guidance, Navigation, and Control Conference,* Monterey, California, USA, 17–19 August 1993, 939–948.

WON, C. C. (1995). Piezoelectric transformer. *Journal of Guidance, Control, and Dynamics,* **18**, 96–101.

WORDEN, K., STANSBY, P., TOMLINSON, G. and BILLINGS, S. A. (1994). Identification of non-linear-wave forces. *Journal of Fluids and Structures,* **8** (1), 19–71.

XU, W., ZHANG, X. and NAIR, S. (2001). Dynamic modelling of a two-link flexible manipulator incorporating experimental data. *Proceedings of the American Control Conference,* Arlington, Texas, USA, 25–27 June 2001, 4508–4513.

YAMADA, K. and TSUCHIYA, K. (1990). Efficient computation algorithms for manipulator control of a space robot. *Transactions of Society of Instrument and Control Engineers,* **26** (7), 765–772 (in Japanese).

YANG, J.-H., LIAN, F.-L., and FU, L.-C. (1995). Adaptive hybrid position/force control for robot manipulators with compliant links. *Proceedings of the IEEE International Conference on Robotics and Automation,* Nagoya, Japan, 21–27 May 1995, 603–608.

YANG, T.-C., JACKSON, C. S. and KUDVA, P. (1991). Adaptive control of a flexible link manipulator with unknown load. *IEE Proceedings–D: Control Theory and Applications,* **138**, 1614–1619.

YANG, Y. P. and GIBSON, J. S. (1989). Adaptive control of a manipulator with a flexible link. *Journal of Robotic Systems,* **6** (3), 217–232.

YESILDIREK, A., VANDEGRIFT, M. W. and LEWIS, F. L. (1994). A neural network controller for flexible-link robots. *Proceedings of the IEEE International Symposium on Intelligent Control*, Columbus, Ohio, USA, 16–18 August 1994, 63–68.

YIGIT, A. S. (1988). Dynamics of a radially rotating beam with impact: Implications for robotics, PhD Dissertation, The University of Michigan, USA.

YIGIT, A. S. (1994*a*). On the stability of PD control for a two-link rigid-flexible manipulator. *Transactions of ASME: Journal of Dynamic Systems, Measurement, and Control*, **116**, 208–215.

YIGIT, A. S. (1994*b*). The effect of flexibility on the impact response of a two-link rigid-flexible manipulator. *Journal of Sound and Vibration*, **177**, 349–361.

YIGIT, A. S., SCOTT, R. A. and ULSOY, A. G. (1988). Flexural motion of a radially rotating beam attached to a rigid body. *Journal of Sound and Vibration*, **121** (2), 201–210.

YIGIT, A. and ULSOY, A. G. (1989). Controller design for rigid-flexible multi-body system. *Proceedings of the 28th IEEE Conference on Decision and Control*, Florida, USA, 13–15 December 1989, 665–673.

YIGIT, A. S., ULSOY, A. G. and SCOTT, R. A. (1990). Dynamics of a radially rotating beam with impact, Part 1:Theoretical and computational model; Part 2: Experimental and simulation results. *Transactions of ASME: Journal of Vibration and Acoustics*, **112**, 65–77.

YIM, W. (1993). End-point trajectory control stabilization, and zero dynamics of a three-link flexible manipulator. *Proceeding of the IEEE International Conference on Robotics and Automation*, Atlanta, Georgia, USA, 2–6 May 1993, 468–473.

YOSHIKAWA, T. (1987). Dynamic hybrid position/force control of robot manipulators – Description of hand constraints and calculation of joint driving force. *IEEE Journal of Robotics and Automation*, **3**, 386–392.

YOSHIKAWA, T., HARADA, K. and MATSUMOTO, A. (1996). Hybrid position/force control of flexible-macro/rigid-micro manipulator systems. *IEEE Transactions on Robotics and Automation*, **12**, 633–640.

YOUNG, K.-K. D. (1977). Asymptotic stability of model reference systems with variable structure control. *IEEE Transactions on Automatic Control*, **22** (2), 279–281.

YUAN, B.-S., BOOK, W. J. and SICILIANO, B. (1989). Direct adaptive control of a one-link flexible arm with tracking. *Journal of Robotic Systems*, **6** (6), 663–680.

ZADEH, L. A. (1994). Fuzzy logic, neural networks, and soft computing. *Communications of ACM (Association for Computing Machinery)*, **37**, 77–84.

ZHUANG, M. and ATHERTON, D. P. (1994). PID controller design for a TITO system. *Proceedings of IEE–D: Control Theory and Applications*, **141** (2), 111–120.

ZIEGLER, J. G. and NICHOLS, N. B. (1942). Optimum settings for automatic controllers. *Transactions of ASME*, **64**, 759–768.

ZIEMER, R. E., TRANTOR, W. H. and FANNIN, D. R. (1998). *Signals and systems: Continuous and Discrete*. Prentice Hall, New Jersey, USA.

ZUO, K. and WANG, D. (1992). Closed loop shaped-input control of a class of manipulators with a single flexible link. *Proceedings of the IEEE International Conference on Robotics and Automation*, Nice, France, 12–14 May 1992, 782–787.

ZVEREV, A. I. (1967). *Handbook of filter synthesis*. Wiley, New York, USA.

Index